Pseudodifferential Operators

Pseudodifferential Operators

Michael E. Taylor

PRINCETON UNIVERSITY PRESS
PRINCETON, NEW JERSEY

Published by Princeton University Press, Princeton, New Jersey
In the United Kingdom: Princeton University Press, Guildford, Surrey

Library of Congress Cataloging in Publication Data will be
found on the last printed page of this book

This book has been composed in Monophoto Times Roman
Clothbound editions of Princeton University Press books
are printed on acid-free paper, and binding materials are
chosen for strength and durability
Printed in the United States of America by Princeton
University Press, Princeton, New Jersey

To Lillian Harden and Sarah Richbourg.

Contents

Acknowledgments

In preparing this book I am indebted to many people. Eric Bedford and Jeffrey Rauch provided a great deal of help on Chapters II through VII, and I also thank Ralph Phillips for his constructive criticism of this part of the book. My students David Yingst, Mark Farris, and Li-Yeng Sung have shown me the error of my ways on numerous points and also helped out with the proof of several theorems. People familiar with the subject will perceive the dominant influence of the work of Lars Hörmander in every chapter. My collaboration with Andrew Majda and frequent conversations with Richard Melrose, Stanley Osher, and James Ralston have been influential in large parts of this work. Finally, I am indebted to the Alfred P. Sloan foundation for support during part of the time this book was being written and to my colleagues at the Courant Institute, UCLA, and Rice University for their stimulating and supportive atmosphere during the preparation of this book.

MICHAEL TAYLOR
Houston, Texas

Pseudodifferential Operators

Introduction

This book develops a circle of techniques used to treat linear partial differential equations

$$\text{(I.1)} \qquad Pu = \sum_{|\alpha| \le m} a_\alpha(x) D^\alpha u = f$$

on a region Ω, generally supplemented by boundary conditions on one or more hypersurfaces in Ω. The three main classical examples of (I.1) are the following:

$$\text{(I.2)} \qquad \text{The Laplace equation} \quad \Delta u = f$$

where

$$\Delta = \frac{\partial^2}{\partial x_1^2} + \cdots + \frac{\partial^2}{\partial x_n^2} \qquad \text{on} \qquad \Omega \subset \mathbf{R}^n.$$

This is typically supplemented by the Dirichlet boundary condition $u|_{\partial\Omega} = g$ or the Neumann boundary condition $(\partial u/\partial \nu)|_{\partial\Omega} = g$, though other boundary conditions also occur.

$$\text{(I.3)} \qquad \text{The heat equation} \quad \frac{\partial}{\partial t} u = \Delta u, \qquad t \in \mathbf{R}^+, \qquad x \in \Omega.$$

This is typically supplemented by an initial condition $u(0, x) = f(x)$ and, if Ω has nonempty boundary, a boundary condition on $\mathbf{R}^+ \times \partial\Omega$ such as a Dirichlet or Neumann boundary condition.

$$\text{(I.4)} \quad \text{The wave equation} \quad \left(\frac{\partial^2}{\partial t^2} - \Delta\right) u = 0, \qquad t \in \mathbf{R}, \qquad x \in \Omega.$$

This is typically supplemented by the initial condition $u(0, x) = f_1(x)$, $u_t(0, x) = f_2(x)$, and if $\partial\Omega \neq \emptyset$ a boundary condition as in the first two cases.

Equations (I.2)–(I.4) are said to be of elliptic, parabolic, and hyperbolic type, respectively. These equations and natural generalizations make up a large portion of the linear partial differential equations of mathematical physics, and their theory suggests questions to be asked about general linear *PDE*. These questions are typically in one of three categories: existence, uniqueness, qualitative behavior.

For the purposes of mathematical physics, the first two questions may be considered preliminary (though not necessarily trivial from a mathematical viewpoint); after all, the point of the application of calculus to physics is not to prove that real processes actually occur but rather to describe the nature of such occurrences. Thus the third category forms the core of classical linear *PDE*. Within this category, many questions arise; one asks what the solutions look like, and ideally one would want to know everything about them. Such properties as regularity, location of singularities, and estimates in various norms are important examples, particularly emphasized in this treatment, and more questions arise such as on the spectral behavior of P, decay of solutions, location of maxima or nodal sets, limiting behavior under (possibly quite singular) perturbations of the equation or the boundary, and many more. Numerous such questions are addressed in these pages. We also derive existence and uniqueness theorems adequate for many linear equations of mathematical physics, and most such results are fairly simple (unfortunately, one cannot say the same for nonlinear *PDE* of mathematical physics).

To put in perspective the role of pseudodifferential operators in the study of linear *PDE*, we list four tools of linear *PDE*.

(1) *Functional analysis.* The use of various Hilbert, Banach, Frechet, and LF spaces is all pervasive in modern linear *PDE*. Sobolev spaces and spaces of distributions, described in Chapter I, form the setting for most of the analysis, though other spaces, particularly Hölder spaces and Besov spaces (discussed in Chapter XI) make an occasional appearance.

(2) *Fourier analysis.* The use of Fourier series and/or the Fourier transform in constant coefficient PDE is intimately connected with separation of variables, and Fourier analysis was used by Daniel Bernoulli in the very beginning of the study of the subject. In the modern approach, Fourier analysis, via the Plancherel theorem particularly, is often used to get estimates of solutions to constant coefficient equations, obtained from variable coefficient equations by freezing the coefficients and, if a boundary is present, flattening out the boundary. One then patches these estimates together to get estimates in the variable coefficient situation.

(3) *Energy estimates.* This includes integral estimates on quadratic forms of a function and its derivatives. One of the basic estimates of this sort is Gårding's inequality

$$Re(Pu, u) \geq c_1\|u\|_{H^m}^2 - c_2\|u\|_{L^2}^2, \qquad u \in C_0^\infty(\Omega) \tag{I.5}$$

for a partial differential operator

$$P = \sum_{|\alpha| \leq 2m} a_\alpha(x) D^\alpha,$$

assuming

$$Re\, P_{2m}(x, \xi) = Re\left(\sum_{|\alpha|=2m} a_\alpha(x)\xi^\alpha\right) \geq c|\xi|^{2m}.$$

Here

$$\|u\|_{H^m}^2 = \sum_{|\alpha|\leq m} \|D^\alpha u\|_{L^2}^2$$

defines a norm on a space H^m known as a Sobolev space (discussed in Chapter I). When P is a second order scalar differential operator, (I.5) is proved simply by integration by parts. For higher order operators, one way to get (I.5) is to freeze coefficients, obtain such an estimate for constant coefficient operators by Fourier analysis, and glue these estimates together. Gårding's inequality for pseudodifferential operators, generalizing (I.5), will be proved in Chapter II, Section 8, using the calculus of pseudodifferential operators. Energy estimates are also used to prove existence, uniqueness, and finite propagation speed for solutions to hyperbolic equations, and it is in the study of second order hyperbolic equations that the term *energy* has a most direct physical interpretation. Gårding's inequality plays a great role in unifying many of these energy estimates, as will be seen in Chapters IV and V. Weighted L^2 estimates, known as Carleman estimates, have also played an important role in linear *PDE*, though they are not much emphasized here, except in Chapter XIV which considers uniqueness in the Cauchy problem.

(4) *Fundamental solutions and parametrices.* A distribution $E(x, y)$ such that

$$P(x, D_x)E = \delta(x - y) \tag{I.6}$$

is called a fundamental solution. If the two sides of (I.6) differ by a smooth function, E is called a parametrix. One can get a lot of information about solutions to (I.1) if one of these is known. For variable coefficient equations it is usually impossible to construct a fundamental solution; one is generally happy to be able to construct a parametrix. One way to do this for elliptic operators is the following. A first order approximation to a parametrix is obtained by freezing coefficients and dropping lower order terms, getting fundamental solutions for such simple equations, via Fourier analysis, and gluing these together. A true parametrix is then obtained by an iterative procedure. This technique is often called the Levi parametrix method, and one of the simplest applications of the calculus of pseudodifferential operators, as we shall see in Chapter III, is to carry out such a procedure. For nonelliptic operators, much more subtle methods are required to construct parametrices. Developing such methods is one of the main themes of this book.

We begin in Chapter I with a summary of the basic facts about distributions and Sobolev spaces most frequently used in *PDE*. There are numerous excellent books giving more leisurely and complete treatments of this subject, and we mention Yosida's *Functional Analysis* [1], Gelfand et al. *Generalized Functions* [1], Donoghue's *Distribution Theory* [1], Adam's *Sobolev Spaces* [1], and particularly the first part of Lions and Magene's work on boundary value problems [1]. The reader is assumed to be familiar with functional analysis and should have some understanding of distribution theory.

Our subject proper starts with Chapter II, where pseudodifferential operators are defined and some of their basic properties are studied, such as the behavior of products and adjoints of such operators, their continuity on L^2 and Sobolev spaces, the fact that they do not increase the singular support of distributions to which they are applied, and the Gårding inequality, generalizing (I.5). In Chapters III through V this calculus of pseudodifferential operators is applied to some basic questions of existence and regularity of solutions to elliptic, hyperbolic, and parabolic equations, and elliptic boundary value problems. The regularity theorem for elliptic differential operators P of order m, defined as those for which $|P_m(x, \xi)| \geq c|\xi|^m$, states that if Pu is C^∞ on an open set U, $u \in C^\infty(U)$, and more generally if Pu possesses a certain degree of smoothness, in an appropriate space, then u possesses m more degrees of smoothness. An operator P, for which u is C^∞ wherever Pu is with perhaps u possessing fewer than m extra orders of smoothness then Pu generally, is called hypoelliptic, and some results on hypoelliptic operators are given in Chapter III, though further, deeper results are given in Chapter XV. In Chapter IV one application is given to a topic in nonlinear *PDE*: the short time existence of solutions to quasi-linear hyperbolic equations.

In Chapter VI a new concept, that of the wave front set of a distribution, is introduced. The wave front set of a distribution on U is a subset of the cotangent bundle T^*U, lying over the singular support of u, and thus is a refinement of the notion of singular support. It turns out to be the natural language for stating theorems on the propagation of singularities of solutions to PDE, and Chapter VI gives a proof of Hörmander's theorem on propagation of singularities. This proof requires a new tool, the sharp Gårding inequality, proved in Chapter VII.

In Chapter VIII a new theme is taken up, the calculus of Fourier integral operators, a class more general than pseudodifferential operators. Such operators are useful for constructing parametrices for many operators that are not hypoelliptic, in particular hyperbolic operators, and their use extends classical methods of geometrical optics. This study makes use of some basic notions of the symplectic form and Hamiltonian vector fields. Some excellent references for this aspect of advanced calculus include Abraham and Marsden's *Foundations of Mechanics* [1] and Arnold's *Mathematical Methods of Classical Physics* [1], and also Caratheodory's classic [1]. The Hamiltonian vector field associated with the principal symbol of a pseudodifferential operator (of classical type) generates a flow on the cotangent bundle called the bicharacteristic flow, and Hörmander's propagation of singularities theorem asserts that if $Pu = f \in C^\infty(U)$, then the wave front set of u is invariant under this bicharacteristic flow. A second proof of this result is given in Chapter VIII, and in Chapters IX and X propagation of singularities for solutions to boundary value problems is discussed, first in the case of bicharacteristics transversal to the boundary and then for bicharacteristics that "graze" the boundary, being convex with respect to the boundary.

In Chapter XI we study the behavior of various classes of pseudodifferential operators on L^p and Hölder spaces and include a treatment of estimates for solutions to regular elliptic boundary value problems (discussed in Chapter V in the L^2 context) within these categories. In this chapter we make use of results of Marcinkiewicz, Mikhlin, and Hörmander on continuity of certain Fourier multipliers on $L^p(\mathbf{R}^n)$. Stein's book *Singular Integrals and Differentiability Properties of Functions* [1] provides a thorough treatment of this material.

Chapter XII studies questions about the eigenvalues and eigenfunctions of an elliptic self adjoint operator on a compact manifold, including eigenvalue asymptotics and convergence of eigenfunction expansions, and some applications to harmonic analysis on compact Lie groups, among other things. Fourier integral operators provide the tool for studying functions of an elliptic self-adjoint operator. This study in turn is the key to a systematic generalization of many topics in classical harmonic analysis on the torus. Several alternative approaches are given to some results, some of which involve using Tauberian theorems, and we include an appendix on Wiener's Tauberian theorem and some consequences. Generally, I have preferred to refer to other books where certain prerequisite material has been well treated, to keep this work from becoming enormous. However, it seems to me that most expositions of Wiener's Tauberian theorem encourage the reader to avoid perceiving the nature of Wiener's inspiration—namely that his closure of translates theorem was a very simple result that wonderfully tied together many seemingly diverse Tauberian theorems, of which the Hardy-Littlewood and Karamata theorems are important and most frequently used examples.

Chapter XIII is devoted to the Calderon-Vaillancourt theorem on L^2 boundedness of pseudodifferential operators in a borderline case not covered in Chapter II, and to Hörmander-Melin inequalities, on the semiboundedness of second order pseudodifferential operators. Chapter XIV gives some results on uniqueness in the Cauchy problem: given a hypersurface Σ dividing U into two parts, U^+ and U^-, when does $Pu = 0$ on U, $u = 0$ on U^+ imply $u = 0$ on a neighborhood of Σ?

These first fourteen chapters deal primarily with operators which are either elliptic or whose characteristics are simple. The final chapter studies operators with double characteristics. Included here are such hypoelliptic operators as arise in the analysis of the $\bar{\partial}$-Neumann problem in strictly pseudoconvex domains and also certain equations of mathematical physics, such as the equations of crystal optics.

Chapters II through VII have been covered by the author in a one semester course at the University of Michigan, for students who the previous semester had covered some basic *PDE*, including the Sobolev space theory and Fourier analysis given in Chapter I. A preliminary version of these chapters was published as a Springer Lecture Note (Taylor [3]). Most of the material in Chapters VIII through XV has been covered by the author in various courses and seminars at Michigan, Stony Brook, the Courant Institute, and Rice University. The entire book contains more material than one would cover in a year's course, and it should be noted that not every chapter depends on all the previous ones. Chapter XI on L^p and Hölder estimates could be read directly after Chapter II. If one wanted to get quickly to Chapter VIII on Fourier integral operators, one could skip Chapter V on elliptic boundary value problems and the proof of propagation of singularities in Chapter VI using the sharp Gårding inequality, and hence also Chapter VII. As for Chapter XIII, the Claderon-Vaillancourt theorem could be treated directly after Chapter II, and the Hörmander-Melin inequality after Chapter VIII. Chapter XIV and almost all of Chapter XV could be read right after Chapter VIII, the Hörmander-Melin inequality playing a role in some results of the first section of Chapter XV.

The treatment of Fourier integral operators emphasizes the local theory. I avoid discussing an invariantly defined symbol of a Fourier integral operator and in particular do not introduce the Keller-Maslov line bundle. For a complementary treatment of Fourier integral operators which does emphasize such global techniques, I recommend Duistermaat's *Fourier Integral Operators* [1]. Also the reader should consult the original paper of Hörmander [14], of Duistermaat and Hörmander [1], and the book of Guillemin and Sternberg [1].

The beginning student of partial differential equations should be aware that many other methods have been brought to bear besides those connected with pseudodifferential operators. This present book is not intended as a general introduction to the subject of partial differential equations, and we want to give the reader some guide to a set of books which, together, would provide a fairly complete introduction. First, some general texts introducing most of the basic problems of the classical theory, on a fairly elementary level, are: R. Courant and D. Hilbert, *Methods of Mathematical Physics*, vol. 2 [1]; and P. Garabedian, *Partial Differential Equations* [1]. More complete references are given in the bibliography.

Some books that cover a more restricted class of problems, using more advanced techniques, particularly functional analysis and energy estimates, include: L. Bers, F. John and M. Schechter, *Partial Differential Equations*, [1]; L. Hörmander, *Linear Partial Differential Operators*, [16]; S. Mizohata, *The Theory of Partial Differential Equations* [1]; and, F. Trèves, *Linear Partial Differential Equations with Constant Coefficients* [1]. For a treatment of elliptic operators, see: S. Agmon, *Lectures on Elliptic Boundary Value Problems* [2]; D. Gilbarg and N. Trudinger, *Elliptic Partial Differential Equations of the Second Order* [1]; J. Lions and E. Magenes, *Inhomogeneous Boundary Value Problems and Applications* [1]; C. B. Morrey, *Multiple Integrals in the Calculus of Variations* [1]; and M. Protter and H. Weinberger, *Maximum Principles in Differential Equations* [1]. The books by Gilbarg and Trudinger and by Morrey give thorough treatments of quasilinear elliptic equations.

For the connection between the heat equation and diffusion processes, see: A. Friedman, *Stochastic Differential Equations and Applications* [2]; and H. McKean, *Stochastic Integrals* [1]. For the study of scattering theory, quantum mechanical and classical, respectively, I refer to: W. Amrein, J. Jauch, and K. Sinha, *Scattering Theory in Quantum Mechanics* [1]; and P. Lax and R. Phillips, *Scattering Theory* [1]. Some results on nonlinear wave equations are given in R. Courant and K. Friedrichs, *Supersonic Flow and Shock Waves* [61]; and G. Whitham, *Linear and Nonlinear Waves* [1]. Finally, one should see an old fashioned treatment of *PDE* using mainly Green's formula and separation of variables. A good one is I. Stakgold, *Boundary Value Problems of Mathematical Physics* [1]; and good treatments of the special function theory created to treat such problems are given in N. Levedev, *Special Functions and their Applications* [1]; and F. Olver, *Asymptotics and Special Functions* [1].

It seems quite likely that special function theory will make a comeback in *PDE* as modern breakthroughs allow one to use such functions in a more sophisticated way than was done during the last century, even as today more sophisticated use is made of the exponential function than in the eighteenth century. The only higher transcendental function used in this book is the Airy function, which appears in Chapter X.

CHAPTER I

Distributions and Sobolev Spaces

Distributions and Sobolev spaces form a most convenient vector by which methods of functional analysis are brought to bear on problems in partial differential equations, and Fourier analysis plays an enormous role. The basic results used in most of this book are summarized in Sections 1–5. Brief proofs of most of the crucial results are given and references are provided for further results. For more leisurely treatments of Sobolev spaces the reader might refer to Trèves [1], Yosida [1], Bergh and Löfström [1], Lions and Magenes [1], or Adams [1]. Of course, Schwartz [1] and Gelfand et al. [1] are the classics for distribution theory; see also Donoghue [1]. The treatment here owes a great deal to the expositions of Yosida and Hörmander [6], except that I emphasize the interpolation method of A. Calderon. The sixth section treats very briefly Sobolev spaces associated with L^p. Such spaces occur only in Chapters XI and XII of this book. I refer particularly to Stein [1] for analysis on L^p.

I agree with Sternberg [1] that any long list of concepts should be accompanied by some nontrivial theorem, so in Section 7 some basic Fourier analysis is applied to prove local solvability of partial differential equations with constant coefficients.

§1. Distributions

Here we define certain spaces of smooth functions, $C^\infty(\Omega)$ and $C_0^\infty(\Omega)$, and their dual spaces, $\mathscr{E}'(\Omega)$ and $\mathscr{D}'(\Omega)$, which are spaces of distributions. We let Ω be an open subset of $\mathbf{R}^n$, or more generally a smooth paracompact manifold. Suppose $\Omega = \bigcup \Omega_j$ where each Ω_j is open and has compact closure in Ω_{j+1}. When $\Omega = \mathbf{R}^n$, we shall define the Schwartz space $\mathscr{S}$ of rapidly decreasing functions, and its dual $\mathscr{S}'$, the space of tempered distributions.

For a compact subset K of Ω and a nonnegative integer j, define a seminorm $p_{K,j}$ on smooth functions by

$$p_{K,j}(u) = \sup_{x \in K} \{|D^\alpha u(x)| : |\alpha| \le j\} \tag{1.1}$$

if $\Omega \subset \mathbf{R}^n$, where $\alpha = (\alpha_1, \ldots, \alpha_n)$ and

$$D^\alpha u(x) = i^{-|\alpha|} \frac{\partial^{\alpha_1}}{\partial x^{\alpha_1}} \cdots \frac{\partial^{\alpha_n}}{\partial x^{\alpha_n}} u(x). \tag{1.2}$$

If Ω is a manifold, cover K by open sets U_j, having compact closure in V_j, with coordinate charts $\varkappa_j: V_j \to \mathbf{R}^n$, and replace (1.1) by

$$\sum_\ell \sup_{x \in \varkappa_\ell(U_\ell \cap K)} \{D^\alpha u \circ \varkappa_\ell^{-1}(x): |\alpha| \le j\}. \tag{1.3}$$

DEFINITION 1.1. *The space $C^\infty(\Omega)$ consists of the infinitely differentiable functions on Ω equipped with the topology such that $u_k \to u$ as $k \to \infty$ if, and only if, for each j, K,*

$$p_{j,K}(u - u_k) \to 0. \tag{1.4}$$

It is easy to see that one need only require (1.4) for $K = \bar{\Omega}_\nu$, $\nu \in \mathbf{Z}^+$. The topology on $C^\infty(\Omega)$ is independent of the choice of Ω_ν and (for Ω a manifold) of the coordinate charts used to give (1.3). With the countable family of norms $p_{j,\nu}(u) = p_{j,\bar{\Omega}_\nu}(u)$, $C^\infty(\Omega)$ is complete, so it is a Fréchet space.

If $K \subset \Omega$ is compact, let $C_0^\infty(K)$ be the closed subspace of $C^\infty(\Omega)$ consisting of functions which vanish outside K, with the induced topology.

DEFINITION 1.2. *The space $C_0^\infty(\Omega)$ consists of the infinitely differentiable functions on Ω with compact support, equipped with the inductive limit topology*

$$C_0^\infty(\Omega) = \bigcup_{j=1}^\infty C_0^\infty(\bar{\Omega}_j), \tag{1.5}$$

that is to say, the topology on $C_0^\infty(\Omega)$ is the strongest locally convex vector space topology such that each injection $C_0^\infty(\bar{\Omega}_j) \to C_0^\infty(\Omega)$ is continuous.

One can verify that the topology on $C_0^\infty(\Omega)$ is independent of the increasing compact cover $\bar{\Omega}_j$ used in (1.5). Equipped with this topology, $C_0^\infty(\Omega)$ is known as an *LF* space.

Given a locally convex topological vector space E, i.e., one with a basis of neighborhoods of the origin consisting of convex sets, we define its *dual* E' to consist of the continuous linear functions $\alpha: E \to \mathbf{C}$. There are several popular topologies for E'; here are three. The weak topology on E' is the topology of pointwise convergence on E. The Mackey topology on E' is the topology of uniform convergence on subsets K of E which are compact, convex, and balanced (i.e., $u \in K$, $\lambda \in \mathbf{C}$, $|\lambda| \le 1 \Rightarrow \lambda u \in K$). The strong topology on E' is the topology of uniform convergence on subsets of E which are bounded, convex, and balanced, where a bounded set K of E is one on which each seminorm defining the topology of E is bounded. In case E is $C^\infty(\Omega)$ or $C_0^\infty(\Omega)$, a consequence of Ascoli's theorem is that every bounded subset of E is precompact, so in these cases the Mackey and strong topologies on E' coincide.

DEFINITION 1.3. *The space $\mathscr{D}'(\Omega)$ is the dual of $C_0^\infty(\Omega)$ and $\mathscr{E}'(\Omega)$ is the dual of $C^\infty(\Omega)$. Elements of $\mathscr{D}'(\Omega)$ are called distributions.*

If u belongs to $C_0^\infty(\Omega)$ or $C^\infty(\Omega)$ and ω is an element of the corresponding dual space, we denote the action of ω on u by

$$\langle u, \omega \rangle. \tag{1.6}$$

Given a Riemannian metric on Ω, there is a natural mapping $C^\infty(\Omega) \to \mathscr{D}'(\Omega)$ and $C_0^\infty(\Omega) \to \mathscr{E}'(\Omega)$ given by

$$\langle u, f \rangle = \int_\Omega u(x) f(x)\, d\,\text{vol}(x). \tag{1.7}$$

The pairing

$$(u, f) = \int_\Omega u(x)\overline{f(x)}\, d\,\text{vol}(x)$$

induces a sesquilinear pairing $(u, \omega) = \langle u, \bar{\omega} \rangle$. The image of $C^\infty(\Omega)$ is dense in $\mathscr{D}'(\Omega)$ and similarly $C_0^\infty(\Omega)$ is dense in $\mathscr{E}'(\Omega)$.

If $\omega \in \mathscr{D}'(\Omega)$, $U \subset \Omega$ is open, we say ω vanishes on U if $\langle u, \omega \rangle = 0$ for all $u \in C_0^\infty(U)$. We call the support of ω, denoted supp ω, the complement of the largest set on which ω vanishes. The inclusion $C_0^\infty(\Omega) \to C^\infty(\Omega)$ induces a natural inclusion $\mathscr{E}'(\Omega) \to \mathscr{D}'(\Omega)$, and it is not hard to see that $\mathscr{E}'(\Omega)$ consists of those distributions on Ω (i.e., elements of $\mathscr{D}'(\Omega)$) with compact support.

We define, on functions on $\mathbf{R}^n$, the seminorms $q_{j,k}$ by

$$q_{j,k}(u) = \sup_{x \in \mathbf{R}^n} \{(1 + |x|^2)^{j/2} |D^\alpha u(x)| : |\alpha| \le k\}. \tag{1.8}$$

DEFINITION 1.4. *The space $\mathscr{S}(\mathbf{R}^n)$ consists of smooth functions u on $\mathbf{R}^n$ for which each $q_{j,k}(u)$ is finite, with the Fréchet space topology determined by these seminorms. Its dual is denoted $\mathscr{S}'(\mathbf{R}^n)$.*

We have defined the differential operators D^α on functions on $\mathbf{R}^n$ by (1.2); alternatively, with

$$D_j = i^{-1}\frac{\partial}{\partial x_j}, \qquad D^\alpha = D_1^{\alpha_1} \cdots D_n^{\alpha_n}.$$

If $P(\xi)$ is a polynomial,

$$P(\xi) = \sum_{|\alpha| \le m} a_\alpha \xi^\alpha,$$

denote by $P(D)$ the differential operator

$$P(D) = \sum_{|\alpha| \le m} a_\alpha D^\alpha. \tag{1.9}$$

More generally, if

$$p(x, \xi) = \sum_{|\alpha| \le m} a_\alpha(x)\xi^\alpha, \qquad a_\alpha \in C^\infty(\Omega),$$

let

$$p(x, D) = \sum_{|\alpha| \le m} a_\alpha(x)D^\alpha. \tag{1.10}$$

Such operators are called differential operators on $\Omega \subset \mathbf{R}^n$. Clearly $p(x, D): C^\infty(\Omega) \to C^\infty(\Omega)$ and $p(x, D): C_0^\infty(\Omega) \to C_0^\infty(\Omega)$, continuously. One can define operators $p(x, D)^t$ and $p(x, D)^*$ by

$$\begin{aligned} \langle p(x, D)u, f \rangle &= \langle u, p(x, D)^t f \rangle, \\ (p(x, D)u, f) &= (u, p(x, D)^* f) \end{aligned} \tag{1.11}$$

where either $u \in C^\infty(\Omega)$, $f \in C_0^\infty(\Omega)$ or $u \in C_0^\infty(\Omega)$, $f \in C^\infty(\Omega)$. We suppose Ω has a volume element, generally Lebesgue measure if $\Omega \subset \mathbf{R}^n$. It is easy to see that such operators $p(x, D)^t$ and $p(x, D)^*$ are also differential operators. If Ω is a manifold, $P: C^\infty(\Omega) \to C^\infty(\Omega)$ is called a differential operator if in any local coordinate system P has the above form, and given a volume element on Ω, P^t and P^* are defined analogously to (1.11) and one can see that these are also differential operators.

$p(x, D)$ is defined on $\mathscr{E}'(\Omega)$ and $\mathscr{D}'(\Omega)$ by

$$\begin{aligned} \langle u, p(x, D)\omega \rangle &= \langle p(x, D)^t u, \omega \rangle, \qquad \text{or equivalently,} \\ (u, p(x, D)\omega) &= (p(x, D)^* u, \omega). \end{aligned} \tag{1.12}$$

With such a definition, one can show that $p(x, D)$ is the unique continuous extension as an operator on $\mathscr{E}'(\Omega)$ or on $\mathscr{D}'(\Omega)$ of the operator $p(x, D)$ on $C_0^\infty(\Omega)$ or on $C^\infty(\Omega)$, respectively.

If the coefficients $a_\alpha(x) \in C^\infty(\mathbf{R}^n)$ satisfy the slow growth condition

$$|D^\beta a_\alpha(x)| \le C_{\alpha\beta}(1 + |x|^2)^N, \qquad N = N(\alpha, \beta), \tag{1.13}$$

then $p(x, D): \mathscr{S}(\mathbf{R}^n) \to \mathscr{S}(\mathbf{R}^n)$, continously, and similarly for $p(x, D)^*$ and $p(x, D)^t$. Thus (1.12) defines $p(x, D): \mathscr{S}'(\mathbf{R}^n) \to \mathscr{S}'(\mathbf{R}^n)$.

For more information on the topology of these spaces, the reader might refer to Schwartz [1], Donoghue [1], Yosida [1], or Trèves [1].

Exercise

1.1. Let $u \in \mathscr{E}'(\mathbf{R}^n)$ and suppose $\operatorname{supp} u = \{0\}$. Show that, for some K, $a_\alpha \in \mathbf{C}$,

$$u = \sum_{|\alpha| \le K} a_\alpha D^\alpha \delta.$$

Hint: First note that $\langle f, u \rangle = 0$ for all $f \in C^\infty(\mathbf{R}^n)$ vanishing to infinite order at 0. Now show there is a K such that $\langle f, u \rangle = 0$ for all $f \in C^\infty(\mathbf{R}^n)$

vanishing to order K at 0. Here δ, called the Dirac delta, or the delta function, is defined by

$$\langle f, \delta \rangle = f(0), \qquad f \in C^\infty(\mathbf{R}^n).$$

§2. The Fourier Transform

If $u \in L^1(\mathbf{R}^n)$, we define its Fourier transform by

$$\hat{u}(\xi) = (2\pi)^{-n/2} \int u(x) e^{-ix \cdot \xi} \, dx \tag{2.1}$$

where $x \cdot \xi = x_1\xi_1 + \cdots + x_n\xi_n$. Clearly $u \in L^1$ implies $\hat{u} \in L^\infty$ and $\|\hat{u}\|_{L^\infty} \le (2\pi)^{-n/2}\|u\|_{L^1}$. We also denote $\hat{u}(\xi)$ by $\mathfrak{F}u(\xi)$. Changing the sign in the exponent in (2.1), we also define

$$\mathfrak{F}^* u(\xi) = \tilde{u}(\xi) = (2\pi)^{-n/2} \int u(x) e^{ix \cdot \xi} \, dx.$$

PROPOSITION 2.1. *If $u \in \mathcal{S}(\mathbf{R}^n)$, then $\hat{u}, \tilde{u} \in \mathcal{S}(\mathbf{R}^n)$.*

Proof. If $u \in \mathcal{S}$, then we can differentiate (2.1) under the integral sign to obtain

$$D_\xi^\alpha \hat{u}(\xi) = (2\pi)^{-n/2} \int u(x)(-x)^\alpha e^{ix \cdot \xi} \, dx. \tag{2.2.}$$

Now, realizing that $\xi^\beta e^{-ix \cdot \xi} = (-1)^{|\beta|} D_x^\beta e^{-ix \cdot \xi}$ and integrating (2.2) by parts we get

$$\xi^\beta D_\xi^\alpha \hat{u}(\xi) = (2\pi)^{-n/2}(-1)^{|\alpha| + |\beta|} \int D_x^\beta(x^\alpha u(x)) e^{-ix \cdot \xi} \, dx, \tag{2.3}$$

which implies that each $\xi^\beta D_\xi^\alpha \hat{u} \in L^\infty(\mathbf{R}^n)$, provided $u \in \mathcal{S}(\mathbf{R}^n)$. This yields $\hat{u} \in \mathcal{S}$, and similarly one has $\tilde{u} \in \mathcal{S}$.

THEOREM 2.2. *The Fourier transform $\mathfrak{F}: \mathcal{S}(\mathbf{R}^n) \to \mathcal{S}(\mathbf{R}^n)$ is an isomorphism; in fact $\mathfrak{F}\mathfrak{F}^* = \mathfrak{F}^*\mathfrak{F} = I$, so in particular, for $u \in \mathcal{S}$,*

$$u(x) = (2\pi)^{-n/2} \int \hat{u}(\xi) e^{ix \cdot \xi} \, d\xi. \tag{2.4}$$

Proof. Note that (2.4) implies $\mathfrak{F}^*\mathfrak{F} = I$; the proof that $\mathfrak{F}\mathfrak{F}^* = I$ is similar, so it suffices to prove (2.4). Write

$$\begin{aligned} (2\pi)^{-n/2} \int \hat{u}(\xi) e^{ix \cdot \xi} \, d\xi &= (2\pi)^{-n} \iint u(y) e^{i(x-y) \cdot \xi} \, dy \, d\xi \\ &= \lim_{\epsilon \downarrow 0} (2\pi)^{-n} \iint u(y) e^{i(x-y) \cdot \xi - \epsilon|\xi|^2} \, dy \, d\xi \\ &= \lim_{\epsilon \downarrow 0} (2\pi)^{-n} \int \left\{ \int e^{i(x-y) \cdot \xi - \epsilon|\xi|^2} \, d\xi \right\} u(y) \, dy \\ &= \lim_{\epsilon \downarrow 0} \int p(\epsilon, x-y) u(y) \, dy \end{aligned} \tag{2.5}$$

where

$$p(\epsilon, x) = (2\pi)^{-n} \int e^{ix \cdot \xi - \epsilon|\xi|^2} \, d\xi. \tag{2.6}$$

In a moment we shall evaluate the integral (2.6) and show that

$$p(\epsilon, x) = (4\pi\epsilon)^{-n/2} e^{-|x|^2/4\epsilon} = \epsilon^{-n/2} q(x/\sqrt{\epsilon}) \tag{2.7}$$

where $q(x) = p(1, x) = (4\pi)^{-n/2} e^{-|x|^2/4}$. Note that $q \in \mathcal{S}(\mathbf{R}^n)$, and our derivation of (2.7) will show that

$$\int_{\mathbf{R}^n} q(x) \, dx = 1. \tag{2.8}$$

Consequently it is an elementary exercise to verify that, for any bounded continuous $u(x)$,

$$\lim_{\epsilon \downarrow 0} \int p(\epsilon, x - y) u(y) \, dy = u(x). \tag{2.9}$$

(2.5) and (2.9) yield (2.4).

To establish (2.7), note that $p(\epsilon, x)$, defined by (2.6), is an entire analytic function of $x \in \mathbf{C}^n$; it is convenient to verify that

$$p(\epsilon, ix) = (4\pi\epsilon)^{-n/2} e^{|x|^2/4\epsilon}, \qquad x \in \mathbf{R}^n \tag{2.10}$$

from which (2.7) follows by analytic continuation. Now,

$$\begin{aligned} p(\epsilon, ix) &= (2\pi)^{-n} \int e^{-x \cdot \xi - \epsilon|\xi|^2} \, d\xi \\ &= (2\pi)^{-n} e^{|x|^2/4\epsilon} \int e^{-|(x/2\sqrt{\epsilon}) + \sqrt{\epsilon}\xi|^2} \, d\xi \\ &= (2\pi)^{-n} e^{|x|^2/4\epsilon} \int e^{-\epsilon\xi^2} \, d\xi \\ &= (2\pi)^{-n} \epsilon^{-n/2} e^{|x|^2/4\epsilon} \int_{\mathbf{R}^n} e^{-|\xi|^2} \, d\xi, \end{aligned}$$

so to prove (2.10), it suffices to show that

$$\int_{\mathbf{R}^n} e^{-|\xi|^2} \, d\xi = \pi^{n/2}.$$

If

$$A = \int_{-\infty}^{\infty} e^{-|\xi|^2} \, d\xi,$$

then

$$\int_{\mathbf{R}^n} e^{-|\xi|^2} \, d\xi = A^n,$$

but

$$\begin{aligned} A^2 &= \int_{\mathbf{R}^2} e^{-|\xi|^2} \, d\xi \\ &= \int_0^{2\pi} \int_0^{\infty} e^{-r^2} \, r \, dr \, d\theta \\ &= 2\pi \int_0^{\infty} e^{-r^2} \, r \, dr = \pi, \end{aligned}$$

so we are done. These calculations also establish (2.8). The proof of Theorem 2.2 is complete.

We now define

$$\mathfrak{F}:\mathcal{S}'(\mathbf{R}^n) \to \mathcal{S}'(\mathbf{R}^n) \tag{2.11}$$

by

$$(u, \mathfrak{F}\omega) = (\mathfrak{F}^*u, \omega). \tag{2.12}$$

Note that (2.12) holds for $\omega \in \mathcal{S}(\mathbf{R}^n)$. This defines the unique continuous extension of $\mathfrak{F}$ from $\mathcal{S}$ to $\mathcal{S}'$. Similarly $\mathfrak{F}^*:\mathcal{S}' \to \mathcal{S}'$ is defined. From Theorem 2.2 it follows that

$$\mathfrak{F}\mathfrak{F}^* = \mathfrak{F}^*\mathfrak{F} = I \quad \text{on} \quad \mathcal{S}'(\mathbf{R}^n),$$

and hence $\mathfrak{F}$ and $\mathfrak{F}^*$ are isomorphisms from $\mathcal{S}'(\mathbf{R}^n)$ to itself.

Differentiating under the integral sign, or integrating by parts, we see that (2.4) implies, for $u \in \mathcal{S}(\mathbf{R}^n)$,

$$\begin{aligned} D^\alpha u(x) &= (2\pi)^{-n/2} \int \xi^\alpha \hat{u}(\xi) e^{ix\cdot\xi}\, d\xi, \\ (-x)^\beta u(x) &= (2\pi)^{-n/2} \int D_\xi^\beta \hat{u}(\xi) e^{ix\cdot\xi}\, d\xi. \end{aligned} \tag{2.13}$$

Thus $D^\alpha = \mathfrak{F}^{-1}\xi^\alpha\mathfrak{F}$, $x^\beta = \mathfrak{F}^{-1}(-D)^\beta\mathfrak{F}$, on $\mathcal{S}(\mathbf{R}^n)$. These formulas consequently hold on $\mathcal{S}'(\mathbf{R}^n)$. From $\mathfrak{F}^{-1} = \mathfrak{F}^*$ on $\mathcal{S}$ it follows that, for $u \in \mathcal{S}$

$$\|\hat{u}\|_{L^2}^2 = (\mathfrak{F}u, \mathfrak{F}u) = (\mathfrak{F}^*\mathfrak{F}u, u) = (u, u) = \|u\|_{L^2}^2. \tag{2.14}$$

From (2.14) it follows that $\mathfrak{F}$ has a unique extension as an isometry

$$\mathfrak{F}:L^2(\mathbf{R}^n) \to L^2(\mathbf{R}^n).$$

Similarly, $\mathfrak{F}^*$ extends uniquely to an isometry on $L^2(\mathbf{R}^n)$, and $\mathfrak{F}^*\mathfrak{F} = \mathfrak{F}\mathfrak{F}^* = I$ on L^2. Hence $\mathfrak{F}$ and $\mathfrak{F}^*$ are *unitary* operators on $L^2(\mathbf{R}^n)$. This fact is called Plancherel's theorem.

If $u, v \in \mathcal{S}(\mathbf{R}^n)$, we define the convolution $u*v$ by

$$u*v(x) = \int_{\mathbf{R}^n} u(y)v(x-y)\, dy. \tag{2.15}$$

Then $u*v \in \mathcal{S}$; we say $\mathcal{S}*\mathcal{S} \subset \mathcal{S}$. By (2.15) we also have $L^1*L^1 \subset L^1$, $C_0^\infty*\mathcal{S} \subset \mathcal{S}$, $C_0^\infty*C^\infty \subset C^\infty$, etc. Also we have, under appropriate circumstances,

$$\langle u*v, w\rangle = \langle v, u*\check{w}\rangle$$

where

$$jw(x) = \check{w}(x) = w(-x).$$

We can naturally define j on $\mathcal{D}'$, $\mathcal{E}'$, $\mathcal{S}'$, by duality. Then, for a distribution ω, $u*\omega$ is defined by

$$\langle v, u*\omega\rangle = \langle u*v, \check{\omega}\rangle.$$

We have $C_0^\infty * \mathcal{D}' \subset C^\infty$, $\mathcal{S} * \mathcal{S}' \subset \mathcal{S}'$, etc. We can get $\mathcal{E}' * \mathcal{D}' \subset \mathcal{D}'$, $\mathcal{E}' * \mathcal{S}' \subset \mathcal{S}'$, $\mathcal{E}' * \mathcal{E}' \subset \mathcal{E}'$ by

$$\langle v, \omega * \alpha \rangle = \langle v * \omega, \check{\alpha} \rangle.$$

Note that, for u, v belonging to appropriate pairs of spaces,

$$(u * v)\hat{}(\xi) = (2\pi)^{n/2} \hat{u}(\xi) \hat{v}(\xi) \tag{2.16}$$

as is most easily shown by first proving (2.16) for $u, v \in \mathcal{S}(\mathbf{R}^n)$, where it is a simple consequence of Fubini's theorem.

The *delta function*, an element $\delta \in \mathcal{E}' \subset \mathcal{S}' \subset \mathcal{D}'$, is defined by

$$\langle u, \delta \rangle = u(0).$$

Note that

$$\hat{\delta}(\xi) = (2\pi)^{-n/2},$$
$$\delta * \omega = \omega,$$
$$P(D)\omega = P(D)\delta * \omega.$$

By (2.13) we see that, for a polynomial $P(\xi)$,

$$(P(D)u)\hat{}(\xi) = P(\xi)\hat{u}(\xi)$$

or

$$P(D) = \mathfrak{F}^{-1} P(\xi) \mathfrak{F}. \tag{2.17}$$

More generally, for a function $p(\xi)$, we define

$$p(D) = \mathfrak{F}^{-1} p(\xi) \mathfrak{F}. \tag{2.18}$$

If $p(\xi)$ satisfies the slow growth condition (1.13), $p(D)$ maps $\mathcal{S}(\mathbf{R}^n)$ and $\mathcal{S}'(\mathbf{R}^n)$ to themselves. By Plancherel's theorem, $p(D): L^2 \to L^2$ if $p \in L^\infty(\mathbf{R}^n)$.

We end this section with a brief discussion of Fourier series, defined for functions (or distributions) on $\mathbf{T}^n = \mathbf{R}^n / 2\pi \mathbf{Z}^n$ by

$$u(x) = \sum_{m \in \mathbf{Z}^n} \hat{u}(m) e^{im \cdot x} \tag{2.19}$$

where

$$\hat{u}(m) = (2\pi)^{-n} \int_{\mathbf{T}^n} u(x) e^{-im \cdot x} \, dx \tag{2.20}$$

for functions, and, for $u \in \mathcal{D}'(\mathbf{T}^n)$,

$$\hat{u}(m) = (2\pi)^{-n} \langle u, e^{-im \cdot x} \rangle. \tag{2.21}$$

Plancherel's theorem here says

$$\sum_m |\hat{u}(m)|^2 = (2\pi)^{-n} \int_{\mathbf{T}^n} |u(x)|^2 \, dx.$$

Note that

$$D^\alpha u(x) = \sum_{m \in \mathbf{Z}^n} m^\alpha \hat{u}(m) e^{im \cdot x}. \tag{2.22}$$

It follows that $u \in C^\infty(\mathbf{T}^n)$ if, and only if, $\hat{u}(m)$ is a rapidly decreasing sequence in $\mathbf{Z}^n$, i.e., for all k,

$$\sup_{m \in \mathbf{Z}^n} (1+|m|)^k |\hat{u}(m)| < \infty. \tag{2.23}$$

By duality, $u \in \mathscr{D}'(\mathbf{T}^n)$ if, and only if, $\hat{u}(m)$ is a polynomially bounded sequence, i.e., for some ℓ,

$$|\hat{u}(m)| \leq C(1+|m|)^\ell. \tag{2.24}$$

As in (2.18), we define, for a function p on $\mathbf{Z}^n$,

$$p(D)u = \sum p(m)\hat{u}(m)e^{im \cdot x}. \tag{2.25}$$

Thus $p(D): C^\infty \to C^\infty$ and $p(D): \mathscr{D}'(\mathbf{T}^n) \to \mathscr{D}'(\mathbf{T}^n)$ provided $p(m)$ is polynomially bounded, while $p(D): L^2(\mathbf{T}^n) \to L^2(\mathbf{T}^n)$ if, and only if, $p(m)$ is bounded.

§3. Sobolev Spaces on $\mathbf{R}^n$

Let $k \geq 0$ be an integer. We define the Sobolev space $H^k(\mathbf{R}^n)$ to consist of all $u \in L^2(\mathbf{R}^n)$ such that $D^\alpha u \in L^2(\mathbf{R}^n)$, $|\alpha| \leq k$. By the Plancherel formula and (2.13), this is equivalent to requiring $\xi^\alpha \hat{u}(\xi) \in L^2$ for $|\alpha| \leq k$, or, what is the same, $(1+|\xi|)^k \hat{u} \in L^2$. This leads to the following more general definition.

DEFINITION 3.1. *For $s \in \mathbf{R}$, $H^s(\mathbf{R}^n)$ is the set of $u \in \mathscr{S}'(\mathbf{R}^n)$ such that $\hat{u}$ is locally square integrable and* $(1+|\xi|)^s \hat{u} \in L^2(\mathbf{R}^n)$.

We use the norm defined by

$$\|u\|_{H^s}^2 = \int (1+|\xi|^2)^s |\hat{u}(\xi)|^2 \, d\xi \tag{3.1}$$

which, when $s = k$ is an integer, is equivalent to

$$\sum_{|\alpha| \leq k} \|D^\alpha u\|_{L^2}^2.$$

$H^s(\mathbf{R}^n)$ is a Hilbert space, and the Fourier transform is an isometric isomorphism

$$\mathfrak{F}: H^s(\mathbf{R}^n) \to L^2(\mathbf{R}^n, (1+|\xi|^2)^s \, d\xi).$$

Since the Schwartz space $\mathscr{S}(\mathbf{R}^n)$ is clearly dense in $L^2(\mathbf{R}^n, (1+|\xi|^2)^s \, d\xi)$, and $\mathfrak{F}^{-1}\mathscr{S} = \mathscr{S}$, it follows that $\mathscr{S}(\mathbf{R}^n)$ is dense in $H^s(\mathbf{R}^n)$.

It is easy to see that, if $p(\xi)$ is a measurable function satisfying the estimate $|p(\xi)| \leq C(1+|\xi|)^m$, then $p(D)$, defined by (2.18), maps $H^s(\mathbf{R}^n) \to H^{s-m}(\mathbf{R}^n)$. In particular a differential operator with constant coefficients, of order m, maps H^s to H^{s-m}. If $a(x) \in \mathscr{S}$, then $au = (2\pi)^{n/2}\mathfrak{F}^{-1}(\hat{a}*\hat{u})$ shows that $au \in H^s(\mathbf{R}^n)$ provided $u \in H^s(\mathbf{R}^n)$. On the other hand, if $s = k \geq 0$ is an integer, then the basic rules of calculus show that $u \mapsto au$ is a map from $H^k(\mathbf{R}^n)$ to itself provided

$$|D^\beta a(x)| \leq c_\beta, \qquad \text{all } \beta \geq 0. \tag{3.2}$$

It follows by duality that $u \mapsto au$ maps $H^{-k}(\mathbf{R}^n)$ to itself. It will follow from the results of the next section that such a map takes $H^s(\mathbf{R}^n)$ to itself, for all $s \in \mathbf{R}$. Consequently, if

$$p(x, \xi) = \sum_{|\alpha| \leq m} a_\alpha(x)\xi^\alpha$$

and each coefficient $a_\alpha(x)$ satisfies (3.2), then $p(x, D): H^s(\mathbf{R}^n) \to H^s(\mathbf{R}^n)$.

We now show that elements of $H^s(\mathbf{R}^n)$ are smooth in the classical sense for sufficiently large positive s.

THEOREM 3.1. *If $s > n/2$, then each $u \in H^s(\mathbf{R}^n)$ is bounded and continuous.*

Proof. By the Fourier inversion formula, it suffices to prove $\hat{u}(\xi)$ is in $L^1(\mathbf{R}^n)$. Indeed, using Cauchy's inequality, we get

$$\int |\hat{u}(\xi)|\, d\xi \leq \left(\int |\hat{u}(\xi)|^2(1+|\xi|^2)^s\, d\xi\right)^{1/2}\left(\int (1+|\xi|^2)^{-s}\, d\xi\right)^{1/2}$$

the latter factor being finite precisely for $s > n/2$. Note that, by the Riemann–Lebesgue lemma, we can even conclude that $u(x)$ vanishes at infinity.

COROLLARY 3.2. *If $s > (n/2) + k$, then $H^s(\mathbf{R}^n) \subset C^k(\mathbf{R}^n)$.*

If $s = (n/2) + \alpha$, $0 < \alpha < 1$, we can obtain Hölder continuity. We say $u \in C^\alpha(\mathbf{R}^n)$, $0 < \alpha < 1$, provided u is bounded and

$$|u(x+y) - u(y)| \leq c|y|^\alpha, \qquad |y| \leq 1.$$

PROPOSITION 3.3. *If $s = (n/2) + \alpha$, $0 < \alpha < 1$, then $H^s(\mathbf{R}^n) \subset C^\alpha(\mathbf{R}^n)$.*

Proof. For $u \in H^s(\mathbf{R}^n)$, use the Fourier inversion formula to write

$$\begin{aligned} |u(x+y) - u(x)| &= (2\pi)^{-n/2}\left|\int \hat{u}(\xi)e^{ix\cdot\xi}(e^{iy\cdot\xi} - 1)\, d\xi\right| \\ &\leq c\left(\int |\hat{u}|^2(1+|\xi|)^{n+2\alpha}\, d\xi\right)^{1/2} \\ &\quad \times \left(\int |e^{iy\cdot\xi} - 1|^2(1+|\xi|)^{-n-2\alpha}\, d\xi\right)^{1/2}. \end{aligned}$$

Now if $|y| \le 1$,

$$\int |e^{iy\cdot\xi} - 1|^2 (1+|\xi|)^{-n-2\alpha}\, d\xi$$

$$\le c \int_{|\xi| \le |y|^{-1}} |y|^2 |\xi|^2 (1+|\xi|)^{-n-2\alpha}\, d\xi + 4 \int_{|\xi| \ge |y|^{-1}} (1+|\xi|)^{-n-2\alpha}\, d\xi$$

$$\le c|y|^2 \int_0^{|y|^{-1}} \frac{r^{n+1}}{(1+r)^{n+2\alpha}}\, dr + 4c \int_{|y|^{-1}}^{\infty} \frac{r^{n-1}}{(1+r)^{n+2\alpha}}\, dr$$

$$\le c|y|^2 + c|y|^2 \begin{Bmatrix} \dfrac{|y|^{2\alpha-2} - 1}{2\alpha - 2} & (0 < \alpha < 1) \\ \log \dfrac{1}{|y|} & (\alpha = 1) \end{Bmatrix} + c|y|^{2\alpha}$$

which implies that, for $|y| \le 1/2$,

$$\text{(3.3)} \qquad \begin{aligned} |u(x+y) - u(x)| &\le c|y|^{\alpha} \qquad \text{if} \quad 0 < \alpha < 1 \\ &\le c|y|\left(\log \frac{1}{|y|}\right)^{1/2} \qquad \text{if} \quad \alpha = 1. \end{aligned}$$

Note the different modulus of continuity for $\alpha = 1$. Elements of $H^{n/2+1}(\mathbf{R}^n)$ need not be Lipschitz, and elements of $H^{n/2}(\mathbf{R}^n)$ need not be bounded. In fact, if $\hat{u}(\xi) = (1+|\xi|)^{-n}/\log(2+|\xi|)$, it is easy to see that $u \in H^{n/2}(\mathbf{R}^n)$ and not hard to show that $u \notin L^{\infty}(\mathbf{R}^n)$. (Generally, given any positive $\hat{u}$, if $\hat{u} \notin L^1$, then $u \notin L^{\infty}$.)

By duality, Theorem 3.1 implies that all finite measures on $\mathbf{R}^n$ belong to $H^s(\mathbf{R}^n)$ for $s < -(n/2)$. In particular, $\delta \in H^{-n/2-\epsilon}(\mathbf{R}^n)$, $\epsilon > 0$. Consequently, $D^{\alpha}\delta \in H^{-n/2-|\alpha|-\epsilon}(\mathbf{R}^n)$.

Theorem 3.1 is one of many Sobolev imbedding theorems. Another important result is that

$$\text{(3.4)} \qquad H^s(\mathbf{R}^n) \subset L^q(\mathbf{R}^n), \qquad q = \frac{2n}{n-2s}, \qquad \text{if } 0 \le s < \tfrac{1}{2}n.$$

We refer the reader to Bergh and Löfström [1] or Stein [1] for a proof of (3.4); a slightly weaker result will be established in the next section.

Next we consider the behavior of the trace map; with $x = (x_1, x')$, $\tau u(x') = u(0, x')$; clearly $\tau : \mathscr{S}(\mathbf{R}^n) \to \mathscr{S}(\mathbf{R}^{n-1})$.

THEOREM 3.4. *The map τ extends uniquely to a continuous linear operator*

$$\tau : H^s(\mathbf{R}^n) \to H^{s-1/2}(\mathbf{R}^{n-1}), \qquad \text{if } s > \tfrac{1}{2}.$$

Proof. It suffices to derive the appropriate norm estimate. Let $f = \tau u$, $u \in \mathcal{S}$. Therefore,

$$\hat{f}(\xi') = \int \hat{u}(\xi)\, d\xi_1$$

and

$$|\hat{f}(\xi')|^2 \le \left(\int |\hat{u}(\xi)|^2(1+|\xi|)^{2s}\, d\xi_1\right)^{1/2}\left(\int (1+|\xi|)^{-2s}\, d\xi_1\right)^{1/2}$$

if $s > 1/2$, and we estimate the last factor by

$$\begin{aligned}\int (1+|\xi|)^{-2s}\, d\xi_1 &\le c\int (1+|\xi'|^2+\xi_1^2)^{-s}\, d\xi_1 \\ &= c'(1+|\xi'|^2)^{-s+1/2}.\end{aligned} \tag{3.5}$$

Therefore,

$$(1+|\xi'|^2)^{s-1/2}|\hat{f}(\xi')|^2 \le c\int |\hat{u}(\xi)|^2(1+|\xi|)^{2s}\, d\xi,$$

and integrating with respect to ξ' gives

$$\|f\|^2_{H^{s-1/2}(\mathbf{R}^n)} \le c\|u\|^2_{H^s(\mathbf{R}^n)} \tag{3.6}$$

as desired.

Iteration shows that, if $u \in H^s(\mathbf{R}^n)$, $s > k/2$, then $u|_{\mathbf{R}^{n-k}}$ belongs to $H^{s-k/2}(\mathbf{R}^{n-k})$. The result of Theorem 3.4 is sharp.

THEOREM 3.5. *The restriction map $\tau: H^s(\mathbf{R}^n) \to H^{s-1/2}(\mathbf{R}^{n-1})$, $s > 1/2$, is onto.*

Proof. Take $g \in H^{s-1/2}(\mathbf{R}^{n-1})$, Fourier transform $\hat{g}(\xi')$. Let

$$\varphi(\xi) = \hat{g}(\xi')\frac{(1+|\xi'|^2)^{s-1/2}}{(1+|\xi|^2)^s}$$

and let $u = \tilde{\varphi}(x)$; i.e., $\hat{u} = \varphi$. We claim $u \in H^s(\mathbf{R}^n)$ and $u(0, x') = cg(x')$, c a nonzero constant. Indeed,

$$(1+|\xi|^2)^s|\varphi(\xi)|^2 = |\hat{g}(\xi')|^2(1+|\xi'|^2)^{s-1/2}\frac{(1+|\xi'|^2)^{s-1/2}}{(1+|\xi|^2)^s} \tag{3.7}$$

and, by (3.5), $\int (1+|\xi|^2)^{-s}\, d\xi_1 \le c(1+|\xi'|^2)^{-s+1/2}$, so the left side of (3.7) has finite integral, giving $u \in H^s(\mathbf{R}^n)$. Meanwhile

$$\begin{aligned}\int \varphi(\xi)\, d\xi_1 &= \hat{g}(\xi')(1+|\xi'|^2)^{s-1/2}\int (1+|\xi|^2)^{-s}\, d\xi_1 \\ &= c\hat{g}(\xi_1), \qquad c \ne 0,\end{aligned}$$

so $u(0, x') = cg(x')$, as asserted.

Note that, if the operator $(1-\Delta)^{s/2}$ is defined by (2.18) with $p(\xi) = (1+|\xi|^2)^{s/2}$, then

$$H^s(\mathbf{R}^n) = (1-\Delta)^{-s/2}L^2(\mathbf{R}^n). \tag{3.8}$$

Also, for $s > 0$, $H^s(\mathbf{R}^n) = \mathscr{D}((1-\Delta)^{s/2})$, the domain of the unbounded self-adjoint operator $(1-\Delta)^{s/2}$.

Also note that $H^{-s}(\mathbf{R}^n)$ is the dual to $H^s(\mathbf{R}^n)$.

§4. The Complex Interpolation Method

It turns out that, for example, if $T: H^m(\mathbf{R}^n) \to H^m(\mathbf{R}^n)$ and $T: H^\mu(\mathbf{R}^n) \to H^\mu(\mathbf{R}^n)$, then $T: H^s(\mathbf{R}^n) \to H^s(\mathbf{R}^n)$, for all s between m and μ. This is a special case of an important circle of results derivable by Calderon's complex interpolation method (Calderon [2]), which we describe here.

Let E and F be two Banach spaces, naturally included in some common locally convex space V. $E + F = \{u + v : u \in E, v \in F\}$ has a natural Banach space topology. If Ω is the vertical strip in the complex plane

$$\Omega = \{z \in \mathbf{C} : 0 < Re\, z < 1\},$$

define

(4.1) $\mathscr{H}_{E,F}(\Omega) = \{u(z)$ bounded continuous on $\bar{\Omega}$ with values in $E+F$; holomorphic on Ω: $\|u(iy)\|_E$ and $\|u(1+iy)\|_F$ bounded$\}$.

We define the intermediate spaces $[E, F]_\theta$ by

$$[E, F]_\theta = \{u(\theta); u \in \mathscr{H}_{E,F}(\Omega)\}, \qquad 0 \le \theta \le 1.$$

We give $[E, F]_\theta$ the Banach space topology making it isomorphic to the quotient $\mathscr{H}_{E,F}(\Omega)/\{u : u(\theta) = 0\}$. Our first important result is the following.

THEOREM 4.1. *Suppose T is defined on $E+F$ to $\tilde{E}+\tilde{F}$ as a linear operator and $T: E \to \tilde{E}$, $T: F \to \tilde{F}$, where $\tilde{E}$, $\tilde{F}$ are contained in some locally convex space $\tilde{V}$. Then, for $0 \le \theta \le 1$,*

$$T: [E, F]_\theta \to [\tilde{E}, \tilde{F}]_\theta. \tag{4.2}$$

Proof. Let $u \in \mathscr{H}_{E,F}(\Omega)$, $u(\theta) = u$. One verifies immediately that $Tu(z) \in \mathscr{H}_{\tilde{E},\tilde{F}}(\Omega)$, so $Tu = Tu(\theta) \in [\tilde{E}, \tilde{F}]_\theta$, as asserted.

Property (4.2) is the interpolation property. To apply it, we need to characterize $[E, F]_\theta$ in important cases. Here is one basic result.

THEOREM 4.2. *If A is a positive self-adjoint operator on a Hilbert space H, then for $0 \le \theta \le 1$,*

$$[H, \mathscr{D}(A)]_\theta = \mathscr{D}(A^\theta). \tag{4.3}$$

Proof. First suppose $v \in \mathscr{D}(A^\theta)$. We want to write $v = u(\theta)$ for some $u \in \mathscr{H}_{H,\mathscr{D}(A)}(\Omega)$. Let

$$u(z) = A^{-z+\theta}v.$$

Then $u(\theta) = v$, $u(iy) = A^{-iy}(A^\theta v)$ is bounded in H, and $u(1+iy) = A^{-1}A^{-iy}(A^\theta v)$ is bounded in $\mathscr{D}(A)$.

Conversely, suppose $u(z) \in \mathscr{H}_{H,\mathscr{D}(A)}(\Omega)$. We need to prove that $u(\theta) \in \mathscr{D}(A^\theta)$. Let $\epsilon > 0$ and note that, by the maximum principle,

$$\begin{aligned}
&\|(1+i\epsilon A)^{-1}A^z u(z)\|_H \\
&\qquad \le \sup_{y\in\mathbf{R}} \max\{\|(1+i\epsilon A)^{-1}A^{iy}u(y)\|_H, \|(1+i\epsilon A)^{-1}A^{1+iy}u(1+iy)\|_H\} \\
&\qquad \le C, \qquad \text{independent of } \epsilon.
\end{aligned}$$

This implies $A^\theta u(\theta) \in H$, so $u(\theta) \in \mathscr{D}(A^\theta)$, as desired.

An immediate corollary is

$$(4.4) \qquad [H^\sigma(\mathbf{R}^n), H^s(\mathbf{R}^n)]_\theta = H^{\theta s+(1-\theta)\sigma}(\mathbf{R}^n).$$

Thus maps that take $H^k(\mathbf{R}^n)$ to itself for all integers (respectively, all nonnegative integers) automatically take $H^s(\mathbf{R}^n)$ to itself for all real s (respectively, for all $s \ge 0$). In particular, let $\chi: \mathbf{R}^n \to \mathbf{R}^n$ be a C^∞ diffeomorphism, and define χ^* by

$$\chi^* u(x) = u(\chi(x)).$$

The chain rule implies $\chi^*: H^k(\mathbf{R}^n) \to H^k(\mathbf{R}^n)$ for all positive integers if χ is *linear* outside some compact set, and dually $\chi^*: H^{-k} \to H^{-k}$. Thus interpolation shows that

$$\chi^*: H^s(\mathbf{R}^n) \to H^s(\mathbf{R}^n), \qquad s \in \mathbf{R}.$$

This invariance of $H^s(\mathbf{R}^n)$ under such coordinate changes will allow us to define $H^s(M)$ where M is a compact manifold, in the next section.

It is also useful to know how L^p spaces interpolate. Let X be a σ-finite measure space, then

$$(4.5) \qquad [L^q(X), L^p(X)]_\theta = L^r(X), \qquad \frac{1}{r} = \frac{\theta}{p} + \frac{1-\theta}{q}$$

provided $p, q < \infty$, $0 \le \theta \le 1$. (4.5) also holds with p or $q = \infty$ if $0 < \theta < 1$, and if X has finite measure it holds for $0 \le \theta \le 1$. For a proof of (4.5), see Calderon [2] or Bergh and Löfström [1], or the reader might try it as an exercise. A consequence of (4.5) for $X = \mathbf{R}^n$ is the following result of Hausdorff and Young on the behavior of the Fourier transform.

PROPOSITION 4.3. *The Fourier transform* $\mathfrak{F}: L^p(\mathbf{R}^n) \to L^{p'}(\mathbf{R}^n)$ *if* $1 \le p \le 2$, $(1/p) + (1/p') = 1$.

Proof. We know that $\mathfrak{F}: L^1 \to L^\infty$ and $\mathfrak{F}: L^2 \to L^2$, so the rest follows from (4.5) by interpolation.

As a consequence of Proposition (4.3) we can give a short proof of a result slightly weaker than (3.4), namely,

$$(4.6) \qquad H^s(\mathbf{R}^n) \subset L^p(\mathbf{R}^n), \qquad 2 \le p < \frac{2n}{n-2s}, \qquad \text{if } 0 < s < \tfrac{1}{2}n.$$

Indeed, if $u \in H^s(\mathbf{R}^n)$, $s > 0$, and if p' is such that $2p's/(2-p') > n$, then

$$\int |\hat{u}(\xi)|^{p'}\, d\xi \le \left(\int (|\hat{u}|^{p'}(1+|\xi|)^{p's})^{q'}\, d\xi \right)^{1/q'} \left(\int (1+|\xi|)^{-p'sq}\, d\xi \right)^{1/q}$$

where $q' = 2/p'$, so $q = 2/(2-p')$. Hence, for $u \in H^s(\mathbf{R}^n)$, $s > 0$,

$$\hat{u}(\xi) \in L^{p'} \qquad \text{for} \qquad 2 \le p < \frac{2n}{n-2s}$$

and, by Proposition 4.3, with the roles of p and p' interchanged, we get (4.6).

There is also a real interpolation, introduced by Lions and Peetre, which we briefly mention, refering to Bergh and Löfström [1] or Lions and Magenes [11] for details. For simplicity, suppose $F \subset E$, the injection being continuous; E and F are Banach spaces, as before. For $u \in E$, let

$$(4.7) \qquad K(u, \epsilon) = \inf_{v \in F} \|u - v\|_E + \epsilon \|v\|_F.$$

We define, for $0 < \theta < 1$, $1 \le p \le \infty$,

$$(4.8) \qquad [E, F]_{\theta,p} = \{u \in E : K(u, \cdot)E^{-\theta} \in L^p(\mathbf{R}^+, \epsilon^{-1}\, d\epsilon)\}.$$

An interpolation theorem analogous to Theorem 4.1 holds. In fact, one of the advantages of these real interpolation spaces is that *nonlinear* mappings can be handled, under certain hypotheses. See Tartar [1], Bona and Scott [1], or Peetre [1].

It turns out that, since $L^2(\mathbf{R}^n)$ is a Hilbert space, for $0 < \theta < 1$,

$$H^{k\theta}(\mathbf{R}^n) = [L^2(\mathbf{R}^n), H^k(\mathbf{R}^n)]_\theta = [L^2(\mathbf{R}^n), H^k(\mathbf{R}^n)]_{\theta,2}.$$

However, for general Banach spaces E and F, none of the spaces (4.8) need coincide with the complex interpolation spaces $[E, F]_\theta$.

§5. Sobolev Spaces on Bounded Domains and Compact Manifolds

Let M be a compact manifold, which we could endow with a Riemannian metric. If $u \in \mathscr{D}'(M)$, we say $u \in H^s(M)$ provided that, on any coordinate patch $U \subset M$, any $\psi \in C_0^\infty(U)$, the element $u\psi \in \mathscr{E}'(U)$ belongs to $H^s(U)$, if U is identified with its image in $\mathbf{R}^n$. In particular, for $s = k$ a nonnegative integer, $H^k(M)$ in the set of $u \in L^2(M)$ such that, for any ℓ smooth vector fields $X_1, \dots, X_\ell$ on M, $\ell \le k$, $X_1 \cdots X_\ell u \in L^2(M)$. By the invariance under coordinate changes derived in Section 4, it suffices to work with any single coordinate cover of M.

A fairly straightforward consequence of (4.4), whose proof we leave to the reader, is that

$$[H^s(M), H^\sigma(M)]_\theta = H^{\theta\sigma + (1-\theta)s}(M). \tag{5.1}$$

If M has a Riemannian metric and Δ is its Laplace operator, it will follow from results of Chapter III, Section 1 (Exercise 1.2) that, if $(-\Delta)^k$ is defined as an unbounded operator on $L^2(M)$ with domain

$$\mathcal{D}((-\Delta)^k) = H^{2k}(M),$$

then $(-\Delta)^k$ is self adjoint. Granted this, (5.1) and (4.3) imply

$$\mathcal{D}((-\Delta)^{s/2}) = H^s(M), \qquad s > 0. \tag{5.2}$$

Actually, the operator $(-\Delta)^{s/2}$ will turn out to satisfy the conditions of that exercise, but that fact will not be proved until Chapter XII.

In case M is the torus $\mathbf{T}^n$, $H^s(\mathbf{T}^n)$ can be characterized by Fourier series; $u \in H^s(\mathbf{T}^n)$ if, and only if,

$$\sum_{m \in \mathbf{Z}^n} |\hat{u}(m)|^2 (1 + |m|)^{2s} < \infty. \tag{5.3}$$

Indeed, for $s = k$ a nonnegative integer, this follows from Plancherel's theorem on the torus, and then it follows in general by duality and interpolation. In particular, (5.2) is transparent for $M = \mathbf{T}^n$. Since the Fourier series representation diagonalizes Δ on $\mathbf{T}^n$, we see that, for any $\sigma > 0$,

$$(1 - \Delta)^{-\sigma/2} : L^2(\mathbf{T}^n) \to L^2(\mathbf{T}^n) \text{ is compact.} \tag{5.4}$$

The operator is self adjoint and its eigenvalues form a sequence tending to zero, all having finite multiplicity. It follows that

$$I : H^{s+\sigma}(\mathbf{T}^n) \to H^s(\mathbf{T}^n) \text{ is compact}, \qquad \sigma > 0, \qquad s \in \mathbf{R}. \tag{5.5}$$

I is the inclusion map. For any compact M, by using a coordinate cover and partition of unity to break things up on M and transfer them to the torus, one can deduce from (5.5) that

$$I : H^{s+\sigma}(M) \to H^s(M) \text{ is compact}, \qquad \sigma > 0, \qquad s \in \mathbf{R}. \tag{5.6}$$

We also remark that $H^{-s}(M)$ is the dual of $H^s(M)$.

Preparatory to studying compact manifolds with boundary, we look at Sobolev spaces on $\mathbf{R}^n_+ = \{x \in \mathbf{R}^n : x_1 > 0\}$. For $k \ge 0$ an integer, we want $H^k(\mathbf{R}^n_+) = \{u \in L^2(\mathbf{R}^n_+) : D^\alpha u \in L^2(\mathbf{R}^n_+), |\alpha| \le k\}$. It is not difficult to show that the space $\mathscr{S}(\mathbf{R}^n_+)$ of restrictions to $\mathbf{R}^n_+$ of $\mathscr{S}(\mathbf{R}^n)$ is dense in $H^k(\mathbf{R}^n_+)$. We claim each $u \in H^k(\mathbf{R}^n_+)$ is the restriction to $\mathbf{R}^n_+$ of an element of $H^k(\mathbf{R}^n)$. To see this, fix an integer N and let

$$Eu(x) = \begin{cases} u(x), & x_1 > 0 \\ \sum_{j=1}^{N} a_j u(-jx_1, x'), & x_1 < 0 \end{cases} \tag{5.7}$$

defined a priori for $u \in \mathscr{S}(\mathbf{R}^n_+)$. The a_j will be picked to yield:

THEOREM 5.1. *The map E has a unique continuous linear extension to*

$$E: H^k(\mathbf{R}^n_+) \to H^k(\mathbf{R}^n), \qquad k \le N-1.$$

Proof. It suffices to derive an H^k estimate on Eu for $u \in \mathscr{S}(\mathbf{R}^n_+)$. Such an estimate will automatically follow if all the derivatives of Eu of order $\le N-1$ match up at $x_1 = 0$, i.e., if

$$\sum_{j=1}^{N} (-j)^\ell a_j = 1, \qquad \ell = 0, 1, \ldots, N-1. \tag{5.8}$$

Formula (5.8) is a linear system of N equations in N unknowns; its determinant is seen to be nonzero, so appropriate a_ℓ can be found.

COROLLARY 5.2. *The restriction $\rho: H^k(\mathbf{R}^n) \to H^k(\mathbf{R}^n_+)$ is surjective.*

Suppose, for $s \ge 0$, we define $H^s(\mathbf{R}^n_+)$ by interpolation:

$$H^s(\mathbf{R}^n_+) = [L^2(\mathbf{R}^n_+), H^k(\mathbf{R}^n_+)]_\theta; \qquad s < k, \qquad s = \theta k. \tag{5.9}$$

Using Theorem 5.1, we can show that this definition is consistent; if s is an integer, it agrees with the first definition, and for general $s \ge 0$, (5.9) is independent of the integer $k > s$ chosen. Such an argument also yields the following:

$$\rho: H^s(\mathbf{R}^n) \to H^s(\mathbf{R}^n_+) \text{ is onto}, \qquad s \ge 0. \tag{5.10}$$

Thus we can identify $H^s(\mathbf{R}^n_+)$ with $H^s(\mathbf{R}^n)/\{u : u = 0 \text{ on } \mathbf{R}^n_+\}$, with the quotient topology. To prove (5.10), note that, by Theorem 5.1 and interpolation, one has, for $0 \le s \le N-1$,

$$E: H^s(\mathbf{R}^n_+) \to H^s(\mathbf{R}^n) \tag{5.11}$$

and (5.10) follows immediately from (5.11), taking N large enough.

Now let Ω be a smooth compact manifold with smooth boundary, possibly a bounded domain in $\mathbf{R}^n$ with smooth boundary. We can suppose Ω is imbedded as a submanifold of a compact manifold M of the same dimension. If $\Omega \subset \mathbf{R}^n$, you can put Ω is a large box and identify opposite sides to get $\Omega \subset \mathbf{T}^n$; in the general case you can let M be the double of Ω (see Milnor [1]).

If $k \geq 0$ is an integer, we define $H^k(\Omega)$ to consist of all $u \in L^2(\Omega)$ such that, for all differential operators P of order $\leq k$ whose coefficients are smooth on $\bar{\Omega}$, $Pu \in L^2(\Omega)$. $C^\infty(\bar{\Omega})$ can be seen to be dense in $H^k(\Omega)$, and, by covering a neighborhood $\partial\Omega \subset M$ with coordinate patches and locally using the extension operator E from above, we can get, for each finite N an extension operator

$$E: H^k(\Omega) \to H^k(M), \qquad 0 \leq k \leq N-1. \tag{5.12}$$

If, for real $s \geq 0$ we define $H^s(\Omega)$ by

$$H^s(\Omega) = [L^2(\Omega), H^k(\Omega)]_\theta, \qquad s = \theta k, \qquad \theta < 1, \tag{5.13}$$

we see that $E: H^s(\Omega) \to H^s(M)$, so the restriction $\rho: H^s(M) \to H^s(\Omega)$ is onto, and

$$H^s(\Omega) \approx H^s(M)/\{u : u|_\Omega = 0\}, \tag{5.14}$$

which shows that (5.13) is independent of k and consistent for s an integer. Factoring the natural inclusion $H^{s+\sigma}(\Omega) \to H^s(\Omega)$ $(\sigma > 0)$ through by

$$\begin{array}{ccc} H^{s+\sigma}(\Omega) & \xrightarrow{I} & H^s(\Omega) \\ {\scriptstyle E}\downarrow & & \uparrow{\scriptstyle \rho} \\ H^{S+\sigma}(M) & \xrightarrow{I} & H^s(M) \end{array}$$

we see that, for $s \geq 0$,

$$I: H^{s+\sigma}(\Omega) \to H^s(\Omega) \text{ is compact}, \qquad \sigma > 0, \tag{5.15}$$

a result known as Rellich's theorem.

$\partial\Omega$ is a smooth compact manifold, on which Sobolev spaces have been defined. Upon using local coordinate systems flattening out $\partial\Omega$, together with a partition of unity, and applying the trace theorem, Theorem 3.4, one can show that the natural trace map $\tau u = u|_{\partial\Omega}$ is a continuous operator

$$\tau: H^s(\Omega) \to H^{s-1/2}(\partial\Omega), \qquad s > \tfrac{1}{2}, \tag{5.16}$$

and furthermore τ is surjective.

For $s \geq 0$, we define $\mathring{H}^s(\Omega)$ to be the closure of $C_0^\infty(\mathring{\Omega})$ in $H^s(\Omega)$, where $\mathring{\Omega}$ denotes the interior of Ω. If $s > 1/2$, $u \in \mathring{H}^s(\Omega)$, it follows from (5.16) that $u|_{\partial\Omega} = 0$, so $\mathring{H}^s(\Omega)$ is strictly contained in $H^s(\Omega)$ for $s > 1/2$. On the other hand, we have

$$\mathring{H}^s(\Omega) = H^s(\Omega) \qquad \text{if} \qquad 0 \leq s \leq \tfrac{1}{2}. \tag{5.17}$$

That is, $C_0^\infty(\mathring{\Omega})$ is dense in $H^s(\Omega)$ if $0 \le s \le 1/2$. By (5.14) we can identify the dual of $H^s(\Omega)$ with $\{\omega \in H^{-s}(M) = \langle u, \omega \rangle = 0 \text{ if } u \in H^s(M),\ u = 0 \text{ on } \Omega\}$, or

$$H^s(\Omega)^* = \{u \in H^{-s}(M) : \operatorname{supp} u \subset \bar{\Omega}\}. \tag{5.18}$$

Using this, we can prove (5.17), for otherwise there would be a nonzero element ω of $H^{-s}(M)$, $0 \le s \le 1/2$, supported on $\partial\Omega$. In fact, generally speaking, if S is a smooth submanifold of M of codimension k, then

$$u \in H^{-k/2}(M), \qquad \operatorname{supp} u \subset S \Rightarrow u = 0. \tag{5.19}$$

By reducing (5.19) to the case $u \in \mathscr{E}'(\mathbf{R}^n)$, $\operatorname{supp} u \subset \mathbf{R}^{n-k}$, and using the partial Fourier transform in the $\mathbf{R}^{n-k}$ variables, such an assertion in this case follows from the fact that

$$u \in H^{-k/2}(\mathbf{R}^k), \qquad \operatorname{supp} u \subset \{0\} \Rightarrow u = 0. \tag{5.20}$$

In fact, such u must be a finite sum of derivatives of δ (see Exercise 1.1) and its Fourier transform must be a polynomial, from which (5.20) follows. Tracing backwards yields (5.20) $\Rightarrow$ (5.19) $\Rightarrow$ (5.17).

It is easy to see that, for $k \ge 0$ an integer,

$$\mathring{H}^k(\Omega) = \{u \in H^k(M) : \operatorname{supp} u \subset \bar{\Omega}\}. \tag{5.21}$$

In fact, it can be shown that, for $s \ge 0$

$$\mathring{H}^s(\Omega) = \{u \in H^s(M) : \operatorname{supp} u \subset \bar{\Omega}\}, \qquad s - \tfrac{1}{2} \notin \mathbf{Z}. \tag{5.22}$$

The fact that (5.22) fails for s a half-integer is intriguing, especially since it turns out that

$$[L^2(\Omega), \mathring{H}^k(\Omega)]_\theta = \{u \in H^{k\theta}(M) : \operatorname{supp} u \subset \bar{\Omega}\}.$$

For information on this we refer the reader to Fujiwara [1].

For more information on $H^s(\Omega)$, $\mathring{H}^s(\Omega)$ and related function spaces, we refer the reader to Lions and Magenes [1], Hörmander [6], and Trèves [1]. Also Adams [1] collects a lot of material on the spaces $H^k(\Omega)$, for $k \ge 0$ an integer, when Ω has a fairly nasty boundary.

We close this section with a simple exercise that brings together several techniques developed so far. Let Ω be the unit disc in $\mathbf{R}^2$:

$$\Omega = \{(x, y) \in \mathbf{R}^2 : x^2 + y^2 \le 1\}.$$

Then $\partial\Omega = S^1$ is the unit circle. Denote $x + iy = z = re^{i\theta}$, $\theta \in \mathbf{R}/2\pi\mathbf{Z} \approx S^1$. For smooth $f(\theta)$, we denote by $u = PIf$ the unique harmonic function on Ω equal to $f(\theta)$ on $\partial\Omega$. We can write out $u(re^{i\theta})$ using Fourier series. If

$$f(\theta) = \sum_{n=-\infty}^{\infty} \hat{f}(n) e^{in\theta}; \qquad \hat{f}(n) = \frac{1}{2\pi} \int_0^{2\pi} f(\theta) e^{-in\theta}\, d\theta, \tag{5.23}$$

let

$$(5.24)\qquad \begin{aligned} u(z) &= \sum_{n=-\infty}^{\infty} \hat{f}(n) r^{|n|} e^{in\theta} \\ &= \sum_{n=0}^{\infty} \hat{f}(n) z^n + \sum_{n=1}^{\infty} \hat{f}(-n) \bar{z}^n. \end{aligned}$$

The last formula makes it apparent that u is harmonic since it is the sum of a holomorphic and an antiholomorphic function. That $u|_{\partial\Omega} = f$ is the Fourier inversion formula. We prove the following.

PROPOSITION 5.3. *The map* $PI: H^s(S^1) \to H^{s+1/2}(\Omega)$ *for* $s \geq -(1/2)$.

Proof. It suffices to prove this for $s = k - (1/2)$, $k = 0, 1, \ldots$, and the result for general $s \geq -(1/2)$ will follow by interpolation. To say $f \in H^{k-1/2}(S^1)$ means

$$\sum_{n=-\infty}^{\infty} |\hat{f}(n)|^2 (1+|n|)^{2k-1} < \infty.$$

Now the functions $\{r^{|n|} e^{in\theta} : n \in \mathbf{Z}\}$ are clearly mutually orthogonal on $L^2(\Omega)$, and

$$(5.25)\qquad \iint_\Omega |r^{|n|} e^{in\theta}|^2 \, dx\, dy = 2\pi \int_0^1 r^{2|n|} r\, dr = \frac{2\pi}{2|n|+1}.$$

In particular, $f \in H^{-1/2}(\Omega)$ implies

$$\sum_{n=-\infty}^{\infty} |\hat{f}(n)|^2 (1+|n|)^{-1} < \infty$$

which yields $PIf \in L^2(\Omega)$, by (5.25). Now if $f \in H^{k-1/2}(S^1)$, then $(\partial/\partial\theta)^\nu f \in H^{-1/2}(S^1)$, $0 \leq \nu \leq k$, so $(\partial/\partial\theta)^\nu PIf = PI(\partial/\partial\theta)^\nu f \in L^2(\Omega)$. We need to show that $(\partial/\partial\theta)^\mu (\partial/\partial\theta)^\nu PIf \in L^2$ near $\partial\Omega$ for $0 \leq \mu + \nu \leq k$. Indeed, if we let

$$\Lambda_0 f = \sum_{n=-\infty}^{\infty} |n| \hat{f}(n) e^{in\theta},$$

it follows from Plancherel's theorem that $(\partial/\partial\theta)^\nu \Lambda_0^\mu f \in H^{-1/2}(S^1)$ for $0 \leq \mu + \nu \leq k$, while a short calculation yields

$$(r(\partial/\partial r))^\mu (\partial/\partial\theta)^\nu PIf = PI((\partial/\partial\theta)^\nu \Lambda_0^\mu f)$$

which hence belongs to $L^2(\Omega)$. Since PIf is smooth at $r = 0$, this finishes the proof.

We remark that Proposition 5.3 is true for all $s \in \mathbf{R}$, which requires a slightly more involved proof.

Exercises

5.1. Let H_0 be a Hilbert space, $H_1 \subset H_0$, the domain of some self-adjoint operator. Let H_{-1} denote the dual to H_1. We have the natural inclusions

$$H_1 \subset H_0 \subset H_{-1}$$

the latter inclusion being induced by the inner product on H_0. Let $H_\theta = [H_0, H_1]_\theta$, $H_{-\theta} = [H_0, H_{-1}]_\theta$, $0 \leq \theta \leq 1$. The dual of H_θ is naturally a subspace of H_{-1}. Show that it is equal to $H_{-\theta}$.

5.2. Using Exercise 5.1, show that, for $0 \leq \theta \leq 1$,

$$[\mathring{H}^k(\Omega), L^2(\Omega)]_\theta = H^{k\theta}_{\bar{\Omega}}(M)$$
$$= \{u \in H^{k\theta}(M) : \text{supp } u \in \bar{\Omega}\}.$$

For the connection between this space and $\mathring{H}^{k\theta}(\Omega)$, see Fujiwara [1].

5.3. Let I be an interval in the circle S^1. Let $u = 1$ on I. Show that $u \in \mathring{H}^{1/2}(I) = H^{1/2}(I)$ but $u \notin H^{1/2}_{\bar{I}}(S^1)$.

§6. Sobolev Spaces, L^p Style

One can replace L^2 by L^p in the concepts introduced in Sections 3 and 5, and study such spaces. We sketch this here. Such spaces will only be used in Chapters XI and XII.

For an integer $k \geq 0$, define $W^k_p(\mathbf{R}^n)$ to consist of $u \in L^p(\mathbf{R}^n)$ such that $D^\alpha u \in L^p(\mathbf{R}^n)$ for $|\alpha| \leq k$. In analogy to the characterization (3.8) of $H^s(\mathbf{R}^n)$, define $\mathscr{L}^s_p(\mathbf{R}^n)$ by

$$\mathscr{L}^s_p(\mathbf{R}^n) = (1 - \Delta)^{-s/2} L^p(\mathbf{R}^n), \qquad s \in \mathbf{R}. \tag{6.1}$$

For $p = 2$, the identification of W^k_2 with $\mathscr{L}^k_2$, as noted in Section 3, follows from Plancherel's theorem. Here, we need a replacement for Plancherel's theorem, which is provided by the Marcinkiewicz multiplier theorem in Marcinkiewicz [1]. The following formulation is due to Hörmander [4]; see also Stein [1] for a proof.

THEOREM 6.1. *Let $p(\xi)$ be a smooth function on $\mathbf{R}^n$ such that*

$$|\xi|^{|\alpha|} |p^{(\alpha)}(\xi)| \leq C_\alpha, \qquad |\alpha| \leq \left[\frac{n}{2}\right] + 1. \tag{6.2}$$

Then, for $1 < p < \infty$,

$$p(D) : L^p(\mathbf{R}^n) \to L^p(\mathbf{R}^n), \tag{6.3}$$

with operator norm bounded by $C(p) \sum_\alpha C_\alpha$.

In fact, one can take $C(p) = Cp$ for $p \geq 2$, $C(p) = C/p - 1$ for $1 < p \leq 2$. With this result, we can establish

$$W_p^k(\mathbf{R}^n) = \mathscr{L}_p^k(\mathbf{R}^n), \qquad 1 < p < \infty. \tag{6.4}$$

Indeed, the fact that, for $|\alpha| \leq k$, $D^\alpha(1-\Delta)^{-k/2}: L^p \to L^p$ follows from Theorem 6.1, and this shows $\mathscr{L}_p^k \subset W_p^k$. Conversely, if $u \in W_p^k$, Theorem 6.1 implies that, for $|\alpha| \leq k$, $D^\alpha(1-\Delta)^{-k/2}D^\alpha u \in L^p$; so $(1-\Delta)^{-k/2}(1-\Delta)^k u \in L^p$, or $(1-\Delta)^{k/2}u \in L^p$, which implies $u \in \mathscr{L}_p^k$. Next, we claim that

$$[L^p(\mathbf{R}^n), \mathscr{L}_p^s(\mathbf{R}^n)]_\theta = \mathscr{L}_p^{s\theta}(\mathbf{R}^n), \qquad 1 < p < \infty. \tag{6.5}$$

The proof of this is parallel to that of Theorem 4.2, with Theorem 6.1 replacing the use of the spectral theorem, whose role in the proof of (4.2) was to bound the operator norm of A^{-iy} by 1. In this case, Theorem 6.1 implies that

$$\|(1-\Delta)^{-iy}\|_{\mathscr{L}(L^p)} \leq C_p(1+|y|)^n. \tag{6.6}$$

Thus, if $v \in \mathscr{L}_p^{s\theta}(\mathbf{R}^n)$, let

$$u(z) = e^{z^2}(1-\Delta)^{(-z+\theta)s/2}v.$$

Then $u(\theta) = e^{\theta^2}v$, $u(iy) = e^{-y^2}(1-\Delta)^{-iys/2}((1-\Delta)^{s\theta/2}v)$ is bounded in $L^p(\mathbf{R}^n)$ and $u(1+iy) = e^{(1+iy)^2}(1-\Delta)^{-s/2}(1-\Delta)^{-is/2y}((1-\Delta)^{s\theta/2}v)$ is bounded in $\mathscr{L}_p^s(\mathbf{R}^n)$, so indeed $u \in \mathscr{H}_{L^p(\mathbf{R}^n), \mathscr{L}_p^s(\mathbf{R}^n)}(\Omega)$ which implies $\mathscr{L}_p^{s\theta}(\mathbf{R}^n) \subset [L^p(\mathbf{R}^n), \mathscr{L}_p^s(\mathbf{R}^n)]_\theta$. The reverse containment is similarly established, in analogy with the proof of Theorem 4.2. Generalizing (6.5), we can establish, for $1 < p < \infty$,

$$[\mathscr{L}_p^\sigma(\mathbf{R}^n), \mathscr{L}_p^s(\mathbf{R}^n)]_\theta = \mathscr{L}_p^{\sigma(1-\theta)+s\theta}(\mathbf{R}^n). \tag{6.7}$$

If M is a compact manifold without boundary, one defines $\mathscr{L}_p^s(M)$ in analogy with $H^s(M)$, via coordinate charts, and proves

$$[\mathscr{L}_p^\sigma(M), \mathscr{L}_p^s(M)]_\theta = \mathscr{L}_p^{\sigma(1-\theta)+s\theta}(M), \qquad 1 < p < \infty. \tag{6.8}$$

If Ω is a compact subdomain of M, with smooth boundary, one defines $W_p^k(\Omega)$ in the obvious fashion and generalizes the extension theorems of Section 5 to get $E: W_p^k(\Omega) \to W_p^k(M) = \mathscr{L}_p^k(M)$, $1 < p < \infty$. If we define $\mathscr{L}_p^s(\Omega)$ for $s \geq 0$ by

$$\mathscr{L}_p^s(\Omega) = [L^p(\Omega), W_p^k(\Omega)]_\theta, \qquad 0 \leq \theta \leq 1, \qquad s = k\theta, \tag{6.9}$$

it follows that $E: \mathscr{L}_p^s(\Omega) \to \mathscr{L}_p^s(M)$ and hence

$$\mathscr{L}_p^s(\Omega) \approx \mathscr{L}_p^s(M)/\{u: u = 0 \text{ on } \Omega\}. \tag{6.10}$$

Also, of course, $\mathscr{L}_p^k(\Omega) = W_p^k(\Omega)$ for $k \geq 0$ an integer.

Generalizing Rellich's theorem, one has, for $s \geq 0$, $1 < p < \infty$,

$$(6.11) \qquad I: \mathscr{L}_p^{s+\sigma}(\Omega) \to \mathscr{L}_p^s(\Omega) \text{ compact}, \qquad \sigma > 0.$$

By arguments as in Section 5, it is possible to reduce (6.11) to showing that, for $\sigma > 0$, $1 < p < \infty$

$$(6.12) \qquad (1-\Delta)^{-\sigma/2}: L^p(\mathbf{T}^n) \to L^p(\mathbf{T}^n) \text{ is compact.}$$

To prove this, one can use the Marcinkiewicz multiplier theorem for the torus, which asserts that, if $p(\xi)$ is a smooth function on $\mathbf{R}^n$ satisfying (6.2) and if $p(D): \mathscr{D}'(\mathbf{T}^n) \to \mathscr{D}'(\mathbf{T}^n)$ is defined by (2.25), then $p(D): L^p(\mathbf{T}^n) \to L^p(\mathbf{T}^n)$, $1 < p < \infty$, with a bound on the operator norm given as in Theorem 6.1. Indeed, it was for Fourier multipliers on the torus that Marcinkiewicz originally proved his theorem. We apply this result as follows. Take $\varphi \in C_0^\infty(\mathbf{R}^n)$, $\varphi(\xi) = 1$ for $|\xi| \leq 1$, and let

$$p_\epsilon(\xi) = (1 + |\xi|^2)^{-\sigma/2} \varphi(\epsilon\xi).$$

Then each $p_\epsilon(D)$ has finite rank, so certainly is compact. On the other hand, the Marcinkiewicz multiplier theorem implies

$$\|(1-\Delta)^{-\sigma/2} - p_\epsilon(D)\|_{\mathscr{L}(L^p)} \to 0 \text{ as } \epsilon \to 0, \qquad 1 < p < \infty,$$

so $(1-\Delta)^{-\sigma/2}$ is a norm limit of compact operators, which implies (6.12).

The Sobolev imbedding theorem, Theorem 3.1, has the following generalization:

$$(6.13) \qquad \mathscr{L}_p^s(\mathbf{R}^n) \subset C(\mathbf{R}^n) \cap L^\infty(\mathbf{R}^n) \qquad \text{if} \qquad s > \frac{n}{p}.$$

To prove this it suffices to show that $(1-\Delta)^{-s/2}\delta \in L^{p'}(\mathbf{R}^n)$ if $s > (n/p)$, since $u = (1-\Delta)^{-s/2}\delta * (1-\Delta)^{s/2}u$. Note that

$$(6.14) \qquad \psi(x) = (1-\Delta)^{-s/2}\delta = (2\pi)^{-n} \int e^{ix\cdot\xi}(1 + |\xi|^2)^{-s/2}\, d\xi.$$

In fact, one can show that $\psi(x) = (1-\Delta)^{-s/2}\delta$ is smooth for $x \neq 0$, rapidly decreasing as $|x| \to \infty$, and satisfies the estimate

$$(6.15) \qquad |\psi(x)| \leq C|x|^{-n+s}, \qquad |x| \leq 1, \qquad s < n$$

which is sufficient. For such an estimate of (6.14), see Titchmarsh [1]; a much more general result will be proved in Chapter XII, Section 3, Lemma 3.1. The generalization of (3.4) of Section 3 is

$$(6.16) \qquad \mathscr{L}_p^s(\mathbf{R}^n) \subset L^q(\mathbf{R}^n), \qquad q = \frac{np}{n-ps}, \qquad 0 \leq s < n/p.$$

For a proof of this, see Bergh and Löfström [1].

The trace theorems for restrictions to $\partial\Omega$ of $u \in \mathscr{L}^s_p(\Omega)$ are more subtle for $p \neq 2$ than for $p = 2$. τu loses $1/p$ derivatives, generally, but does not belong to $\mathscr{L}^{s-1/p}(\partial\Omega)$, necessarily, but rather to a Besov space:

$$\tau: \mathscr{L}^s_p(\Omega) \to B^{s-1/p}_p(\partial\Omega), \qquad s > \frac{1}{p}.$$

We give a brief account of the Besov spaces $B^s_p(M)$ in Chapter XI; one characterization of $B^s_p(M)$ involves the real interpolation method:

$$B^s_p(M) = [L^p(M), W^k_p(M)]_{\theta,p}, \qquad s = \theta k < k.$$

See in particular Bergh and Löfström [1]. More generally, $B^s_p(\Omega)$ is defined, $s > 0$.

In much of the literature of Sobolev spaces, one sees the spaces $W^s_p(\Omega)$. For $s = k$, a nonnegative integer, they are the spaces $W^k_p(\Omega) = \mathscr{L}^k_p(\Omega)$ discussed here. However, if $s > 0$ is not an integer, $W^s_p(\Omega) \neq \mathscr{L}^s_p(\Omega)$; rather, $W^s_p(\Omega)$ is isomorphic to the Besov space $B^s_p(\Omega)$ when s is not an integer (but not when s is an integer). This seemingly bizarre turn of events arose because terminology was fixed before the interpolation and trace theorems were completely understood.

For further material on $\mathscr{L}^s_p(\Omega)$ and $B^s_p(\Omega)$, we refer particularly to Bergh and Löfström [1] and, for $W^k_p(\Omega)$, k an integer, where Ω may have a nonsmooth boundary, to Adams [1].

§7. Local Solvability of Constant Coefficient *PDE*

Here we take the space to apply some basic Fourier analysis and distribution theory to the question of solving the constant coefficient partial differential equation

$$P(D)u = f \tag{7.1}$$

on a neighborhood U of 0, given $f \in \mathscr{D}'(\mathbf{R}^n)$, $P(\xi)$ a polynomial. We show below that, for any bounded neighborhood U of 0, there exists $u \in \mathscr{D}'(\mathbf{R}^n)$ solving (7.1) on U. If $f \in C^\infty$, we produce $u \in C^\infty$. Such a result was first proved by Malgrange [1] and Ehrenpreis [1]; see also Hörmander [6] and Trèves [1]. They have more precise results on regularity of u than the proof here will provide. In particular, if $f \in L^2_{\text{loc}}$, there exists a solution to (7.1) with $u \in L^2_{\text{loc}}$, and $P^{(\alpha)}(D)u \in L^2_{\text{loc}}$, where $P^{(\alpha)}(\xi) = D^\alpha_\xi P(\xi)$. The following proof of local solvability was worked out by the author and Jiri Dadok.

Note that, for $\alpha \in \mathbf{R}^n$, solving (7.1) is equivalent to solving

$$P(D+\alpha)v = g, \qquad \text{near } 0 \tag{7.2}$$

with $v = e^{i\alpha\cdot x}u$, $g = e^{i\alpha\cdot x}f$. Now to solve (7.2) near 0, we can cut g off to be supported near 0 and work on $\mathbf{T}^n = \mathbf{R}^n/\mathbf{Z}^n$. Thus (7.2) is certainly locally solvable, given the following result.

THEOREM 7.1. *For almost all* $\alpha \in A = \{(\alpha_1, \ldots, \alpha_n): 0 \le \alpha_\nu \le 1\}$,

$$P(D+\alpha): \mathscr{D}'(\mathbf{T}^n) \to \mathscr{D}'(\mathbf{T}^n)$$

is an isomorphism, as is $P(D+\alpha): C^\infty(\mathbf{T}^n) \to C^\infty(\mathbf{T}^n)$.

This theorem is a consequence of the following.

PROPOSITION 7.2. *Take a polynominal* $P(\xi)$ *of order m on* $\mathbf{R}^n$. *For almost all* $\alpha \in A$, *there are constants C, N such that*

$$|P(k+\alpha)|^{-1} \le C(1+|k|^2)^N, \qquad k \in \mathbf{Z}^n. \tag{7.3}$$

In order to prove (7.3), we shall need the following elementary fact about the behavior of a polynomial near its zero set.

PROPOSITION 7.3. *Let* $P(\xi)$ *be a polynomial of order m on* $\mathbf{R}^n$. *Then there is a* $\delta = \delta(m, n)$ *such that*

$$|P(\xi)|^{-\delta} \in L^1_{\text{loc}}.$$

We shall postpone the proof of Proposition 7.3, and first show how it yields (7.2). First of all, we claim that, for any polymial $P(\xi)$ of order m on $\mathbf{R}^n$, there exist $\delta > 0$ and M such that

$$\int |P(\xi)|^{-\delta}(1+|\xi|^2)^{-M}\, d\xi < \infty. \tag{7.4}$$

Indeed, Proposition 7.3 guarantees

$$\int_{|\xi|\le 1} |P(\xi)|^{-\delta}\, d\xi < \infty,$$

while for M sufficiently large

$$\int_{|\xi|\ge 1} |P(\xi)|^{-\delta}|\xi|^{-2M}\, d\xi \le C \int_{|\xi|\le 1} \left|P\left(\frac{\xi}{|\xi|^2}\right)\right|^{-\delta} d\xi,$$

and Proposition 7.3 also implies that, for $\delta > 0$ small enough,

$$\left|P\left(\frac{\xi}{|\xi|^2}\right)\right|^{-\delta} \in L^1_{\text{loc}}.$$

Now, using (7.4), note that

$$\int_A \sum_{k\in\mathbf{Z}^n} |P(k+\alpha)|^{-\delta}(1+|k|^2)^{-M}\, d\alpha \le C\int_{\mathbf{R}^n} |P(\xi)|^{-\delta}(1+|\xi|^2)^{-M}\, d\xi < \infty. \tag{7.5}$$

Thus, for almost all $\alpha \in A$,

$$\sum_{k \in \mathbf{Z}^n} |P(k+\alpha)|^{-\delta}(1+|k|^2)^{-M} < \infty. \tag{7.6}$$

Of course, (7.2) is an immediate consequence of (7.6).

It remains only to prove Proposition 7.3. This is a simple consequence of the Weierstrass preparation theorem, which states that if $F(z)$ is a holomorphic function $F(0) = 0$, $F(z_1, 0) = g(z_1)z_1^k$, $g(0) \neq 0$, then there is a neighborhood Ω of 0 in $\mathbf{C}^n$ and a holomorphic function $q(z)$ on Ω, $q(0) \neq 0$, such that

$$q(z)F(z) = z_1^k + \sum_{j=0}^{k-1} \lambda_j(z')z_1^j \tag{7.7}$$

with $\lambda_j(z')$ holomorphic for z' near 0. For a proof of this classical result, see Gunning and Rossi [1] or Hörmander [8]. We apply it to the special case of polynomials, and restricting our complex variables to $\mathbf{R}^n$.

We must prove $|P(\xi)|^{-\delta}$ is locally integrable, in some neighborhood of any given point $\xi^0 \in \mathbf{R}^n$; may as well suppose $P(\xi^0)=0$. Let $F(\xi)=P(\xi+\xi^0)$, so we show $|F(\xi)|^{-\delta}$ is integrable near $\xi = 0$. Making an orthogonal change of coordinates, we can suppose $F(\xi_1, 0) = g(\xi_1)\xi_1^k$ with $g(0) \neq 0$, $k \leq n$. Thus (7.7) applies: $q(0) \neq 0$ and

$$q(\xi)F(\xi) = \xi_1^k + \sum_{j=0}^{k-1} \lambda_j(\xi')\xi_1^j. \tag{7.8}$$

Now it is clear that, for $|\xi'|$ small, $\delta < 1/n$, there is a $c_1 > 0$ such that

$$\int_{-c_1}^{c_1} |F(\xi)|^{-\delta}\, d\xi_1 \leq C < \infty \qquad (\forall |\xi'| \text{ small})$$

and this implies

$$\int_{|\xi| \leq c_2} |F(\xi)|^{-\delta}\, d\xi < \infty,$$

as desired. Our proof of Theorem 7.1, and hence of local solvability, is now complete.

CHAPTER II

Pseudodifferential Operators

In this chapter the basic classes of operators that will serve up to Chapter VIII are introduced and discussed. The classes of symbols $S^m_{\rho,\delta}$ defined in Section 1 were introduced in Hörmander [12] and my presentation follows that paper and also part of Hörmander's [14] fairly closely. We define pseudodifferential operators and investigate their behavior on smooth functions and distributions and investigate adjoints and products of such operators and their continuity on Sobolev spaces. Behavior of certain families of pseudodifferential operators, in particular Friedrich's mollifiers, is considered and finally Gårding's inequality is proved.

§1. The Fourier Integral Representation and Symbol Classes

Recall that the Fourier inversion formula is

$$f(x) = \int \hat{f}(\xi) e^{ix\cdot\xi}\, d\xi \tag{1.1}$$

where $\hat{f}(\xi)$ is the Fourier transform of f, given by

$$\hat{f}(\xi) = (2\pi)^{-n} \int f(x) e^{-ix\cdot\xi}\, dx. \tag{1.2}$$

If we differentiate (1.1), using the notation $D_j = (1/i)(\partial/\partial x_j)$, we get

$$D^\alpha f(x) = \int \xi^\alpha \hat{f}(\xi) e^{ix\cdot\xi}\, d\xi.$$

Hence if

$$p(x, D) = \sum_{|\alpha| \le k} a_\alpha(x) D^\alpha$$

is a differential operator,

$$p(x, D) f(x) = \int p(x, \xi) \hat{f}(\xi) e^{ix\cdot\xi}\, d\xi \tag{1.3}$$

where

$$p(x, \xi) = \sum_{|\alpha| \le k} a_\alpha(x) \xi^\alpha.$$

We shall use the Fourier integral representation (1.3) to define pseudodifferential operators, taking the function $p(x, \xi)$ to belong to a general class of symbols, which we now define.

DEFINITION 1.1. *Let Ω be an open subset of $\mathbf{R}^n$, $m, \rho, \delta \in \mathbf{R}$, and suppose $0 \le \rho, \delta \le 1$. We define the symbol class*

$$S^m_{\rho,\delta}(\Omega)$$

to consist of the set of $p \in C^\infty(\Omega \times \mathbf{R}^n)$ with the property that, for any compact $K \subset \Omega$, any multi-indices α, β, there exists a constant $C_{K,\alpha,\beta}$ such that

$$|D^\beta_x D^\alpha_\xi p(x, \xi)| \le C_{K,\alpha,\beta}(1+|\xi|)^{m-\rho|\alpha|+\delta|\beta|} \tag{1.4}$$

for all $x \in K$, $\xi \in \mathbf{R}^n$. We drop the Ω and use $S^m_{\rho,\delta}$ when the context is clear.

The class $S^m_{\rho,\delta}$ was introduced by Hörmander in [12]. The subclass $S^m_{1,0}$, defined by

$$|D^\beta_x D^\alpha_\xi p(x, \xi)| \le C_{K\alpha\beta}(1+|\xi|)^{m-|\alpha|}$$

had been defined by Kohn and Nirenberg in [1]. Prior to that, a subclass of $S^m_{1,0}$ had been studied, the class S^m, defined as follows.

DEFINITION 1.2. *The symbol $p(x, \xi)$ belongs to $S^m(\Omega)$ if $p \in S^m_{1,0}(\Omega)$ and there are smooth $p_{m-j}(x, \xi)$, homogeneous of degree $m-j$ in ξ for $|\xi| \ge 1$, i.e.,*

$$p_{m-j}(x, r\xi) = r^{m-j} p_{m-j}(x, \xi), \qquad |\xi| \ge 1, r \ge 1$$

such that

$$p(x, \xi) \sim \sum_{j \ge 0} p_{m-j}(x, \xi)$$

where the asymptotic condition means that

$$p(x, \xi) - \sum_{j=0}^{N} p_{m-j}(x, \xi) \in S^{m-N-1}_{1,0}(\Omega).$$

Clearly if $p(x, \xi)$ is homogeneous of degree m in ξ and if $\varphi(\xi) = 0$ for $|\xi| \le C_1$, $\varphi(\xi) = 1$ for $|\xi| \ge C_2 > C_1$, $\varphi \in C^\infty$, then $\varphi(\xi)p(x, \xi) \in S^m \subset S^m_{1,0}$. We also have the following simple facts, whose proofs follow easily from the chain rule.

PROPOSITION 1.3. *Let $p \in S^m_{\rho,\delta}(\Omega)$, $q \in S^\mu_{\rho',\delta'}$. Then*

$$p^{(\alpha)}_{(\beta)} = D^\beta_x D^\alpha_\xi p \in S^{m-\rho|\alpha|+\delta|\beta|}_{\rho,\delta}, \tag{1.5}$$

and

$$p(x,\xi)q(x,\xi) \in S^{m+\mu}_{\rho'',\delta''} \quad \textit{where} \quad \rho'' = \min(\rho, \rho'), \quad \delta'' = \max(\delta, \delta'). \tag{1.6}$$

$$\textit{If } |p(x, \xi)^{-1}| \le C(1 + |\xi|)^{-m}, \quad \textit{then } p(x, \xi)^{-1} \in S^{-m}_{\rho,\delta}. \tag{1.7}$$

Such symbols $p(x, \xi)$, as we have said, lead to operators $p(x, D)$ defined by (1.3).

DEFINITION 1.4. *If $p(x, \xi) \in S^m_{\rho,\delta}$, the operator $p(x, D)$ is said to belong to $OPS^m_{\rho,\delta}$. More generally, if Σ is any symbol class and $p(x, \xi) \in \Sigma$, we say $p(x, D) \in OP\Sigma$.*

It is now necessary to investigate $p(x, D)$ as an operator.

THEOREM 1.5. *If $p \in S^m_{\rho,\delta}(\Omega)$, then $p(x, D)$ defined by (1.3) is a continuous operator*

$$p(x, D): C_0^\infty(\Omega) \to C^\infty(\Omega). \tag{1.8}$$

If $\delta < 1$, then the map can be extended to a continuous map

$$p(x, D): \mathscr{E}'(\Omega) \to \mathscr{D}'(\Omega). \tag{1.9}$$

Proof. If $p \in S^m_{\rho,\delta}(\Omega)$, $u \in C_0^\infty(\Omega)$, then the integral

$$p(x, D)u = \int p(x, \xi)\hat{u}(\xi)e^{ix \cdot \xi}\, d\xi$$

is absolutely convergent, and one can differentiate under the integral sign, obtaining always absolutely convergent integrals. Thus we have (1.8). To obtain (1.9), we need a lemma.

LEMMA 1.6. *Let $p \in S^m_{\rho,\delta}(\Omega)$, $v \in C_0^\infty(\Omega)$. Then for all $\xi, \eta \in \mathbf{R}^n$,*

$$\left|\int v(x)p(x, \xi)e^{ix \cdot \eta}\, dx\right| \le C_N(1+|\xi|)^{m+\delta N}(1+|\eta|)^{-N}. \tag{1.10}$$

Proof. Integration by parts yields

$$\left|\eta^\alpha \int v(x)p(x, \xi)e^{ix \cdot \eta}\, dx\right| = \left|\int D_x^\alpha(v(x)p(x, \xi))e^{ix \cdot \eta}\, dx\right| \le C_\alpha(1+|\xi|)^{m+\delta|\alpha|}$$

by (1.4), from which (1.10) follows.

To complete the proof of Theorem 1.5, we show that the functional

$$v \to \langle p(x, D)u, v\rangle, \qquad v \in C_0^\infty(\Omega) \tag{1.11}$$

is well defined if $u \in \mathscr{E}'(\Omega)$. Indeed, formally, write

$$\begin{aligned}\langle p(x, D)u, v\rangle &= \int v(x)p(x, \xi)\hat{u}(\xi)e^{ix \cdot \xi}\, d\xi\, dx \\ &= \int p_v(\xi)\hat{u}(\xi)\, d\xi\end{aligned} \tag{1.12}$$

where $p_v(\xi) = \int v(x)p(x, \xi)e^{ix \cdot \xi}\, dx$. For (1.12) to be well defined, for any $u \in \mathscr{E}'(\Omega)$, we only need that $p_v(\xi)$ be rapidly decreasing. But the lemma implies (taking $\eta = \xi$) that

$$|p_v(\xi)| \le C_N(1+|\xi|)^{m-(1-\delta)N}. \tag{1.13}$$

If we assume $\delta < 1$, this implies the rapid decrease. Hence (1.12) can be taken to define (1.11), and the proof of Theorem 1.5 is complete.

Exercises

1.1. Define $S^{-\infty}$ to be

$$\bigcap_{m>0} S^{-m}_{1,0} = \bigcap_{m>0} S^{-m}_{\rho,\delta}.$$

If $p(x, \xi) \in S^{-\infty}$, show that

$$p(x, D): \mathscr{E}'(\Omega) \to C^\infty(\Omega).$$

1.2. Suppose $p(x, \xi) \in S^m_{1,0}$. Using (1.3), show that

$$p(x, D): C_0^{m+\ell+k}(\Omega) \to C^\ell(\Omega)$$

provided $k > n/2$, $n = \dim \Omega$. (This result will be immensely improved.)

1.3. If $p(x, \xi) \in S^0_{\rho,\delta}(\Omega)$ and φ is smooth on a neighborhood of the closure of the set of values assumed by p, show that $\varphi(p(x, \xi)) \in S^0_{\rho,\delta}(\Omega)$.

§2. The Pseudolocal Property

The purpose of this section is to prove the following:

THEOREM 2.1. *If* $p(x, \xi)$ $S^m_{\rho,\delta}(\Omega)$, $\delta < 1$ *and if* $\rho > 0$, *we have, for* $u \in \mathscr{E}'(\Omega)$,

$$\text{sing supp } p(x, D)u \subset \text{sing supp } u. \tag{2.1}$$

Here the singular support of a distribution u, denoted sing supp u, is the complement of the open set on which u is smooth. We refer to (2.1) as the pseudolocal property. Before proving Theorem 2.1, we give a few preliminaries.

If $K \in \mathscr{D}'(\Omega \times \Omega)$, then there is associated a map $K: C_0^\infty(\Omega) \to \mathscr{D}'(\Omega)$ defined by $\langle Ku, v\rangle = \langle K, u(x)v(y)\rangle$. The converse is also true, and is known as the Schwartz kernel theorem. We shall use the following simple assertion, whose proof we leave as an exercise to the reader.

LEMMA 2.2. (Singular support lemma) *Suppose* $K \in \mathscr{D}'(\Omega \times \Omega)$ *satisfies*

$$K: C_0^\infty(\Omega) \to C^\infty(\Omega) \tag{2.2}$$

and

$$K: \mathscr{E}'(\Omega) \to \mathscr{D}'(\Omega). \tag{2.3}$$

$$K \text{ is } C^\infty \text{ off the diagonal in } \Omega \times \Omega. \tag{2.4}$$

Then sing supp $Ku \subset$ sing supp u, *for* $u \in \mathscr{E}'(\Omega)$.

To prove Theorem 2.1, we need to analyze the kernel K of $p(x, D)$ and verify (2.4). We have

$$
\begin{aligned}
(2.5)\qquad \langle K, uv\rangle &= \langle p(x, D)u, v\rangle \\
&= \int v(x)p(x, D)u(x)\, dx \\
&= \iint p(x, \xi)e^{ix\cdot\xi}v(x)\hat{u}(\xi)\, d\xi\, dx \\
&= (2\pi)^{-n}\iiint p(x, \xi)e^{i(x-y)\cdot\xi}v(x)u(y)\, dy\, d\xi\, dx.
\end{aligned}
$$

Thus, with the appropriate interpretation as a distribution integral,

$$(2.6)\qquad K = (2\pi)^{-n}\int p(x, \xi)e^{i(x-y)\cdot\xi}\, d\xi.$$

Consequently,

$$(2.7)\qquad (x-y)^{\alpha}K = \int e^{i(x-y)\cdot\xi}D_{\xi}^{\alpha}p(x, \xi)\, d\xi.$$

From this representation one sees that the integral is absolutely convergent for α so large that $m-\rho|\alpha| < -n$ ($n = \dim\Omega$), and more generally applying j derivatives yields an absolutely convergent integral provided $m-\rho|\alpha| < -n-j$, so in that case $(x-y)^{\alpha}K \subset C^{j}(\Omega\times\Omega)$. Hence K is smooth off the diagonal $x = y$, and Theorem 2.1 is proved.

We remark that this proof leads to the estimate, for x and y in compact subsets of Ω,

$$|D_{x,y}^{\beta}K| \le C|x-y|^{-k}$$

where $k \ge 0$ is any integer strictly greater than $(1/\rho)(m+n+|\beta|)$. This is not sharp. For example, if $p(x, \xi) \in S_{1,0}^{m}$, it is true that

$$|K(x, y)| \le \begin{cases} C|x-y|^{-(m+n)} & \text{if} \quad m > -n \\ C|\log|x-y|| & \text{if} \quad m = -n. \end{cases}$$

We shall return to this point in Chapter XII, Lemma 3.1.

§3. Asymptotic Expansions of a Symbol

THEOREM 3.1. *Suppose $p_j \in S_{\rho,\delta}^{m_j}(\Omega)$, $m_j \downarrow -\infty$. Then there exists $p \in S_{\rho,\delta}^{m_0}(\Omega)$ such that, for all $N > 0$,*

$$(3.1)\qquad p - \sum_{j=0}^{N-1} p_j \in S_{\rho,\delta}^{m_N}(\Omega).$$

If (3.1) *holds we say*

$$p \sim \sum_{j \geq 0} p_j. \tag{3.2}$$

Proof. Pick compact sets K_j, $K_1 \subset K_2 \subset \cdots \to \Omega$. Pick $\varphi \in C^\infty(\mathbf{R}^n)$ with $\varphi(\xi) = 0$ for $|\xi| \leq 1/2$, $\varphi(\xi) = 1$ for $|\xi| \geq 1$. Construct $p(x, \xi)$ of the form

$$p(x, \xi) = \sum_{j=0}^{\infty} \varphi(\epsilon_j \xi) p_j(x, \xi) \tag{3.3}$$

where ϵ_j are picked so small that

$$|D_x^\beta D_\xi^\alpha \varphi(\epsilon_j \xi) p_j(x, \xi)| \leq 2^{-j}(1 + |\xi|)^{m_j + 1 - \rho|\alpha| + \delta|\beta|}$$

for $|\alpha| + |\beta| + i \leq j$ and $x \in K_i$. It is easy to verify that (3.3) is convergent and that $p(x, \xi)$ satisfies (3.1).

We remark that an analogous proof shows that for any sequence a_j there is a function $f \in C^\infty(\mathbf{R})$ such that $f^{(j)}(0) = j!a_j$, a classical result due to Borel.

The next theorem shows that the asymptotic relation (3.2) is valid if an apparently weaker relation than (3.1) is assumed to hold. It also gives a useful tool for proving that a function $p \in C^\infty(\Omega \times \mathbf{R}^n)$ belongs to a certain symbol class.

THEOREM 3.2. *Let* $P_j \in S^{m_j}_{\rho,\delta}(\Omega)$, $m_j \downarrow -\infty$, $j \geq 0$. *Let* $p \in C^\infty(\Omega \times \mathbf{R}^n)$ *and assume there are* $C_{\alpha\beta}$, $\mu = \mu(\alpha, \beta)$ *such that*

$$|D_x^\beta D_\xi^\alpha p(x, \xi)| \leq C_{\alpha\beta}(1 + |\xi|)^\mu. \tag{3.4}$$

If there exist $\mu_k \to \infty$ *such that*

$$\left| p(x, \xi) - \sum_{j=0}^{k} p_j(x, \xi) \right| \leq C_k(1 + |\xi|)^{-\mu_k}, \tag{3.5}$$

then $p \in S^{m_0}_{\rho,\delta}(\Omega)$ *and* $p \sim \Sigma p_j$, *in the sense that* (3.1) *holds.*

Proof. By Theorem 3.1 we can find $q \in S^{m_0}_{\rho,\delta}(\Omega)$ such that $q \sim \Sigma p_j$. It only remains to show that $p - q \in S^{-\infty}$. (3.5) implies that

$$|p(x, \xi) - q(x, \xi)| \leq C_{K,N}(1 + |\xi|)^{-N}, \qquad x \in K.$$

We need only verify that such an inequality holds for $D_x^\beta D_\xi^\alpha(p - q)$. For this, we use the inequality

$$\sum_{|\alpha|=1} \sup_{K_1} |D^\alpha f|^2 \leq C \sup_{K_2} |f| \sum_{|\alpha| \leq 2} \sup_{K_2} |D^\alpha f|, \tag{3.6}$$

where $K_1 \subset \operatorname{int} K_2 \subset K_2$, K_j compact. We briefly defer the proof of (3.6). If we apply this inequality to the functions

$$F_\xi(x, \eta) = p(x, \xi+\eta) - q(x, \xi+\eta)$$

taking $K_1 = K \times \{0\}$, K_2 a small neighborhood of K_1, we get

$$\sup_{x \in K} |\nabla_{x,\xi}(p-q)(x, \xi)|^2$$

$$\leq C \sup_{(x,\eta)\in K_2} |p(x, \xi+\eta) - q(x, \xi+\eta)| \left(\sum_{|\alpha| \leq 2} \sup_{(x,\eta)\in K_2} |D^\alpha_{x,\eta}(p-q)(x, \xi+\eta)| \right)$$

$$\leq C'_\mu(1+|\xi|)^{-\mu},$$

since the first factor is rapidly decreasing and the second factor has polynomial growth. Inductively it follows that $D_x^\beta D_\xi^\alpha(p-q)$ is rapidly decreasing, and the proof is complete, modulo establishing (3.6).

(3.6) can be proved directly from the mean value theorem, but we shall do it via the following general result about semigroups.

PROPOSITION 3.3. *Let the closed linear operator A generate a contraction semigroup on a Banach space X. Then, for $u \in \mathcal{D}(A^2)$, we have*

$$\|Au\|^2 \leq 4\|u\| \, \|A^2u\|.$$

Proof. From the identity

$$-tAu = t(t-A)^{-1}A^2u + t^2u - t^2t(t-A)^{-1}u$$

and the inequality

$$\|t(t-A)^{-1}\| \leq 1,$$

valid for the generator of a contraction semigroup, we get, for $t > 0$,

$$t\|Au\| \leq \|A^2u\| + 2t^2\|u\|, \qquad \text{so}$$

$$\|Au\| \leq \inf_{t>0}((1/t)\|A^2u\| + 2t\|u\|) = 2\|A^2u\|^{1/2}\|u\|^{1/2}.$$

COROLLARY 3.4. *For all u $C_0^\infty(\mathbf{R}^n)$,*

$$\left\| \frac{\partial}{\partial x_j} u \right\|^2_{L^\infty} \leq 4\|u\|_{L^\infty} \left\| \frac{\partial^2}{\partial x_j^2} u \right\|_{L^\infty}.$$

Proof. Apply Proposition 3.3 to the Banach space $C_0(\mathbf{R}^n)$ and the translation group $U^tf = f(x+te_j)$.

Now (3.6) is a simple consequence of Corollary 3.4.

Our next goal is to introduce a class of operators, apparently more general than the class of pseudodifferential operators defined in Section 1, and then to show that the two classes in fact coincide, under some circumstances. This will make for greater freedom in constructing pseudodifferential operators, which will be very useful in succeeding sections.

The operators we now consider are of the form

$$Au(x) = (2\pi)^{-n} \iint a(x, y, \xi)u(y)e^{i(x-y)\cdot\xi}\, dy\, d\xi. \tag{3.7}$$

The *amplitude* $a(x, y, \xi)$ is assumed to belong to one of the following classes.

DEFINITION 3.5. *Let* $0 \leq \rho, \delta_1, \delta_2$. *We say* $a(x, y, \xi) \in S^m_{\rho,\delta_1,\delta_2}(\Omega \times \Omega \times \mathbf{R}^n)$ *if, on compact subsets of* $\Omega \times \Omega$, *we have*

$$|D_y^\gamma D_x^\beta D_\xi^\alpha a(x, y, \xi)| \leq C(1 + |\xi|)^{m - \rho|\alpha| + \delta_1|\beta| + \delta_2|\gamma|}. \tag{3.8}$$

Granted (3.8), the proof of Lemma 1.6 shows that, if $u \in C_0^\infty(\Omega)$, then

$$\left|\int u(y)a(x, y, \xi)e^{-iy\cdot\xi}\, dy\right| \leq C_N(1 + |\xi|)^{m-(1-\delta_2)N},$$

so if $\delta_2 < 1$, (3.7) is absolutely integrable, for $u \in C_0^\infty(\Omega)$. Similarly you can differentiate under the integral sign and conclude that

$$A: C_0^\infty(\Omega) \to C^\infty(\Omega) \qquad \text{provided} \qquad \delta_2 < 1. \tag{3.9}$$

In order to reduce A to the form $p(x, D)$, it is convenient to assume that A is properly supported, a concept we now define.

DEFINITION 3.6. *A distribution* $A \in \mathscr{D}'(\Omega \times \Omega)$ *is said to be properly supported if* supp A *has compact intersection with* $K \times \Omega$ *and with* $\Omega \times K$, *for any compact* $K \subset \Omega$.

Alternatively A is properly supported provided $A: C_0^\infty(\Omega) \to \mathscr{E}'(\Omega)$ and $A^t: C_0^\infty(\Omega) \to \mathscr{E}'(\Omega)$, hence $A: C^\infty(\Omega) \to \mathscr{D}'(\Omega)$. Note that, if A is given by (3.7) and if $b(x, y)$ has proper support, then the operator $\tilde{A}$ given by (3.7) with $a(x, y, \xi)$ replaced by $b(x, y)a(x, y, \xi)$ is properly supported.

DEFINITION 3.7. *If* A *is given by* (3.7) *with* $a(x, y, \xi) \in S^m_{\rho,\delta_1,\delta_2}$ *and if* A *is properly supported, we say*

$$A \in OPS^m_{\rho,\delta_1,\delta_2}.$$

Note that, if $A \in OPS^m_{\rho,\delta_1,\delta_2}$, $\delta_2 < 1$, then

$$A: C^\infty(\Omega) \to C^\infty(\Omega). \tag{3.10}$$

Indeed, we know that $A: C_0^\infty(\Omega) \to C^\infty(\Omega)$. Now if $u \in C^\infty(\Omega)$ and $K \subset \Omega$ is compact, pick $v \in C_0^\infty(\Omega)$ such that supp $A \cap (\Omega \times K)$ is contained in $\hat{K} \times \hat{K}$ with $\hat{K}$ compact and $v = 1$ on a neighborhood of $\hat{K}$. It follows that $Au = A(vu)$ on K, so Au, which a priori belongs to $\mathscr{D}'(\Omega)$, is smooth on the interior of K. This proves (3.10). We now have the following result.

THEOREM 3.8. *Let $A \in OPS^m_{\rho,\delta_1,\delta_2}$. Assume $0 \le \delta_2 < \rho \le 1$. Then there is a $p(x, \xi) \in S^m_{\rho,\delta}$, with $\delta = \max(\delta_1, \delta_2)$, such that*

$$A = p(x, D).$$

In fact, $p(x, \xi) = e^{-ix\cdot\xi}A(e^{ix\cdot\xi})$, and we have the asymptotic expansion

$$p(x, \xi) \sim \sum_{\alpha \ge 0} \frac{i^{|\alpha|}}{\alpha!} D^\alpha_\xi D^\alpha_y a(x, y, \xi)\Big|_{y=x}. \tag{3.11}$$

Proof. We see that $p(x, \xi) = e^{-ix\cdot\xi}A(e^{ix\cdot\xi})$ is a well-defined smooth function of its arguments, and if we apply the linear operator A to

$$u(x) = \int \hat{u}(\xi)e^{ix\cdot\xi}\, d\xi,$$

we get

$$Au(x) = \int \hat{u}(\xi)p(x, \xi)e^{ix\cdot\xi}\, d\xi.$$

It remains to show that $p \in S^m_{\rho,\delta}$ and that (3.11) holds. Note that the general term in the sum in (3.11) belongs to $S^{m-(\rho-\delta_2)|\alpha|}_{\rho,\delta}$.

Let $b(x, y, \eta) = a(x, x+y, \eta)$ and $\hat{b}(x, \xi, \eta) = (2\pi)^{-n}\int b(x, y, \eta)e^{-iy\cdot\xi}\, dy$, so

$$p(x, \eta) = \int \hat{b}(x, \xi, \eta+\xi)\, d\xi. \tag{3.12}$$

The hypotheses on $a(x, y, \xi)$ imply

$$|D^\gamma_y D^\beta_x D^\alpha_\eta b(x, y, \eta)| \le C(1+|\eta|)^{m+\delta|\beta|+\delta_2|\gamma|-\rho|\alpha|}, \qquad \delta = \delta_1 \vee \delta_2. \tag{3.13}$$

Note that $a(x, y, \xi)$ can be replaced by an amplitude of the form $\tilde{a}(x, y)a(x, y, \xi)$ where $\tilde{a}(x, y)$ has proper support in $\Omega\times\Omega$, and $\tilde{a} = 1$ on an appropriate neighborhood of the diagonal, without changing A, if A is properly supported, so we can suppose without loss of generality that $a(x, y, \xi)$ is properly supported. Thus, as x runs over any compact subset of Ω, $b(x, y, \eta)$ vanishes for y outside some compact set. Now (3.13) implies

$$|D^\beta_x D^\alpha_\eta \hat{b}(x, \xi, \eta)| \le C_\nu(1+|\eta|)^{m+\delta|\beta|+\delta_2\nu-\rho|\alpha|}(1+|\xi|)^{-\nu}. \tag{3.14}$$

Since $\delta_2 < 1$, it follows that $p(x, \eta)$ and any of its derivatives can be bounded by some power of $1+|\eta|$. We now set things up to apply Theorem 3.2.

If we take the Taylor expansion of $\hat{b}(x, \xi, \eta + \xi)$ in the last argument, about the point η, (3.14) yields

$$\Big|\hat{b}(x, \xi, \eta+\xi) - \sum_{|\alpha|<N} \frac{1}{\alpha!}(iD_\eta)^\alpha \hat{b}(x, \xi, \eta)\xi^\alpha\Big|$$
$$\le C_\nu|\xi|^N(1+|\xi|)^{-\nu} \sup_{0\le t\le 1} (1 + |\eta+t\xi|)^{m+\delta_2\nu-\rho N}$$

where ν can be 0 or any positive number. With $\nu = N$ we obtain a bound

$$C(1+|\eta|)^{m-(\rho-\delta_2)N} \qquad \text{if} \qquad |\xi| \leq \tfrac{1}{2}|\eta|,$$

and if N is large, we get a bound by any power of $(1+|\xi|)^{-1}$ for $|\eta| < 2|\xi|$. Hence

$$\left| p(x, \eta) - \sum_{|\alpha|<N} \frac{1}{\alpha!} (iD_\eta)^\alpha D_y^\alpha b(x, y, \eta)|_{y=0} \right| \leq C(1+|\eta|)^{m+n-(\rho-\delta_2)N}.$$

Theorem 3.8 now follows immediately from theorem 3.2.

Properly supported operators can be composed, and the composition is properly supported. Since the distribution kernel of any pseudodifferential operator is smooth off the diagonal, we may write such an operator as the sum of a properly supported operator and a smoothing operator. In the future, pseudodifferential operators will usually be assumed to be properly supported, often without comment.

§4. Adjoints and Products

THEOREM 4.1. *If $p(x, D) \in OPS^m_{\rho,\delta}$ is properly supported, $\delta < 1$, then*

$$p(x, D)^* \in OPS^m_{\rho,0,\delta}.$$

Proof. One has

$$(p(x, D)u, v) = (2\pi)^{-n} \int \overline{v(y)} \iint e^{i(y-x)\cdot\xi} p(y, \xi) u(x)\, dx\, d\xi\, dy$$

$$= (2\pi)^{-n} \left(\int \overline{u(x)} \iint e^{i(x-y)\cdot\xi} p(y, \xi)^* v(y)\, dy\, d\xi\, dx \right)^*,$$

so

$$p(x, D)^* v = (2\pi)^{-n} \iint p(y, \xi)^* e^{i(x-y)\cdot\xi} v(y)\, dy\, d\xi \tag{4.1}$$

which is in the form (3.7) with $a(x, y, \xi) = p(y, \xi)^*$.

Applying Theorem 3.8, we have the following.

THEOREM 4.2. *If $p(x, D) \in OPS^m_{\rho,\delta}$ is properly supported, $0 \leq \delta < \rho \leq 1$, then*

$$p(x, D)^* \in OPS^m_{\rho,\delta}$$

and indeed $p(x, D)^* = p^*(x, D)$ with

$$p^*(x, \xi) \sim \sum_{\alpha \geq 0} \frac{i^{|\alpha|}}{\alpha!} D_\xi^\alpha D_x^\alpha p(x, \xi)^*. \tag{4.2}$$

Proof. Immediate. Equation (3.11) yields (4.2).

THEOREM 4.3. *Let* $p(x, D) \in OPS^m_{\rho',\delta'}$ and $q(x, D) \in OPS^\mu_{\rho'',\delta''}$ *be properly supported,* $0 \le \delta'' < \rho'' \le 1$. *Then*

$$p(x, D)q(x, D) \in OPS^{m+\mu}_{\rho,\delta',\delta''}, \qquad \rho = \min(\rho', \rho'').$$

Proof. First apply Theorems 4.1 and 4.2 to the operator $q(x, D)^* = q^*(x, D)$.

$$q(x, D)u(x) = q(x, D)^{**}u = (2\pi)^{-n} \iint q^*(y, \xi)^* e^{i(x-y)\cdot\xi} u(y)\, dy\, d\xi.$$

This implies

$$\widehat{q(x, D)u}(\xi) = (2\pi)^{-n} \int e^{-iy\cdot\xi} q^*(y, \xi)^* u(y)\, dy \tag{4.3}$$

provided $0 \le \delta'' < \rho'' \le 1$. Thus

$$\begin{aligned} p(x, D)q(x, D)u &= \int e^{ix\cdot\xi} p(x, \xi)\widehat{q(x, D)u}(\xi)\, d\xi \\ &= (2\pi)^{-n} \iint e^{i(x-y)\cdot\xi} p(x, \xi) q^*(y, \xi)^* u(y)\, dy\, d\xi. \end{aligned} \tag{4.4}$$

Thus $p(x, D)q(x, D)$ is of the form (3.7) with $a(x, y, \xi) = p(x, \xi)q^*(y, \xi)^*$, and the proof is complete.

THEOREM 4.4. *Let* $p(x, D) \in OPS^m_{\rho',\delta'}$ *and* $q(x, D) \in OPS^\mu_{\rho'',\delta''}$ *be properly supported. Suppose* $0 \le \delta'' < \rho \le 1$ *with* $\rho = \min(\rho', \rho'')$. *Then*

$$p(x, D)q(x, D) \in OPS^{m+\mu}_{\rho,\delta}, \qquad \delta = \max(\delta', \delta'')$$

and $p(x, D)q(x, D) = r(x, D)$ *with*

$$r(x, \xi) \sim \sum_{\alpha \ge 0} \frac{i^{|\alpha|}}{\alpha!} D^\alpha_\xi p(x, \xi) D^\alpha_x q(x, \xi). \tag{4.5}$$

Proof. We apply Theorem 3.8 to the conclusion of Theorem 4.3. We need only check (4.5). From (3.11) we have

$$\begin{aligned} r(x, \xi) &\sim \sum_{\alpha \ge 0} \frac{i^{|\alpha|}}{\alpha!} D^\alpha_\xi D^\alpha_y (p(x, \xi) q^*(y, \xi)^*)\Big|_{y=x} \\ &\sim \sum_{\alpha,\beta \ge 0} \frac{i^{|\alpha|-|\beta|}}{\alpha!\beta!} D^\alpha_\xi (p(x, \xi) D^\beta_\xi D^{\alpha+\beta}_y q(y, \xi))\Big|_{y=x}. \end{aligned}$$

If one expands this out, numerous terms cancel and (4.5) is produced.

We note the following consequences of the symbol expansions (4.2) and (4.5):

$$p^*(x, \xi) - p(x, \xi)^* \in S^{m-(\rho-\delta)}_{\rho,\delta}; \tag{4.6}$$

$$r(x, \xi) - p(x, \xi)q(x, \xi) \in S^{m+\mu-(\rho'-\delta'')}_{\rho,\delta}. \tag{4.7}$$

In particular, if $p(x, \xi) \in S^m_{1,0}$, $q(x, \xi) \in S^\mu_{1,0}$, we have

$$p^*(x, \xi) - p(x, \xi)^* \in S^{m-1}_{1,0},$$
$$r(x, \xi) - p(x, \xi)q(x, \xi) \in S^{m+\mu-1}_{1,0}.$$

This case will be of most frequent use, but when other cases are called upon they can be used to powerful effect. In Chapter X we shall find it very useful that

$$OPS^m_{1/3,2/3} \cdot OPS^\mu_{1,0} \subset OPS^{m+\mu}_{1/3,2/3}.$$

We remark that Theorem 4.4 remains valid when the hypothesis $\delta'' < \min(\rho', \rho'')$ is relaxed to $\delta'' < \rho'$, which should seem reasonable since the terms in (4.5) still have order tending to $-\infty$. The trick using $q(x, D)^{**}$, due to Hörmander [14], does not work in this more general case, and for a proof the reader is referred to Hörmander [12].

Exercises

4.1. Suppose the adjoint of $P \in OPS^m_{\rho,\delta}(\Omega)$ is defined with respect to a volume form $\rho(x)\,dx$: $(P^*u, v) = \int P^*u\bar{v}\rho\,dx = (u, Pv) = \int u\overline{Pv}\rho\,dx$. Assume ρ smooth and >0. Compute the symbol of P^* now. Show that (4.6) remains valid in this more general context.

This arises if one is on a manifold (see the next section) and chooses coordinates that take the volume form to $\rho\,dx$.

4.2. Suppose $P \in OPS^m$, $Q \in OPS^\mu$. Show that $P^* \in OPS^m$ and $PQ \in OPS^{m+\mu}$.

§5. Coordinate Changes: Operators on a Manifold

We shall now see what happens to a pseudodifferential operator under a change of coordinates. Let Ω and $\mathcal{O}$ be regions in $\mathbf{R}^n$ and let $\chi: \Omega \to \mathcal{O}$ be a diffemorphism. If $P = p(x, D) \in OPS^m_{\rho,\delta}(\Omega)$, so $P: C_0^\infty(\Omega) \to C^\infty(\Omega)$, set

$$\tilde{P}u = P(u \circ \chi) \circ \chi^{-1} \tag{5.1}$$

so $\tilde{P}: C_0^\infty(\mathcal{O}) \to C^\infty(\mathcal{O})$.

THEOREM 5.1. *If $P \in OPS^m_{\rho,\delta}(\Omega)$ is properly supported and if $\rho > 1/2$ and $\rho + \delta \geq 1$, then $\tilde{P} \in OPS^m_{\rho,\delta}(\mathcal{O})$ and $\tilde{P} = \tilde{p}(x, D)$ with*

$$\tilde{p}(\chi(x), \xi) \sim \sum_{\alpha \geq 0} \frac{1}{\alpha!} \varphi_\alpha(x, \xi) p^{(\alpha)}(x, {}^t\chi'(x)\xi), \tag{5.2}$$

where $\varphi_\alpha(x, \xi)$ is a polynomial in ξ of degree $\leq (1/2)|\alpha|$, with $\varphi_0(x, \xi) = 1$.

We have used the shorthand notation $p^{(\alpha)}(x, \xi) = D_\xi^\alpha p(x, \xi)$. More generally, the notation

(5.3) $$p_{(\beta)}^{(\alpha)}(x, \xi) = D_x^\beta D_\xi^\alpha p(x, \xi)$$

is useful. Also, $\chi'(x)$ denotes the Jacobian of χ, and ${}^t\chi'(x)$, its transpose.

Proof. Write $\chi_1 = \chi^{-1}$ and

$$\begin{aligned}\tilde{P}u(x) &= (2\pi)^{-n} \int e^{i(\chi_1(x)-y)\cdot\xi} p(\chi_1(x), \xi) u(\chi(y))\, dy\, d\xi \\ &= (2\pi)^{-n} \int e^{i(\chi_1(x)-\chi_1(y))\cdot\xi} p(\chi_1(x), \xi) u(y) \det(\chi_1'(y))\, dy\, d\xi \\ &= (2\pi)^{-n} \int e^{i(x-y)\cdot{}^t\Phi(x,y)\xi} p(\chi_1(x), \xi) \det \chi_1'(y) u(y)\, dy\, d\xi\end{aligned}$$

where Φ is defined by setting

$$\begin{aligned}(\chi_1(x)-\chi_1(y))\cdot\xi &= \sum_j (\chi_1^j(x)-\chi_1^j(y))\xi_j \\ &= \sum_{j,k} \Phi_{kj}(x, y)(x_k - y_k)\xi_j.\end{aligned}$$

Thus Φ is defined and smooth near the diagonal in $\mathcal{O} \times \mathcal{O}$, and

(5.4) $$\tilde{P}u(x) = (2\pi)^{-n} \int e^{i(x-y)\cdot\xi} p(\chi_1(x), \psi(x, y)\xi) D(x, y) u(y)\, dy\, d\xi + Ku,$$

with $\psi(x, y) = \Phi(x, y)^{-1}$ and $D(x, y) = \det \chi_1'(y) \det \psi(x, y)\Xi(x, y)$ where $\Xi(x, y)$ is identically 1 in a neighborhood of the diagonal in $\mathcal{O} \times \mathcal{O}$ and is thrown in because $\psi(x, y)$ might not be defined everywhere. The resulting error term K is a smoothing operator which, being properly supported, belongs to $OPS^{-\infty}$.

Now in Formula (5.4) we have one familiar representation of a pseudodifferential operator, namely the form (3.7), with

(5.5) $$a(x, y, \xi) = p(\chi_1(x), \psi(x, y)\xi) D(x, y).$$

A straightforward application of the chain rule shows that

$$\begin{aligned}|D_y^\gamma D_x^\beta D_\xi^\alpha a(x, y, \xi)| &\le C(1+|\xi|)^{m-\rho|\alpha|} \sum_{\mu+\nu\le|\beta|} (1+|\xi|)^{(1-\rho)\nu+\delta\mu+(1-\rho)|\gamma|} \\ &\le C(1+|\xi|)^{m-\rho|\alpha|+|\beta|\max(\delta, 1-\rho)+(1-\rho)|\gamma|}.\end{aligned}$$

Consequently we have the following intermediate result on the way to Theorem 5.1.

PROPOSITION 5.2. *If $P \in OPS_{\rho,\delta}^m(\Omega)$ is properly supported, then*

$$\tilde{P} \in OPS_{\rho,\delta',1-\rho}^m(\mathcal{O}), \qquad \delta' = \max(\delta, 1-\rho).$$

To continue with the proof of Theorem 5.1, the hypotheses imply that $\delta' = \delta$ and that $1-\rho < \rho$, so Theorem 3.8 is applicable. We have $\tilde{P} \in OPS^m_{\rho,\delta}(\mathcal{O})$ and

$$\tilde{p}(x, \xi) \sim \sum_{\alpha \leq 0} \frac{i^{|\alpha|}}{\alpha!} D^\alpha_\xi D^\alpha_y(p(\chi_1(x), \psi(x, y)\xi)D(x, y))\Big|_{y=x}. \tag{5.6}$$

Now $\psi(x, x) = {}^t\chi'(\chi_1(x))$ and $D(x, x) = 1$, from which we see that (5.6) leads to (5.2).

This proves Theorem 5.1. Note that

$$\tilde{p}(x, \xi) - p(\chi_1(x), {}^t\chi'(\chi_1(x))\xi) \in S^{m-(2\rho-1)}_{\rho,\delta} \tag{5.7}$$

since the general term in (5.6) belongs to $S^{m-\rho|\alpha|+(1-\rho)|\alpha|}_{\rho,\delta}$. Making a more careful analysis of (5.6), one can show that

$$\varphi_\alpha(x, \xi) = D^\alpha_y \exp i(\chi(y) - \chi(x) - \chi'(x)(y-x)) \cdot \xi|_{y=x},$$

but we shall not go into the details, but refer the reader to Hörmander [6].

We can now define the concept of a pseudodifferential operator on a manifold M. Namely $A: C^\infty_0(M) \to C^\infty(M)$ belongs to $OPS^m_{\rho,\delta}(M)$ if the kernel of A is smooth off the diagonal in $M \times M$ and if for any coordinate neighborhood U in M with $\chi: U \to \mathcal{O}$ a diffeomorphism onto an open subset $\mathcal{O}$ of $\mathbf{R}^n$, the map of $C^\infty_0(\mathcal{O})$ into $C^\infty(\mathcal{O})$ given by $u \mapsto A(u \circ \chi) \circ \chi^{-1}$ belongs to $OPS^m_{\rho,\delta}(\mathcal{O})$. We assume $\rho > 1/2$ and $\rho + \delta \geq 1$.

If $p(x, D) \in OPS^m_{\rho,\delta}(\mathcal{O})$, we define the principal symbol of $p(x, D)$ to be the equivalence class in $S^m_{\rho,\delta}(\mathcal{O})/S^{m-(2\rho-1)}_{\rho,\delta}(\mathcal{O})$. We shall also call any member of this equivalence class a principal symbol of $p(x, D)$. By (5.7) it follows that if A is a pseudodifferential operator on a manifold M, its principal symbol is a well-defined function on the cotangent bundle $T^*(M)$.

§6. L^2 and Sobolev Space Continuity

The basic aim of this section is to prove that if $A \in OPS^m_{\rho,\delta}(\Omega)$ and $\delta < \rho$, then $A: H^s_{\text{comp}}(\Omega) \to H^{s-m}_{\text{loc}}(\Omega)$. Since the proof for $\delta = 0$ is more straightforward than the proof in the more general case, we give that first.

PROPOSITION 6.1. *If* $p(x, \xi) \in S^0_{0,0}(\mathbf{R}^n)$ *has support in* $|x| \leq C_0$, *then* $p(x, D): L^2(\mathbf{R}^n) \to L^2(\mathbf{R}^n)$, *continuously.*

Proof. Write $p(x, \xi) = \int p_\eta(\xi) e^{ix\cdot\eta}\, d\eta$ where

$$p_\eta(\xi) = (2\pi)^{-n} \int p(x, \xi) e^{-ix\cdot\eta}\, dx.$$

The assumptions on $p(x, \xi)$ imply that $|p_\eta(\xi)| \leq C_N(1 + |\eta|)^{-N}$, since

$$\eta^\alpha p_\eta(\xi) = (2\pi)^{-n} \int D^\alpha_x p(x, \xi) e^{-ix\cdot\eta}\, dx.$$

Thus

$$\|p_\eta(D)u\|_{L^2} \le C_N(1+|\eta|)^{-N}\|u\|_{L^2}.$$

Since $p(x, D) = \int e^{ix\cdot\eta}p_\eta(D)\, d\eta$ and $|e^{ix\cdot\eta}| = 1$, we see that

$$\|p(x, D)u\|_{L^2} \le C_N \int (1+|\eta|)^{-N}\, d\eta\|u\|_{L^2}$$
$$\le C_1\|u\|_{L^2}$$

if we take $N > n$.

Note that this proof breaks down if $\delta > 0$, since in that case we could only conclude that $|p_\eta(\xi)| \le C_N(1+|\xi|)^{\delta N}(1+|\eta|)^{-N}$. The following ingenious argument is due to Hörmander [14]. It is a very beautiful variation on the idea that a positive linear functional λ on $C(K)$, the space of continuous functions on a compact Hausdorff space K, is automatically continuous, with norm $\|\lambda\| = \lambda(1)$. Furthermore, the proof builds up machinery that will also prove Gårding's inequality, in Section 8.

Lemma 6.2. *If $p(x, \xi) \in S^0_{\rho,\delta}(\Omega)$, $\delta < \rho$, and if $Re\, p(x, \xi) \ge C > 0$, then there exists a $B \in OPS^0_{\rho,\delta}$ such that, with $Re\, P = (1/2)(P + P^*)$,*

$$Re\, p(x, D) - B^*B \in OPS^{-\infty}. \tag{6.1}$$

Proof. We shall construct the symbol $b(x, \xi) \sim \Sigma b_j(x, \xi)$ with $b_j \in S^{-j(\rho-\delta)}_{\rho,\delta}$. Start with

$$b_0(x, \xi) = (Re\, p(x, \xi))^{1/2}.$$

It is easy to infer (see Exercise 1.3) that $b_0(x, \xi) \in S^0_{\rho,\delta}$. Furthermore, the formulas for adjoints and products show that

$$Re\, p(x, D) - b_0(x, D)^*b_0(x, D) = R_1 \in OPS^{-(\rho-\delta)}_{\rho,\delta}.$$

Proceeding by induction, suppose we have the terms $b_0, \dots, b_j$ in the asymptotic expansion. We need $b_{j+1} \in S^{-(j+1)(\rho-\delta)}_{\rho,\delta}$ such that

$$Re\, p(x, D) = ((b_0^* + \cdots + b_j^*) + b_{j+1}^*)((b_0 + \cdots + b_j) + b_{j+1}) + R_{j+1} \tag{6.2}$$

with $R_{j+1} \in OPS^{-(j+1)(\rho-\delta)}_{\rho,\delta}$. The right-hand side of (6.2) is equal to

$$Re\, p(x, D) + R_j + b_{j+1}^*(b_0 + \cdots + b_{j+1}) + (b_0^* + \cdots + b_{j+1}^*)b_{j+1} + R_{j+1}$$
$$= Re\, p(x, D) + R_j + b_{j+1}^*b_0 + b_0^*b_{j+1} \bmod OPS^{-(j+1)(\rho-\delta)}_{\rho,\delta},$$

where $R_j \in OPS^{-j(\rho-\delta)}_{\rho,\delta}$ is the analogous remainder term in the previous stage. Note that $R_j = R_j^*$ so its principal symbol is real or, if a matrix, self adjoint. We require of the *symbol* of b_{j+1} that

$$b_{j+1}^*b_0 + b_0b_{j+1} = -R_j. \tag{6.3}$$

Consequently, we need merely pick $b_{j+1} = -(1/2)b_0^{-1}R_j$ in the scalar case.

This finishes the construction of $b(x, \xi)$ in the scalar case, which suffices to obtain the results of this section, but we shall pause to clean up the case where $p(x, \xi)$ is a $k \times k$ system, with $Re\, p(x, \xi) = (1/2)(p(x, \xi) + p(x, \xi)^*) \geq C > 0$. Thus $b_0(x, \xi)$ is a positive self-adjoint matrix. It follows that (6.3) has a *unique* self-adjoint solution $b_{j+1}(x, \xi) = b_{j+1}(x, \xi)^*$. Indeed, the map $\Phi: S_k \to S_k$ on the space of $k \times k$ self-adjoint matrices defined by $\Phi(A) = Ab_0 + b_0A$ is easily seen to have eigenvalues $\{\lambda_j + \lambda_i\}$ where $\lambda_j > 0$ are the eigenvalues of b_0. This completes the proof of Lemma 6.2 in the case of systems.

We obtain the following L^2 estimate.

THEOREM 6.3. *Let $A \in OPS^0_{\rho,\delta}(\Omega)$, $0 \leq \delta < \rho \leq 1$; assume that*

$$\limsup_{|\xi| \to \infty} |A(x, \xi)| < M < \infty. \tag{6.4}$$

If $K \subset\subset \Omega$, there is an $R \in OPS^{-\infty}$ such that

$$\|Au\|^2_{L^2(K)} \leq M^2\|u\|^2 + (Ru, u). \tag{6.5}$$

Proof. The operator $C = M^2 - A^*A$ has principal symbol $C(x, \xi) = M^2 - |A(x, \xi)|^2 > 0$, so by Lemma 6.2 there is a $B \in OPS^0_{\rho,\delta}$ such that

$$C - B^*B = M^2 - A^*A - B^*B = -R \in OPS^{-\infty}.$$

Thus

$$\begin{aligned}\|Au\|^2_{L^2} &\leq (Au, Au) + (Bu, Bu) \\ &\leq M^2\|u\|^2_{L^2} + (Ru, u),\end{aligned}$$

and the proof is complete.

The following corollary is immediate.

COROLLARY 6.4. *If $\lim_{|\xi| \to \infty} A(x, \xi) = 0$, then $A: L^2(K) \to L^2_{\text{loc}}(\Omega)$ is compact.*

The result that $OPS^m_{\rho,\delta}: H^s \to H^{s-m}$ if $0 \leq \delta < \rho \leq 1$ follows from the L^2 continuity result of Theorem 6.3, via use of the operators $\Lambda^\sigma \in OPS^\sigma_{1,0}(\mathbf{R}^n)$ defined by $\Lambda^\sigma u = \int (1 + |\xi|^2)^{\sigma/2} e^{ix \cdot \xi}\hat{u}(\xi)\, d\xi$. Clearly $\Lambda^\sigma: H^s \to H^{s-\sigma}$ isomorphically. On $\Omega \subset \mathbf{R}^n$, alter Λ^σ so it is properly supported.

THEOREM 6.5. *If $A \in OPS^m_{\rho,\delta}(\Omega)$ is properly supported, $0 \leq \delta < \rho \leq 1$, then*

$$A: H^s_{\text{loc}}(\Omega) \to H^{s-m}_{\text{loc}}(\Omega).$$

Proof. It suffices to show that $\Lambda^{s-m}A\Lambda^{-s} \in OPS^0_{\rho,\delta}$ takes $L^2_{\text{loc}}(\Omega)$ to $L^2_{\text{loc}}(\Omega)$ which follows immediately from Theorem 6.3.

Note that, in the proof of Lemma 6.2 above, the assumption $\delta < \rho$ was necessary in order to ensure that in the asymptotic series defining B the order of the terms went to $-\infty$. Calderon and Vaillancourt have shown that $A \in OPS^0_{\rho,\rho}$ is continuous on L^2, $0 \le \rho < 1$. We shall prove this result in Chapter XIII. One key ingredient in the proof of this is the L^2 continuity of $p(x, D)$ on $L^2(\mathbf{R}^n)$ when one assumes

$$|D_x^\beta D_\xi^\alpha p(x, \xi)| \le C_{\alpha\beta}, \qquad x, \xi \in \mathbf{R}^n$$

but does not assume p has compact support in x. Unlike Proposition 6.1, L^2 continuity in this case is not trivial. For $OPS^0_{1,0}$ one also has continuity on L^p, $1 < p < \infty$, and on Hölder spaces. This will be taken Chapter XI.

Exercises

In the following exercises, let M be a compact manifold and $P \in OPS^0_{\rho,\delta}(M)$, $1 - \rho \le \delta < \rho$.

6.1. Show that the quantity $\sup_M \limsup_{|\xi| \to \infty} |p(x, \xi)|$ is a well-defined quantity, where $p(x, \xi)$ is a principal symbol of P. Call it $N(P)$.

6.2. Show that $\inf \{\|P + K\| : K \text{ compact on } L^2(M)\} \le N(P)$.

Suppose that $N(P) > 0$, say there are $x_j \to x_0$ in M (surrounded by some coordinate system, in which we express the symbol $p(x, \xi)$ of P) and $\xi_j \to \infty$ such that $|p(x_j, \xi_j)| \ge C_0 > 0$. We may as well assume $x_0 = 0$, the origin. We want to show that P is *not compact*. Before doing the following exercises, look ahead at Section 7.

6.3. Pick $\bar{\delta} > \delta$, $\bar{\rho} < \rho$, slightly, so that $1 - \bar{\rho} \le \bar{\delta} < \bar{\rho}$; pick ψ, $\varphi \in C_0^\infty(\mathbf{R}^n)$, both equal to 1 inside the unit ball and vanishing outside the ball of radius 2. Let

$$q_j(x, \xi) = \psi(|\xi_j|^{\bar{\delta}}(x - x_j))\varphi(|\xi_j|^{-\bar{\rho}}(\xi - \xi_j)).$$

Show that $q_j(x, \xi)$ is bounded in $S^0_{\bar{\rho},\bar{\delta}}$.

6.4. With $q_j(x, D) = Q_j$, show that $\|Q_j\|_{\mathscr{L}(L^2)}$ does not tend to 0, as $j \to \infty$.

6.5. Show that $Q_j u \to 0$ in L^2 for each $u \in L^2$. If P were compact, we could deduce that $\|PQ_j\| \to 0$.

6.6. Let $a_j = p(x_j, \xi_j)$. Show that $PQ_j - a_j Q_j$ is bounded in $OPS^{-\epsilon}_{\bar{\rho},\bar{\delta}}$ for some $\epsilon > 0$.

6.7. Let $T_j \to 0$ strongly on $L^2(M)$ and suppose T_j is bounded in $OPS^{-\epsilon}_{\bar{\rho},\bar{\delta}}(M)$ for some $\epsilon > 0$. Show that $\|T_j\| \to 0$. Hint: Show that $T_j \Lambda^\epsilon \to 0$ strongly and write $T_j = (T_j \Lambda^\epsilon)\Lambda^{-\epsilon}$, with $\Lambda^{-\epsilon}$ compact on $L^2(M)$.

6.8. Deduce that, if P were compact, $\|a_j Q_j\| \to 0$. Conclude from this that P is not compact if $N(P) > 0$.

6.9. Show that $N(P) = \inf \{\|P + K\| : K \text{ compact on } L^2(M)\}$. Hint: Find a sequence u_j with $\|u_j\| = 1$ and $\|Q_j u_j - u_j\| \to 0$.

§7. Families of Pseudodifferential Operators: Friedrichs' Mollifiers

We make $S^m_{\rho,\delta}(\Omega)$ into a Frechet space by means of the seminorms

$$|p|_{K,\alpha,\beta} = \sup_{x\in K} |D^\beta_x D^\alpha_\xi p(x,\xi)|(1+|\xi|)^{-m+\rho|\alpha|-\delta|\beta|}.$$

It easily follows from the reasoning of Section 6 that the map $p(x,\xi) \mapsto p(x,D)$ is a continuous map from $S^m_{\rho,\delta}(\Omega)$ to $\mathscr{L}(H^s_{\text{comp}}(\Omega), H^{s-m}_{\text{loc}}(\Omega))$ if $\delta < \rho$.

Similarly, if M is a compact manifold, $0 \le 1-\rho \le \delta < \rho \le 1$, we can give $OPS^m_{\rho,\delta}(M)$ a natural Frechet space topology by using coordinate neighborhoods, and all the natural maps

$$OPS^m_{\rho,\delta}(M) \to \mathscr{L}(H^s(M), H^{s-m}(M))$$

are continuous.

In particular, bounded families of symbols give rise to bounded families of operators. In particular, let $p(\xi) \in S^\sigma_{1,0}(\mathbf{R}^n)$, $\sigma \le 0$, and let

$$p_\epsilon(\xi) = p(\epsilon\xi).$$

A simple application of the chain rule shows that $\{p_\epsilon : 0 < \epsilon \le 1\}$ is bounded in $S^0_{1,0}(\mathbf{R}^n)$. In particular, one could take $p(\xi) \in S^{-\infty}(\mathbf{R}^n)$, for example,

$$p(\xi) = e^{-|\xi|^2}.$$

If coordinate patches and partitions of unity are used, it is easy to build on any compact manifold M a family of operators J_ϵ satisfying the following conditions.

DEFINITION 7.1. *A Friedrichs' mollifier on M is a family J_ϵ of scalar pseudodifferential operators, $0 < \epsilon \le 1$, such that*

(7.1) $J_\epsilon \in OPS^{-\infty}(M)$ *for each* $\epsilon \in (0, 1]$.

(7.2) $\{J_\epsilon : 0 < \epsilon \le 1\}$ *is a bounded subset of* $OPS^0_{1,0}(M)$.

(7.3) $J_\epsilon u \to u$ *in* $L^2(M)$ *as* $\epsilon \to 0$, *for each* $u \in L^2(M)$.

Friedrich's mollifiers are useful operators to have around, as we shall see in Chapters IV and V. We note the following simple consequences of the arguments of Section 5.

PROPOSITION 7.2. *Let $A \in OPS^m_{\rho,\delta}(M)$, $1-\rho \le \delta < \rho$. If J_ϵ is a Friedrich's mollifier on M, then $[A, J_\epsilon] = AJ_\epsilon - J_\epsilon A$ has the following properties*

(7.4) $[A, J_\epsilon] \in OPS^{-\infty}(M)$, $0 < \epsilon \le 1$, *and*

(7.5) $\{[A, J_\epsilon] : 0 < \epsilon \le 1\}$ *is a bounded subset of* $OPS^{m-\rho\wedge(1-\delta)}_{\rho,\delta}(M)$.

We now give an application of Friedrichs' mollifiers to proving the identity of weak and strong solutions to pseudodifferential equations.

DEFINITION 7.3. *Let M be a compact manifold, $A: C^\infty(M) \to C^\infty(M)$ and $A: \mathcal{D}'(M) \to \mathcal{D}'(M)$. Take $f \in L^2(M)$. A function $u \in L^2(M)$ is said to be a weak solution of the equation*

$$Au = f \tag{7.6}$$

if this equation holds when A is applied to u in the distribution sense. On the other hand, u is said to be a strong solution of (7.6) *if there exists a sequence $u_j \to u$ in $L^2(M)$, with $u_j \in C^\infty(M)$, such that $Au_j = f_j \to f$ in $L^2(M)$.*

Clearly every strong solution is a weak solution. Conversely:

PROPOSITION 7.4. *If $A \in OPS^1_{1,0}(M)$, then every weak solution to* (7.6) *is a strong solution.*

Proof. With J_ϵ a Friedrichs' mollifier, let $\epsilon_j \to 0$ and set $u_j = J_{\epsilon_j} u$. To show that $\|Au_j - f\|_{L^2} \to 0$, write

$$Au_j = J_{\epsilon_j} Au + [A, J_{\epsilon_j}]u = J_{\epsilon_j} f + [A, J_{\epsilon_j}]u,$$

and use Proposition 7.2.

If $A \in OPS^m_{1,0}(M)$ with $m > 1$, weak and strong solutions need not coincide, though they do if A is elliptic (since by elliptic regularity u must belong to $H^m(M)$).

The Friedrichs' mollifier technique was introduced by Friedrichs [2] in order to prove such "weak = strong" results. For weak = strong results on manifolds with boundary, see Sarason [1], Ralston [2], or Tartakoff [1].

Exercises

7.1. Build a Friedrichs' mollifier on M.

7.2. Prove Proposition 7.2.

7.3. Let $p_{\epsilon,\eta}(\xi) = e^{-\epsilon\xi^2 + i\eta\xi}$. Show that $\{p_{\epsilon,\eta}: 0 < \epsilon, |\eta| \leq 1, \eta^2 \leq c\epsilon\}$ is bounded in $S^0_{1,0}(\mathbf{R}^n)$. More generally, show that, if $0 < \rho \leq 1$, $\{p_{\epsilon,\eta}: 0 < \epsilon, |\eta| \leq 1; \eta^2 \leq c\epsilon^\rho\}$ is bounded in $S^0_{\rho,0}(\mathbf{R}^n)$.

§8. Gårding's Inequality

This inequality, first proved by Gårding [1] for differential operators and by Calderon and Zygmund for singular integral operators, is one of the fundamental results of the theory. We shall see it applied to good effect (via Exercise 8.1) in proving energy estimates for symmetrizable hyperbolic equations, a task to which the tool was originally applied by

Gårding and it will also play an important role in deriving a priori estimates for elliptic boundary value problems in Chapter V.

THEOREM 8.1. *Let $p(x, D) \in OPS^m_{\rho,\delta}(\Omega)$ and assume $0 \leq \delta < \rho \leq 1$. Suppose $Re\, p(x, \xi) \geq C|\xi|^m$ for $|\xi|$ large, with $C > 0$. Then, for any $s \in \mathbf{R}$, for any compact $K \subset \Omega$, and all $u \in C_0^\infty(K)$, we have*

$$Re(p(x, D)u, u) \geq C_0\|u\|^2_{H^{m/2}} - C_1\|u\|^2_{H^s} \tag{8.1}$$

Proof. Replacing $p(x, D)$ by $q(x, D) = \Lambda^{-m/2}p(x, D)\Lambda^{-m/2}$, we can suppose without loss of generality that $m = 0$. So suppose $Re\, p(x, \xi) \geq C > 0$, $p(x, \xi) \in S^0_{\rho,\delta}$. Now Lemma 6.8 applies to $r(x, \xi) = Re\ p(x, \xi) - (1/2)C$, to yield $B \in OPS^0_{\rho,\delta}$ with $r(x, D) - B^*B = S \in OPS^{-\infty}$, and hence

$$Re(p(x, D)u, u) - \tfrac{1}{2}C(u, u) = (Bu, Bu) + Re(Su, u)$$

which immediately implies (8.1) in the case $m = 0$.

There is a sharp form of Gårding's inequality which says that under the weaker hypothesis that $Re\, p(x, \xi) \geq 0$, if $p(x, D) \in OPS^m_{1,0}$, it follows that

$$Re(p(x, D)u, u) \geq -C_1\|u\|^2_{H^{(m-1)/2}}, \qquad u \in C_0^\infty(K).$$

We shall return to this in Chapters VI and VII.

An operator $p(x, D) \in OPS^m_{\rho,\delta}$ (with $0 \leq \delta < \rho \leq 1$) which satisfies the condition

$$Re\, p(x, \xi) \geq C|\xi|^m \text{ for } |\xi| \text{ large}$$

for some positive C, is called strongly elliptic.

Exercises

8.1. Let M be a compact manifold, $p(x, D) \in OPS^m_{\rho,\delta}(M)$, $1 - \rho \leq \delta < \rho$, $m > 0$. Suppose $p(x, D)$ is self adjoint and that $Re\, p(x, \xi) \geq C|\xi|^m$ for $|\xi|$ large. Show that there exists a self adjoint $P \in OPS^m_{\rho,\delta}(M)$ such that

$$P - p(x, D) \in OPS^{-\infty}$$

and

$$P \geq cI > 0.$$

Hint: Examine the spectrum of $p(x, D)$.

§9. References to Further Work

Other classes of operators, some of which will be discussed later on in this book, have played an important role in linear *PDE*. Here I briefly describe classes of pseudodifferential operators of Beals and Fefferman

[2], Beals [1], [2] and Hörmander [23], referring to these sources for details.

Generalizing the definition of $S^m_{\rho,\delta}$, one can consider symbols $p(x, \xi)$ satisfying estimates of the form

$$|D_x^\beta D_\xi^\alpha p(x, \xi)| \le c_{\alpha\beta} e^{\lambda(x,\xi)} \Phi(x, \xi)^{-|\alpha|} \varphi(x, \xi)^{-|\beta|} \tag{9.1}$$

where φ, Φ, λ have properties to be described shortly. If (9.1) holds, $p(x, \xi)$ is said to belong to the symbol class

$$S^\lambda_{\Phi,\varphi}.$$

In case $\varphi(x, \xi) = (1+|\xi|^2)^{-\delta/2}$, $\Phi(x, \xi) = (1+|\xi|^2)^{\rho/2}$ and $e^{\lambda(x,\xi)} = (1+|\xi|^2)^{m/2}$, then the Beals-Fefferman class $S^\lambda_{\Phi,\varphi}$ coincides with Hörmander's class $S^m_{\rho,\delta}$. The *weight functions* Φ, φ are assumed to satisfy the following conditions:

$$\varphi \le C; \tag{9.2}$$

$$\Phi\varphi \ge C; \tag{9.3}$$

$$c \le \Phi(x, \xi)\Phi(y, \eta)^{-1} \le C \quad \text{and} \quad c \le \varphi(x, \xi)\varphi(y, \eta)^{-1} \le C \tag{9.4}$$
$$\text{if} \quad |x-y| \le c\varphi(x, \xi) \quad \text{and} \quad |\xi - \eta| \le c\Phi(x, \xi).$$

$$\text{With } R = \Phi\varphi^{-1}, \quad R(x, 0) \le C(1+|x|)^C. \tag{9.5}$$

$$c \le R(x, \xi)R(y, \eta)^{-1} \le C, \quad \text{if } |\xi-\eta| \le cR(x, \xi)^{\delta+1/2} \text{ and} \tag{9.6}$$
$$|x-y| \le cR(x, \xi)^\delta R(y, \eta)^{-1/2} \quad (\text{for some } \delta > 0).$$

The functions Φ, φ are regarded as assigning to each point (x, ξ) two characteristic units of length, $\varphi(x, \xi)$ in the x-direction and $\Phi(x, \xi)$ in the ξ-direction. Symbols (and the weight functions themselves) do not vary too rapidly over such distances. The pair Φ, φ is called *localizable* if, in addition to (9.2) through (9.6), one has, for some $\varepsilon > 0$,

$$\Phi \ge c_1(1+|\xi|)^\varepsilon. \tag{9.7}$$

We note that in the special case $\Phi = (1+|\xi|^2)^{\rho/2}$, $\varphi = (1+|\xi|^2)^{-\delta/2}$, (9.3) says $\delta \le \rho$ and (9.6) implies $\delta < 1$.

Given such a pair of weight functions, one assumes λ belongs to the class $\mathcal{O}(\Phi, \varphi)$, defined by λ continuous on $\mathbf{R}^{2n}$ and

$$|\lambda(x, \xi) - \lambda(y, \eta)| \le C \quad \text{if} \quad |x-y| \le c\varphi(x, \xi) \quad \text{and} \tag{9.8}$$
$$|\xi - \eta| \le \Phi(x, \xi),$$

$$c(\Phi\varphi)^{-m} \le e^\lambda \Phi^{-K} \varphi^{-k} \le C(\Phi\varphi)^m \text{ for some } K, k, m. \tag{9.9}$$

One element of $\mathcal{O}(\Phi, \varphi)$, denoted (K, k), is

$$(K, k) = K \log \Phi + k \log \varphi. \tag{9.10}$$

If $p(x, \xi) \in S^{\lambda}_{\Phi,\varphi}$ and $q(x, \xi) \in S^{\mu}_{\Phi,\varphi}$, then $p(x, D)q(x, D) = r(x, D) \in OPS^{\lambda+\mu}_{\Phi,\varphi}$, and

$$r(x, \xi) - \sum_{|\alpha|<N} \frac{1}{\alpha!} p^{(\alpha)}(x, \xi)q_{(\alpha)}(x, \xi) \in S^{\lambda+\mu-(N,N)}_{\Phi,\varphi}. \tag{9.11}$$

Also $p(x, D)^* = s(x, D) \in OPS^{\lambda}_{\Phi,\varphi}$ and

$$s(x, \xi) - \sum_{|\alpha|<N} \frac{1}{\alpha!} \overline{p}^{(\alpha)}_{(\alpha)} \in S^{\lambda-(N,N)}_{\Phi,\varphi}. \tag{9.12}$$

It may happen that, for $N \to \infty$, elements of $S^{\lambda-(N,N)}_{\Phi,\varphi}$ do not have progressively lower order, such as when $\Phi = \varphi^{-1}$; in particular, the case $S^m_{1/2,1/2}$ is included in results (9.11) and (9.12). The case of $OPS^m_{1/2,1/2}$ will be treated in this book in Chapter VIII, Section 8. As for L^2-continuity, one has

$$P \in OPS^{(0,0)}_{\Phi,\varphi} \Rightarrow P: L^2(\mathbf{R}^n) \to L^2(\mathbf{R}^n). \tag{9.13}$$

This includes the case $P \in OPS^0_{\rho,\rho}$, and thus (9.13) is an extension of the Calderon-Vaillancourt theorem, which will be proved for $P \in OPS^0_{\rho,\rho}$ $(0 < \rho < 1)$ in Chapter XIII.

The genesis of the Beals-Fefferman classes of pseudodifferential operators lies in their paper [1] on local solvability of *PDE*. This paper uses a Calderon-Zygmund style decomposition of symbols of classical operators and L^2 estimates on bounded subsets of $OPS^0_{1/2,1/2}$ to establish inequalities leading to local solvability. The classes $OPS^{\lambda}_{\Phi,\varphi}$ incorporate such techniques and hence lead to a more straightforward proof of such local solvability; see Beals and Fefferman [2]. $OPS^{\lambda}_{\Phi,\varphi}$ has also been used to construct inverses of hypoelliptic operators, generalizing results that will be established in Chapter III, Section 3, via $OPS^m_{\rho,\delta}$.

The operator classes of Hörmander [23] are more general still. If $p \in C^\infty(T^*(\mathbf{R}^n)) = C^\infty(\mathbf{R}^{2n})$, we denote by

$$p^{(k)}(x_0, \xi_0)(t_1, \ldots, t_k)$$

the kth order Fréchet derivative of p at (x_0, ξ_0); t_j are tangent vectors to $T^*(\mathbf{R}^n)$ at (x_0, ξ_0); we can identify t_j with elements of $\mathbf{R}^{2n}$. One considers symbols $p(x, \xi)$ satisfying estimates of the form

$$|p^{(k)}(x, \xi)(t_1, \ldots, t_k)| \le C_k m(x, \xi) \prod_{j=1}^{k} g_{x,\xi}(t_j)^{1/2}. \tag{9.14}$$

$g_{x,\xi}$ is a family of metrics and m is a weight for g, specified below. The class of such symbols is denoted by

$$S(m, g).$$

The assumptions for the metric g, a positive-definite quadratic form in t for each (x, ξ), are the following.

(9.15) (Slowing varying) For $w, y, t \in \mathbf{R}^{2n}$,

$$C^{-1}g_w(t) \leq g_y(t) \leq Cg_w(t) \text{ if } g_w(y-w) \leq c.$$

(9.16) (σ-temperate) $g_w^\sigma(t)/g_y^\sigma(t) \leq c(1 + g_y^\sigma(w-y))^N$, where

$$g_w^\sigma(y) = \sup_{z \in \mathbf{R}^{2n}} \frac{\sigma(y, z)^2}{g_w(z)},$$

σ denoting the symplectic form on $\mathbf{R}^{2n} = T^*(\mathbf{R}^n)$.

(9.17) (Uncertainty principle) $g_w(t) \leq g_w^\sigma(t)$.

The weight function m of (9.14) is assumed to satisfy the following conditions:

(9.18) $$C^{-1}m(w) \leq m(y) \leq Cm(w) \quad \text{if} \quad g_w(y-w) \leq c;$$

(9.19) $$\frac{m(w)}{m(y)} \leq C(1 + g_w^\sigma(w-y))^N.$$

To such general symbols, Hörmander associates not only the operator $p(x, D)$ defined by (1.3) but also the prescription of Weyl:

$$p_w(x, D) = (2\pi)^{-n} \int \hat{p}(\chi, \omega) e^{i(\chi \cdot x + \omega \cdot D)} \, d\chi \, d\omega$$

where

$$\hat{p}(\chi, \omega) = (2\pi)^{-n} \int p(x, \xi) e^{-i\chi \cdot x - \omega \cdot \xi} dx \, d\xi$$

which can be restated as

(9.20) $$p_w(x, D)u = (2\pi)^{-n} \iint e^{i(x-y)\cdot\xi} p(\tfrac{1}{2}(x+y), \xi) u(y) \, dy \, d\xi.$$

Denote by $OP_WS(m, g)$ the set of such operators. If g satisfies (9.15) through (9.17) and m_j are weights satisfying (9.18) and (9.19), and if $p_j(x, \xi) \in S(m_j, g)$, then $p_{1W}(x, D)p_{2W}(x, D) = r_W(x, D) \in OP_WS(m_1m_2, g)$, with

$$r(x, \xi) = \int p_1(\tfrac{1}{2}(x+z+(t/2)), \eta) p_2(\tfrac{1}{2}(x+z-(t/2)), \xi) e^{iE} \, dz \, d\eta \, dt \, d\tau,$$

$$E = (x - z + (t/2)) \cdot (\eta - \xi) + (z - x + (t/2)) \cdot (\tau - \xi).$$

Furthermore,

$$r(x,\xi)-\sum_{j<N}\frac{1}{j!}\left(\frac{-i}{2}\right)^j\left(\sum_{i=1}^n\frac{\partial^2}{\partial y_i\,\partial\xi_i}-\frac{\partial^2}{\partial x_i\,\partial\eta_i}\right)^j p_1(x,\xi)p_2(y,\eta)\Big|_{x=y,\,\xi=\eta} \in S(h^N m_1 m_2, g)$$

where

$$h(w)=\sup_{t\neq 0}\left(\frac{g_w(t)}{g_w^\sigma(t)}\right)^{1/2}. \tag{9.22}$$

The L^2 continuity result is (for g satisfying (9.15) through (9.17))

$$P\in OP_W S(1,g)\Rightarrow P: L^2(\mathbf{R}^n)\to L^2(\mathbf{R}^n). \tag{9.23}$$

We remark that, if

$$g_{x,\xi}(y,\eta)=(1+|\xi|^2)^\delta|y|^2+(1+|\xi|^2)^{-\rho}|\eta|^2,$$

then $S((1+|\xi|^2)^{m/2},g)=S^m_{\rho,\delta}$, and more generally if

$$g_{x,\xi}(y,\eta)=\varphi(x,\xi)^{-2}|y|^2+\Phi(x,\xi)^{-2}|\eta|^2,$$

then $S(e^\lambda,g)=\dot{S}^\lambda_{\Phi,\varphi}$. For details on this general class of operators, see Hörmander [23] and also Beals [5].

CHAPTER III

Elliptic and Hypoelliptic Operators

Our first application of the machinery of pseudodifferential operators developed in Chapter II will be to obtain regularity theorems for solutions to elliptic equations and certain generalizations, known as hypoelliptic equations, $Pu = f$, which have the property that u is smooth wherever f is. The most important example of a nonelliptic hypoelliptic operator covered by the techniques of this chapter is the heat operator, or more generally the class of parabolic operators. A more sophisticated hypoelliptic operator arises in the study of strictly pseudoconvex domains, in several complex variables. This cannot be treated by the methods of this chapter. We return to this in Chapter XV. In the third section we treat hypoelliptic operators with "slowly varying strength," following Hörmander [12], where the symbol classes $S^m_{\rho,\delta}$ were first introduced.

§1. Elliptic Operators

In this section we examine regularity of elliptic operators in $OPS^m_{\rho,\delta}$, defined on a region Ω, assuming $\rho > \delta$. All operators will be assumed to be properly supported. Elliptic operators seem to arise "naturally" only for $(\rho, \delta) = (1, 0)$, but their analysis in the more general case is a convenient tool for the study of certain classes of hypoelliptic operators that we shall introduce in the next section.

DEFINITION 1.1 *The operator $p(x, D) \in OPS^m_{\rho,\delta}$ is elliptic of order m if on each compact $K \subset \Omega$ there are constants C_K and R such that*

$$|p(x, \xi)| \geq C_K(1+|\xi|)^m \quad \text{if} \quad x \in K, |\xi| \geq R. \tag{1.1}$$

Thus the Laplace operator $\Delta = (\partial^2/\partial x_1^2) + \cdots + (\partial^2/\partial x_n^2)$ is elliptic of order two, since $\sigma_\Delta(x, \xi) = -\xi_1^2 - \cdots - \xi_n^2 = -|\xi|^2$, but the wave operator $\Box = \Delta_{n-1} - (\partial^2/\partial x_n^2)$, with symbol $\sigma_\Box(x, \xi) = -|\xi'|^2 + \xi_n^2$, is not elliptic.

DEFINITION 1.2. *A parametrix Q for the operator $P \in OPS^m_{\rho,\delta}$ is a properly supported operator which is a two-sided inverse for P modulo smoothing operators;*

$$PQ - I = K_1 \in OPS^{-\infty},$$
$$QP - I = K_2 \in OPS^{-\infty}.$$

THEOREM 1.3. *If $P = p(x, D) \in OPS^m_{\rho,\delta}$ is elliptic, $\rho > \delta$, then there is a properly supported $Q \in OPS^{-m}_{\rho,\delta}$ which is a parametrix for P.*

Proof. We shall construct Q by successive approximations. First set

$$q_0(x, \xi) = \zeta(x, \xi)p(x, \xi)^{-1} \in S^{-m}_{\rho,\delta} \tag{1.2}$$

where $\zeta(x, \xi)$ vanishes in a neighborhood of the zeros of p, and is identically one for large ξ, say for $|\xi| \geq C$. $q_0 \in S^{-m}_{\rho,\delta}$ is a simple consequence of the chain rule and the assumed estimate $|p(x, \xi)^{-1}| \leq C(1+|\xi|)^{-m}$, $|\xi| \geq C$, since

$$D^\beta_x D^\alpha_\xi q_0(x, \xi) = \sum_{\substack{\alpha_1+\cdots+\alpha_\mu=\alpha \\ \beta_1+\cdots+\beta_\mu=\beta}} C(D^{\beta_1}_x D^{\alpha_1}_\xi p) \cdots (D^{\beta_\mu}_x D^{\alpha_\mu}_\xi p)p^{-1-\mu}, \qquad |\xi| \geq C.$$

Let $Q_0 = q_0(x, D)$. Thus Q_0P has symbol

$$\sigma_{Q_0P}(x, \xi) \sim \sum_{\alpha \geq 0} \frac{1}{\alpha!} q_0^{(\alpha)}(x, \xi)p_{(\alpha)}(x, \xi) = 1 + r(x, \xi).$$

Hence

$$Q_0P = I + R, \qquad R \in OPS^{-(\rho-\delta)}_{\rho,\delta}. \tag{1.3}$$

Consequently, we can define $E \in OPS^0_{\rho,\delta}$ to have the asymptotic expansion

$$E \sim I - R + R^2 - R^3 + \cdots,$$

and then

$$(EQ_0)P = I + K_2, \qquad K_2 \in OPS^{-\infty}. \tag{1.4}$$

Consequently, we take $Q = EQ_0$, so Q is a left parametrix of P.

Similarly we can construct a right parametrix $\tilde{Q}$ of P, namely, with $PQ_0 = I + \tilde{R}$, $\tilde{R} \in OPS^{-(\rho-\delta)}_{\rho,\delta}$, take $\tilde{E} \sim I - \tilde{R} + \tilde{R}^2 - \cdots$ and let $\tilde{Q} = Q_0\tilde{E}$. We claim that $Q = \tilde{Q}$ mod $OPS^{-\infty}$, which implies Q is a two-sided parametrix. Indeed, note that, with $K_j \in OPS^{-\infty}$,

$$QP\tilde{Q} = (I + K_2)\tilde{Q} = \tilde{Q} + K_2\tilde{Q}, \qquad \text{and}$$
$$QP\tilde{Q} = Q(I + K_1) = Q + QK_1.$$

Hence $Q - \tilde{Q} = K_2\tilde{Q} - QK_1 \in OPS^{-\infty}$ as asserted, and the theorem is proved.

This construction of a parametrix leads immediately to our elliptic regularity theorem.

THEOREM 1.4. *If $P \in OPS^m_{\rho,\delta}$ is a properly supported elliptic operator, $\rho > \delta$, then, for any $u \in \mathscr{D}'(\Omega)$,*

$$\text{sing supp } u = \text{sing supp } Pu. \tag{1.5}$$

Thus, u is smooth wherever Pu is.

Proof. Take a parametrix $Q \in OPS^{-m}_{\rho,\delta}$ as in Theorem 1.3. Then $u = Q(Pu)$ mod C^∞ while the pseudolocal property for Q implies that sing supp $Q(Pu) \subset$ sing supp Pu. Thus sing supp $u \subset$ sing supp Pu in this case. The converse inclusion sing supp $Pu \subset$ sing supp u is simply the pseudolocal property of P, so (1.5) is established.

Similarly one obtains the following corollary, whose proof is left as an exercise.

COROLLARY 1.5. *With P as above, $\omega \subset \Omega$ an open subset, if $Pu|_\omega \in H^s_{\text{loc}}(\omega)$, then $u|_\omega \in H^{s+m}_{\text{loc}}(\omega)$.*

Exercises

1.1. Let M be a compact Riemannian manifold, $A_0 \in OPS^m_{1,0}$ an elliptic operator on M, $m > 0$. Assume A_0 is formally self adjoint, i.e.,

$$(A_0 u, v) = (u, A_0 v), \qquad u, v \in C^\infty(M).$$

Show that

$$\pm i + A_0 : H^m(M) \to L^2(M)$$

is an isomorphism.

1.2. With notation as above, define an unbounded operator A on $L^2(M)$ by

$$\mathcal{D}(A) = H^m(M), \qquad Au = A_0 u \text{ for } u \in \mathcal{D}(A).$$

Show that A is self adjoint.

§2. Hypoelliptic Operators with Constant Strength

In the last section, we showed that elliptic operators satisfy a certain regularity condition, namely (1.5). More generally, we define an operator P to be *hypoelliptic* if this property holds, i.e.,

$$\text{sing supp } u = \text{sing supp } Pu.$$

We shall begin by characterizing those hypoelliptic partial differential operators with constant coefficients. We shall not give a complete proof. We shall show that the condition (2.2) or (2.4) is *sufficient* for hypoellipticity, and that hypoellipticity implies the (a priori) weaker condition (2.3); the equivalence between these two conditions involves some technical facts about zeros of polynomials in $\mathbf{C}^n$, for which the reader is referred to Hörmander [6] or Trèves [1].

THEOREM 2.1. *Let $P(\xi)$ be a polynomial. The following are equivalent.*

$$P(D) \text{ is hypoelliptic.} \tag{2.1}$$

$$\left|\frac{P^{(\alpha)}(\xi)}{P(\xi)}\right| \le C(1+|\xi|)^{-\rho|\alpha|} \qquad \text{for } |\xi| \text{ large.} \tag{2.2}$$

(2.3) *If $V = \{\zeta \in \mathbf{C}^n : P(\zeta) = 0\}$, then $\zeta \in V$, $|Re\ \zeta| \to \infty \Rightarrow |Im\ \zeta| \to \infty$.*

(2.4) *There exists $\rho > 0$, $C > 0$ such that, for $\zeta \in V$, $|\zeta|$ large, $|Im\, \zeta| \geq C|\zeta|^{\rho}$.*

Proof. First we show that (2.2) implies $P(D)$ is hypoelliptic. Indeed, from (2.2) it follows that

$$Q(\xi) = \varphi(\xi)P(\xi)^{-1} \in S^0_{\rho,0}$$

if $\varphi(\xi)$ vanishes in a neighborhood of the zeros of $P(\xi)$ and is identically 1 for large $|\xi|$. In fact, this follows from the formula

$$D_\xi^\alpha Q(\xi) = \sum_{\alpha_1+\cdots+\alpha_\mu=\alpha} CP^{(\alpha_1)}(\xi)\cdots P^{(\alpha_\mu)}(\xi)P(\xi)^{-1-\mu}, \qquad |\xi| \text{ large},$$

granted (2.2). Since it is clear that $Q(D) \in OPS^0_{\rho,0}$ provides a two-sided parametrix for $P(D)$, the proof of Theorem 1.4 shows that $P(D)$ is hypoelliptic.

Next, we show that (2.1) implies (2.3). For this, let

$$\begin{aligned} \mathcal{N} &= \{u \in C(\Omega): P(D)u = 0\} \subset C(\Omega) \\ &= \{u \in C^1(\Omega): P(D)u = 0\} \subset C^1(\Omega), \qquad \text{if } P(D) \text{ is hypoelliptic.} \end{aligned}$$

Since $\mathcal{N}$ is a Fréchet space under topologies induced from either $C(\Omega)$ or $C^1(\Omega)$ and since one is stronger than the other, the open mapping theorem implies these two topologies coincide. Thus, if $\zeta_\nu \in V$, $\{Im\, \zeta_\nu\}$ bounded, then $\{e^{i\zeta_\nu \cdot x}\}$ is a subset of $\mathcal{N}$, bounded in $C(\Omega)$ and hence bounded in $C^1(\Omega)$. Since

$$\frac{\partial}{\partial x_j} e^{i\zeta_\nu \cdot x} = i\zeta_\nu^j e^{i\zeta_\nu \cdot x},$$

where $\zeta_\nu = (\zeta_\nu^1, \ldots, \zeta_\nu^n)$, it follows that $\{\zeta_\nu\}$ is bounded, so we have the implication (2.1) $\Rightarrow$ (2.3).

To finish our argument, we shall show that (2.2) and (2.4) are equivalent. Clearly (2.4) implies (2.3), but we refer to Hörmander [6] or Trèves [1] for the proof that (2.3) implies (2.4), using an argument involving the Seidenberg-Tarski theorem. The equivalence of (2.2) and (2.4) is an immediate consequence of the following lemma.

LEMMA 2.2. *Let $P(\xi)$ be a polynomial and let $d(\xi) = \text{dist}\ (\xi, V)$, the distance from $\xi \in \mathbf{R}^n$ to $V = \{\zeta \in \mathbf{C}^n: P(\zeta) = 0\}$. Then*

$$C_1 \leq d(\xi) \sum_{\alpha \neq 0} \left| \frac{P^{(\alpha)}(\xi)}{P(\xi)} \right|^{1/|\alpha|} \leq C_2. \tag{2.5}$$

Proof. We have

$$P(\xi + \eta) - P(\xi) = \sum_{\alpha \neq 0} \frac{\eta^\alpha}{\alpha!} P^{(\alpha)}(\xi).$$

Hence

$$\frac{P(\xi+\eta)}{P(\xi)} - 1 = \sum_{\alpha\neq 0} \frac{\eta^\alpha}{\alpha!}\frac{P^{(\alpha)}(\xi)}{P(\xi)}. \tag{2.6}$$

Pick C such that

$$\sum_{\alpha\neq 0} \frac{C^{|\alpha|}}{\alpha!} < 1.$$

Suppose

$$|\eta| < C \min_{\alpha\neq 0} \left|\frac{P^{(\alpha)}(\xi)}{P(\xi)}\right|^{-1/|\alpha|},$$

with ξ fixed. Then (2.6) implies $P(\xi+\eta) \neq 0$. Hence

$$d(\xi) \geq C \min_{\alpha\neq 0} \left|\frac{P^{(\alpha)}(\xi)}{P(\xi)}\right|^{-1/|\alpha|}$$

which proves half the lemma

For the second half, let $|\eta| \leq d(\xi)$. Then if $g(t) = P(\xi+t\eta)$, we have

$$\left|\frac{P(\xi+\eta)}{P(\xi)}\right| = \left|\frac{g(1)}{g(0)}\right| = \prod_{j=1}^{m}\left|\frac{1-t_j}{t_j}\right| \leq 2^m$$

where t_j are the roots of $g(t)$; $t_j \geq 1$. Thus

$$|P(\xi+\eta)| \leq 2^m |P(\xi)| \qquad \text{if} \qquad |\eta| \leq d(\xi). \tag{2.7}$$

If we apply this to Cauchy's formula

$$P^{(\alpha)}(\xi) = c \int_{\Gamma_n} \frac{P(\xi+\zeta)}{\zeta^\alpha \zeta_1 \cdots \zeta_n}\, d\zeta$$

where $\Gamma_n = \{\zeta \in \mathbf{C}^n : |\zeta_j - \xi_j| = 2^{-1/2n}\}$, we get

$$|P^{(\alpha)}(\xi)| \leq \frac{C2^m|P(\xi)|}{d(\xi)^{|\alpha|}}$$

which completes the proof.

Thus we have completely characterized those hypoelliptic differential operators with constant coefficients. For operators with variable coefficients, results are more subtle and less complete. There are several directions to go in when looking for large families of hypoelliptic operators. Here, we examine one of them.

DEFINITION 2.3. *The operator*

$$p(x, D) = \sum_{|\alpha|\leq m} a_\alpha(x) D^\alpha$$

is a formally hypoelliptic operator of constant strength if there exists a polynomial $P(\xi)$ *such that*

(2.8) $$P(D) \text{ is hypoelliptic, and}$$

(2.9) $$\frac{p(x, \xi)}{P(\xi)} \text{ and } \frac{P(\xi)}{p(x, \xi)} \text{ are bounded, for large } \xi, \text{ locally uniformly in } x.$$

We shall show that such operators are indeed hypoelliptic, but first we prove some convenient technical lemmas about hypoelliptic polynomials.

PROPOSITION 2.4. *If* $P(D)$ *is hypoelliptic and if* $Q(\xi)$ *is a polynomial such that* $P(\xi)/Q(\xi)$ *and* $Q(\xi)/P(\xi)$ *are bounded, for large* ξ, *then* $Q(D)$ *is also hypoelliptic.*

Proof. Let $d_1(\xi)$ be the distance from ξ to $V = \{\zeta \in \mathbf{C}^n : P(\zeta) = 0\}$. Lemma 2.2 implies

(2.10) $$d_1(\xi)^{|\alpha|}|P^{(\alpha)}(\xi)| \leq C^{|\alpha|}|P(\xi)|.$$

To complete the proof, we shall need the following fact, which we shall prove shortly: if $Q(\xi)/P(\xi)$ is bounded for large ξ, then with

(2.11) $$\tilde{R}(\xi, t) = \left(\sum_{\alpha} |R^{(\alpha)}(\xi)|^2 t^{2|\alpha|}\right)^{1/2}$$

for any polynomial $R(\xi)$, we have

(2.12) $$\tilde{Q}(\xi, t) \leq C\tilde{P}(\xi, t) \qquad \text{for} \qquad \xi \in \mathbf{R}^n,\ t \geq 1.$$

Granted this, we proceed. When ξ is so large that $d_1(\xi) \geq 1$, we apply (2.12) with $t = d_1(\xi)$, and use (2.10) to get

(2.13) $$\sum_{\alpha} |Q^{(\alpha)}(\xi)|^2 d_1(\xi)^{2|\alpha|} \leq C|P(\xi)|^2.$$

Since $P(\xi)/Q(\xi)$ is bounded for large ξ, we have

$$\sum_{\alpha} |Q^{(\alpha)}(\xi)|^2 d_1(\xi)^{2|\alpha|} \leq C|Q(\xi)|^2.$$

Since $d_1(\xi) \geq C'|\xi|^{\rho}$ for large ξ, with $\rho > 0$, it follows that $Q(\xi)$ satisfies condition (2.2) of Theorem 2.1, so $Q(D)$ is hypoelliptic.

It remains to prove the inequality (2.12). We obtain this as the second of the following two lemmas.

LEMMA 2.5. *For each* K *there are constants* C_1, C_2 *such that, for all polynomials* R *of degree* $\leq K$, *we have, with* $\zeta \in \mathbf{R}^n$, $\xi \in \mathbf{R}^n$, $t \geq 0$,

(2.14) $$C_1\tilde{R}(\xi, t) \leq \sup_{|\zeta| \leq t} |R(\xi + \zeta)| \leq C_2\tilde{R}(\xi, t).$$

Proof. The right-hand inequality follows from the Taylor expansion of $R(\xi+\zeta)$ about ξ. Since, if we set $R_t(\xi) = R(t\xi)$ we get $\tilde{R}(\xi, t) = \tilde{R}_t(\xi/t, 1)$, it suffices to prove the other inequality with $t = 1$.

To do this, pick N distinct points $\zeta_1, \ldots, \zeta_N$ in the unit ball in $\mathbf{R}^n$, where N is the dimension of the space $\mathscr{P}_K$ of polynomials in $\mathbf{C}[X_1, \ldots, X_n]$ of degree $\leq K$, such that the linear functionals $P \to P(\zeta_j)$ form a *basis* of $\mathscr{P}_K^*$. To see that this is possible, it suffices to show that the linear functionals $P \to P(\zeta)$, $\zeta \in B = \{\zeta \in \mathbf{R}^n : |\zeta| \leq 1\}$, *span* $\mathscr{P}_K$. If not though, there would be a $P_0 \in \mathscr{P}_K$ annihilated by all such functionals, i.e., $P(\zeta) = 0$ for $\zeta \in \mathbf{R}^n$, $|\zeta| \leq 1$, which implies $P = 0$. Consequently, you can pick a *dual basis* $P_1, \ldots, P_N$ for $\mathscr{P}_K$, so

$$P_j(\zeta_k) = \delta_{jk}.$$

Consequently, for all polynomials $R(\xi)$ of degree K,

$$R(\eta) = \sum_{j=1}^{N} R(\zeta_j)P_j(\eta), \tag{2.15}$$

a result known as the Lagrange interpolation formula. To continue after (2.14), we have

$$R(\xi+\eta) = \sum_{j=1}^{N} R(\xi+\zeta_j)P_j(\eta);$$

therefore,

$$R^{(\alpha)}(\xi) = \sum_{j=1}^{N} P_j^{(\alpha)}(0)R(\xi+\zeta_j).$$

The left-hand inequality in (2.14), with $t = 1$, follows.

LEMMA 2.6. *Inequality* (2.12) *above is valid*, i.e., $\tilde{Q}(\xi, t) \leq C\tilde{P}(\xi, t), t \geq 1$.

Proof. Using Lemma 2.5 and the boundedness of $Q(\xi)/P(\xi)$ for large ξ, we find

$$\begin{aligned}\tilde{Q}(\xi, t) &\leq C_1 \sup_{|\zeta|\leq t} |Q(\xi+\zeta)| \\ &\leq C_2\Big(1 + \sup_{|\zeta|\leq t} |P(\xi+\zeta)|\Big) \\ &\leq C_2 \sup_{|\zeta|\leq t} \tilde{P}(\xi+\zeta, 1) \\ &\leq C_3 \sup_{|\zeta|\leq t+1} |P(\xi+\zeta)| \leq C_4\tilde{P}(\xi, t+1) \\ &\leq C_4(1+t^{-1})^m\tilde{P}(\xi, t).\end{aligned}$$

Thus the proof of Proposition 2.4 is complete. In particular, it follows that if $p(x, D)$ is formally hypoelliptic of constant strength, then any constant coefficient operator $P_0(D) = p(x_0, D)$ obtained by freezing the coefficients of $p(x, D)$ at some point, is hypoelliptic.

Within the proof of Proposition 2.4, we derived the following result, which it will be convenient to point out.

LEMMA 2.7. *If $P(D)$ is hypoelliptic and $Q(\xi)/P(\xi)$ is bounded for large ξ, then*

$$\left|\frac{Q^{(\alpha)}(\xi)}{P(\xi)}\right| \leq C|\xi|^{-\rho|\alpha|} \qquad \textit{for large } \xi.$$

Proof. This follows from the estimate (2.13).

Suppose now that $p(x, D)$ is a formally hypoelliptic operator with constant strength. Freeze the coefficients at one point x_0 to obtain a hypoelliptic constant coefficient operator $P(D) = p(x_0, D)$. Let $E(\xi) = P(\xi)^{-1}$ for large ξ, so as was shown in the proof of Theorem 2.1, $E \in OPS^0_{\rho,0}$ and E is a parametrix of $P(D)$.

PROPOSITION 2.8. *The operators $Ep(x, D)$ and $p(x, D)E$ belong to $OPS^0_{\rho,0}$ and are elliptic.*

Proof. The fact that $p(x, \xi)E(\xi) \in S^0_{\rho,0}$ is a routine estimation, using Lemma 2.7, and we also see that $|p(x, \xi)E(\xi)| \geq C > 0$ for large ξ. Thus $p(x, D)E$ is an elliptic operator in $OPS^0_{\rho,0}$. Since the symbol of $Ep(x, D)$ is asymptotic to

$$E(\xi)p(x, \xi) + \sum_{\alpha>0} \frac{1}{\alpha!} E^{(\alpha)}(\xi)p_{(\alpha)}(x, \xi) = E(\xi)p(x, \xi) \bmod S^{-\rho}_{\rho,0},$$

the same conclusion holds for $Ep(x, D)$.

Now for the main result of this section.

THEOREM 2.9. *If $p(x, D)$ is a formally hypoelliptic operator of constant strength, then $p(x, D)$ is hypoelliptic.*

Proof. If $A \in OPS^0_{\rho,0}$ is a parametrix for the elliptic operator $Ep(x, D)$, then AE is a left parametrix for $p(x, D)$, so the proof of Theorem 1.4 shows that $p(x, D)$ is hypoelliptic.

Exercises

2.1. Let

$$A(t, x, D_x) = \sum_{|\alpha|\leq 2} a(t, x)D_x^\alpha$$

be a second order operator, with $D_j = (1/i)(\partial/\partial x_j)$, which is strongly elliptic in the sense that

$$Re \sum_{|\alpha|=2} a_\alpha(t, x)\xi^\alpha \geq C|\xi|^2.$$

Prove that $(\partial/\partial t) - A(t, x, D_x)$ is formally hypoelliptic with constant strength.

2.2. Consider the hypoelliptic operator $P(D) = (\partial/\partial t) - (\partial^2/\partial x^2)$ on $\mathbf{R}^2$. Show that there exists a nonlinear change of coordinates of $\mathbf{R}^2$ with respect to which P becomes an operator which is not formally hypoelliptic. Is this new operator hypoelliptic?

2.3. Let X and Y be smooth real vector fields on $\Omega \subset \mathbf{R}^2$ which are linearly independent at each point. Prove that the second order operator $X^2 - Y$ is hypoelliptic.

2.4. Let $p(x, D)$ be a formally hypoelliptic operator with constant strength. Prove that its adjoint $p(x, D)^*$ also enjoys this property.

In general, the adjoint of a hypoelliptic operator has the local solvability property. A similar argument, in another context, is given in Chapter VI of this volume.

§3. Hypoelliptic Operators with Slowly Varying Strength

In this section we prove the hypoellipticity of operators $p(x, D) \in OPS^m_{1,0}$ satisfying the following condition.

DEFINITION 3.1. *The operator $p(x, D)$ is formally hypoelliptic of slowly varying strength if, with $0 \le \delta < \rho \le 1$ and some $\mu < \infty$, we have, for ξ large,*

$$|D_x^\beta D_\xi^\alpha p(x, \xi)|\,|p(x, \xi)^{-1}| \le C(1+|\xi|)^{-\rho|\alpha|+\delta|\beta|}, \quad \textit{and} \tag{3.1}$$

$$|p(x, \xi)^{-1}| \le C(1+|\xi|)^{\mu}. \tag{3.2}$$

In this section we shall assume $p(x, \xi)$ is scalar.

To tie this in with the concept of Section 2, note the following.

PROPOSITION 3.2. *A differential operator $p(x, D)$ is formally hypoelliptic of constant strength if, and only if, for ξ large,*

$$|p(x, \xi)^{-1}|\,|D_x^\beta D_\xi^\alpha p(x, \xi)| \le C(1+|\xi|)^{-\rho|\alpha|}. \tag{3.3}$$

Proof. Clearly $p(x, D)$ formally hypoelliptic of constant strength implies (3.3). Also, (3.3) clearly implies any frozen coefficient operator $P_0(D) = p(x_0, D)$ is hypoelliptic. It remains to show that, for any $x, y \in \Omega$, ξ large,

$$\frac{p(x, \xi)}{p(y, \xi)} \text{ is bounded.} \tag{3.4}$$

Indeed, for $\alpha = 0$, $|\beta| = 1$, (3.3) is equivalent to

$$D_x^\beta \log p(x, \xi) \text{ bounded,} \qquad \xi \text{ large}$$

which implies

$$\log p(x, \xi) - \log p(y, \xi) \text{ bounded}, \qquad \xi \text{ large}$$

or

$$\log \frac{p(x, \xi)}{p(y, \xi)} \text{ bounded}.$$

This implies (3.4) and completes the proof.

Given $p(x, \xi)$ satisfying (3.1) and (3.2), if $e_0(x, \xi) = \varphi(\xi)p(x, \xi)^{-1}$ with $\varphi(\xi) = 0$ in a neighborhood of the zeros of $p(x, \xi)$, $\varphi(\xi) = 1$ for ξ large, we see that $e_0(x, \xi) \in S^{\mu}_{\rho,\delta}$. The symbol of $e_0(x, D)p(x, D)$ is asymptotic to

$$1 + \sum_{\alpha > 0} \frac{1}{\alpha!} e_0^{(\alpha)}(x, \xi)p_{(\alpha)}(x, \xi) = 1 \bmod S^{-(\rho-\delta)}_{\rho,\delta}$$

since, for ξ large,

$$e_0^{(\alpha)}p_{(\alpha)} = \sum c(p^{-1}p^{(\alpha_1)}) \cdots (p^{-1}p^{(\alpha_n)})p^{-1}p_{(\alpha)}.$$

So

$$e_0(x, D)p(x, D) = I + R, \qquad R \in OPS^{-(\rho-\delta)}_{\rho,\delta}.$$

It follows that $I + R$ has a two sided parametrix $F \in OPS^0_{\rho,\delta}$, so $Fe_0(x, D)$ is a left parametrix of $p(x, D)$. (Similarly one can construct a right parametrix of $p(x, D)$, and prove it coincides, mod $OPS^{-\infty}$, with $Fe_0(x, D)$.) Again, the proof of Theorem 1.4 yields the following conclusion.

THEOREM 3.3. *If $p(x, D)$ is formally hypoelliptic of slowly varying strength, then $p(x, D)$ is hypoelliptic.*

As we have remarked, the formally hypoelliptic operators of constant strength satisfy (3.1), (3.2) with $\delta = 0$. On the other hand,

$$p(x, \xi) = 1 + |x|^{2\nu}|\xi|^{2\mu} \qquad (\nu \text{ integer})$$

satisfies (3.1), (3.2) with $\rho = 1$, $\delta = \mu/\nu$, provided $\mu < \nu$. This yields the following.

COROLLARY 3.4. *The operator $1 + |x|^{2\nu}(-\Delta)^{\mu}$ is hypoelliptic, provided $\mu < \nu$. In particular for all integers $\nu \geq 2$ ($\nu = 0$ is okay, of course) $1 - |x|^{2\nu}\Delta$ is hypoelliptic.*

Exercise

3.1. Show that $1 - |x|^2\Delta$ is not hypoelliptic.

CHAPTER IV

The Initial Value Problem and Hyperbolic Operators

In this chapter our main aim is to study the Cauchy problem for hyperbolic equations:

$$Lu = f, \qquad \frac{\partial^j}{\partial t^j} u\big|_{t=0} = g_{j+1}, \qquad j = 0, \ldots, m-1.$$

In the first section we reduce this to a first order system and begin to explore the meaning of hyperbolicity. Several notions of hyperbolicity are dealt with in Sections 2 and 3. In Section 4 we study the phenomenon of finite propagation speed, and in Section 5 we study some quasilinear hyperbolic equations a little bit. Some special cases of hyperbolic equations that arise in mathematical physics are studied in Section 6, and Section 7 is devoted to parabolic equations.

§1. Reduction to a First Order System

With

$$L = \frac{\partial^m}{\partial t^m} - \sum_{j=0}^{m-1} A_{m-j}(t, x, D_x) \frac{\partial^j}{\partial t^j},$$

A_{m-j} being a differential operator of order $m-j$, with top order symbol $\tilde{A}_{m-j}(t, x, \xi)$, we derive a first order system equivalent to the Cauchy problem

(1.1) $$Lu = f, \qquad \frac{\partial^j}{\partial t^j} u = g_{j+1}, \qquad j = 0, \ldots, m-1.$$

Suppose for convenience that $x \in \mathbf{R}^n$ or $\mathbf{T}^n$, and with $\Lambda = (1-\Delta)^{1/2} \in OPS^1$, write

(1.2) $$u_j = \left(\frac{\partial}{\partial t}\right)^{j-1} \Lambda^{m-j} u, \qquad j = 1, \ldots, m.$$

Then (1.1) is equivalent to

$$\frac{\partial}{\partial t}\begin{pmatrix} u_1 \\ \vdots \\ \\ u_m \end{pmatrix} = \begin{pmatrix} 0 & \Lambda & 0 & \cdots & 0 \\ & 0 & \Lambda & & \vdots \\ \vdots & & & \ddots & \\ & & & & \Lambda \\ b_1 & b_2 & b_3 & \cdots & b_m \end{pmatrix}\begin{pmatrix} u_1 \\ \vdots \\ \\ u_m \end{pmatrix} + \begin{pmatrix} 0 \\ \vdots \\ 0 \\ f \end{pmatrix} \tag{1.3}$$

where $b_j = A_{m-j+1}(t, x, D_x)\Lambda^{j-m}$, with initial condition $u_j = \Lambda^{m-j}g_j$ at $t = 0$. Changing notation slightly, calling the column vector $(u_1, \ldots, u_m)^t$ u and the matrix in (1.3) K, write (1.3) as

$$\frac{\partial}{\partial t}u = Ku + f, \qquad u(0) = g. \tag{1.4}$$

Note that $K = K(t, x, D_x) \in OPS^1$. Call its principal part $K_1(t, x, D_x)$ and its full symbol $K(t, x, \xi)$. It is easy to check that the eigenvalues of $K_1(t, x, \xi)$ are precisely $i\tau_j(t, x, \xi)$, $j = 1, \ldots, m$, where $\tau_j(t, x, \xi)$ are the roots of

$$L_m(t, x, \tau, \xi) = 0.$$

$$L_m(t, x, \tau, \xi) = (i\tau)^m - \sum_{j=0}^{m-1} \tilde{A}_{m-j}(t, x, \xi)(i\tau)^j.$$

In order to formulate conditions for (1.4) to be hyperbolic, we shall consider the constant coefficient case, $x \in \mathbf{R}^n$. Then the solution to (1.4) is obtained simply by Fourier analysis: with $\hat{u}(t, \xi) = (2\pi)^{-n}\int u(t, x)e^{ix\cdot\xi}\, dx$,

$$\hat{u}(t, \xi) = e^{tK(\xi)}\hat{g}(\xi) \tag{1.5}$$

(taking $f = 0$). One reasonable requirement to make is to demand that, for $g \in L^2(\mathbf{R}^n)$, (1.5) yields $u(t) \in L^2(\mathbf{R}^n)$ for each $t \in \mathbf{R}$. This is equivalent to demanding that $e^{tK(\xi)}$ be bounded as a function of ξ, for each $t \in \mathbf{R}$. The following are some conditions that would guarantee this property.

(1.6) $K(\xi)$ skew symmetric. Then $\|e^{tK(\xi)}\| = 1$.

(1.7) $K(\xi)$ skew symmetrizable, i.e., there is a measurable matrix function $S(\xi)$, with $\|S(\xi)\|, \|S(\xi)^{-1}\| \le C$, such that SKS^{-1} is skew symmetric. Then $\|e^{tK(\xi)}\| \le C^2$.

(1.8) $K(\xi)$ has distinct, pure imaginary eigenvalues. Then K is skew-symmetrizable.

(1.9) $K(\xi) = K_1(\xi) + K_0(\xi)$ where $K_1(\xi)$ satisfies one of the above conditions, or more generally $\|e^{tK_1(\xi)}\| \le C(t)$, and $\|K_0(\xi)\| \le C_2$.

That case (1.9) works out is a special case of the more general assertion that a bounded perturbation of a generator of a semigroup is also a generator of a semigroup. (For a proof, see Hille and Phillips [1].)

If $K(\xi) \in S^1$, so $K(\xi) \sim K_1(\xi) + K_0(\xi) + \cdots$, the requirement that $u(t) \in L^2$ for each $g \in L^2$ is equivalent to demanding that

$$\|e^{rK_1(\omega)}\| \le C, \qquad \text{all } r \in \mathbf{R}, \qquad \omega \in S^{n-1}. \tag{1.10}$$

In particular it is certainly necessary that all the eigenvalues of $K_1(\xi)$ be purely imaginary. Some necessary and sufficient conditions that (1.10) hold are given by the Kreiss matrix theorem. In particular, each of the following two conditions is equivalent to (1.10).

$$\|(e^{\pm K_1(\omega)} - z)^{-1}\| \le \frac{C_0}{|z| - 1}, \qquad |z| > 1 \tag{1.11}$$

with C_0 independent of $\omega \in S^{n-1}$.

(1.12) There exist a constant $C_1 > 0$ and positive definite matrices $H_\pm(\omega)$ such that

$$C_1^{-1} I \le H_\pm \le C_1 I \qquad \text{and} \qquad (H_\pm e^{\pm K_1(\omega)} u, e^{\pm K_1(\omega)} u) \le (H_\pm u, u).$$

For the equivalence of (1.10) through (1.12), see Richtmyer and Morton [1, pp. 74–80].

There is a weaker notion of hyperbolicity, where one demands a solution $u(t)$ to (1.4) for all $t \in \mathbf{R}$, given $g \in H^s(\mathbf{R}^n)$, but only demands that $u(t) \in H^\sigma(\mathbf{R}^n)$, perhaps with $\sigma < s$. We shall not deal with any weakly hyperbolic operators with variable coefficients in this chapter, but since the constant coefficient situation is fairly simple, we shall state and prove a result. (This particular proof was worked out by myself and Jeff Rauch in a conversation at Dominic's in Ann Arbor.)

THEOREM 1.1. *If $K(t, x, \xi) = K(\xi) \in S^1_{1,0}$, then* (1.4) (*with* $f = 0$) *has a solution $u \in C(\mathbf{R}, H^\sigma(\mathbf{R}^n))$, for some $\sigma \in \mathbf{R}$, given $g \in H^s(\mathbf{R}^n)$, provided all eigenvalues $\mu_j(\xi)$ of the complete symbol $K(\xi)$ satisfy*

$$|Re\ \mu_j(\xi)| \le C. \tag{1.13}$$

Proof. We use the following linear algebra result: there exists a measurable *unitary* matrix $U(\xi)$ such that, if

$$M(\xi) = U(\xi) K(\xi) U(\xi)^{-1},$$

then $M(\xi)$ is in upper triangular form. The diagonal elements of $M(\xi)$ must then be the eigenvalues of $K(\xi)$.

$$M(\xi) = \begin{pmatrix} \mu_1(\xi) & & \\ & \ddots & \tilde{M}(\xi) \\ 0 & & \mu_m(\xi) \end{pmatrix}.$$

Since the Fourier multiplier $U(D)$ takes each $H^s(\mathbf{R}^n)$ to itself, it suffices to solve $(\partial/\partial t)v = M(D)v$, $v(0) = h \in H^s(\mathbf{R}^n)$, and get $v(t) \in H^\sigma(\mathbf{R}^n)$. Indeed, we shall construct $e^{tM(D)}$ as a one parameter group on the Hilbert space

$$\mathscr{H} = H^{s-m+1} \oplus H^{s-m+2} \oplus \cdots \oplus H^{s-1} \oplus H^s = \{v : v_j \in H^{s-m+j}\}.$$

Clearly the diagonal part of $M(D)$ generates a one parameter group on $\mathscr{H}$, by hypothesis (1.9). Meanwhile, since the offdiagonal part $\tilde{M}(\xi)$ is nonzero only above the diagonal and satisfies the estimate $\|\tilde{M}(\xi)\| \le C_2(1+|\xi|)$, it follows that $M(D)$ is a bounded perturbation, on $\mathscr{H}$, of its diagonal part, so $M(D)$ also generates a one parameter group on $\mathscr{H}$, and the proof is complete.

Remark. For the necessity of (1.9), see Hörmander [6], Chapter V.

§2. Symmetric Hyperbolic Systems

We shall begin with an analysis of existence and uniqueness for solutions of the simplest type of first order hyperbolic systems, symmetric hyperbolic systems.

DEFINITION 2.1. *The operator $(\partial/\partial t) - K(t, x, D_x)$ is symmetric hyperbolic if $K + K^* \in OPS^0_{1,0}$, $K \in OPS^1_{1,0}$.*

We assume $K = K(t, x, D_x)$ is a smooth one-parameter family of operators in $OPS^1_{1,0}$, and $x \in M$, a compact manifold.

We shall prove that a solution of this initial value problem for $u = u(t, x)$,

$$\frac{\partial}{\partial t} u = Ku + f, \tag{2.1}$$

$$u(0) = g \tag{2.2}$$

exists and is unique by proving an a priori inequality for such solutions. Existence will then follow by a little functional analysis. The following inequality from the theory of ordinary differential equations will be a useful tool.

LEMMA 2.2. (Gronwall's inequality). If $y \in C^1$ *and* $y'(t) + f(t)y \le g(t)$, *then*

$$y(t) \le e^{-\int_0^t f(\tau)\,d\tau}\left[y_0 + \int_0^t g(\tau)e^{\int_0^\tau f(\sigma)\,d\sigma}\,d\tau\right]. \tag{2.3}$$

Proof. The hypothesis is equivalent to the inequality

$$\frac{d}{dt}(ye^{\int_0^t f}) \le g(t)e^{\int_0^t f}$$

and integrating this yields (2.3).

Suppose now that $u \in C^\infty(\mathbf{R} \times M)$ and $(\partial/\partial t)u = Ku + f$. If we differentiate

$$\|u(t)\|^2 = \int_M |u(t, x)|^2 \, dx$$

with respect to t, we get

$$\begin{aligned}\frac{d}{dt}(u, u) &= (u', u) + (u, u') \\ &= (Ku + f, u) + (u, Ku + f) \\ &= ((K + K^*)u, u) + 2\, Re(f, u) \\ &\le C\|u\|^2 + C\|f\|^2\end{aligned}$$

since $K + K^* \in OPS^0_{1,0}$. Applying Gronwall's inequality to this yields

$$\|u(t)\|^2 \le C\|u(0)\|^2 + C\int_0^t \|f(\tau)\|^2 \, d\tau \tag{2.4}$$

with C bounded for t in a bounded interval $-T \le t \le T$. More generally, differentiating

$$\|u(t)\|^2_{H^s(M)} = \|\Lambda^s u(t)\|^2_{L^2(M)}$$

yields

$$\|u(t)\|^2_{H^s} \le C\|u(0)\|^2_{H^s} + C\int_0^t \|f(\tau)\|^2_{H^s(M)} \, d\tau. \tag{2.5}$$

Indeed, you can take $C = 1 + c_1|t|$, though we shall not use this observation until (5.4). This inequality is valid whenever $u \in H^1([-T, T], H^{s+1}(M))$, $f \in L^2([-T, T], H^s(M))$, $g \in H^{s+1}(M)$, and (2.1), (2.2) are satisfied. C is a function of T, M, and some norms of $K(t, x, D_x)$ in $C^\infty(\mathbf{R}, OPS^1_{1,0})$.

We shall obtain the solution of the initial value problem (2.1), (2.2) as a limit of solutions to the problem

$$\begin{aligned}&\frac{\partial}{\partial t}u = KJ_\epsilon u + f, \\ &u(0) = g\end{aligned} \tag{2.6}$$

where J_ϵ is a Friedrichs' mollifier on M, which was discussed in Chapter II. The point of this is that, for each $\epsilon > 0$, $K_\epsilon = KJ_\epsilon$ is a continuous linear operator on $H^s(M)$, so (2.6) can be regarded as a Banach-space valued ordinary differential equation, to which the Picard iteration method applies (see e.g., Dieudonne [1]). Thus, given $g \in H^{s+1}(M)$, $f \in L^2([-T, T], H^{s+1}(M))$, we can solve (2.6), producing a solution $u_\epsilon \in H^1([-T, T], H^{s+1}(M))$. Note that, if $f \in C([-T, T], H^{s+1}(M))$, then $u_\epsilon \in C^1([-T, T], H^{s+1}(M))$.

Now since $\{K_\epsilon : 0 < \epsilon \leq 1\}$ is a bounded subset of $OPS^1_{1,0}$ and $\{K_\epsilon + K_\epsilon^* : 0 < \epsilon \leq 1\}$ is a bounded subset of $OPS^0_{1,0}$, we get the estimate

$$\|u_\epsilon(t)\|^2_{H^s} \leq C\|g\|^2_{H^s} + C\int_0^t \|f(\tau)\|^2_{H^s}\, d\tau \tag{2.7}$$

with C *independent of* ϵ, $0 < \epsilon \leq 1$. Of course, we want to let $\epsilon \to 0$.

First note that, by the estimate (2.7), $\{u_\epsilon : 0 < \epsilon \leq 1\}$ is a bounded subset of $C([-T, T], H^s(M))$ given

$$g \in H^{s+1}(M), \qquad f \in C([-T, T], H^{s+1}(M)).$$

Since $u'_\epsilon = K_\epsilon u_\epsilon + f$, it follows that $\{u'_\epsilon : 0 < \epsilon \leq 1\}$ is a bounded subset of $C([-T, T], H^{s-1}(M))$. Hence $\{u_\epsilon\}$ is a bounded subset of $C^1([-T, T], H^{s-1}(M))$. Furthermore, for each $t_0 \in [-T, T]$, $\{u_\epsilon(t_0) : 0 < \epsilon \leq 1\}$, being a bounded subset of $H^s(M)$, is a relatively compact subset of $H^{s-1}(M)$. Hence, by Ascoli's theorem (see Dieudonné [1] or Dugundji [1]), there is a sequence $\epsilon_n \to 0$ such that u_{ϵ_n} converges, in $C([-T, T], H^{s-1}(M))$, to a limit we call u. Clearly u satisfies (2.1), (2.2) in the distribution sense.

We are almost through with the proof of our first main result:

THEOREM 2.3. *Let $(\partial/\partial t) - K$ be first order symmetric hyperbolic system. Then, given $g \in H^s(M)$ and $f \in L^2([-T, T], H^s(M))$, the Cauchy problem* (2.1), (2.2) *has a unique solution $u \in C([-T, T], H^s(M))$.*

Proof. Let $g_j \in H^{s+4}(M)$, $g_j \to g$ in $H^s(M)$, and let $f_j \in C([-T, T], H^{s+4}(M))$, $f_j \to f$ in $L^2([-T, T], H^s(M))$. The argument above also produces solutions u_j to (2.1), (2.2) with g, f replaced by g_i, f_j, with $u_j \in C([-T, T], H^{s+2}(M))$. Since $u'_j = Ku_j + f_j$, it follows that $u_j \in C^1([-T, T], H^{s+1}(M))$. Hence we can apply the energy inequality (2.5) and conclude that $\{u_j\}$ is a Cauchy sequence in $C([-T, T], H^s(M))$. The limit u solves our system.

As for uniqueness, since any $u \in C([-T, T], H^s(M))$ solving (2.1), (2.2) must belong to $H^1([-T, T], H^{s-1}(M))$, the energy inequality (2.1), with s replaced by $s-2$, applies, and we see that the solution is unique.

Exercises

2.1. Suppose u satisfies (2.1), (2.2). If $g \in H^s(M), f \in H^\ell([-T, T], H^s(M))$, $\ell \geq 0$, show that

$$u \in C^k([-T, T], H^{s-k}(M)), \qquad 0 \leq k \leq \ell. \tag{2.8}$$

If $f \in C^\ell([-T, T], H^s(M))$, show that (2.8) holds for $0 \leq k \leq \ell + 1$.

2.2. If $g \in H^s(M)$, $f \in H^s([-T, T] \times M)$, show that, if $s \geq 0$

$$u \in H^s([-T, T] \times M).$$

Hint: Use Exercise 2.1 to do it for $s \geq 0$ an integer, and then interpolate.

2.3. Let $u(t)$ solve (2.1) with $f = 0$, $u(s) = \delta_y I$ (I is the $k \times K$ identity matrix). Let $R(t, s)$ be defined by

$$R(t, s) = \begin{cases} u(t) & t \geq s \\ 0 & t \leq s \end{cases}$$

Thus $R(t, s) \in L^\infty_{\text{loc}}(\mathbf{R}^2, H^\sigma(M))$, $\sigma < -(n/2)$. Show that

$$\left(\frac{\partial}{\partial t} - K\right) R = \delta_{(s,y)} I.$$

R is called the Fundamental solution.

§3. Strictly Hyperbolic Equations

As (1.3) shows, the first order systems arising from higher order equations tend not to be symmetric. (See, however, exercise 3.1.) We introduce a large convenient class here.

DEFINITION 3.1. $(\partial/\partial t) - K$ *is called strictly hyperbolic if* $K \in OPS^1$ *has principal symbol* $K_1(t, x, \xi)$ *whose eigenvalues, for each fixed* (t, x, ξ), $\xi \neq 0$, *are pure imaginary and distinct.*

In a similar fashion, an operator L of order m as considered in Section 1 is said to be strictly hyperbolic if, for each (t, x, ξ), the roots $\tau_1, \ldots, \tau_m$ of $L_m(t, x, \tau, \xi) = 0$ are real and distinct, provided $\xi \neq 0$. Clearly such strictly hyperbolic operators produce strictly hyperbolic systems via (1.3).

The way we shall treat the Cauchy problem

$$\frac{\partial}{\partial t} u = Ku + f, \qquad u(0) = g \tag{3.1}$$

when $(\partial/\partial t) - K$ is strictly hyperbolic is to construct what is called a *symmetrizer* for K, which will allow us to prove the energy estimates (2.4) and (2.5) of the last section. As before, we suppose $x \in M$, a compact manifold.

DEFINITION 3.2. *A symmetrizer for $(\partial/\partial t) - K$ is a smooth one parameter family of operators $R = R(t) \in OPS^0$ such that*

$$R_0(t, x, \xi) \text{ is a positive definite matrix, for } |\xi| \geq 1, \tag{3.2}$$

$$RK + (RK)^* \in OPS^0_{1,0}. \tag{3.3}$$

If such a symmetrizer exists, one says $(\partial/\partial t) - K$ is symmetrizable.

PROPOSITION 3.1. *Any strictly hyperbolic first order system $(\partial/\partial t) - K$ has a symmetrizer.*

Proof. If we let the eigenvalues of $K_1(t, x, \xi)$ be $i\lambda_\nu(t, x, \xi)$, $\lambda_1(t, x, \xi) < \lambda_2(t, x, \xi) < \cdots < \lambda_k(t, x, \xi)$, K being a $k \times k$ system, then λ_ν are well-defined C^∞ functions of (t, x, ξ), homogeneous of degree 1 in ξ. Similarly, if $P_\nu(t, x, \xi)$ are the projections onto the associated eigenspaces of $i\lambda_\nu(t, x, \xi)$,

$$P_\nu = \frac{1}{2\pi i} \int_{\gamma_\nu} (\zeta - K_1(t, x, \xi))^{-1} \, d\zeta,$$

then P_ν is smooth and homogeneous of degree 0 in ξ. Cut off near $\xi = 0$, to get $P_\nu \in S^0$. Now let

$$R(t, x, D_x) = \sum_{j=1}^{k} P_j(t, x, D)^* P_j(t, x, D).$$

$R \in OPS^0$ and (3.2) is satisfied. Furthermore, it is easy to see that

$$\sigma_{RK}(t, x, \xi) = i \sum_{j=1}^{k} \lambda_j(t, x, \xi) P_j(t, x, \xi)^* P_j(t, x, \xi) \bmod S^0$$

so (3.3) holds. Thus R is a symmetrizer.

We remark that, whenever $(\partial/\partial t) - K$ is symmetrizable, a symmetrizer exists which is positive definite on $L^2(M)$. In fact, if R_1 is any symmetrizer satisfying (3.2), (3.3), there is an $R = R_1 \bmod OPS^{-1}$ with $R \geq \eta I > 0$ on $L^2(M)$, as follows from Gårding's inequality (see Chapter II, Exercise 8.1.) With such a positive definite symmetrizer, we have the following modification of the energy estimates of Section 2, for u solving (3.1).

$$\begin{aligned}
\frac{d}{dt}(Ru, u) &= (Ru', u) + (Ru, u') + (R'u, u) \\
&= (RKu + Rf, u) + (Ru, Ku + f) + (R'u, u) \\
&= ((RK + K^*R)u, u) + (Rf, u) + (Ru, f) + (R'u, u) \\
&\leq C\|u\|^2 + C\|f\|^2 \\
&\leq C'(Ru, u) + C\|f\|^2.
\end{aligned}$$

Gronwall's inequality applies as before, to yield

$$\|u(t)\|^2 \leq C\|u(0)\|^2 + C \int_0^t \|f(\tau)\|^2 \, d_\tau, \qquad |t| \leq T,$$

and similarly, for all real s,

$$\|u(t)\|_{H^s}^2 \leq C\|u(0)\|_{H^s}^2 + C \int_0^t \|f(\tau)\|_{H^s}^2 \, d\tau.$$

From here one derives, in exactly the same fashion as in Section 2, the following existence and uniqueness theorem.

THEOREM 3.2. *If $(\partial/\partial t) - K$ is symmetrizable (in particular, if strictly hyperbolic), then the Cauchy problem (3.1) has a unique solution $u \in C([-T, T], H^s(M))$, given $g \in H^s(M)$ and $f \in L^2([-T, T], H^s(M))$. If $f \in H^\ell([-T, T], H^s(M))$, $\ell \geq 0$, then $u \in C^k([-T, T], H^{s-k}(M))$, $0 \leq k \leq \ell$.*

The next theorem follows from the reduction technique of Section 1.

THEOREM 3.3 *If L is a strictly hyperbolic operator of order m, then the Cauchy problem*

$$Lu = f, \qquad \frac{\partial^j}{\partial t^j} u(0) = g_{j+1}, \qquad j = 0, \ldots, m-1$$

has a unique solution $u \in C([-T, T], H^s(M))$, given $g_{j+1} \in H^{s-j}(M)$, $f \in L^2([-T, T], H^{s-m+1}(M))$. Indeed,

$$\left(\frac{\partial}{\partial t}\right)^j u \in C([-T, T], H^{s-j}(M)), \qquad j = 0, \ldots, m-1. \tag{3.4}$$

More generally, if $f \in H^\ell([-T, T], H^{s-m+1}(M))$, $\ell \geq 0$, then (3.4) holds for $j = 0, \ldots, m - 1 + \ell$.

Exercises

3.1. Consider the second order equation on $\mathbf{R} \times M$

$$\frac{\partial^2}{\partial t^2} u - Au + B \frac{\partial}{\partial t} u = 0 \tag{3.5}$$

where $A \in OPS^2$ has principal symbol $A_2(t, x, \xi) \leq -C|\xi|^2$ and $A - A^* \in OPS^0$, and $B \in OPS^0$. Let $P \in OPS^1$ be any positive operator on $L^2(M)$ with principal symbol $P_1(t, x, \xi) = (-A_2(t, x, \xi))^{1/2}$, and convert (3.5) into a first order system for (u_1, u_2) via

$$u_1 = Pu, \qquad u_2 = \frac{\partial}{\partial t} u.$$

Show that you get a symmetric hyperbolic system for (u_1, u_2).

3.2. Let $u \in \mathscr{D}'(\mathbf{R} \times M)$ and suppose u satisfies $Lu = 0$ with L strictly hyperbolic. If $u = 0$ for $t < t_0$, prove that $u \equiv 0$.

3.3. Let L be a strictly hyperbolic operator with smooth coefficients on $\mathbf{R} \times \mathbf{R}^n$. By altering, outside some large subset of $\mathbf{R}^n$, the coefficients of L and the Cauchy data $f, g_1, \ldots, g_m$ to be periodic in x, with some large period, and then applying Theorem 3.3, deduce the following local existence theorem: Given any relatively compact open set $(-T, T) \times U \subset \mathbf{R} \times \mathbf{R}^n$, there is a $u \in C((-T, T), H^s(\mathbf{R}^n))$ such that $(\partial^j/\partial t^j)u(0) = g_{j+1}$, $j = 0, \ldots, m-1$, and such that, on $(-T, T) \times U$, $Lu = f$, given $g_{j+1} \in H^{s-j}(\mathbf{R}^n)$ and $f \in L^2(\mathbf{R}, H^{s-m+1}(\mathbf{R}^n))$.

3.4. Discuss the Cauchy problem for a differential operator

$$L = \frac{\partial^m}{\partial t^m} - \sum_{j=0}^{m-1} A_{m-j}(t, x, D_x) \frac{\partial^j}{\partial t^j}$$

where $A_{m-j}(t, x, D_x)$ are themselves $k \times k$ systems of operators.

§4. Finite Propagation Speed: Finite Domain of Dependence

We have seen that a strictly hyperbolic differential equation

$$Lu = f \tag{4.1}$$

on $\mathbf{R} \times \mathbf{R}^n$ can be solved, on any compact set, given Cauchy data

$$u\big|_{t=0} = g_1, \ldots, \frac{\partial^{m-1}}{\partial t^{m-1}} u\big|_{t=0} = g_m. \tag{4.2}$$

We want to examine local uniqueness of such solutions, and consider restrictions on the support of u when f and the g_j are supported in given compact sets. In order to accomplish this, it is useful to generalize the Cauchy problem, replacing (4.2) by

$$u\big|_S = g_1, \quad \frac{\partial}{\partial \nu} u\big|_S = g_2, \ldots, \frac{\partial^{m-1}}{\partial \nu^{m-1}} u\big|_S = g_m \tag{4.3}$$

where S is a hypersurface in $\mathbf{R}^{n+1}$ and $\partial/\partial\nu$ the normal vector field to S. The appropriate condition to place on S for (4.1), (4.3) to be a good Cauchy problem is described as follows.

DEFINITION 4.1. *A covariant vector $V = (V_0, V')$ at (t_0, x_0) is timelike with respect to L if the equation*

$$L_m(t_0, x_0, X + \tau V) = 0$$

has m distinct real roots $\tau_1, \ldots, \tau_m$ when X is not proportional to V.

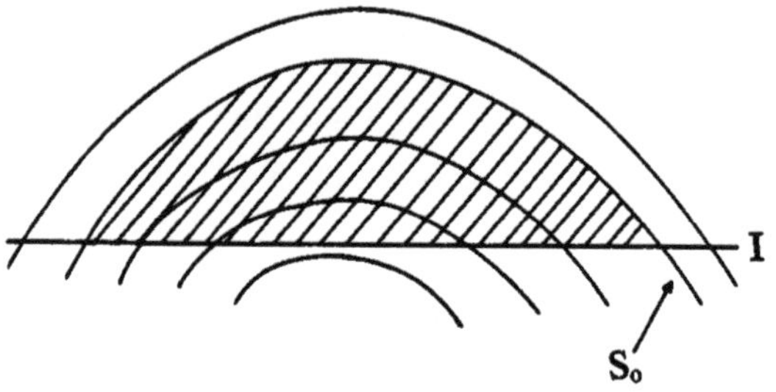

Figure IV.1

DEFINITION 4.2. *A hypersurface S is spacelike with respect to L if its normals are all timelike.*

In particular, the condition of strict hyperbolicity of L given at the beginning of Section 3 is equivalent to saying each hyperplane $\{t = \text{const.}\}$ is spacelike, or that each "vertical" vector $(\tau, \xi) = (\tau, 0)$ is timelike. Clearly vectors sufficiently close to "vertical" will remain timelike. Indeed, if $V = (V_0, V')$ and if

$$\frac{|V_0|}{|V'|} > \frac{|\tau|}{|\xi|}$$

for all (τ, ξ) with $L_m(t_0, x_0, \tau, \xi) = 0$, $\xi \neq 0$, then V is timelike.

Suppose S_0 is a spacelike hypersurface and a neighborhood U of S_0 is foliated by a one parameter family S_σ of spacelike hypersurfaces. Then one can locally change coordinates, putting on (s, y) coordinates so that these hypersurfaces coincide with the surfaces $\{s = \text{const.}\}$, and in this new coordinate system, by Definition 4.2, L becomes a new operator $\tilde{L}(s, y, D_{s,y})$ which is strictly hyperbolic, so the Cauchy problem

$$\tilde{L}v = \tilde{f}, \qquad \frac{\partial^j}{\partial s^j} v\Big|_{s=0} = \tilde{g}_j$$

can be locally solved. It easily follows that the Cauchy problem (4.1), (4.3) can be solved on compact subsets of U.

We shall use this local existence theorem to prove a uniqueness theorem, by duality. Let $L = L(t, x, D_{t,x})$ be strictly hyperbolic. Let G be a bounded domain in $\mathbf{R}^{n+1}$, whose boundary consists of two parts, $\partial G = S_0 \cup I$ where $I = \partial G \cap \{t = 0\}$. Suppose that S_0 is spacelike, and furthermore that S_0 is part of a one parameter family of spacelike surfaces S_σ foliating a neighborhood of $\bar{G}$. See Figure IV.1.

THEOREM 4.3. *Under the above hypotheses, suppose $u \in C(\mathbf{R}, H^s(\mathbf{R}^n))$, that the Cauchy data of u vanish on I and $Lu = 0$ on G. Then $u = 0$ on G.*

Proof. We shall show that $\langle u, \varphi \rangle = 0$ for all $\varphi \in C_0^\infty(G)$. To do this, first note that the surfaces S_σ are also spacelike for L^*, since the principal symbols of L and L^* must be equal, up to sign. Now let Φ be a solution to the Cauchy problem

$$L^*\Phi = \varphi, \qquad \frac{\partial^j}{\partial \nu^j} \Phi|_{S_0} = 0, \qquad j = 0, \ldots, m-1$$

in a neighborhood of $\bar{G}$. We can suppose Φ is zero on the opposite side of G from S_0. Now the following identity follows by integration by parts:

$$0 = \langle \Phi, Lu \rangle = \langle L^*\Phi, u \rangle = \langle \varphi, u \rangle.$$

This concludes the proof.

This type of argument was first used by Holmgren to prove uniqueness in the Cauchy problem for differential operators with analytic coefficients, and noncharacteristic boundary. See Garabedian [1] or John [1]; in that case, the existence theorem used was the Cauchy-Kowalevski theorem.

An immediate consequence of Theorem 4.3 is the following. Let u be a solution to (4.1), (4.2), with f and g_j supported on $|x| \le R$. Suppose any surface S whose normal $\nu = (\nu_0, \nu')$ satisfies

$$\frac{|\nu'|}{|\nu_0|} \le C_0^{-1}$$

is *spacelike*. Then $u(t, x)$ must be zero for $|x| \ge R + C_0|t|$. This expresses the assertion that signals travel at finite speed. We now introduce the concept of a domain of dependence.

DEFINITION 4.4. *Let $(t_0, x_0) \in \mathbf{R}^{n+1}$. If there is a domain G containing (t_0, x_0), with boundary $\partial G = S_0 \cup I$ satisfying the hypotheses preceding Theorem 4.3, we say I is a domain of dependence for (t_0, x_0), and we say (t_0, x_0) has a finite domain of dependence.*

The following theorem is a simple consequence of the above discussion.

THEOREM 4.5. *Let L be a strictly hyperbolic operator as above.*

(i) *If the vector (ν_0, ν') is timelike whenever*

$$\frac{|\nu'|}{|\nu_0|} \le C_0^{-1},$$

then any $(t_0, x_0) \in \mathbf{R}^{n+1}$ has the bounded domain of dependence

$$B_{c_0|t_0|}(x_0) = \{x \in \mathbf{R}^n : |x - x_0| \le C_0|t_0|\}.$$

(ii) *Given $g_j \in \mathscr{D}'(\mathbf{R}^n)$, $f \in C(\mathbf{R}^n))$, the Cauchy problem*

$$Lu = f, \qquad \frac{\partial^j}{\partial t^j} u|_{t=0} = g_{j+1}, \qquad j = 0, \ldots, m$$

has a unique solution $u \in C(\mathbf{R}, \mathscr{D}'(\mathbf{R}^n))$, *if each point* $(t_0, x_0) \in \mathbf{R}^{n+1}$ *has a finite domain of dependence.*

(iii) *If the conditions of* (i) *are satisfied and if* $x \in K$ *for* $x \in \text{supp } g_j$, $(t, x) \in \text{supp } f$, *then*, $\text{supp } u \subset \{(t, x): \text{dist } (x, K) \leq C_0|t|\}$.

Exercises

4.1. Suppose

$$L = \frac{\partial^m}{\partial t^m} + \sum_{j=0}^{m-1} A_{m-j}(t, x, D_x) \frac{\partial^j}{\partial t^j}$$

is a differential operator of order m, strictly hyperbolic, and suppose every $(t_0, x_0) \in \mathbf{R}^{n+1}$ has a finite domain of dependence. Then there is a unique $R \in C(\mathbf{R}, H^{n/2+m-1-\epsilon}_{\text{loc}}(\mathbf{R}^n))$ such that

$$LR = 0, \qquad R(0) = 0, \ldots, \qquad \frac{\partial^{m-2}}{\partial t^{m-2}} R(0) = 0, \qquad \frac{\partial^{m-1}}{\partial t^{m-1}} R(0) = \delta(x).$$

Let $\Phi \in \mathscr{D}'(\mathbf{R}^{n+1})$ be defined by

$$\Phi(t) = \begin{matrix} R(t) & t > 0 \\ 0 & t < 0 \end{matrix}.$$

Prove that $L\Phi = \delta$. What can you say about the support of Φ?

4.2. Concoct an example of a strictly hyperbolic operator L whose coefficients near infinity behave so badly that not all points $(t_0, x_0) \in \mathbf{R}^{n+1}$ have a finite domain of dependence.

4.3. Let $K(t, x, D_x) \in OPS^1$ on a compact manifold M and suppose $(\partial/\partial t) - K$ is strictly hyperbolic. Suppose solutions to $(\partial/\partial t - K)u = 0$ enjoy finite propagation speed. Show that K must be a *differential operator.*

§5. Quasilinear Hyperbolic Systems

We consider in this section local solvability of the Cauchy problem for a quasilinear hyperbolic system

$$\frac{\partial}{\partial t} u = K(t, u, x, D_x)u, \tag{5.1}$$

$$u(0) = g \tag{5.2}$$

where $K = K(t, u)$ is a smooth family of pseudodifferential operators in $OPS^1(\mathbf{T}^n)$, depending on t and u. We suppose that either K_1 is skew symmetric or K_1 has distinct, purely imaginary eigenvalues, i.e., $(\partial/\partial t) - K$ is either symmetric hyperbolic or strictly hyperbolic. Under such a hypothesis, we shall show that, given $g \in C^\infty(\mathbf{T}^n)$, or more generally given $g \in H^M(\mathbf{T}^n)$ for M sufficiently large, there is an interval $(-T, T)$ on which (5.1), (5.2) has a solution. For some reason, the literature seems to concentrate on symmetric hyperbolic systems. The following argument was worked out during several conversations between the author and Andrew Majda along the beach of Venice, California.

The proof will use an iterative method common in the solution of such nonlinear problems. Given u on $\mathbf{R} \times \mathbf{T}^n$ with $u(0) = g$, we define $Fu = v$ to be the solution to the system

$$\frac{\partial}{\partial t} v = K(t, u, x, D_x)v, \qquad v(0) = g. \tag{5.3}$$

Solving (5.1), (5.2) amounts to finding a fixed point of F, i.e., a function u such that $Fu = u$ on $(-T, T) \times \mathbf{T}^n$.

In order to treat this problem, it will be necessary to deal with pseudo-differential operators with less than C^∞ symbols. The following symbol classes will be useful.

DEFINITION 5.1. *We say $p(x, \xi) \in H^M S^m_{1,0}$ provided*

$$\|D_\xi^\alpha p(\cdot, \xi)\|_{H^M(\mathbf{T}^n)} \le C(1 + |\xi|)^{m - |\alpha|}, \qquad \text{for } |\alpha| \le M.$$

It is clear that for continuity on Sobolev spaces $H^s(\mathbf{T}^n)$, when s is bounded, and to perform any finite number of operations, one needs only finitely such differentiability of a symbol. We state a few specific results of this nature, leaving the simple proofs (modifications of the proofs of Chapter II) to the reader.

LEMMA 5.2. *If $p(x, \xi) \in H^M S^0_{1,0}$ and $M > n/2$, then $p(x, D): L^2(\mathbf{T}^n) \to L^2(\mathbf{T}^n)$. More generally, $p(x, D): H^s(\mathbf{T}^n) \to H^s(\mathbf{T}^n)$ for $|s| \le \mu$, provided $M > (n/2) + \mu$.*

LEMMA 5.3. *Given any M, m_1, m_2 there is a μ such that, if $p_j(x, D) \in OPH^\mu S^{m_j}_{1,0}$, then $p_1(x, D)p_2(x, D) \in OPH^M S^{m_1+m_2}_{1,0}$, $p_1(x, D)^* \in OPH^M S^{m_1}_{1,0}$, and*

$$[p_1(x, D), p_2(x, D)] \in OPH^M S^{m_1+m_2-1}_{1,0}.$$

PROPOSITION 5.4. *Given any μ there is an M such that if $K \in H^M S^1_{1,0}$ and R and $(\partial/\partial t)R \in H^M S^0_{1,0}$ are such that*

$$(R(t, x, D_x)u, u) \ge c_0\|u\|^2_{L^2},$$
$$RK + K^*R \in H^M S^0_{1,0},$$

then the system

$$\frac{\partial}{\partial t} u = Ku + f,$$

$$u(0) = g$$

has a unique solution $u \in C([-T, T], H^s(\mathbf{T}^n))$, *given* $g \in H^s(\mathbf{T}^n)$, $f \in C([-T, T], H^s(\mathbf{T}^n))$, *and supposing* $|s| \le \mu$. *Such a solution satisfies the estimate, for* $|s| \le \mu$

$$\|u(t)\|_{H^s}^2 \le (1 + C|t|)\left[\|g\|_{H^s}^2 + \int_0^t \|f(\tau)\|_{H^s}^2 \, d\tau\right] \qquad |t| \le T \tag{5.4}$$

where C *depends on finitely many seminorms of* K, R, *and* $(\partial/\partial t)R$, *in* $H^M S_{1,0}^1$ *and* $H^M S_{1,0}^0$, *and of* $RK + K^*R$ *in* $H^M S_{1,0}^0$, *on* T *and on* μ, C_0, *and on* n. M *depends on* μ *and* n, *but not on the order of the system.*

To return to our iterative method (5.3), suppose $g \in H^M(\mathbf{T}^n)$, with the size of M to be determined. Suppose also that $u \in C([-T, T], H^M(\mathbf{T}^n))$ and $(\partial/\partial t)u \in C([-T, T], H^{M-1}(\mathbf{T}^n))$. First of all, take M large enough so there is a unique solution $v \in C^1([-T, T], H^n(\mathbf{T}^n))$. Say this happens if

$$M \ge M_1. \tag{5.5}$$

To obtain a more precise estimate on v, it is convenient to obtain equations for various derivatives of v. Indeed, set

$$v_{0\alpha} = D_x^\alpha v, \qquad v_{1\alpha} = \frac{\partial}{\partial t} D_x^\alpha v.$$

Similarly define $u_{0\alpha}$, $u_{1\alpha}$. Applying the chain rule to (5.3) yields

$$\frac{\partial}{\partial t} v_{0\alpha} = K(t, u_{00})v_{0\alpha} + \sum_{\substack{\gamma+\delta+\sigma=\alpha,\ \sigma<\alpha \\ \delta_1+\cdots+\delta_\mu=\delta}} c_{\sigma\gamma\delta_1\cdots\delta_\mu} u_{0\delta_1} \cdots u_{0\delta_\mu} K_{\gamma\mu}(t, u_{00}) v_{0\sigma}, \tag{5.6}$$

$$\begin{aligned} \frac{\partial}{\partial t} v_{1\beta} = {} & K(t, u_{00})v_{1\beta} + u_{10}K_\mu(t, u_{00})v_{0\beta} + K_t(t, u_{00})v_{0\beta} \\ & + \sum_{\substack{\gamma+\delta+\sigma=\beta \\ \text{etc.}}} c_{\sigma\cdots\delta_\mu}\{u_{1\delta_1} u_{0\delta_2} \cdots u_{0\delta_\mu} K_{\gamma\mu}(t, u_{00}) v_{0\sigma} \\ & + u_{0\delta_1} \cdots u_{0\delta_\mu} \cdot u_{10} \cdot K_{\gamma,\mu+1} v_{0\sigma} \\ & + u_{0\delta_1} \cdots u_{0\delta_\mu} K_{\gamma\mu t}(t, u_{00}) v_{0\sigma} \\ & + u_{0\delta_1} \cdots u_{0\delta_\mu} K_{\gamma\mu}(t, u_{00}) v_{1\sigma}\}. \end{aligned} \tag{5.7}$$

Here $K_{\gamma\mu} = (D_x^\gamma D_u^\mu K)$ and $K_{\gamma\mu t} = (D_t D_x^\gamma D_u^\mu K)$. Now replace $v_{0\sigma}$ by $P_{0\sigma}(\tilde{v})$ where

$$\tilde{v} = \{v_{0\alpha}, v_{1\beta}: 0 \le |\alpha| \le M, 0 \le |\beta| \le M-1\} \tag{5.8}$$

and

$$P_{j\sigma}(\tilde{v}) = \Lambda^{-(M-j-|\sigma|)} \sum_{|\beta|=M-j} c_{\sigma\beta}(x, D_x) v_{j\beta} \tag{5.9}$$

$c_{\sigma\beta}(x, D_x) \in OPS^0$ being appropriately chosen. Thus $P_{j\sigma} \in OPS^{-(M-j-|\sigma|)}$. Similarly, replace $u_{j\delta}$ on the right sides of (5.6) and (5.7) by $P_{j\delta}(\tilde{u})$. The system for $\tilde{v}$ becomes

$$\begin{aligned} \frac{\partial}{\partial t} v_{0\alpha} &= K(t, P_{00}(\tilde{u}))v_{0\alpha} \\ &\quad + \sum_{\sigma<\alpha} c_{\sigma\cdots\delta_\mu}(P_{0\delta_1}\tilde{u}) \cdots (P_{0\delta_\mu}\tilde{u}) K_{\gamma\mu}(t, P_{00}\tilde{u}) P_{0\sigma}\tilde{v}, \end{aligned} \tag{5.10}$$

$$\begin{aligned} \frac{\partial}{\partial t} v_{1\beta} &= K(t, P_{00}(\tilde{u}))v_{1\beta} \\ &\quad + P_{10}(\tilde{u})K(t, P_{00}\tilde{u})P_{0\beta}\tilde{v} + K_t(t, P_{00}\tilde{u})P_{0\beta}\tilde{v} \\ &\quad + \sum_{\sigma<\beta} c_{\sigma\cdots\delta_\mu}\{(P_{1\delta_1}\tilde{u})(P_{0\delta_2}\tilde{u}) \cdots (P_{0\delta_\mu}\tilde{u})K_{\gamma\mu}(t, P_{00}\tilde{u})P_{0\sigma}\tilde{v} \\ &\qquad + (P_{0\delta_1}\tilde{u}) \cdots (P_{0\delta_\mu}\tilde{u})(P_{10}\tilde{u})K_{\gamma,\mu+1}P_{0\sigma}\tilde{v} \\ &\qquad + (P_{0\delta_1}\tilde{u}) \cdots (P_{0\delta_\mu}\tilde{u})K_{\gamma\mu t}(t, P_{00}\tilde{u})P_{0\sigma}\tilde{v} \\ &\qquad + (P_{0\delta_1}\tilde{u}) \cdots (P_{0\delta_\mu}\tilde{u})K_{\gamma\mu}(t, P_{00}\tilde{u})P_{1\sigma}\tilde{v}\}. \end{aligned} \tag{5.11}$$

We rewrite this system as

$$\frac{\partial}{\partial t}\tilde{v} = K(t, P_{00}\tilde{u})\tilde{v} + \Phi(t, x, \tilde{u}, \tilde{v}). \tag{5.12}$$

As a first justification of regarding the terms grouped together in Φ as "zero order," we point out that all the operators $K_{\gamma\delta}(t, P_{00}\tilde{u})P_{0\sigma}$ appearing on the right side of (5.10) are of order ≤ 0, as are all the operators $K(t, P_{00}\tilde{u})P_{0\beta}$ and $K_t(t, P_{00}\tilde{u})P_{0\beta}$ appearing on the right side of (5.11).

Since $K(t, P_{00}\tilde{u}) \in OPH^M S^1_{1,0}$, etc., we shall impose another restriction on the size of M. Namely,

$$M \ge M_2 \tag{5.13}$$

where M_2 is chosen so that, if (5.13) holds, then

$$OPH^{M-1}S^m_{1,0}: H^s(\mathbf{T}^n) \to H^{s-m}(\mathbf{T}^n) \text{ for } |m| \le 1, \qquad |s| \le n.$$

Indeed, we have the following. Let $K(M)$ be the number of multiindices $(\alpha_1, \ldots, \alpha_n)$ of length $|\alpha| \leq M$, and let $K = K(M) + K(M-1)$. Suppose (5.1) is a $k \times k$ system, so $\tilde{v}$ is a kK-tuple of functions.

LEMMA 5.5. *Assuming M satisfies* (5.5) *and* (5.13), *we find that*

$$\Phi: \mathbf{R} \times \mathbf{T}^n \times [L^2(\mathbf{T}^n)]^{2kK} \to [L^2(\mathbf{T}^n)]^{kK} \tag{5.14}$$

is a Lipschitz continuous map.

Proof. First look at the components of Φ of the form

$$\Psi = (P_{0\delta_1}\tilde{u}) \cdots (P_{0\delta_\mu}\tilde{u}) K_{\gamma\delta}(t, P_{00}\tilde{u}) P_{0\sigma}\tilde{v}$$

where $|\delta_1 + \cdots + \delta_\mu + \sigma| \leq M$, $|\sigma| < M$. The map $\tilde{u} \to K_{\gamma\delta}(t, P_{00}\tilde{u})$ is a Lipschitz map of $L^2(\mathbf{T}^n)$ into $OPH^{M-1}S^1_{1,0}$. Consequently, $K_{\gamma\delta}(t, P_{00}\tilde{u})$: $H^s \to H^{s-1}$ for $|s| \leq n$, and for $s > n$ it certainly maps $H^s(\mathbf{T}^n)$ to $L^\infty(\mathbf{T}^n)$. By the Sobolev imbedding theorem, we have

$$H^s(\mathbf{T}^n) \subset L^{2n/n-2s}(\mathbf{T}^n), \qquad 0 \leq s < n/2$$

and $H^s(\mathbf{T}^n) \subset L^\infty(\mathbf{T}^n)$ for $s > n/2$. Since $P_{0\delta_j}\tilde{u} \in H^{M-|\delta_j|}(\mathbf{T}^n)$ while $K_{\gamma\delta}(t, P_{00}\tilde{u})P_{0\sigma}\tilde{v}$ belongs to $H^{M-|\sigma|-1}(\mathbf{T}^n)$ if $M - |\sigma| \leq n$ and belongs to $L^\infty(\mathbf{T}^n)$ if $M - |\sigma| > n$, it follows that Ψ is in $L^2(\mathbf{T}^n)$ and, being multilinear, is a Lipschitz function of its arguments. A similar argument controls all the other terms which make up Φ.

We are now ready to apply the method of Section 4 to the equation (5.12), in order to estimate $\tilde{v}$. Constructing a positive definite symmetrizer $R(t, w, x, D_x)$ for $K(t, w, x, D_x)$, substituting $P_{00}\tilde{u}$ for w, gives $R(t, P_{00}\tilde{u}) \in OPH^M S^0_{1,0}$. Now we can write

$$\begin{aligned} \frac{d}{dt}(R\tilde{v}, \tilde{v}) &= (R(K\tilde{v}+\Phi), \tilde{v}) + (R\tilde{v}, K\tilde{v}+\Phi) + \left(\left(\frac{\partial}{\partial t}R\right)\tilde{v}, \tilde{v}\right) \\ &= ((RK + K^*R)\tilde{v}, \tilde{v}) + (R\Phi, \tilde{v}) + (R\tilde{v}, \Phi) \\ &\quad + ((R_t + (P_{10}\tilde{u}) \cdot R_u)\tilde{v}, \tilde{v}). \end{aligned} \tag{5.15}$$

The last three terms in (5.15) are clearly bounded by

$$B(\|\tilde{u}\|_{L^2})\|\tilde{v}\|^2_{L^2} + B(\|\tilde{u}\|_{L^2})$$

for $t \in [-T, T]$, where B is some function of its argument which we need not specify. Similarly we obtain the bound

$$((RK + K^*R)\tilde{v}, \tilde{v}) \leq B(\|\tilde{u}\|_{L^2})\|\tilde{v}\|^2_{L^2}$$

provided that M is sufficiently large that $RK + K^*R \in OPH^L S^0_{1,0}$ with $L \geq M_2$. This is possible by Lemma 5.3; say this holds provided

$$M \geq M_3. \tag{5.16}$$

Consequently, (5.15) becomes

$$\frac{d}{dt}(R\tilde{v}, \tilde{v}) \le B'(\|\tilde{u}\|_{L^2})(R\tilde{v}, \tilde{v}) + B'(\|\tilde{u}\|_{L^2}), \tag{5.17}$$

and Gronwall's inequality yields

$$\|\tilde{v}(t)\|_{L^2}^2 \le e^{\int_0^t B'(\|u(\tau)\|_{L^2})\,d\tau}\left[\|\tilde{v}(0)\|_{L^2}^2 + \int_0^t B'(\|u(\tau)\|_{L^2})\,d\tau\right]. \tag{5.18}$$

Here $\tilde{v}(0)$ is defined by

$$v_{0\alpha}(0) = D_x^\alpha g, \qquad v_{1\beta} = D_x^\beta(K(t, g)g), \qquad |\alpha| \le M, \qquad |\beta| \le M-1.$$

Note that

$$D_x^\beta(K(t, g)g) = \sum_{\substack{\mu+\nu+\rho=\beta \\ \nu_1+\cdots+\nu_s=\nu}} cg^{(\nu_1)}\cdots g^{(\nu_s)}((D_x^\mu D_g^s K)(t, g))D_x^\rho g$$

which belongs to $L^2(\mathbf{T}^n)$ provided $g \in H^M(\mathbf{T}^n)$, by the argument used in the proof of lemma 5.5. So we can rewrite (5.15) as

$$\|\tilde{v}(t)\|_{L^2}^2 \le e^{\int_0^t B'(\|\tilde{u}(\tau)\|_{L^2})\,d\tau}\left[C_0^2\|g\|_{H^M}^2 + \int_0^t B'(\|\tilde{u}(\tau)\|)\,d\tau\right]. \tag{5.19}$$

Suppose $u \in C([-T, T], H^M(\mathbf{T}^n)) \cap C^1([-T, T], H^{M-1}(\mathbf{T}^n))$, so $\tilde{u} \in C([-T, T], L^2(\mathbf{T}^n))$. Suppose the norm of $\tilde{u}$ in this space is $\le A_0$, where we pick $A_0 \ge 2C_0\|g\|_{H^M}+1$. Pick the t interval fairly small, as follows. Suppose $T \le T_0$ where T_0 is so small that, with $B_1 = \sup\{|B'(\lambda)|: |\lambda| \le A_0\}$,

$$e^{T_0 B_1}[C_0^2\|g\|_{H^M}^2 + T_0 B_1] \le A_0^2. \tag{5.20}$$

It follows from (5.19) that, under this assumption

$$\|\tilde{v}(t)\|_{L^2} \le A_0, \qquad |t| \le T_0.$$

Consequently, for such a small t interval, the mapping $\tilde{v} = \tilde{F}\tilde{u}$ arising from (5.3) maps the set

$$\{\tilde{u} \in C([-T, T], L^2(\mathbf{T}^n)): \|\tilde{u}\| \le A_0\}$$

into itself.

To check convergence of $v_n = F^n u$, we need to estimate the difference between $\tilde{v} = \tilde{F}(\tilde{u})$ and $\tilde{v}_1 = \tilde{F}(\tilde{u}_1)$. From (5.12) we get

$$\frac{\partial}{\partial t}(\tilde{v}-\tilde{v}_1) = K(t, P_{00}\tilde{u})\tilde{v} - K(t, P_{00}\tilde{u}_1)\tilde{v}_1 + \Phi(t, x, \tilde{u}, \tilde{v}) - \Phi(t, x, \tilde{u}_1, \tilde{v}_1)$$

or, with $\tilde{w} = \tilde{v} - \tilde{v}_1$,

$$(5.21) \qquad \frac{\partial}{\partial t}\tilde{w} = K(t, P_{00}\tilde{u})\tilde{w} + [K(t, P_{00}\tilde{u}) - K(t, P_{00}\tilde{u}_1)]\tilde{v}_1 + \Phi(t, x, \tilde{u}, \tilde{v}) - \Phi(t, x, \tilde{u}_1, \tilde{v}_1) = K(t, P_{00}\tilde{u})\tilde{w} + \Delta$$

where, with the aid of Lemma 5.5, we have

$$(5.22) \qquad \|\Delta\|_{L^2(\mathbf{T}^n)} \le C\|\tilde{u}-\tilde{u}_1\|_{L^2}[\|\tilde{v}_1\|_{H^1} + C_1].$$

Since the H^1 norm of $\tilde{v}_1$ is involved, we have no hope of invoking the standard contraction mapping principle, but will have to argue further. If we set $\hat{u} = \{u_{0\alpha}, u_{1\beta}: |\alpha| \le M+1, |\beta| \le M\}$ and define $\hat{v}$ similarly, (5.3) gives rise to $\hat{v} = \hat{F}\hat{u}$, and the proof of (5.19) extends to

$$\|\hat{v}(t)\|_{L^2}^2 \le e^{\int_0^t B'(\|\hat{u}\|)\,d\tau}\left[C_1^2\|g\|_{H^{M+1}}^2 + \int_0^t B'(\|\hat{u}\|)\,d\tau\right]$$

with perhaps a bigger function $B'(\lambda)$. Now, pick $A_1 \ge 2C_1\|g\|_{H^{M+1}} + 1$, supposing $g \in H^{M+1}(\mathbf{T}^n)$, and we see that, perhaps further shrinking the t-interval, there is a T_1 such that, if $\|\hat{u}(t)\|_{L^2} \le A_1$ for $|t| \le T_1$, then

$$(5.23) \qquad \|\hat{v}(t)\|_{L^2} \le A_1, \qquad |t| \le T_1.$$

This implies

$$(5.24) \qquad \|\tilde{v}(t)\|_{H^1} \le CA_1, \qquad |t| \le T_1.$$

And furthermore $\tilde{F}^\nu\tilde{u}$ satisfies the estimate (5.24) for $\nu = 1, 2, 3, \ldots$. Now (5.21), (5.22), and (5.24) yield

$$(5.25) \qquad \|\tilde{w}(t)\|_{L^2} \le A_3 \int_0^t \|\tilde{u}(\tau) - \tilde{u}_1(\tau)\|_{L^2}\,d\tau, \qquad |t| \le T_1$$
$$\le A_3 T_1 \sup_{|\tau| \le T_1} \|\tilde{u}(\tau) - \tilde{u}_1(\tau)\|_{L^2}.$$

It is clear from (5.25) that $\tilde{F}^\nu\tilde{u}$ will converge to a limit as $\nu \to \infty$, in $C([-T_1, T_1], L^2(\mathbf{T}^n))$, provided T_1 is picked so small, in addition to the above requirements, that $A_3 T_1 < 1$. The limit $\tilde{w}$ must be of the form $\{w_{0\alpha}, w_{1\beta}: |\alpha| \le M, |\beta| \le M-1\}$ for some $w \in C([-T_1, T_1], H^M(\mathbf{T}^n)) \cap C^1([-T_1, T_1], H^{M-1}(\mathbf{T}^n))$, and w must solve (5.1), (5.2). Since the terms $\tilde{F}^\nu\tilde{u}$ are bounded in $L^\infty([-T_1, T_1], L^2(\mathbf{T}^n))$, whose unit ball is weak* compact, considering this space as the dual of $L^1([-T, T], L^2(\mathbf{T}^n))$, it follows that $\tilde{w} \in L^\infty([-T, T], L^2(\mathbf{T}^n))$, i.e., $w \in L^\infty([-T, T], H^{M+1}(\mathbf{T}^n))$, and $(\partial/\partial t)w \in L^\infty([-T, T], H^M(\mathbf{T}^n))$. We summarize our existence theorem.

THEOREM 5.6. *Let $g \in H^{M+1}(\mathbf{T}^n)$, where M is sufficiently large. Then, for T_1 sufficiently small, the iterative method* (5.3) *converges to a solution u of* (5.1), (5.2), *with*

$$u \in C([-T, T], H^M) \cap L^\infty([-T, T], H^{M+1}), \tag{5.26}$$

$$\frac{\partial}{\partial t} u \in C([-T, T], H^{M-1}) \cap L^\infty([-T, T], H^M), \tag{5.27}$$

provided (5.1) *is either symmetric hyperbolic or strictly hyperbolic.*

We remark that, for M sufficiently large, such a solution u is unique. Indeed, if u_1 is another solution to (5.1), (5.2) with similar regularity, (5.25) implies

$$\|\tilde{u}(t) - \tilde{u}_1(t)\|_{L^2} \le A_3 \int_0^t \|\tilde{u}(\tau) - \tilde{u}_1(\tau)\|_{L^2}\, d\tau,$$

which immediately implies $\tilde{u}(t) \equiv \tilde{u}_1(t)$, or $u \equiv u_1$. More generally, if u_1 solves (5.1) with $u_1(0) = g_1$, then (5.25) easily generalizes to

$$\|\tilde{u}(t) - \tilde{u}_1(t)\|_{L^2} \le A_3 \int_0^t \|\tilde{u}(\tau) - \tilde{u}_1(\tau)\|_{L^2}\, d\tau + A_4 \|g - g_1\|_{H^M}$$

where A_4 depends on $\|\tilde{u}\|_{L^2}$, $\|\tilde{u}_1\|_{L^2}$ and A_3 depends on $\|\tilde{u}\|_{H^1}$ and $\|\tilde{u}_1\|_{H^1}$. If $T_2 \le 1/2A_3$, we have

$$\sup_{|t| \le T_2} \|\tilde{u}(t) - \tilde{u}_1(t)\|_{L^2} \le 2A_4 \|g - g_1\|_{H^M},$$

or

$$\sup_{|t| \le T_2} \|u(t) - u_1(t)\|_{H^M} \le A_5 \|g - g_1\|_{H^M} \tag{5.28}$$

where we must be careful to note that the size of the interval on which (5.28) is true depends on $\|g\|_{H^{M+1}}$, $\|g_1\|_{H^{M+1}}$.

Using Theorem 5.6, we can handle quasilinear hyperbolic differential operators of the form

$$Lu = \frac{\partial^m u}{\partial t^m} - \sum_{j=0}^{m-1} A_{m-j}(t, u, x, D_x) \frac{\partial^j u}{\partial t^j} \tag{5.29}$$

where A_{m-j} is a differential operator, with smooth dependence on the arguments t, u, x, and we suppose for convenience that $|A_{m-j}| \le C_{m-j} < \infty$. If (5.28) is strictly hyperbolic, then the reduction method of Section 1 produces a strictly hyperbolic first order system, (1.3), with

$$b_j = A_{m-j+1}(t, u, x, D_x)\Lambda^{j-m}.$$

Consequently, Theorem 5.6 applies, and we have the following.

THEOREM 5.7. *Let (5.28) be strictly hyperbolic and let Cauchy data* $u(0) = g_1, \ldots, (\partial^{m-1}/\partial t^{m-1})u(0) = g_m$ *be given,* $g_j \in H^{M-j+1}(\mathbf{T}^n)$. *If M is large enough, $Lu = 0$ has a unique solution u on $(-T, T) \times \mathbf{T}^n$, for T sufficiently small, and*

$$\frac{\partial^j}{\partial t^j} u \in L^\infty([-T, T], H^{M-j}(\mathbf{T}^n)) \cap C([-T, T], H^{M-j-1}(\mathbf{T}^n)),$$

$$0 \le j \le m-1.$$

More generally than (5.28), there arise operators of the form

$$(5.30)\quad Lu = \frac{\partial^m u}{\partial t^m} - \sum_{j=0}^{m-1} A_{m-j}(t, \{D_t^\ell D_x^\beta u : \ell + |\beta| \le m-1\}, x, D_x)\frac{\partial^j u}{\partial t^j}.$$

A typical case of this is the equation

$$u_{tt} - [2 - (1 + u_x^2)^{-3/2}]u_{xx} = 0$$

which occurs in the study of vibrating strings. Using the method of Section 1, we derive the first order system (1.3) with

$$(5.31)\qquad b_j = A_{m-j+1}(t, P_{j1}u, \ldots, P_{j\nu}u, x, D_x)\Lambda^{j-m}$$

with $P_{j\mu} \in OPS^0$, and u standing for $(u_1, \ldots, u_m)^t$ in (5.31). Thus it is necessary to generalize (5.1) to

$$(5.32)\qquad \frac{\partial}{\partial t} u = K(t, P_1 u, \ldots, P_\ell u, x, D_x)u$$

with $P_j \in OPS^0$ and $K(t, v_1, \ldots, v_\ell, x, \xi) \in S^1$, K_1 being either skew symmetric or possessed of distinct purely imaginary eigenvalues. This generalization leads to some minor notational changes in the argument leading up to Theorem 5.6, starting with the formula (5.6), but the argument goes through with very little change, and Theorem 5.6 holds with (5.1) generalized to (5.32). Similarly, Theorem 5.7 holds with (5.29) generalized to (5.30).

Note that the iterative method for solving (5.32) is equivalent to using the iteration $v = F_0 u$ defined by $(\partial^j/\partial t^j)v|_{t=0} = g_{j+1}$ and

$$(5.33)\quad \frac{\partial^m v}{\partial t^m} = \sum_{j=0}^{m-1} A_{m-j}(t, \{D_x^\ell D_x^\beta u : \ell + |\beta| \le m-1\}, x, D_x)\frac{\partial^j v}{\partial t^j}.$$

We have not derived the optimal results in Theorems 5.6 and 5.7. For example, we have made no attempt to find the smallest value of M for which these results hold. In Hughes, Kato, and Marsden [1] it is shown that local solutions exist, at least for symmetric hyperbolic first order differential equations, for $M > (n/2) + 1$, and this result may well hold generally.

These results are interesting, but they are overshadowed by the formation of shocks in solutions to quasilinear hyperbolic equations that might start out smooth.

Consider the following example. Let u solve the equation

$$u_t + uu_x = 0. \tag{5.34}$$

Let $u(0) = -xe^{1-x^2}$, for example. (5.34) says u is constant on the line of slope $1/u(x)$ through $(t, x) = (0, x)$. This implies (5.34) cannot have a solution that is continuous on $(0, T) \times \mathbf{R}$ if $T > 1$, since several such characteristic lines intersect at $x = 0$, $t = 1$. Nevertheless, there is a bounded measurable function defined on $\mathbf{R}^+ \times \mathbf{R}$ which solves (5.34) in the weak sense that $(\partial/\partial t)u + (1/2)(\partial/\partial x)(u^2) = 0$, where the derivatives are taken in the distribution sense. There has been a bit of work done on the global existence and properties of weak solutions to quasilinear equations, especially those said to be in conservation form. See in particular the papers of Lax [2], Glimm [1], Glimm and Lax [1], DiPerna [1], and many other papers mentioned in their bibliographies, for work on systems. It turns out that weak solutions are not unique, but certain solutions are singled out by so-called entropy conditions. There are very many unsolved problems in this area, which remains one of the deepest and most mysterious areas of mathematics. Particularly, the problems are all wide open in more than one space variable, except for scalar first order equations, for which we refer to Volpert [1], Johnson and Smoller [1], and Kotlow [1].

One knows that (5.34) has a unique global weak solution, for $t \geq 0$, satisfying a certain entropy condition, given initial data $u(0) \in L^1(\mathbf{R}) \cap L^\infty(\mathbf{R})$. Now integration by parts shows that, for nice smooth solutions to (5.34) which die at infinity, $\|u(t)\|_{L^2(\mathbf{R})}$ is constant. Indeed, it is the case that, for $u(0) \in L^2(\mathbf{R})$, $u(t) \in L^2(\mathbf{R})$ for $t \geq 0$. See the exercises after Section 7 for more on this. The methods of this section, suitably sharpened, show that there is a local solution for $u(0) \in H^{3/2+\epsilon}(\mathbf{R})$, and the solution operator $S(t): u(0) \mapsto u(t)$ is a continuous operator on bounded subsets of $H^{3/2+\epsilon}(\mathbf{R})$. The following curious phenomenon, however, arises. As proved by DiPerna [1] in a much more general context, continuous solutions to such a "strictly nonlinear" equation as (5.34) must be Lipschitz on the interior of their regions of continuity. This implies that $S(t)$ cannot be a bounded evolution operator on $H^\sigma(\mathbf{R})$ for $1/2 < \sigma \leq 3/2$. Since it turns out that $S(t)$ is bounded on $L^2(\mathbf{R})$, there is a gap in the set of $\sigma \geq 0$ such that $S(t)$ (for t in some small interval) is bounded on bounded subsets of $H^\sigma(\mathbf{R}^n)$. This is an interesting phenomenon, which does not contradict the existing theorems (see, e.g., Bona and Scott [1]) on interpolation for nonlinear operators.

Exercises

5.1. Let L be strictly hyperbolic on $\mathbf{R}\times\mathbf{T}^n$ of the form (5.29) and suppose that any vector $V = (V_0, V')$ with

$$\frac{|V'|}{|V_0|} \le C_0^{-1}$$

is timelike for

$$\frac{\partial^m}{\partial t^m} - \sum_{j=0}^{m-1} A_{m-j}(t, v, x, D_x)\frac{\partial^j}{\partial t^j},$$

for any value of v, or more generally suppose L is of the form (5.30) and that any such V is timelike for

$$\frac{\partial^m}{\partial t^m} - \sum_{j=0}^{m-1} A_{m-j}(t, \{v_{\ell\beta}: \ell + |\beta| \le m-1\}, x, D_x)\frac{\partial^j}{\partial t^j},$$

for any values of $v_{\ell\beta}$. Show that solutions enjoy the following version of finite propagation speed: Let u_1 solve $Lu_1 = 0$, $(\partial^j/\partial t^j)u_1|_{t=0} = g_{j+1}$ while u_2 solves $Lu_2 = 0$, $(\partial^j/\partial t^j)u_2|_{t=0} = h_{j+1}, j = 0, \ldots, m-1$. Suppose u_1 and u_2 are sufficiently smooth. Suppose that $g_{j+1} = h_{j+1}$ outside of $x \in K$, $j = 0, \ldots, m-1$. Show that $u_1 = u_2$ outside of $\{(t, x): \text{dist}(x, K) \le C_0|t|\}$. Hint: Obtain such an identity for the approximations to u_j given by (5.33), using the linear theory.

5.2. Using Exercise 5.1, discuss the Cauchy problem on $(-T, T)\times M$ for M noncompact.

5.3. Let M be sufficiently large and let $u \in C([-T, T], H^M(\mathbf{T}^n))$ be a solution to $Lu = 0$ as in Theorem 5.7. Suppose $g_j \in C^\infty(\mathbf{T}^n)$. Show that $u \in C^\infty((-T, T)\times\mathbf{T}^n)$.

§6. The Vibrating Membrane Problem

In this section we shall indicate how to obtain a hyperbolic equation describing one physical process. For a further introduction to the equations governing continua, see Goldstein [1].

We suppose we have a thin membrane, pulled tight, in a region Ω in the x-y plane, which vibrates in the direction perpendicular to the x-y plane. If $u(t, x, y)$ is the displacement from the x-y plane, Hooke's law says that the potential energy in the membrane due to such a displacement is of the form

$$V = \int_\Omega \int f(x, y, |\nabla_{x,y} u|^2)\, dx\, dy \tag{6.1}$$

where $f(x, y, \lambda)$ depends on the elastic properties of the membrane. In the simplest approximation, one sets

$$f(x, y, |\nabla u|^2) = \sigma(x, y)|\nabla u|^2. \tag{6.2}$$

Generally speaking, one supposes $(\partial/\partial\lambda)f \geq c_0 > 0$. On the other hand, the kinetic energy due to this vibration is

$$T = \frac{1}{2}\int_\Omega \int \mu(x, y)|u_t|^2\, dx\, dy \tag{6.3}$$

where $\mu(x, y)$ is the mass density, assumed to be smooth and strictly positive. Now Hamilton's principle says that the action

$$\int_{t_0}^{t_1} (T - V)\, dt$$

is stationary. Thus, we are led to the calculus of variations problem

$$\frac{d}{ds} I(u + sv)|_{s=0} = 0 \qquad \text{for all} \qquad v \in C_0^\infty((t_0, t_1) \times \Omega) \tag{6.4}$$

where

$$I(u) = \int_{t_0}^{t_1} \int_\Omega \left\{ f(x, y, |\nabla u|^2) - \frac{\mu}{2}|u_t|^2 \right\} dx\, dy\, dt. \tag{6.5}$$

Now

$$\begin{aligned} \frac{d}{ds} I(u + sv)|_{s=0} &= \int_{t_0}^{t_1} \int_\Omega \left[2\frac{\partial f}{\partial \lambda}(x, y, |\nabla u|^2)\, \nabla u \cdot \nabla v - \mu u_t v_t \right] dx\, dy\, dt \\ &= \int_{t_0}^{t_1} \int_\Omega v \left[\frac{\partial}{\partial t}(\mu u_t) - 2 \operatorname{div}\left(\frac{\partial f}{\partial \lambda}(x, y, |\nabla u|^2)\, \nabla u \right) \right] dx\, dy\, dt, \end{aligned}$$

so (6.4) yields the partial differential equation for u:

$$\frac{\partial}{\partial t}(\mu u_t) - 2 \operatorname{div}\left(\frac{\partial f}{\partial \lambda}(x, y, |\nabla u|^2)\, \nabla u \right) = 0. \tag{6.6}$$

In the special case when f is given by (6.2), the equation becomes

$$\frac{\partial}{\partial t}(\mu u_t) - 2 \operatorname{div}(\sigma \nabla u) = 0 \tag{6.7}$$

or

$$\mu u_{tt} - 2\sigma \Delta u - 2\nabla\sigma \cdot \nabla u = 0.$$

It is clear that (6.7) is hyperbolic, assuming μ and σ positive. More generally, (6.6) is a quasilinear hyperbolic equation, whose existence theory, either locally on Ω when $\Omega \subset \mathbf{R}^2$ or on $(-T, T) \times \Omega$ if $\Omega = \mathbf{T}^2$, follows from the discussion of Section 5 provided f is smooth and, with $F(x, y, p) = f(x, y, |p|^2)$,

$$0 < c_0|\xi|^2 \leq \sum_{j,k} \frac{\partial^2 F}{\partial p_k \, \partial p_j} \xi_j \xi_k \leq c_1 |\xi|^2.$$

When $\Omega \subset \mathbf{R}^2$, it is necessary to specify boundary data of u as well as its initial data. For instance, if our membrane is securely fastened to the head of a drum, we have $u = 0$ on $\partial\Omega$, for all t. In this chapter we have not touched on such mixed initial boundary value problems, though some special cases will be treated in the following exercises, in the linear case. The primary "energy identity" is the fact that the total energy

$$E = T + V = \int_\Omega \left\{ \frac{\mu}{2} |u_t|^2 + f(x, y, |\nabla u|^2) \right\} dx \, dy \tag{6.8}$$

is constant.

Exercises

6.1. If u is sufficiently smooth and satisfies (6.6) with $u = 0$ on $\partial\Omega$, show that $(d/dt)E = 0$, with E given by (6.8).

6.2. Let L be any negative semidefinite self-adjoint operator on a Hilbert space H, with domain $\mathcal{D}(L)$. Discuss solutions to the abstract Cauchy problem

$$\frac{\partial^2}{\partial t^2} v - Lv = 0 \tag{6.9}$$

$$v(0) = g_1 \in \mathcal{D}((-L)^{1/2}), \qquad \frac{\partial v}{\partial t}(0) = g_2 \in H. \tag{6.10}$$

6.3. Consider the linear equation (6.7). Let $v = \mu^{1/2} u$, $H = L^2(\Omega)$. Let

$$Lv = 2\mu^{-1/2} \sum_j \frac{\partial}{\partial x_j} \sigma \frac{\partial}{\partial x_j} (\mu^{-1/2} v).$$

Note that (6.7) formally becomes (6.9), and note that, if $v, w \in C^\infty(\bar{\Omega})$, vanishing on $\partial\Omega$, then $(Lv, w) = (v, Lw)$. Look ahead at Chapter V and show that L is self adjoint, and hence consider existence and smoothness of solution to (6.7) with Cauchy data as in (6.10), and the Dirichlet condition $u = 0$ on $\partial\Omega$.

§7. Parabolic Evolution Equations

DEFINITION 7.1. *If $-K \in OPS^m_{1,0}$ is strongly elliptic of order $m > 0$ on M, a compact manifold, then the operator $(\partial/\partial t) - K$ is a strongly parabolic operator on $\mathbf{R} \times M$.*

Here we take $K = K(t)$ to be a smooth family of operators on M, as before. If u is sufficiently smooth, $(\partial/\partial t)u = Ku + f$, we have the a priori estimate

$$\begin{aligned}\frac{\partial}{\partial t}\|u(t)\|^2_{L^2} &= (Ku + f, u) + (u, Ku + f)\\ &= ((K+K^*)u, u) + 2\,Re(u, f)\\ &\le c\|u\|^2 + c\|f\|^2\end{aligned}$$

where Gårding's inequality is applied to the strongly elliptic operator $-K$ to give this last inequality. Thus

$$\|u(t)\|^2_{L^2} \le c\|u(0)\|^2_{L^2} + c\int_0^t \|f(\tau)\|^2_{L^2}\,d\tau, \qquad 0 \le t \le T_1$$

for $u \in C^1([0, T], H^m(M))$. From this energy inequality, existence and uniqueness of solutions to the initial value problem follows, by the same sort of arguments as given in Section 2.

Note that, unlike hyperbolic equations, strongly parabolic equations can only be solved for positive t, not for negative t.

To solve the initial value problem for $(\partial/\partial t) - K$, $K \in OPS^m_{1,0}$, we only need $-Re\,\tau(t, x, \xi) \ge c|\xi|^m$ for any eigenvalue τ of $K(t, x, \xi)$, $|\xi| \ge c$. To show this involves constructing a symmetrizer, in the spirit of Section 3. We exhibit a more general construction in the next chapter.

Exercises 7.1 through 7.5 give another approach to parabolic equations in the temporally homogeneous case, using the theory of semigroups. For the theory of semigroups, see Yosida [1], Chapter IX.

Exercises

7.1. Suppose $A \in OPS^m_{1,0}$, on a compact manifold M, is such that $-A$ is strongly elliptic of order m. Prove that the spectrum and numerical range of $A - \lambda_0$ is in a cone in the negative half plane in $\mathbf{C}$ if λ_0 is sufficiently large. Deduce that A generates a strongly continuous semigroup e^{tA}; in fact, A generates a holomorphic semigroup. Hint: Look at the argument in Agmon [2, p. 102].

7.2. Let $f \in L^2(M)$. Show that $e^{tA}f \in C^\infty(M)$ for each $t > 0$. If $u(t, x) = e^{tA}f(x)$, show that $u \in C^\infty((0, \infty) \times M)$. Hint: $e^{tA}f \in \mathscr{D}(A^k)$ for all k since e^{tA} is a holomorphic semigroup.

7.3. Prove that e^{tA} is a strongly continuous semigroup of operators on each Sobolev space $H^s(M)$, $s \in \mathbf{R}$.

7.4. Let A generate a holomorphic semigroup on a Hilbert space H. Prove that, for all $f \in H$,

$$\int_0^1 Re(-Ae^{tA}f, e^{tA}f)\, dt = \tfrac{1}{2}\|f\|^2 - \tfrac{1}{2}\|e^{tA}f\|^2.$$

Deduce that, when A is as in Exercises 7.1 through 7.3, $f \in H^s(M)$, $u = e^{tA}f$, then $u \in L^2([0, 1], H^{s+m/2}(M))$.

7.5. In the situation of Exercise 7.4, if $m = 1$, show that $f \in H^s(M)$ implies $u \in H^{s+1/2}([0, 1] \times M)$. Hint: See Theorem 4.3.1 of Hörmander [6].

In the following exercises we consider the equation

$$u_t + \frac{\partial}{\partial x} g(u) + Lu = \epsilon u_{xx}, \qquad u(0) = f(x) \tag{7.1}$$

on $\mathbf{R}^+ \times S^1$. We suppose g is smooth and $|g(s)| \le cs^2$ for s large, for example

$$g(u) = \tfrac{1}{2}u^2 \tag{7.2}$$

and we suppose L is a skew adjoint operator on $L^2(S^1)$, commuting with $(\partial^2/\partial x^2) = \Delta$, for example, $L = 0$, or with some real constant a,

$$L = a\frac{\partial^3}{\partial x^3}. \tag{7.3}$$

We suppose $\epsilon > 0$.

7.6. Show that (7.1) is identical to the integral equation

$$u(t) = e^{t(\epsilon\Delta - L)}f + \int_0^t e^{(t-s)(\epsilon\Delta - L)} \frac{\partial}{\partial x} g(u(s))\, ds.$$

7.7. Suppose that g satisfies

$$|g(u)| \le c_0(1 + u^2)$$

and

$$|(u-v)^{-1}(g(u)-g(v))| \le c_1(|u|+|v|)$$

which follows from g smooth and $|g'(u)| \le c_1$. Show that $u \mapsto g(u)$ is a Lipschitz map of $L^2 \to L^1$, and in fact

$$\|g(u)-g(v)\|_{L^1} \le c_1\|u-v\|_{L^2}.$$

7.8. Show that the operator norm of $e^{t(\epsilon\Delta - L)}(\partial/\partial x): L^1 \to L^2$ is the same as that of $(\partial/\partial x)e^{t\epsilon\Delta}: L^2 \to L^\infty$, and this is bounded by

$$c(\epsilon t)^{-3/4}.$$

Hint: Estimate the kernel of these operators.

7.9. If Φ is a map on the set of $u \in C([0, T_0], L^2(S^1))$ with $u(0) = f$, given by

$$\Phi u(t, x) = e^{t(\epsilon\Delta - L)}f + \int_0^t e^{(t-s)(\epsilon\Delta - L)} \frac{\partial}{\partial x} g(u)\, ds,$$

show that

$$\|\Phi u - \Phi \bar{u}\|_{C(I,L^2)} \le c(\epsilon)\|u - \bar{u}\|_{C(I,L^2)} \int_0^{T_0} s^{-3/4}\, ds$$

where $I = [0, T_0]$. Deduce that, for T_0 sufficiently small, Φ is a contraction map, and so has a unique fixed point. Furthermore, the size of T_0 depends only on $\|f\|_{L^2}$, and on ϵ.

7.10. Show that, for the solution $u \in C([O, T_0], L^2(S^1))$ constructed above, $\|u(t)\|_{L^2}$ is nonincreasing, and deduce that, for any $\epsilon > 0$, $f \in L^2(S^1)$, (7.1) has a *global* solution $u \in C([0, \infty), L^2(S^1))$. Hint: Rewrite (7.1) as

$$\frac{\partial}{\partial t} u + \left[\frac{\partial}{\partial x} \circ h(u) + h(u) \circ \frac{\partial}{\partial x}\right] u + Lu = \epsilon u_{xx}$$

where $h(u)$ solves

$$uh'(u) + 2h(u) = g'(u)$$

and show that such a smooth h can be found.

7.11. Now what? Does anything come from letting $\epsilon \downarrow 0$?

In (7.1), suppose $L = 0$, for the rest of the exercises.

7.12. Suppose $f \in L^\infty(S^1)$. Let $u_\epsilon(t, x)$ denote the solution to (7.1), $\epsilon > 0$. Show that, on $\mathbf{R}^+ \times S^1$,

$$|u_\epsilon(t, x)| \le \|f\|_{L^\infty}.$$

Hint: Apply the maximum principle.

7.13. Show that, for any smooth function u on S^1,

$$\int_{S^1} -\Delta\left(\frac{\partial u}{\partial x}\right) \operatorname{sgn} \frac{\partial u}{\partial x}\, dx \ge 0 \quad \text{and} \tag{7.4}$$

$$\int_{S^1} \frac{\partial}{\partial x}(g(u)_x) \operatorname{sgn} \frac{\partial u}{\partial x}\, dx = 0. \tag{7.5}$$

7.14. Show that, for a solution u_ϵ to (7.1) with $u_\epsilon(0)$ of bounded variation (we say $u^0 = u_\epsilon(0) \in BV$),

$$\frac{d}{dt}\int \left|\frac{\partial u_\epsilon}{\partial x}\right| dx \leq 0. \tag{7.6}$$

(Hint: Differentiate (7.1) with respect to x, multiply by sgn $(\partial u/\partial x)$, and integrate.) Hence, if $u^0 \in BV$, $\{u_\epsilon : 0 < \epsilon \leq 1\}$ is *bounded* in $L^\infty(\mathbf{R}^+, BV)$.

7.15. Now show that $\{u_\epsilon : 0 < \epsilon \leq 1\}$ is bounded in $\mathbf{C}^{1/2}(\mathbf{R}^+, L^1)$. Hint: Use the integral equation (for $0 \leq t < t'$)

$$u_\epsilon(t) - u_\epsilon(t') = e^{\epsilon t\Delta}u^0 - e^{\epsilon t'\Delta}u^0 + \int_{t'}^{t} e^{(t-s)\epsilon\Delta} D_x g(u_\epsilon(s))\, ds$$
$$+ \int_0^{t'} [e^{(t-s)\epsilon\Delta} - e^{(t'-s)\epsilon\Delta}] D_x g(u_\epsilon(s))\, ds$$

and derive the estimate

$$\|e^{\sigma\Delta} - I\|_{\mathscr{L}(BV, L^1(S^1))} \leq c\sigma^{1/2}, \qquad \sigma > 0$$

to show that

$$\|u_\epsilon(t) - u_\epsilon(t')\|_{L^1} \leq [c\epsilon^{1/2}|t - t'|^{1/2} + c|t - t'|]\|u^0\|_{BV}.$$

7.16. Using the results of Exercises 7.14 and 7.15, show that, as $\epsilon \to 0$, provided $u^0 \in BV$, a subsequence of u_ϵ converges to

$$u_0 \in L^\infty(\mathbf{R}^+, BV) \cap C(\mathbf{R}^+, L^1)$$

solving

$$\frac{\partial}{\partial t} u_0 + g(u_0)_x = 0,$$

$$u_0(0) = u^0(x) \in BV.$$

7.17. Let Φ be a smooth *convex* function. Multiply (7.1) by $\Phi'(u_\epsilon)\varphi$, for any $\varphi \in C_0^\infty(\mathbf{R}^+ \times S^1)$ which is *positive*, and integrate to get

$$\int_{\mathbf{R}^+}\int_{S^1}\left\{-\Phi(u_\epsilon)\frac{\partial \varphi}{\partial t} + \epsilon\left|\frac{\partial u_\epsilon}{\partial x}\right|^2 \Phi''(u_\epsilon)\varphi + \epsilon\frac{\partial u_\epsilon}{\partial x}\cdot\frac{\partial \varphi}{\partial x}\Phi'(u_\epsilon)\right.$$
$$\left. + \varphi\frac{\partial}{\partial x}\left(\int_k^{u_\epsilon} a_i(s)\Phi'(s)\, ds\right)\right\} dx\, dt = 0,$$

with k any constant. Note that

$$\epsilon\left|\frac{\partial u_\epsilon}{\partial x}\right|^2 \Phi''(u_\epsilon)\varphi \geq 0$$

and show that

$$\lim_{\epsilon\to 0} \iint \epsilon \frac{\partial u_\epsilon}{\partial x} \cdot \frac{\partial \varphi}{\partial x} \Phi'(u_\epsilon)\, dx\, dt = 0$$

assuming u_ϵ solves (7.1) with $u_\epsilon(0) = u^0 \in L^2$. If $u^0 \in BV$, deduce that for any limit point u_0 of u_ϵ, $\epsilon \to 0$,

$$\int_{\mathbf{R}^+} \int_{S^1} \left\{ \Phi(u_0) \frac{\partial \varphi}{\partial t} + \left(\int_k^{u_0} g'(s)\Phi'(s)\, ds \right) \frac{\partial \varphi}{\partial x} \right\} dx\, dt \geq 0. \tag{7.7}$$

We remark that (7.7) is called the entropy condition.

7.18. Taking a sequence of convex functions $\Phi_j(\xi) \to |\xi - k|$, $\Phi_j'(\xi) \to \operatorname{sgn}(\xi - k)$, show that any limit u_0 of u_ϵ, $\epsilon \to 0$, satisfies, for positive test function φ,

$$\int_{\mathbf{R}^+} \int_{S^1} \left\{ |u_0 - k| \frac{\partial \varphi}{\partial t} + \operatorname{sgn}(u_0 - k)(g(u_0) - g(k)) \frac{\partial \varphi}{\partial x} \right\} dx\, dt \geq 0 \tag{7.8}$$

for all $k \in \mathbf{R}$. Somewhat more preversely, (7.8) is called the entropy condition.

7.19. If u_0 and v_0 solve (7.1), with $\epsilon = 0$, and are limits of solutions u_ϵ, v_ϵ, $\epsilon \to 0$, with $u_\epsilon(0) = u^0$, $v_\epsilon(0) = v^0 \in BV$, show that

$$\|u_0(t) - v_0(t)\|_{L^1} \leq \|u^0 - v^0\|_{L^1}.$$

Hint: This can be obtained by a very tricky application of (7.8). See Krushkov [1]. Deduce *uniqueness* of solutions to (7.1) with $\epsilon = 0$ which are obtained by such a limiting process as $\epsilon \downarrow 0$, for $u^0 \in BV$. (Recall $L = 0$.) Denote the solution operator by

$$S(t)u^0 = u_0(t).$$

7.20. With $S(t)$ defined as above on BV, Exercises 7.12 and 7.19 above show that, for $f, g \in BV(S^1)$,

$$\|S(t)f - S(t)g\|_{L^1} \leq \|f - g\|_{L^1},$$
$$\|S(t)f\|_{L^\infty} \leq \|f\|_{L^\infty}.$$

Deduce that $S(t)$ has a unique continuous extension to $L^1(S^1)$, which maps $L^\infty(S^1)$ to itself. Deduce further that $S(t)$ is bounded from $L^p(S^1)$ to itself, $1 \leq p \leq \infty$, and in particular $S(t)$ is bounded from $L^2(S^1)$ to itself. Hint: This is an interpolation theorem for a nonlinear operator. See Tartar [1] or Bona and Scott [1]. The key is to use the Lions-Peetre real method (K method) of interpolation, rather than a complex interpolation method (see Bergh and Löfström [1].)

For further material on (7.1) with $L = 0$, see Johnson and Smoller [1], Vol'pert [1], Kotlow [1], and Lax [2].

§8. References to Further Work

One big subject we have not touched on is mixed problems for hyperbolic equations. For symmetric hyperbolic equations, see Courant and Hilbert [1] and Lax and Phillips [1] and for another general class see Kreiss [1], and Rauch [1]. One basic class is treated in Chapter IX.

A topic which is interesting both for mixed problems and the pure initial value problem is scattering theory; see Lax and Phillips [1]. Closely allied to this is the method of geometrical optics, with which parametrices for hyperbolic systems have been constructed by Lax [1], Ludwig [1], Hörmander [13] and many others. One application of these methods is to determine the singularities of solutions to hyperbolic (and more general) equations. Using the theory of pseudodifferential operators, we shall obtain some such results in Chapter VI. In Chapter VIII we introduce Fourier integral operators and go much more deeply into this.

Another important topic we have not touched on is difference schemes approximating solutions to initial value problems. See in particular Richtmyer and Morton [1]. A calculus of pseudodifferential operators has been developed to treat some problems in this area. See Yamaguti and Nogi [1], Lax and Nirenberg [1], and Vaillancourt [1].

Certain mixed problems for second order hyperbolic equations can be studied by means of functional analysis and the elliptic theory. One approach, due to Phillips, is outlined in the beginning of [1].

CHAPTER V

Elliptic Boundary Value Problems

Let M be a smooth compact manifold with boundary ∂M. Suppose L is an elliptic operator of order m on M, and $B_j, j = 1, \ldots, \nu$, are differential operators of order $m_j \leq m - 1$ defined in a neighborhood of ∂M. I want to investigate the boundary value problem

$$(0.1) \qquad \begin{aligned} Lu &= f, \\ B_j u|_{\partial M} &= g_j, \qquad 1 \leq j \leq \nu. \end{aligned}$$

Two properties we shall investigate are existence of solutions and smoothness. As we shall see in Section 2, existence will follow from the a priori estimates needed to prove smoothness results, via functional analysis.

I use a type of decoupling argument, due to Calderon, to derive basic a priori estimates for the first order system to which we reduce (0.1). The presentation of this is inspired by the thesis of Polking [1]. This leads to regularity theorems, with loss of δ derivatives, under appropriate hypotheses ($0 \leq \delta < 1$). The case $\delta = 0$ corresponds to regular elliptic boundary value problems, a class of which we make a detailed study in Section 4. In Section 5, I present a slightly different method (than in Section 2) for proving boundary regularity, which yields results on regularity with loss of δ derivatives, with no restriction on δ. This is useful, since the $\bar{\partial}$ Neumann problem, the most important nonregular problem, works with $\delta = 1$, on strongly pseudoconvex domains. In Chapter XV we shall study a class of hypoelliptic operators which includes the pseudodifferential operators on ∂M one would encounter upon treating the $\bar{\partial}$-Neumann problem by the method of Section 5 of this chapter.

Section 5 uses a "complete decoupling" procedure developed by the author in [4], and also by Kumano-go in [5]. We shall see it again in Chapter IX, where it will treat reflection of singularities problems and will also furnish more information on solutions to elliptic boundary value problems.

There are many topics on elliptic equations I have not had the space to cover. The reader is referred especially to Morrey [1], Agmon [2], and Lions and Magenes [1] for more material. Particularly, one should study the method of quadratic forms to solve boundary value problems, emphasized in Agmon [2]. As an exercise, the reader should try to obtain regularity theorems for such operators from the results of this chapter.

§1. Reduction to First Order Systems and Decoupling

We begin our analysis of (0.1). To obtain estimates on a solution u, it is only necessary to worry about the behavior of u near the boundary, since Chapter III yields the desired estimates and smoothness of u on the interior of M.

We proceed to reduce the problem near the boundary to a standard form. First, the collar neighborhood theorem says that there is a neighborhood of ∂M in M diffeomorphic to $I \times \partial M$, where $I = [0, 1]$. For a proof of this see Milnor's [1]. If we choose coordinates on $I \times \partial M$, (y, x), $y \in [0, 1]$, L takes the form

$$L = \frac{\partial^m}{\partial y^m} + \sum_{j=0}^{m-1} A_j(y, x, D_x) \frac{\partial^j}{\partial y^j}$$

where A_j is a differential operator of order $m - j$. This can be reduced to a first order system $(\partial/\partial y)u = Ku$, $K = K(y, x, D_x) \in OPS^1$, by the same technique as used in Chapter IV. Thus, if we set

$$u_j = \left(\frac{\partial}{\partial y}\right)^{j-1} \Lambda^{m-j} u, \qquad j = 1, \ldots, m,$$

$Lu = f$ becomes the system (1.3) of Chapter IV, which we rewrite as there:

$$\frac{\partial}{\partial y} u = Ku \tag{1.1}$$

upon relabeling $(u_1, \ldots, u_m)$ as u, and if the operators B_j of (0.1) have the form

$$B_j = \sum_{k=0}^{m_j} b_{jk}(x, D_x) \frac{\partial^k}{\partial y^k},$$

the boundary condition at $y = 0$ becomes

$$\Lambda^{m-m_j-1} \sum_{k=0}^{m_j} b_{jk}(x, D_x)\Lambda^{k+1-m} u_{k+1}(0) = \Lambda^{m-m_j-1} g_j$$

which we write as

$$Bu(0) = h \tag{1.2}$$

with $B \in OPS^0(M)$. We now discuss (1.1), (1.2).

DEFINITION 1.1. *The system* (1.1) *is elliptic if, and only if, the principal symbol* $K_1(y, x, \xi)$ *of* $K \in OPS^1$ *is a matrix with no purely imaginary eigenvalues for* $(y, x) \in I \times \partial M$, $\xi \neq 0$.

It is easy to see that if L is elliptic, then (1.1) will satisfy this criterion.

We now construct some pseudodifferential operators which will be useful for the proof of a priori estimates. Some will be analogous to the symmetrizers of Chapter IV, but first we want to "decouple" (1.1), as follows. Define

$$E_0(y, x, \xi)$$

to be the projection onto the direct sum of the generalized eigenspaces of $K_1(y, x, \xi)$, corresponding to all eigenvalues with positive real part, defined by

$$E_0(y, x, \xi) = \frac{1}{2\pi i} \int_\gamma (\zeta - K_1(y, x, \xi))^{-1} \, d\zeta \tag{1.3}$$

where γ is a curve in the right-hand plane containing the positive eigenvalues of $K_1(y, x, \xi)$. Clearly E_0 is homogeneous of degree 0 in ξ, so is the principal symbol of a smooth one parameter family of operators $E = E(y) \in OPS^0(\partial M)$. Now define

$$A = (2E-1)K \in OPS^1. \tag{1.4}$$

Then the principal symbol $A_1(y, x, \xi)$ of A has only eigenvalues with positive real part, for $\xi \neq 0$. We now want to construct operators $P = P(y) \in OPS^0(\partial M)$ such that PA is strongly elliptic. This plays the role of the "symmetrizer."

LEMMA 1.2. *Let* A *be an* $m \times m$ *matrix and suppose the eigenvalues of* A *have positive real part. Then there is a positive matrix* P *such that* $PA + A^*P$ *is also positive.*

Proof. For any $\epsilon > 0$, A is similar to $D_0 + \epsilon D_1$ where D_0 is diagonal and $D_0 + D_1$ is the Jordan canonical form of A. If $QAQ^{-1} = D_0 + \epsilon D_1$, write $Q = UP_1$ with U unitary and P_1 positive. Then $P_1AP_1^{-1} = U^*D_0U + \epsilon U^*D_1U = N + \epsilon B$ where N is normal. The eigenvalues of N and A are the same, so there is an $\eta > 0$ such that $N + N^* \geq \eta$. Taking $\epsilon < (1/3)\eta$, the matrix $N + \epsilon B$ is accretive. If $P = P_1^2$, if follows that $PA + A^*P = P_1(N+\epsilon B)P_1 + P_1(N^*+\epsilon B^*)P_1$ is positive.

Before we proceed, two observations are in order. First, for a given positive matrix P, the set of A such that $PA + A^*P$ is positive, is open. Second, for a given compact set K of $m \times m$ matrices, the set $\{P : P > 0, PA + A^*P > 0 \text{ for all } A \in K\}$ is an open convex set of matrices.

LEMMA 1.3. *If* $A = (2E-1)K$ *is as defined by* (1.4), *so the eigenvalues of* $A_1(y, x, \xi)$ *have positive real part, there is a smooth one parameter* *family $P = P(y) \in OPS^0$ such that P is positive definite and PA is strongly elliptic.*

Proof. With Lemma 1.2 and subsequent observations, it is easy using a partition of unity to construct $P_0(y, x, \xi)$ on the cosphere bundle of ∂M such that the symbols P_0 and $P_0 A_1 + A_1^* P_0$ are positive definite. Extend P_0 to be homogeneous of degree 0 in ξ. As a consequence of Gårding's inequality (see Chapter II, Section 8, Exercise 8.1) we can construct positive definite $P(y) \in OPS^0$ with principal symbol P_0. The proof is complete.

This preliminary work done, we are now ready to prove some a priori estimates.

§2. A Priori Estimates and Regularity Theorems

We are working on the space $\Omega = I \times \partial M$. It is convenient to use the Sobolev-type spaces $H_{(k,s)}(\Omega)$ with norms

$$\|u\|_{(k,s)}^2 = \sum_{j=0}^{k} \|D_y^j \Lambda^{k-j+s} u(y)\|_{L^2(\Omega)}^2. \tag{2.1}$$

Here $k \geq 0$ is an integer and $s \in \mathbf{R}$ is any real number. It is easy to see that $H_{(k,0)}(\Omega)$ coincides with the usual Sobolev space $H^k(\Omega)$. We shall use $|u|_s$ to denote $\|u\|_{H^s(\partial M)}$.

The derivation of a priori inequalities here is based on the notion that the projection E splits K up into a forward and a backward evolution operator. We shall differentiate $(P\Lambda^{s/2}Eu, \Lambda^{s/2}Eu)$ and $(P\Lambda^{s/2}(1-E)u, \Lambda^{s/2}(1-E)u)$ with respect to y, in the spirit of the energy inequalities of Chapter IV, and add the two up. Some new features, naturally, will appear. Let

$$H = \frac{\partial}{\partial y} - K,$$

and

$$\begin{aligned} \frac{\partial}{\partial y}(P\Lambda^{1/2}Eu, \Lambda^{1/2}Eu) \tag{2.2} \\ &= (P\Lambda^{1/2}Eu', \Lambda^{1/2}Eu) + (P\Lambda^{1/2}Eu, \Lambda^{1/2}Eu') + (P'\Lambda^{1/2}Eu, \Lambda^{1/2}Eu) + \cdots \\ &= (P\Lambda^{1/2}E(Hu+Ku), \Lambda^{1/2}Eu) + (P\Lambda^{1/2}Eu, \Lambda^{1/2}E(Hu+Ku)) + R_1(u). \end{aligned}$$

Here and below $R_j(u)$ denotes a remainder term with the property that

$$|R_j(u)| \leq C|u|_1|u|_0.$$

Rearranging the first term in (2.2), we get

$$(PEHu, \Lambda Eu) + (PEKu, \Lambda Eu) + R_2(u)$$
$$= (PEHu, \Lambda Eu) + (PAEu, \Lambda Eu) + R_3(u)$$

where we have used $E = (2E-1)E \bmod OPS^{-1}$ together with the commutativity of the principal symbols of E and K, to get $PEK = PAE \bmod OPS^0$. Similarly we rewrite the second term in (2.2) as $(PAEu, EHu) + (P\Lambda Eu, EKu) + R_4(u) = (\Lambda Eu, PEHu) + (\Lambda Eu, PAEu) + R_5(u)$. Thus,

$$(2.3)\quad \frac{\partial}{\partial y}(P\Lambda^{1/2}Eu, \Lambda^{1/2}Eu)$$
$$= 2\,Re(PEHu, \Lambda u) + ([PA + A^*P]Eu, \Lambda Eu) + R_6(u).$$

Similarly we get

$$(2.4)\quad \frac{\partial}{\partial y}(P\Lambda^{1/2}(1-E)u, \Lambda^{1/2}(1-E)u)$$
$$= 2\,Re(P(1-E)Hu, \Lambda(1-E)u)$$
$$- ([PA + A^*P](1-E)u, \Lambda(1-E)u) + R_7(u).$$

Adding (2.3) and (2.4) and applying Gårding's inequality to $(PA+A^*P)\Lambda^{-1} \in OPS^0$ yields

$$C|\Lambda u|^2 \le C|\Lambda Eu|^2 + C|\Lambda(1-E)u|^2$$
$$\le ([PA+A^*P]Eu, \Lambda Eu) + ([PA+A^*P](1-E)u, \Lambda(1-E)u) + R_8(u)$$
$$\le \frac{d}{dy}(P\Lambda^{1/2}Eu, \Lambda^{1/2}Eu) - \frac{d}{dy}(P\Lambda^{1/2}(1-E)u, \Lambda^{1/2}(1-E)u)$$
$$+ C|Hu|\,|\Lambda u| + R_9(u).$$

Integrating with respect to y, we get

$$(2.5)\quad \|u\|^2_{(0,1)} + |E(0)u(0)|^2_{1/2} + |(1-E(1))u(1)|^2_{1/2}$$
$$\le C\|Hu\|^2_{(0,0)} + C|E(1)u(1)|^2_{1/2} + C|(1-E(0))u(0)|^2_{1/2} + C\|u\|^2_{(0,0)}.$$

Since

$$\|u\|^2_{(1,0)} = \|D_y u\|^2_{(0,0)} + \|u\|^2_{(0,1)} \le \|Hu\|^2_{(0,0)} + C\|u\|^2_{(0,1)},$$

we get

$$(2.6)\quad \|u\|^2_{(1,0)} + |E(0)u(0)|^2_{1/2} + |(1 - E(1))u(1)|^2_{1/2}$$
$$\le C\|Hu\|^2_{(0,0)} + C|E(1)u(1)|^2_{1/2} + C|(1-E(0))u(0)|^2_{1/2} + C\|u\|^2_{(0,\sigma)},$$

for any $\sigma \in \mathbf{R}$, using Poincaré's inequality.

From now on, let $R(y) = y + (1-2y)(1-E(y))$, so $R(0) = 1-E(0)$ and $R(1) = E(1)$. Also let $|u|_\tau^2 = |u(1)|_\tau^2 + u(0)|_\tau^2$. We generalize (2.6) by substituting $\Lambda^t u$ for u.

$$\begin{aligned} \|u\|_{(1,t)}^2 &= \|\Lambda^t u\|_{(1,0)}^2 \\ &\le C\|H\Lambda^t u\|_{(0,0)}^2 + C|R\Lambda^t u|_{1/2}^2 + C\|\Lambda^t u\|_{(0,\sigma)}^2 \\ &\le C\|Hu\|_{(0,t)}^2 + C|Ru|_{1/2+t}^2 + C\|u\|_{(0,\sigma+t)}^2 + C|u|_{t-1/2}^2. \end{aligned}$$

Now

$$|u|_{t-1/2}^2 \le C\|u\|_{(1,t-1)}^2 \le C\|Hu\|_{(0,t-1)}^2 + C\|u\|_{(0,t)}^2,$$

so

$$\|u\|_{(1,t)}^2 \le C\|Hu\|_{(0,t)}^2 + C|Ru|_{1/2+t}^2 + C\|u\|_{(0,t)}^2,$$

and again using Poincaré's inequality we get, for any $\sigma \in \mathbf{R}$,

$$\|u\|_{(1,t)}^2 \le C\|Hu\|_{(0,t)}^2 + C|Ru|_{1/2+t}^2 + C\|u\|_{(0,\sigma)}^2. \tag{2.7}$$

Now (2.6) also implies

$$|(1-R)u|_{1/2}^2 \le C\|Hu\|_{(0,0)}^2 + C|Ru|_{1/2}^2 + C\|u\|_{(0,\sigma)}^2,$$

and in analogy with the argument above this yields

$$|(1-R)u|_{1/2+t}^2 \le C\|Hu\|_{(0,t)}^2 + C|Ru|_{1/2+t}^2 + C\|u\|_{(0,\sigma)}^2.$$

If we substitute $(1-R)u$ for u in this inequality and use the fact that the principal symbols of $R(y)$ and $K(y)$ commute, for each y, so $H(1-R) = (1-R)H + OPS^0$, together with the fact that $R(1-R) \in OPS^{-1}$ at $y = 0$ and 1, we obtain the important inequality

$$|(1-R)u|_{1/2+t}^2 \le C\|Hu\|_{(0,t)}^2 + C\|u\|_{(0,t)}^2 + C|u|_{t-1/2}^2. \tag{2.8}$$

Now suppose the boundary operator $B \in OPS^1(\partial M)$ satisfies the estimate, for each $\tau \in \mathbf{R}$,

$$|g|_{\tau-\delta}^2 \le C|(1-R)g|_\tau^2 + C|Bg|_\tau^2 + C|g|_{\tau-1}^2, \qquad g \in C^\infty(\partial M) \tag{2.9}$$

for some δ with $0 \le \delta < 1$. Inequality (2.8) implies

$$|u|_{1/2+t-\delta}^2 \le C\|Hu\|_{(0,t)}^2 + C\|u\|_{(0,t)}^2 + C|Bu|_{1/2+t}^2 + C|u|_{t-1/2}^2.$$

Now, (2.7) yields

$$\|u\|_{(1,t-\delta)}^2 \le C\|Hu\|_{(0,t-\delta)}^2 + C|u|_{1/2+t-\delta}^2 + C\|u\|_{(0,t)}^2,$$

so

$$\|u\|_{(1,t-\delta)}^2 \le C\|Hu\|_{(0,t)}^2 + C|Bu|_{1/2+t}^2 + C\|u\|_{(0,\sigma)}^2 + C|u|_\sigma^2 \tag{2.10}$$

for any $\sigma \in \mathbf{R}$, where again Poincaré's inequality has been used. Indeed, we can eliminate the term $|u|_\sigma^2$ on the right as was done leading up to (2.7).

We are now in sight of our first main goal, the following a priori estimate.

THEOREM 2.1. *Suppose (2.9) is satisfied by the boundary operator B. Then for any integer $k \geq 1$, and for all $t \in \mathbf{R}$,*

$$\|u\|^2_{(k,t-\delta)} \leq C\|Hu\|^2_{(k-1,t)} + C|Bu|^2_{t-1/2+k} + C\|u\|^2_{0,\sigma} \tag{2.11}$$

for all $u \in H_{(k,t)}(I \times \partial M)$.

Proof. We proceed by induction, the case $k = 1$ being inequality (2.10). Suppose (2.11) is valid for k. Now

$$\begin{aligned}\|u\|^2_{(k+1,t-\delta)} &\leq C\|D_y u\|^2_{(k,t-\delta)} + C\|u\|^2_{(k,t+1-\delta)}\\ &\leq C\|Hu\|^2_{(k,t-\delta)} + C\|u\|^2_{(k,t+1-\delta)}.\end{aligned}$$

We estimate the right side, using (2.11) for k, with t replaced by $t+1$, and the induction step follows.

To derive the regularity theorem corresponding to estimate (2.11), we make use of a Friedrich's mollifier J_ϵ on ∂M, discussed in Chapter II, Section 7.

LEMMA 2.2 *If $u \in H_{(1,r-1)}$ and $Hu \in H_{(0,r)}$, $Bu \in H^{r+1/2}(\partial M)$, then $u \in H_{(1,r-\delta)}$, provided (2.9) holds.*

Proof. For each $\epsilon > 0$,

$$J_\epsilon u \in H_{(1,\infty)} = \bigcap_{s<\infty} H_{(1,s)},$$

so (2.11) applies, and we get

$$\begin{aligned}\|J_\epsilon u\|^2_{(1,r-\delta)} &\leq C\|HJ_\epsilon u\|^2_{(0,r)} + C|BJ_\epsilon u|^2_{1/2+r} + C\|J_\epsilon u\|^2_{(0,\sigma)}\\ &\leq C\|Hu\|^2_{(0,r)} + C|Bu|^2_{1/2+r} + C\|u\|^2_{(0,r)} + C|u|^2_{r-1/2}\end{aligned}$$

using the facts that $[J_\epsilon, K]$ is bounded in $OPS^0_{1,0}$ and $[J_\epsilon, B]$ is bounded in $OPS^{-1}_{1,0}$. Hence

$$\|J_\epsilon u\|^2_{(1,r-\delta)} \leq \|Hu\|^2_{(0,r)} + C|Bu|^2_{1/2+r} + C\|u\|^2_{(1,r-1)}$$

a bound which is *independent* of ϵ. Thus $\{J_\epsilon u : 0 < \epsilon \leq 1\}$ is a bounded subset of $H_{(1,r-\delta)}$, so some subsequence $J_{\epsilon_n} u$ converges weakly in $H_{(1,r-\delta)}$, as $n \to \infty$. But $J_\epsilon u \to u$ in $H_{(1,r-1)}$, so we conclude that $u \in H_{(1,r-\delta)}$, as asserted.

LEMMA 2.3. *The hypothesis $u \in H_{(1,r-1)}$ in Lemma 2.1 can be replaced by*

$$u \in H_{(1,-\infty)} = \bigcap_{s>-\infty} H_{(1,s)}.$$

Proof. By induction.

PROPOSITION 2.4. *Let $u \in H_{(0,\sigma)}$, $Hu \in H_{(0,r)}$, and $Bu \in H^{r+1/2}(\partial M)$. Then $u \in H_{(1,r-\delta)}$, if the hypothesis (2.9) on B holds.*

Proof. Since $u \in H_{(0,\sigma)}$, $(\partial/\partial y)u = Hu + Ku \in H_{(0,\sigma')}$ with $\sigma' = \min(r, \sigma - 1)$. Thus $u \in H_{(1,\sigma')}$, and the result follows from Lemma 2.3.

The following is our regularity theorem.

THEOREM 2.5. *If* $u \in H_{(0,-\infty)}$, $Hu \in H_{(k,t)}$, *and* $Bu \in H^{1/2+t+k}(\partial M)$, *then, provided the hypothesis* (2.9) *on* B *holds,*

$$u \in H_{(k+1,t-\delta)}.$$

Proof. First note that $u \in H_{(0,-\infty)}$, $Hu \in H_{(k,t)}$, $Bu \in H^{1/2+t+k}(\partial M)$ implies $u \in H_{(1,t-\delta+k)}$, by Proposition 2.4. To deduce that $u \in H_{(k+1,t-\delta)}$, it will suffice to infer this from the hypothesis $u \in H_{(k,t-\delta+1)}$; an induction argument will finish off the proof. But $u \in H_{(k,t-\delta+1)}$ implies $(\partial/\partial y)u = Hu + Ku \in H_{(k,t-\delta)}$, which implies $u \in H_{(k+1,t-\delta)}$.

COROLLARY 2.6. *If* $u \in H_{(0,-\infty)}(I \times M)$, $Hu \in C^\infty(I \times \partial M)$, $Bu \in C^\infty(\partial M)$, *then* $u \in C^\infty(I \times \partial M)$, *provided* (2.9) *holds.*

It is desirable to generalize Theorem 2.5 to cases where $\delta \geq 1$. We shall return to this in Section 5.

Exercises

2.1. Translate Theorems 2.1 and 2.5 into estimates and regularity theorems for the elliptic boundary value problem $Lu = f$, $B_j u = g_j$, under appropriate hypotheses. The estimate should look like

$$\|u\|^2_{H^{m+k}(\Omega)} \leq C\|Lu\|^2_{H^k(\Omega)} + C\sum_{j=1}^{\nu} |B_j u|^2_{m+k-m_j-1/2} + C\|u\|^2_{H^0(\Omega)}.$$

2.2. Suppose $u \in H^{-k}(I \times \partial M)$ and $Hu = f \in C^\infty(I \times \partial M)$. Prove that $u \in H_{(0,\sigma)}$ for some σ. Hint: $(\partial/\partial y)u = f + Ku \in H^{-k}(I, H^{-k-1}(\partial M))$, $u'' = f' + K(f + Ku) + K'u$, etc.

§3. Closed Range and Fredholm Properties

We begin by proving a couple of functional analytic results which belong to the standard theory of Fredholm operators, but which are stated in a slightly different form than usual, a form most useful for our purposes.

PROPOSITION 3.1. *Let* $T: E \to X$ *be a closed linear operator between Banach spaces, and suppose* $K: E \to Y$ *is compact.* If

$$\|u\|_E \leq C\|Tu\|_X + C\|Ku\|_Y, \qquad u \in \mathcal{D}(T), \tag{3.1}$$

then T *has closed range.*

Proof. Since we can replace E by $\mathscr{D}(T)$ with its graph topology, we may assume $T: E \to X$ is continuous. Let $Tu_n \to f$ in X. We need $u \in E$ with $Tu = f$. Let $V = \ker T$. We divide up the argument into two cases.

(i) If dist $(u_n, V) \leq \alpha < \infty$, we can take $v_n = u_n \bmod V$, $\|v_n\| \leq 2\alpha$, and $Tv_n = Tu_n \to f$. Passing to a subsequence, we may assume $Kv_n \to g$ in Y. The inequality (3.1) yields

$$\|v_m - v_n\| \leq C\|Tv_n - Tv_m\| + C\|Kv_m - Kv_n\| \to 0,$$

so $v_n \to v$ and $Tv = f$.

(ii) If dist $(u_n, V) \to \infty$, assume dist $(u_n, V) > 2$ for all n. Let $v_n = u_n \bmod V$ be such that dist $(u_n, V) \leq \|v_n\| \leq$ dist $(u_n, V) + 1$. Of course, $Tv_n = Tu_n$. If

$$w_n = \frac{v_n}{\|v_n\|},$$

then

$$\text{dist}\,(w_n, V) \geq \frac{\text{dist}\,(v_n, V)}{\text{dist}\,(v_n, V) + 1} \geq \frac{1}{2}.$$

Also $\|w_n\| = 1$, so we can assume $Kw_n \to g$ in Y. $Tw_n \to 0$. Thus inequality (3.1) implies

$$\|w_n - w_m\| \leq C\|Tw_n - Tw_m\| + C\|Kw_m - Kw_n\| \to 0.$$

Thus $w_n \to w$ in E, and we see simultaneously that dist $(w, V) \geq 1/2$ and that $Tw = 0$, a contradiction. Hence case (ii) is impossible and the proposition is proved.

PROPOSITION 3.2. *If $T: E \to X$ is a closed linear operator between Banach spaces, which has closed range, of finite codimension, and if $K: E \to X$ is compact, then $T + K$ has closed range, of finite codimension.*

Proof. Without loss of generality, T is continuous. Also, it suffices to prove the proposition assuming T is onto, for if V is a finite dimensional complementary subspace to the range $R(T)$, $T \oplus j: E \oplus V \to X$ is onto, and we need only apply such a special case of our proposition to $(T+K) \oplus 0 = (T \oplus j) + (K \oplus (-j))$.

But if T is onto, T^* is injective and has closed range. Hence for $w \in X'$, the dual of X, $\|w\|_{X'} \leq C\|T^*w\|_{E'}$. Thus we have the inequality

$$\|w\|_{X'} \leq C\|(T^* + K^*)w\|_{E'} + C\|K^*w\|_{E'}.$$

From Proposition 3.1 it follows that $(T+K)^*$ has closed range; clearly its kernel is finite dimensional. This implies that $T+K$ has closed range of finite codimension, as desired.

The next assertion follows easily from Proposition 3.2 and will be very convenient.

PROPOSITION 3.3. *Let $T:E \to X$ be a closed linear operator, $j:V \to X$ a compact injection. Suppose that, for each $x \in \tilde{X}$, a closed linear subspace of X of finite codimension, there is a $u \in \mathscr{D}(T)$ such that $Tu - x \in j(V)$. Then T has closed range, of finite codimension.*

Proof. The hypothesis implies that $T \oplus j: E \oplus V \to X$ has closed range, of finite codimension. Now apply Proposition 3.2, with T replaced by $T \oplus j$ and $K = 0 \oplus (-j)$.

The following corollary gives the flavor of Proposition 3.3., while not being quite general enough for all our needs. The point of such a proposition is that the problem of solvability modulo such a "smooth" error is often localizable. The proposition enables one to pass from a local analysis to a global result.

COROLLARY 3.4. *Let $T:H^m(M) \to H^\mu(M)$ be a closed linear map. Suppose that for each $f \in H^\mu(M)$, there is a $u \in H^m(M)$ such that*

$$Tu = f \bmod C^\infty(\bar{M}).$$

Then T has closed range, of finite codimension.

Returning to differential equations, we show that the system

$$\begin{aligned} Hu &= f, \\ Ru|_{\partial M} &= 0 \end{aligned} \tag{3.2}$$

can be solved, if f satisfies a finite number of linear conditions. Let

$$E = \{u \in H_{(1,t)}: Ru|_{\partial M} = 0\}$$

and define

$$H_0: E \to H_{(0,1)}(I \times \partial M)$$

by

$$H_0 u = Hu.$$

The a priori estimates of Section 2 show that ker H_0 is finite dimensional. (Note that $Ru|_{\partial M} = 0$ implies $(1 - E(0))u(0) = 0$ and $E(1)u(1) = 0$). It will be convenient to assume that

$$E(t)^2 = E(t), \qquad t = 0, \text{ or } 1 \tag{3.3}$$

i.e., each $E(t)$ is actually a projection. Indeed, it is easy to see that if γ is a small curve in $\mathbf{C}$ encircling 1, which avoids the spectrum of $E(t)$, which is discrete in $\mathbf{C} - \{0, 1\}$, then

$$\frac{1}{2\pi i} \int_\gamma (\zeta - E(t))^{-1} \, d\zeta$$

is a projection which differs from $E(t)$ by an element of OPS^{-1}.

THEOREM 3.5. *The map H_0 is Fredholm.*

Proof. We need only show that the range $R(H_0)$ is closed and of finite codimension. By the estimate (2.7) we have

$$\|u\|_E^2 \leq C\|H_0 u\|_{(0,t)}^2 + C\|u\|_{(0,\sigma)}^2$$

so Proposition 3.1 implies that $R(H_0)$ is closed. Now suppose $w \in H_{(0,-t)}$, the dual of $H_{(0,t)}$, and suppose $w \perp R(H_0)$, i.e.,

$$\langle H_0 u, w\rangle = 0 \qquad \text{for all} \qquad u \in E. \tag{3.4}$$

Then (3.4) holds in particular for all $u \in C_0^\infty((0, 1) \times \partial M)$, so $H^*w = 0$ where

$$H^* = -\frac{\partial}{\partial y} - K^*.$$

Thus $(\partial/\partial y)w = H^*w + K^*w = K^*w \in H_{(0,-t-1)}$, so $w \in H_{(1,-t-1)}$. Hence $w|_{\partial M}$ is well defined and belongs to $H^{-t-1/2}(\partial M)$, and we have

$$\begin{aligned} 0 = \langle H_0 u, w\rangle &= \langle Hu, w\rangle - \langle u, H^*w\rangle \\ &= (u(1), w(1)) - (u(0), w(0)) \end{aligned}$$

for all $u \in E$, if w satisfies (3.4).

Now for any $v \in H^{t+1/2}(\partial M)$, there is a $u \in H_{(1,t)}$ such that $u|_{\partial M} = (1-R)v$. In particular $Ru|_{\partial M} = 0$, so $u \in E$. Thus

$$((1-R(0))v_0, w(0)) = ((1-R(1))v_1, w(1))$$

for all $v_0, v_1 \in H^{t+1/2}(\partial M)$, if w satisfies (3.4). Taking $v_1 = 0$ yields $(1-R(1)^*)w(1) = 0$ and taking $v_1 = 0$ yields $(1-R(0)^*)w(0) = 0$. Thus all $w \perp R(H_0)$ must satisfy the boundary value problem

$$H^*w = 0, \qquad (1-R^*)w|_{\partial M} = 0. \tag{3.5}$$

But $1-R^*$ plays the same role for the elliptic operator H^* as R plays for H, so the set of $w \in H_{(1,-t-1)}$ satisfying (3.5) is finite dimensional, and hence $R(H_0)$ has finite codimension. This proves the theorem.

Next we consider the inhomogeneous boundary value problem

$$Hu = f, \qquad Ru|_{\partial M} = g. \tag{3.6}$$

Given $f \in H_{(0,t)}(I \times \partial M)$, $g \in H^{t+1/2}(\partial M)$, we would like to find a solution $u \in H_{(1,t)}$. Since $R(j)^2 = R(j)$ for $j = 0, 1$, we must require that $(1 - R)g = 0$. Thus let

$$H_{1-R}^{t+1/2}(\partial M) = \{g \in H^{t+1/2}(\partial M) : (1-R)g = 0\}. \tag{3.7}$$

Define

$$H_1 : H_{(1,t)} \to H_{(0,t)} \oplus H_{1-R}^{t+1/2}(\partial M) \tag{3.8}$$

by

$$H_1 u = \{Hu, Ru|_{\partial M}\}. \tag{3.9}$$

THEOREM 3.6. *The map H_1 is Fredholm.*

Proof. Take $F = Ag \in H_{(1,t)}$, $F|_{\partial M} = g$. Consider the equation

$$Hv = f - HF, \qquad Rv|_{\partial M} = 0. \tag{3.10}$$

By Theorem 3.5, there is a solution v to (3.10) provided $f - HAg$ satisfies some finite number of linear conditions. In such a case, set $u = v + F$. The Fredholm property of H_1 is now immediate.

Next we consider the boundary value problem

$$Hu = 0, \qquad Bu|_{\partial M} = g \tag{3.11}$$

where $B \in OPS^0(\partial M)$, $B: H^{t+1/2}(\partial M) \to H^{t+1/2}(\partial M)$. Here the domain and range spaces, even though given the same name, can be considered to be Sobolev spaces of sections of different vector bundles. For (3.11) to be solvable, $u|_{\partial M}$ must satisfy a certain condition, as required by the following.

PROPOSITION 3.7. *If $u \in H_{(1,t)}$ and $Hu = 0$, then $(1-R)u|_{\partial M} \in H^{t+3/2}(\partial M)$.*

Proof. Since the hypothesis implies $u \in H_{(0,t+1)}$, this is a simple consequence of the estimate (2.8), using Friedrichs' molifiers.

We use Proposition 3.7 to motivate the hypothesis we will make on the boundary operator B in order to guarantee that (3.11) is Fredholm. Crudely, suppose that we require $(1-R)u = 0$ on ∂M. Then (3.11) could be solved if we could find h such that $BRh = g$, since then we need only solve $Hu = 0$, $Ru|_{\partial M} = h$, which can be done if h (and hence g) satisfies a certain finite number of linear conditions. We take this as motivation for the following. Define

$$BR: H^{\tau-\delta}(\partial M) \to H^{\tau}(\partial M)$$

to be the closed linear operator with domain $\mathcal{D}(BR) = \{u \in H^{\tau-\delta}: BRu \in H^{\tau}\}$. We make the following hypothesis on the boundary operator B:

(3.12) $BR: H^{\tau-\delta}(\partial M) \to H^{\tau}(\partial M)$ has closed range, of finite codimension.

THEOREM 3.8. *Suppose the pair $\{H, B\}$ satisfies the hypothesis* (3.12). *Then for any $g \in H^{t+1/2}(\partial M)$ satisfying a certain finite number of linear conditions, there is a solution $u \in H_{(1,t-\delta)}$ of* (3.11).

Proof. If $E = \{u \in H_{(1,t-\delta)}: Hu = 0\}$, we are to show that the closed linear map

$$\beta: E \to H^{t+1/2}(\partial M)$$

given by $\beta u = Bu|_{\partial M}$ has closed range of finite codimension. Let the range of BR be F, a closed finite codimensional subspace of $H^{t+1/2}(\partial M)$. There exists a continuous map $T: F \to H^{t-\delta}$ such that $BRTg = g$, $g \in F$. Now if $g \in F$ and if we can solve the equation

$$Hv = 0, \qquad Rv|_{\partial M} = RTg \in H^{t-\delta+1/2}(\partial M) \tag{3.13}$$

for $v \in H_{(1,t-\delta)}$, which we can if RTg satisfies a certain finite number of linear conditions, by Theorem 3.6, then $v \in E$ and

$$\begin{aligned} Bv|_{\partial M} &= BRv + B(1-R)v \\ &= BRTg + B(1-R)v \\ &= g + B(1-R)v. \end{aligned}$$

Thus $\beta v - g \in H^{3/2+t-\delta}$ (by Proposition 3.7), so by proposition 3.3, β has closed range of finite codimension, as desired.

By the same argument leading to the proof of Theorem 3.6, Theorem 3.8 extends immediately to the following.

THEOREM 3.9. *Suppose B satisfies condition* (3.12). *Then the system*

$$Hu = f, \qquad Bu = g$$

can be solved for $u \in H_{(1,t-\delta)}$, given $f \in H_{(0,t)}$, $g \in H^{t+1/2}(\partial M)$, provided $\{f, g\}$ satisfies a certain finite number of linear conditions.

Note that Theorems 3.8 and 3.9 do not require $\delta < 1$. If both condition (2.9) and condition (3.12) are satisfied, for some $\delta \in [0, 1)$, then $u \mapsto \{Hu, Bu\}$ leads to a closed, densely defined Fredholm map of $H_{(1,t-\delta)}$ into $H_{(0,t)} \oplus H^{t+1/2}(\partial M)$. Actually, this coincidence only seems to occur for $\delta = 0$ (in cases where $\delta < 1$), the most important case, which we discuss in the next section.

Finally, we translate Theorem 3.9 into an existence theorem, modulo finitely many linear conditions, for boundary value problems for the elliptic boundary problem (0.1) on a general compact manifold M with boundary.

THEOREM 3.10. *Let L be an elliptic operator of order m and consider the boundary value problem*

$$Lu = f \text{ on } M, \qquad B_j u|_{\partial M} = g_j, \qquad 1 \le j \le \nu. \tag{3.14}$$

Suppose the operators H, R and B are determined by the process of Section 1, *and suppose Hypothesis* (3.12) *holds. Let $f \in H^k(M)$, $g_j \in H^{m+k-m_j-1/2}(\partial M)$. Then, provided $\{f, g_j\}$ satisfy a certain finite number of linear conditions, there exists $u \in H^{m+k-\delta}(M)$ satisfying* (3.14).

Proof. Write $f = f_1 + f_2$ where f_1 is supported in the interior of M and f_2 is supported in a small neighborhood of ∂M, using a partition of unity $\varphi_1 + \varphi_2 = 1$; $f_j = \varphi_j f$. The ellipticity of L alone shows that there is a properly supported parametrix, so one can find $u_1 \in H^{m+k}(M)$, supported away from ∂M, such that

$$Lu_1 = f_1 \bmod C_0^\infty(\text{int } M). \tag{3.15}$$

Now we claim that, given finitely many linear conditions on $f_2 = \varphi_2 f$ and on g_j, there is a $v \in H^{m+k-\delta}(M)$ such that, on a neighborhood of ∂M reparametrized as $I \times \partial M$,

$$\begin{aligned} Lv &= f_2 \text{ on } (0, 1) \times \partial M, \\ B_j v(0) &= g_j. \end{aligned} \tag{3.16}$$

Indeed, an auxiliary condition can be placed on $v(1)$, gotten by reducing (3.14) to a first order system for $\tilde{v}$ and imposing the condition $R(1)\tilde{v}(1) = 0$. The existence of such v then follows from Theorem 3.9. The regularity result, Corollary 2.6, implies that such v is C^∞ near $\{y = 1\}$, so if $\psi \in C^\infty(M)$ is equal to 1 near ∂M and vanishes near $\{y = 1\}$, set $u_2 = \psi v$. Then let $u = u_1 + u_2$. It follows that $u \in H^{m+k-\delta}(M)$ and

$$Lu = f \bmod C^\infty(\bar{\Omega}), \qquad B_j u = g_j. \tag{3.17}$$

Theorem 3.10 is now a consequence of Proposition 3.3.

§4. Regular Boundary Value Problems

Let L be an elliptic operator on M. We may suppose that L maps sections of some vector bundle E_0 over M to sections of a vector bundle F, of the same fiber dimension as E_0. The process described in Section 1 leads to a pseudodifferential operator R and the boundary operators B_j lead to an operator $B \in OPS^0(\partial M)$ mapping sections of a vector bundle E to sections of a vector bundle G. Consider the following two conditions:

$$|g|_\tau^2 \le C|(1-R)g|_\tau^2 + C|Bg|_\tau^2 + C|g|_{\tau-1}^2, \qquad g \in C^\infty(\partial M); \tag{4.1}$$

(4.2) $BR: H^\tau(\partial M) \to H^\tau(\partial M)$ has closed range, of finite codimension.

These conditions are Hypotheses (2.9) and (3.12), respectively, with $\delta = 0$. As we have seen, (4.1) leads to a regularity theorem and (4.2) leads to an existence theorem, modulo finitely many linear conditions. Actually, the boundary operators B_j only lead to B defined on $u(0)$. At $y = 1$ it is convenient to impose the boundary condition

$$R(1)u(1) = 0. \tag{4.3}$$

DEFINITION 4.1. *If* (4.1) *and* (4.2) *are both satisfied, the boundary value problem*

$$Lu = f, \qquad B_j u = g_j, \qquad j = 1, \ldots, \nu$$

is called a regular elliptic boundary value problem or a coercive boundary value problem.

The purpose of this section is to give more explicit conditions for regularity. Let b be the principal symbol of B, r that of R, homogeneous of degree zero in ξ.

PROPOSITION 4.1. *Consider the following conditions.*

(4.4) *For each $(x_0, \xi_0) \in T^*(\partial M)\backslash 0$, there is no $v \in E_{x_0}$ such that*

$$v - r(x_0, \xi_0)v = 0,$$
$$b(x_0, \xi_0)v = 0$$

(4.5) $$b(x_0, \xi_0)r(x_0, \xi_0): E_{x_0} \to G_{x_0} \text{ is surjective, for each}$$
$$(x_0, \xi_0) \in T^*(\partial M)\backslash 0.$$

Then (4.4) *implies* (4.1) *and* (4.5) *implies* (4.2).

Proof. Condition (4.4) says that $(1-R)^*(1-R) + B^*B$ is strongly elliptic, from which (4.1) follows easily. Condition (4.5) says that $BR(BR)^*$ is elliptic, as an operator on $H^\tau(\partial M, G)$, and hence Fredholm, which implies (4.2).

We want to make these conditions even more explicit by relating them directly to the symbols of L and B_j. The next two propositions will be proved simultaneously.

PROPOSITION 4.2. *For given $(x_0, \xi_0) \in T^*(\partial M)\backslash 0$, the following three conditions are equivalent.*

(4.6) *There is no $v \in E_{x_0}$ such that $v - r(x_0, \xi_0)v = 0$ and*

$$b(x_0, \xi_0)v = 0.$$

(4.7) *There is no nonzero bounded solution on $[0, \infty)$ of the ODE*

$$\frac{d}{dy}\varphi - K_1(0, x_0, \xi_0)\varphi = 0,$$
$$b(x_0, \xi_0)\varphi(0) = 0.$$

(4.8) *There is no nonzero bounded solution on $[0, \infty)$ of the ODE*

$$\frac{d^m}{dy^m}\Phi + \sum_{j=0}^{m-1} \tilde{A}_j(0, x_0, \xi_0)\frac{d^j}{dy^j}\Phi = 0,$$
$$\tilde{B}_j\left(x_0, \xi_0, \frac{d}{dy}\right)\Phi(0) = 0.$$

Here $\tilde{A}_j(0, x, \xi)$ *is the principal symbol of* $\tilde{A}_j(0, x, D_x)$, $\tilde{B}_j(x, \xi_0, \eta)$ *that of* B_j, *with* d/dy *substituted for* η *in this polynomial.*

PROPOSITION 4.3. *For any* $(x_0, \xi_0) \in T^*(\partial M)\backslash 0$, *the following three conditions are equivalent.*

$$b(x_0, \xi_0)r(x_0, \xi_0): E_{x_0} \to G_{x_0} \text{ is onto.} \tag{4.9}$$

(4.10) *There exists a bounded solution on* $[0, \infty)$ *to the ODE*

$$\frac{d}{dy}\varphi - K_1(0, x_0, \xi_0)\varphi = 0,$$

$$b(x_0, \xi_0)\varphi = \eta$$

for any given $\eta \in G_{x_0}$.

(4.11) *There exists a bounded solution on* $[0, \infty)$ *to the ODE*

$$\frac{d^m}{dy^m}\Phi + \sum_{j=0}^{m-1} \tilde{A}_j(0, x_0, \xi_0)\frac{d^j}{dy^j}\Phi = 0,$$

$$\tilde{B}_j\left(x_0, \xi_0, \frac{d}{dy}\right)\Phi(0) = \eta_j$$

for any $\eta_j \in E_{j,x_0}$, *where* B_j *takes sections of* E *to sections of* E_j, *over* ∂M.

Proof. In each case, the first condition is equivalent to the second by virtue of the exponential representation of all the solutions to

$$\frac{d}{dy}\varphi - K_1(0, x_0, \xi_0)\varphi = 0 \tag{4.12}$$

if we recall that $r(x_0, \xi_0)$ is defined to be the projection onto the sum of the generalized eigenspaces of $K_1(0, x_0, \xi_0)$ corresponding to eigenvalues with negative real part, annihilating the other eigenspaces. Also, the second and third conditions in each case are equivalent because of the equivalence between the mth order *ODE*

$$\frac{d^m}{dy^m}\Phi + \sum_{j=0}^{m-1} \tilde{A}_j(0, x_0, \xi_0)\frac{d^j}{dy^j}\Phi = 0 \tag{4.13}$$

and the first order system to which it is reduced, which happens to be (4.12).

If the operators $\tilde{A}_j(y, x, D_x)$ all have scalar principal part, we shall give another, more algebraic, characterization of regularity (which can also be generalized to systems). If $L_m(0, x_0, \eta, \xi)$ is the principal symbol of L, also assumed to be scalar, then none of the roots $\tau_1, \ldots, \tau_m$ of $L_m(0,$

$x_0, \tau, \xi) = 0$ are real, if $\xi \neq 0$. Let

$$M^+(x_0, \xi_0, \tau) = \prod_{k=1}^{\ell} (\tau - \tau_k(x_0, \xi_0)) \tag{4.14}$$

where we assume $\tau_1(x_0, \xi_0), \ldots, \tau_\ell(x_0, \xi_0)$ are all the roots with positive imaginary part. Let $\tilde{B}_j(y, x, \eta, \xi)$ be the principal symbol of B_j. If $\mathbf{C}[\tau]$ denotes the ring of polynomials in τ, with complex coefficients and if $(M^+(x_0, \xi_0, \tau))$ denotes the ideal generated by $M^+(x_0, \xi_0, \tau)$, then $\mathbf{C}[\tau]/(M^+(x_0, \xi_0, \tau))$ is a finite dimensional vector space over $\mathbf{C}$, for each $(x_0, \xi_0) \in T^*(\partial M)\backslash 0$.

PROPOSITION 4.4. *Assume the principal symbols of L and B_j are all scalar. All the conditions of Proposition* 4.2 *are equivalent to*

$$\{\tilde{B}_j(x_0, \xi_0, \tau) : 1 \leq j \leq \nu\} \text{ spans } \mathbf{C}[\tau]/(M^+(x_0, \xi_0, \tau)), \tag{4.15}$$

and all the conditions of Proposition 4.3 *are equivalent to*

$$\{\tilde{B}_j(x_0, \xi_0, \tau) : 1 \leq j \leq \nu\} \textit{ is a linearly independent set in } \mathbf{C}[\tau]/(M^+(x_0, \xi_0, \tau)). \tag{4.16}$$

Proof. The two equivalences follow from the observation that a solution on $[0, \infty)$ to (4.13) is bounded if, and only if, $M^+(x_0, \xi_0, d/dy)\Phi = 0$ on $[0, \infty)$.

In practice, it is usually as easy to use properties (4.8) and (4.11) to verify regularity as to check properties (4.15) and (4.16). There is, however, one interesting result which Proposition 4.4 yields, which we now discuss.

First of all, suppose L and B_j are scalar differential operators leading to a regular boundary value problem. Since $\tilde{B}_j(x_0, -\xi_0, -\tau) = \pm\tilde{B}_j(x_0, \xi_0, \tau)$, it follows that $\{\tilde{B}_j(x_0, \xi_0, \tau) : 1 \leq j \leq \nu\}$ is equivalently a basis of $\mathbf{C}[\tau]/(M^+(x_0, \xi_0, \tau))$ or a basis of $\mathbf{C}[\tau]/(M^+(x_0, -\xi_0, -\tau))$. In particular, for such to happen $M^+(x_0, \xi_0, \tau)$ and $M^+(x_0, -\xi_0, \tau)$ must have the same degree in τ, for each $\xi_0 \neq 0$. More specifically, we have the following concept.

DEFINITION 4.5. *A scalar L is properly elliptic if, and only if, the degree of $M^+(x_0, \xi_0, \tau)$ is independent of $(x_0, \xi_0) \in T^*(\partial M)\backslash 0$.*

Since the roots with positive imaginary part of $L_m(0, x_0, \xi, \tau) = 0$ correspond to the roots with negative imaginary part of $L_m(0, x_0, -\xi, \tau) = 0$, it follows that a properly elliptic operator L has even order $m = 2\mu$, and $M^+(x_0, \xi_0, \tau)$ has order μ, for each $(x_0, \xi_0) \in T^*(\partial M)\backslash 0$. If $\dim M \geq 3$, the assumption that a scalar operator L is properly elliptic is no restriction (see Exercise 4.1).

PROPOSITION 4.6. *If L is a properly elliptic operator of order* 2μ, *the Dirichlet problem*

$$(4.17)\quad Lu = f, \qquad u|_{\partial M} = g_0, \qquad \left.\frac{\partial u}{\partial \nu}\right|_{\partial M} = g_1, \ldots, \left(\frac{\partial}{\partial \nu}\right)^{\mu-1} u|_{\partial M} = g_{\mu-1}$$

is a regular elliptic boundary value problem. Here ν *can be any real vector field which is transversal to* ∂M.

Proof. In appropriate coordinates, $\nu = \partial/\partial y$. The proof amounts to observing that $\{1, \tau, \ldots, \tau^{\mu-1}\}$ must be a basis of $\mathbf{C}[\tau]/(M^+(x_0, \xi_0, \tau))$ under these circumstances.

Another simple consequence of Proposition 4.4 is the following.

PROPOSITION 4.7. *Let L be a second order properly elliptic operator on M, with scalar principal symbol. With* ν *any transversal vector field to* ∂M, *and* $\beta \in C^\infty(\partial M)$, *the boundary value problem*

$$(4.18)\qquad Lu = f, \qquad \left(\frac{\partial}{\partial \nu} + \beta\right) u|_{\partial M} = g$$

is a regular elliptic problem.

Proof. Picking appropriate coordinates, $\nu = \partial/\partial y$, as above, while $M^+(x, \xi, \tau)$ must be a first order polynomial in τ, of the form $\tau - \tau_1(x, \xi)$, with $Im\, \tau_1(x, \xi) > 0$ on $T^*(\partial M)\backslash 0$. It follows that $\{\tau\}$ is a basis of the one dimensional vector space $\mathbf{C}[\tau]/(\tau - \tau_1(x, \xi))$, which proves the theorem.

When $L = \Delta$, the Laplace operator on M, with respect to some Riemannian metric, $\partial/\partial\nu$ is the normal vector and $\beta = 0$, (4.18) is called the Neumann problem. For nonzero β but $L = \Delta$, $\partial/\partial\nu$ normal, it is called the Robin problem. In general it is the *oblique derivative problem.*

Exercises

4.1. Show that any elliptic operator L with complex (scalar) coefficients on a manifold M of dimension $n \geq 3$ is properly elliptic. Hint: The number of roots τ with positive imaginary part of $L_m(y, x, \xi, \tau)$ is locally constant, for $(y, x) \in M$, $\xi \in \mathbf{R}^{n-1}\backslash 0$. But this set is connected, if $n \geq 3$.

4.2. If dim $M = 2$ and there is a boundary value problem for L satisfying the conditions of Proposition 4.4, show that L must be properly elliptic.

4.3. Consider $L = (\partial/\partial z)^2 = ((\partial/\partial x) + i(\partial/\partial y))^2$, a second order elliptic operator on $\mathbf{R}^2$. Show that L is not properly elliptic. Verify that the Dirichlet problem on the disc for L is not regular by constructing an infinite dimensional space of solutions to $Lu = 0$, $u|_{S^1} = 0$.

4.4. Let E and B be functions on $\Omega \subset \mathbf{R}^3$ taking values in $\mathbf{C}^3$, ν be the unit normal to $\partial\Omega$. Show that the following two elliptic boundary value problems are regular.

$$\text{(i)} \begin{cases} \Delta E = F \\ E \times \nu = 0 \text{ on } \partial\Omega \text{ and } \operatorname{div} E = 0 \text{ on } \partial\Omega \end{cases}$$

$$\text{(ii)} \begin{cases} \Delta B = F \\ B \cdot \nu = 0 \text{ on } \partial\Omega \text{ and } \nu \times \operatorname{curl} B = 0 \text{ on } \partial\Omega. \end{cases}$$

The reader who has seen some physics might note a connection between these boundary value problems and Maxwell's equations for electromagnetic waves, in a region bounded by a perfect conductor.

4.5. Let (L, B_j) and (M, C_j) be regular elliptic boundary value problems. Let $\mathscr{D}(L) = \{u \in H^m(M): B_j u = 0 \text{ on } \partial M\}$ and let $\mathscr{D}(M)$ be defined similarly, so we have two closed densely defined linear operators on $L^2(M)$. Suppose L and M are formally adjoint in the sense that

$$(Lu, v) = (u, Mv)$$

for all smooth u with $B_j u = 0$ on ∂M and smooth v with $C_j v = 0$ on ∂M. Prove that $L^* = M$. In particular, a formally self-adjoint boundary problem leads to a self adjoint operator on $L^2(M)$. (See Lions and Magenes [1].)

4.6. If (L, B_j) defines a regular elliptic boundary value problem, let $x_0 \in \partial M$ and suppose $u \in L^2(M)$ is such that Lu and $B_j u|_{\partial M}$ are smooth in a neighborhood of x_0. Prove that u is smooth in a neighborhood of x_0. Hint: If φ is a smooth function supported in a small neighborhood of x_0, note that $L(\varphi u) = \varphi Lu + [L, \varphi]u$ and $[L, \varphi]$ has lower order. Thus a priori estimates lead to $u \in H^1$ in a neighborhood of x_0. Repeat this reasoning. (See Lions and Magenes [1].)

§5. Reduction of a Boundary Value Problem to a Regular One

Consider our first order elliptic system on $I \times \partial M = \bar{\Omega}$;

$$\frac{\partial}{\partial y} u = Ku. \tag{5.1}$$

Suppose we know that the boundary value problem

$$Bu = f \quad \text{on} \quad \partial\Omega \tag{5.2}$$

is regular. We want to use information on this as a tool to analyze another boundary value problem: $Cu = g$ on $\partial\Omega$. First, we will analyze $u|_{\partial\Omega}$ as a function of f.

Indeed, the results of Sections 1 through 4 show that, for $f \in H^{s+1/2}(\partial\Omega)$, satisfying conditions $(f, g_i) = 0, j = 1, \ldots, \ell$, for certain $g_j \in C^\infty(\partial\Omega)$, (5.1), (5.2) has a solution u, and the difference of any two solutions belongs to a finite dimensional subspace of $C^\infty(\bar{\Omega})$. Consequently, there exists an operator

$$T: H^{s+1/2}(\partial\Omega) \to H_{(1,s)}(\Omega) \tag{5.3}$$

such that if $u = Tf$, then u satisfies (5.1), (5.2), mod C^∞. Let

$$Af = Tu|_{\partial\Omega}. \tag{5.4}$$

A is consequently well defined modulo an element of $OPS^{-\infty}(\partial\Omega)$, and clearly $A: H^{s+1/2}(\partial M) \to H^{s+1/2}(\partial M)$, the domain and range spaces generally being sections of different vector bundles. Our main assertion regarding A is:

THEOREM 5.1. $A \in OPS^0(\partial M)$.

In order to prove this theorem, we shall introduce some machinery that will play a further role in Chapter IX, a sort of "total decoupling" of (5.1) into forward and backward evolution equations, done in a context that, after a couple of more chapters, the reader will recognize as "microlocal."

We claim there are operators $U_j(y, x, D_x)$, V_j, and $\tilde{K}_j$ such that, if $\psi_j(x, \xi)$ has order $-\infty$ outside some sufficiently small "conic set" Γ_j, i.e., a subset of $T^*(\partial M)\backslash 0$ such that $(x, r\xi) \in \Gamma_j$, $r > 0$ whenever $(x, \xi) \in \Gamma_j$, then

$$\left(\frac{\partial}{\partial y} - \tilde{K}_j\right)\psi_j = U_j\left(\frac{\partial}{\partial y} - K\right)V_j\psi_j \text{ mod } OPS^{-\infty} \tag{5.5}$$

and $\tilde{K}_j$ has the totally decoupled form

$$K_j = \begin{pmatrix} E_j & \\ & F_j \end{pmatrix} \tag{5.6}$$

where E_j, $F_j \in OPS^1(\partial M)$, the principal symbol of E_j has negative real part, and the principal symbol of F_j has positive real part. In fact, the principal symbol of K is similar to a matrix of the form (5.6), so on any sufficiently small conic set, principal symbols U_j, $V_j = U_j^{-1}$ can be constructed so that, on the principal symbol level

$$\tilde{K}_j = U_j K U_j^{-1} \quad \text{on} \quad \Gamma_j.$$

Complete asymptotic expansions of U_j, V_j, $\tilde{K}_j$ need to be developed so that (5.5), (5.6) hold, and to do this, we proceed as follows.

So far, we have U_{j0}, V_{j0}, $\tilde{K}_{j0}$ such that,

$$\left(\frac{\partial}{\partial y} - \tilde{K}_{j0}\right)\psi_j = U_{j0}\left(\frac{\partial}{\partial y} - K\right)V_{j0}\psi_j$$

and

$$\tilde{K}_{j0} = \begin{pmatrix} E_j & \\ & F_j \end{pmatrix} + \begin{pmatrix} A_{11} & A_{12} \\ A_{21} & A_{22} \end{pmatrix}$$

with $A_{ij} \in OPS^0$. We now try to find $U_{j1} = (1+R_1)U_{j0}$, $V_{j1} = V_{j0}(1+R)^{-1}$ such that a better remainder term occurs. Indeed, we see that

$$U_{j1}\left(\frac{\partial}{\partial y} - K\right)V_{j1}\psi_j = \left(\frac{\partial}{\partial y} - \tilde{K}_{j1}\right)\psi_j$$

with

$$\tilde{K}_{j1} = \begin{pmatrix} E_j & \\ & F_j \end{pmatrix} + (A + \tilde{K}_{j0}R_1 - R_1\tilde{K}_{j0}) + B,$$

$B \in OPS^0$. We want to pick $R_1 \in OPS^{-1}$ so that the off-diagonal terms in $A + \tilde{K}_{j0}R_1 - R_1\tilde{K}_{j0}$ vanish. We want to choose

$$R_1 = \begin{pmatrix} 0 & R_{12} \\ R_{21} & 0 \end{pmatrix}$$

and require that, on the principal symbol level,

$$\begin{aligned} R_{12}F_j - E_jR_{12} &= A_{12}, \\ F_jR_{21} - R_{21}E_j &= -A_{21}. \end{aligned} \tag{5.7}$$

Since the spectra of the matrices E_j and F_j are disjoint, it is a simple linear algebra exercise to obtain *unique* solutions R_{12} and R_{21}. This specifies R_1 and leads to

$$\tilde{K}_{j1} = \begin{pmatrix} E_j & \\ & F_j \end{pmatrix} + \begin{pmatrix} A'_{11} & 0 \\ 0 & A'_{22} \end{pmatrix} + \begin{pmatrix} B_{11} & B_{12} \\ B_{21} & B_{22} \end{pmatrix}$$

with $B_{ij} \in OPS^{-1}$.

One can continue in this fashion, setting $U_{j2} = (1+R_2)U_{j1}$, $V_{j2} = V_{j1}(1+R_2)^{-1}$, with $R_2 \in OPS^{-2}$, etc. This leads to more equations analogous to (5.7). Finally, with

$$\begin{aligned} U_j &\sim \cdots (1 + R_2)(1 + R_1)U_{j0}, \\ V_j &\sim V_{j0}(1 + R_1)^{-1}(1 + R_2)^{-1} \cdots \end{aligned}$$

it is easily verified that (5.5), (5.6) holds.

We now take up the proof of Theorem 5.1. Introduce cut-off's $\varphi_j(x, D) \in OPS^0$, with each $\varphi_j(x, \xi)$ having order $-\infty$ outside a small conic set Γ_j, on which the above construction works, such that

$$I = \sum_j \varphi_j(x, D).$$

For the $f \in \mathscr{D}'(\partial M)$ of (5.2), write $f_j = \varphi_j(x, D)f$, so

$$f = \sum_j f_j.$$

Let $u_j = Tf_j$, so $u = Tf = \sum u_j$. We may suppose (5.5) holds with $\psi_j(x, \xi) = 1$ on a conic neighborhood of the support of $\varphi_j(x, \xi)$. Define w_j to solve the boundary value problem:

$$\left(\frac{\partial}{\partial y} - \tilde{K}_j\right) w_j = 0, \tag{5.8}$$
$$BV_j w_j\big|_{y=0} = f_j.$$

Due to the nature of $\tilde{K}_j$, near $y = 0$ we must have

$$w_j = \begin{pmatrix} \tilde{w}_j \\ 0 \end{pmatrix} \bmod C^\infty.$$

Thus $BV_j w_j(0) = C_j \tilde{w}_j$, $C_j \in OPS^0(\partial M)$. Now the hypothesis that the boundary problem (5.2) is a *regular* boundary value problem implies that C_j is an *elliptic* operator, so it has a parametrix $\tilde{A}_j$, and

$$\tilde{w}_j = \tilde{A}_j f_j, \bmod C^\infty.$$

Thus, for some $A_j \in OPS^0(\partial M)$,

$$w_j(0) = A_j f_j \bmod C^\infty.$$

Let $v_j = V_j w_j$. It is not hard to show that

$$\psi_j(x, D) w_j = w_j \bmod C^\infty$$

(we leave it as an exercise), and this implies that

$$u_j = v_j \bmod C^\infty.$$

Consequently, mod $C^\infty(\partial M)$, we have

$$u(0) = \sum v_j(0) = \sum V_j A_j f_j = \left(\sum_j V_j A_j \varphi_j\right) f = Af$$

which proves Theorem 5.1, at $y = 0$; the proof at $y = 1$ is similar. The method of proof of Theorem 5.1 enables one to write out a complete asymptotic expansion of the symbol of A, but we shall not carry out the details.

Having proved Theorem 5.1, we return to an analysis of the new boundary value problem, with $C \in OPS^0(\partial M)$,

$$\frac{\partial}{\partial y} u = Ku,$$

(5.9)
$$Cu = g \quad \text{on} \quad \partial\Omega.$$

Setting $f = Bu$, we see that solving (5.9) mod C^∞ is equivalent to solving

(5.10)
$$CAf = g \bmod C^\infty(\partial M).$$

The conclusion is immediate:

THEOREM 5.2. *Suppose CA is hypoelliptic. Then any solution u to (5.9) with $g \in C^\infty(\partial M)$ must belong to $C^\infty(\bar{\Omega})$. In particular, if CA is hypoelliptic with loss of δ derivatives, i.e., $CAv \in H^s \Rightarrow v \in H^{s-\delta}$, then*

$$g \in H^{s+1/2}(\partial M) \qquad \text{implies} \qquad u \in H_{(1,s-\delta)}(\Omega).$$

One need not require $\delta < 1$.

In a similar fashion, solvability of (5.10) leads to solvability of (5.9). We leave it to the reader to formulate the results in this case.

In the remainder of this section we shall discuss an example of a boundary value problem for an elliptic equation which is not regular. The system will lead to a pseudodifferential operator which is hypoelliptic with loss of 1/2 derivatives. We first discuss some conditions which lead to such a "subelliptic estimate," for $q(x, D) \in OPS^1$:

(5.11)
$$\|u\|_{H^{1/2}} \le C\|q(x, D)u\|_{L^2} + C\|u\|_{L^2}, \qquad u \in C^\infty(\partial M).$$

Assume the principal symbol $q(x, \xi)$ is scalar. Then $C = [q^*, q] = q^*q - qq^*$ has principal symbol

$$C_1(x, \xi) = \frac{1}{i} \sum_{j=1}^{n} \left(\frac{\partial q_1}{\partial x_j} \frac{\partial \bar{q}_1}{\partial \xi_j} - \frac{\partial \bar{q}_1}{\partial x_i} \frac{\partial q_1}{\partial \xi_i} \right)$$

in local coordinates. The following theorem is due to Hörmander [10]; our proof follows Neri [1].

PROPOSITION 5.3. *Suppose $q(x, D) \in OPS^1(\partial M)$ and suppose $C_1(x, \xi) > 0$ whenever $q_1(x, \xi) = 0$. Then the subelliptic estimate* (5.11) *holds, and furthermore if $u \in \mathcal{D}'(\partial M)$ and $q(x, D)u \in H^s(\partial M)$, then $u \in H^{s+1/2}(\partial M)$.*

Proof. The hypothesis guarantees that $\eta\Lambda^{-1}q^*q + [q^*q]$ is a strongly elliptic operator of order 1, if $\eta > 0$ is large enough. Now

$$\begin{aligned}
\|qu\|_{L^2}^2 &= (qu, qu) = (q^*qu, u) \\
&= (qq^*u, u) + ([q^*, q]u, u) \\
&\ge ([q^*, q]u, u) \\
&= (\{\eta q^*\Lambda^{-1}q + [q^*, q]\}u, u) - \eta(\Lambda^{-1/2}qu, \Lambda^{-1/2}qu) \\
&\ge C\|u\|_{H^{1/2}}^2 - C\|u\|_{L^2}^2 - C\|qu\|_{H^{-1/2}}^2
\end{aligned}$$

by Gårding's inequality. Since we have

$$\|qu\|_{H^{-1/2}}^2 \le \epsilon\|qu\|_{L^2}^2 + C(\epsilon)\|qu\|_{H^{-1}}^2$$
$$\le \epsilon\|qu\|_{L^2}^2 + C(\epsilon)\|u\|_{L^2}^2,$$

the estimate (5.11) is proved. The regularity assertion then follows by the standard argument, using Friedrich mollifiers.

The reader is referred to Hörmander [10] for results on subelliptic estimates for *systems*. By duality, one easily obtains that $q(x, D)^*$ defines a closed linear operator $Q^*: H^{\tau+1/2}(\partial M) \to H^\tau(\partial M)$ with $\mathcal{D}(Q^*) = \{u \in H^{\tau+1/2}(\partial M): q(x, D)^*u \in H^\tau\}$, and Q^* has closed range of finite codimension. Such an operator $q(x, D)^*$ is characterized by having commutator $C' = [q, q^*]$ whose symbol $C_1'(x, \xi)$ is *negative* where $q_1^*(x, \xi) = 0$.

We apply these results to the study of

$$\Delta u = 0, \qquad Bu = g \text{ on boundary} \tag{5.12}$$

where B is a real vector field. For convenience we shall work on the half space $\mathbf{R}_+^{n+1} = \{(y, x): y \ge 0, x \in \mathbf{R}^n\}$. Rather than using the general theory, we shall use the Poisson integral to reduce this to a pseudodifferential equation on the boundary.

In $f \in H^s(\mathbf{R}^n)$, the Poisson integral

$$u(y, x) = PI(f) = (2\pi)^{-n} \int_{\mathbf{R}^n} e^{ix\cdot\xi - y|\xi|}\hat{f}(\xi)\, d\xi \tag{5.13}$$

solves the Dirichlet problem

$$\Delta u = 0 \text{ on } \mathbf{R}_+^{n+1}, \qquad u|_{y=0} = f.$$

Now to solve (5.12), with

$$Bu = a_0(x)\frac{\partial}{\partial y}u + \sum_{j=1}^n a_j(x)\frac{\partial}{\partial x_j}u \tag{5.14}$$

suppose $u = PI(f)$. Then we have

$$Bu|_{y=0} = (2\pi)^{-n}\int \{-a_0(x)|\xi| + i\sum a_j(x)\xi_j\}e^{ix\cdot\xi}\hat{f}(\xi)\, d\xi$$
$$= Tf$$

where $T \in OPS^1$, namely $T = T(x, D)$ with

$$T(x, \xi) = -a_0(x)|\xi| + i\sum_j a_j(x)\xi_j.$$

If $a_0(x)$ never vanishes, then T is elliptic. If $n = 1$, a_j real, it suffices that $a_0(x)$, $a_1(x)$ have no common zeros, for T to be elliptic. However, if $n \ge 2$ and $a_0(x)$ vanishes at places, then obviously T cannot be elliptic. But if we have the estimate

$$|Tg|^2 \geq C|g|^2_{\tau-1/2} - C|g|^2_{\tau-1}, \tag{5.15}$$

then we can derive the a priori inequality

$$\|u\|^2_{s-1/2} \leq C\|\Delta u\|^2_{s-2} + C\|Bu\|^2_{s-3/2} + C\|u\|^2_{s-1}$$

and the corresponding regularity theorem, as before. As we have seen, (5.15) will hold if $T(x, \xi) = 0$ implies $C_1(x, \xi) > 0$ where $C(x, D) = T^*T - TT^*$. On the other hand, an existence theorem will hold if $C_1(x, \xi) < 0$ at such points. In the special case where

$$a_0(x) = x_1, \qquad a_j = \text{real constant for } 1 \leq j \leq n,$$

we see that

$$C_1(x, \xi) = 2a_1|\xi|.$$

Thus in this case we get smoothness if $a_1 > 0$ and existence if $a_1 < 0$.

For such a problem, one does not simultaneously have existence and smoothness. In fact, such problems are not well posed and it is appropriate to impose additional conditions.

Exercises

5.1. Do the linear algebra exercise leading to solvability of (5.7).

5.2. Work out the symbol, mod $S^{-\infty}$, of $A \in OPS^0$ of Theorem 5.1.

CHAPTER VI

Wave Front Sets and Propagation of Singularities

In this chapter a theorem of Hörmander on the propagation of singularities of solutions to certain classes of nonelliptic equations is proved. The appropriate language in which to state this result involves a refinement of the notion of singular support, namely the wave front set of a distribution, considered in Section 1. This notion was introduced by Sato for hyperfunctions, for his "micro-local analysis" (see Sato, Kawai, and Kashiwara [1]), and in the setting described here for distributions by Hörmander. The most basic equations in classical *PDE* for which propagation of singularities results furnish important qualitative information are the hyperbolic equations, and in Section 4 a simple application to a problem in energy decay of solutions to hyperbolic equations is presented.

§1. The Wave Front Set of a Distribution

In this section we introduce the notion of the wave front set of a distribution u, denoted by $WF(u)$, which will refine the notion of the singular support of u. There are several characterizations of $WF(u)$, and we shall start with a definition involving pseudodifferential operators acting on u.

If $p(x, \xi) \in S^m$ has principal symbol $p_m(x, \xi)$ (homogenous in ξ of degree m), we define the characteristic set of $P = p(x, D)$ by

$$\text{char } P = \{(x, \xi) \in T^*(\Omega)\backslash 0 : p_m(x, \xi) = 0\}. \tag{1.1}$$

Such a set is a *conic* subset of $T^*(\Omega)$, where generally a conic set C of $T^*(\Omega)$ has the property that $(x, \xi) \in C$ implies $(x, r\xi) \in C$ for all $r > 0$. We now define $WF(u)$ as

$$WF(u) = \bigcap \{\text{char } P : P \in OPS^0, Pu \in C^\infty\}. \tag{1.2}$$

Clearly $WF(u)$ is a closed conic subset of $T^*(\Omega)\backslash 0$. We relate this to the familiar notion of sing supp u, the smallest closed subset of Ω outside which u is smooth. Let $\pi: T^*(\Omega) \to \Omega$ denote the usual projection of $T^*(\Omega)$ onto its base space.

PROPOSITION 1.1. $\pi(WF(u)) = \text{sing supp } u$.

Proof. If $x_0 \notin$ sing supp u, there is a $\varphi \in C_0^\infty(\Omega)$, $\varphi = 1$ near x_0, such that $\varphi u \in C_0^\infty(\Omega)$. Clearly $(x_0, \xi) \notin$ char φ for any $\xi \neq 0$, so $\pi(WF(u)) \subset$ sing supp u.

Conversely, if $x_0 \notin \pi(WF(u))$, then for any $\xi \neq 0$ there is a $q \in OPS^0$ such that $(x_0, \xi) \notin$ char q and $qu \in C^\infty$. Thus we can construct finitely many $Q_j \in OPS^0$ such that $Q_j u \in C^\infty$ and each (x_0, ξ), $|\xi| = 1$, is noncharacteristic for some Q_j. Let $Q = \sum Q_j^* Q_j \in OPS^0$. Then Q is elliptic near x_0 and $Qu \in C^\infty$, so u is C^∞ near x_0.

We now define an analogous concept for pseudodifferential operators.

DEFINITION 1.2. *Let U be an open conic subset of $T^*(\Omega)\backslash 0$. We say $p \in S^m_{\rho,\delta}$ has order $-\infty$ on U if for each closed conic subset K of U with $\pi(K)$ compact, we have estimates, for each integer N,*

$$|D_x^\beta D_\xi^\alpha p(x, \xi)| \leq C_{\alpha,\beta,N,K}(1 + |\xi|)^{-N}, \qquad (x, \xi) \in K.$$

DEFINITION 1.3. *If $P = p(x, D) \in OPS^m_{\rho,\delta}$, we define the essential support of P (and of p) to be the smallest closed conic subset of $T^*(\Omega)\backslash 0$ on the complement of which p has order $-\infty$. We denote this set by $ES(P) = ES(p)$.*

PROPOSITION 1.4. *Let $u \in \mathcal{D}'(\Omega)$ and suppose U is a conic open subset of $T^*(\Omega)\backslash 0$ with $U \cap WF(u) = \varnothing$. If $P \in OPS^m_{\rho,\delta}$, $\rho > 0$, and $ES(P) \subset U$, then $Pu \in C^\infty$.*

Proof. Taking $P_0 \in OPS^0$ with symbol $\equiv 1$ on a conic neighborhood of $ES(P)$, so $P = P_0 P = P P_0$ mod $OPS^{-\infty}$, it suffices to conclude that $P_0 u \in C^\infty$, so we can specialize the hypothesis to $P \in OPS^0$.

Now we can find $Q_j \in OPS^0$ such that $Q_j u \in C^\infty$ and each $(x, \xi) \in ES(P)$ is noncharacteristic for some Q_j, by (1.2), and if $Q = \sum Q_j^* Q_j$, then $Qu \in C^\infty$ and char $Q \cap ES(P) = \varnothing$. We claim there is $A \in OPS^0$ such that $AQ = P$ mod $OPS^{-\infty}$. Indeed, let $\tilde{Q}$ be an elliptic operator whose symbol equals that of Q on a conic neighborhood of $ES(P)$, and let $\tilde{Q}^{-1}$ denote a parametrix for $\tilde{Q}$. Set $A = P\tilde{Q}^{-1}$. Then $AQ = P\tilde{Q}^{-1}(\tilde{Q} + Q - \tilde{Q}) = P + P\tilde{Q}^{-1}(Q - \tilde{Q}) = P$ mod $OPS^{-\infty}$. Consequently $Pu = AQu \in C^\infty$, so the proposition is proved.

The end of this argument made use of the following simple fact, which we generalize and formalize.

PROPOSITION 1.5. *One has $ES(PQ) \subset ES(P) \cap ES(Q)$ if $P \in OPS^m_{\rho_1,\delta_1}, Q \in OPS^\mu_{\rho_2,\delta_2}$, $\rho_1 > \delta_2$.*

Proof. Note that

$$\sigma_{PQ} \sim \sum_{\alpha \geq 0} \frac{1}{\alpha!} p^{(\alpha)}(x, \xi) q_{(\alpha)}(x, \xi)$$

if $P = p(x, D)$, $Q = q(x, D)$. From this the proposition is immediate.

More generally, if $P_j = p_j(x, D) \in OPS^{m_j}_{\rho_j,\delta_j}$, with $\rho_j > \delta_{j+1}$, we have

$$ES(P_1 \dots P_k) \subset \bigcap_{j=1}^{k} ES(P_j).$$

We are now ready for the basic proposition on the preservation of wave front set under application of pseudodifferential operators, refining the pseudolocal property proved in Chapter II.

THEOREM 1.6. *One has* $WF(Pu) \subset WF(u) \cap ES(P)$ *if* $P \in OPS^m_{\rho,\delta}$, $\rho > 0$.

Proof. First we show $WF(Pu) \subset ES(P)$. Indeed, if $(x_0, \xi_0) \notin ES(P)$, choose $q \in S^0$ so that $q(x, \xi) = 1$ on a conic neighborhood of (x_0, ξ_0) and $ES(Q) \cap ES(P) = \emptyset$. Thus $QP \in OPS^{-\infty}$, so $QPu \in C^\infty$. By (1.2) it follows that $(x_0, \xi_0) \notin WF(Pu)$.

In order to show that $WF(Pu) \subset WF(u)$, let Γ be any conic neighborhood of $WF(u)$ and write $P = P_1 + P_2$, $P_j \in OPS^m_{\rho,\delta}$, with $ES(P_1) \subset \Gamma$ and $ES(P_2) \cap WF(u) = \emptyset$. By Proposition 1.4, $P_2 u \in C^\infty$. Thus $WF(Pu) = WF(P_1 u) \subset \Gamma$, which shows that $WF(Pu) \subset WF(u)$.

As a corollary, we have the following sharper form of the regularity theorem for elliptic operators.

COROLLARY 1.7. *If* $P \in OPS^m_{\rho,\delta}$ *is elliptic*, $\rho > \delta$, *then* $WF(Pu) = WF(u)$.

Proof. We have seen that $WF(Pu) \subset WF(u)$. On the other hand, if E is a parametrix for P, $E \in OPS^{-m}_{\rho,\delta}$, we see that $WF(u) = WF(EPu) \subset WF(Pu)$.

We now introduce another characterization of $WF(u)$, based on decay of the Fourier transform. Recall that if $v \in \mathcal{E}'(\mathbf{R}^n)$, v is smooth if, and only if, $\hat{v}(\xi)$ is rapidly decreasing as $|\xi| \to \infty$. Thus $x_0 \notin \text{sing supp}\, u$ if, and only if, there exists $\varphi \in C_0^\infty$, $\varphi(x_0) \neq 0$, such that $(\varphi u)\hat{}(\xi)$ is rapidly decreasing. More generally, we have the following.

THEOREM 1.8. *The point* $(x_0, \xi_0) \notin WF(u)$ *if, and only if, there is* $\varphi \in C_0^\infty(\Omega)$, $\varphi(x_0) \neq 0$, *and a conic neighborhood* Γ *of* ξ_0 *such that, for every* N,

$$|(\varphi u)\hat{}(\xi)| \leq C_N(1 + |\xi|)^{-N}, \qquad \xi \in \Gamma. \tag{1.3}$$

Proof. Suppose (1.3) holds. Let $\chi(\xi)$ be homogeneous of degree 0 in ξ, for $|\xi| \geq c_0 > 0$, $\chi(\xi_0) \neq 0$, and be supported in Γ. It follows that

$$\chi(D)\varphi u \in C^\infty.$$

$\chi(D)\varphi \in OPS^0$ and $(x_0, \xi_0) \notin \text{char}\,(\chi(D)\varphi)$, so $(x_0, \xi_0) \notin WF(u)$.

Conversely, suppose $(x_0, \xi_0) \notin WF(u)$. Then $\varphi(x)$ and $\chi(\xi)$ can be constructed with $\varphi(x_0) \neq 0$, $\chi(\xi_0) \neq 0$, such that $ES(\chi(D)\varphi)$ is disjoint from

$WF(u)$. Regularizing χ, one can suppose that $\hat{\chi}$ has compact support. Thus $\chi(D)\varphi u \in C_0^\infty$, so

$$|\chi(\xi)(\varphi u)\hat{}\,| \le C_N(1 + |\xi|)^{-N}$$

which implies that (1.3) holds on some conic neighborhood of ξ_0.

We also localize the concept of H^s regularity, as follows.

DEFINITION 1.9. *Let* $(x_0, \xi_0) \in T^*(\Omega)\backslash 0$. *We say* $u \in H^s$ at (x_0, ξ_0) *if we can write* $u = u_1 + u_2$ *such that*

$$u_1 \in H^s \quad \text{and}$$
$$(x_0, \xi_0) \notin WF(u_2).$$

The following refinement of Corollary 1.7 is left as an exercise.

PROPOSITION 1.10. *Let* $u \in \mathscr{D}'(\Omega)$, $P = p(x, D) \in OPS^m_{\rho,\delta}$, $\rho > \delta$, *and suppose* $Pu \in H^s$ at (x_0, ξ_0). *Suppose* $|p(x, \xi)| \ge C|\xi|^m$, $|\xi| \ge C$, *for* (x, ξ) *in a conic neighborhood of* (x_0, ξ_0). *Then*

$$u \in H^{s+m} \qquad \text{at} \qquad (x_0, \xi_0).$$

If u satisfies the condition of Definition 1.9, sometimes we say $u \in H^s$ microlocally near (x_0, ξ_0). Similarly, if $(x_0, \xi_0) \notin ES(P)$, we say $P \in OPS^{-\infty}$ microlocally near (x_0, ξ_0). More generally, if $P = A + B$, $(x_0, \xi_0) \notin ES(A)$, $B \in OPS^m_{\rho,\delta}$, we say $P \in OPS^m_{\rho,\delta}$ microlocally near (x_0, ξ_0).

Exercises

1.1. Let $S \subset \mathbf{R}^n$ be a region with smooth boundary ∂S. Let $u = \chi_S$. Show that $WF(u) = \{(x, \xi): x \in \partial S, \xi \text{ normal to } \partial S\}$.

1.2. Let $\eta \in \mathbf{R}^n$ be a unit vector and let

$$p_\eta(\xi) = e^{\eta \cdot \xi - |\xi|}.$$

Show that $p_\eta \in S^0_{1/2,0}$ for $|\xi| \ge 1$.

1.3. Let $u = p_\eta(D)\delta$. Show that

$$WF(u) = \{(0, r\eta): r > 0\}.$$

1.4. Define $E: \mathscr{D}'(\Omega) \to \mathscr{D}'(S^*(\Omega))$ by

$$Eu(x, \eta) = p_\eta(D)u(x).$$

Show that

$$\text{sing supp } Eu = WF(u) \cap S^*(\Omega).$$

1.5. Extend Corollary 1.7 and Proposition 1.10 to hypoelliptic operators with constant strength.

§2. Propagation of Singularities: the Hamilton Flow

If p is a real valued function on $T^*(M)$, the vector field

$$H_p = \sum_{j=1}^{n} \left(\frac{\partial p}{\partial x_j} \frac{\partial}{\partial \xi_j} - \frac{\partial p}{\partial \xi_j} \frac{\partial}{\partial x_j} \right)$$

on $T^*(M)$ is called the Hamiltonian vector field of p. We shall be particularly interested in the case where $p(x, \xi)$ is homogeneous of degree s in ξ, so is the principal symbol of a pseudodifferential operator.

DEFINITION. *If $p \in S^m$ has principal symbol p_m, the integral curves of H_{p_m} are called the bicharacteristics of p. The integral curves on which $p_m = 0$ are called the null bicharacteristics. Note that $H_{p_m} p_m = 0$, so p_m is constant on each integral curve. The function $H_a b$ is also denoted by $\{a, b\}$, using the Poisson bracket.*

It will be useful to note that the asymptotic expansion of the symbol of a product of pseudodifferential operators shows that, if $p(x, D) \in OPS^m$, $q(x, D) \in OPS^\mu$ have principal symbols p_m, q_μ which are scalar, then the principal symbol of $[p(x, D), q(x, D)]$ is $(1/i)H_{p_m} q_\mu$.

The following is Hörmander's theorem on propagation of singularities.

THEOREM 2.1. *Let $P = p(x, D) \in OPS^m$ have scalar principal symbol. Let $\gamma:[t_0, t_1] \to T^*(M)\backslash 0$ be a null bicharacteristic strip for Re p_m, and assume Im $p_m \geq 0$ on a neighborhood of γ. Suppose*

$$Pu = f \in H^s \quad \text{on} \quad \gamma.$$

If $u \in H^{s+m-1}$ at $\gamma(t_1)$, then $u \in H^{s+m-1}$ on γ.

In particular if $WF(f) \cap \gamma = \emptyset$ and $\gamma(t_0) \notin WF(u)$, then $\gamma \cap WF(u) = \emptyset$. If P has real principal symbol, we can reverse the time direction in this proposition and conclude that $WF(u) \backslash WF(f)$ is contained in char P *and is invariant under the flow of H_{p_m}.*

Proof. We begin by making some simplifications. We can multiply P on the left by an elliptic operator $E = e(x, D) \in OPS^{1-m}$. Since char (EP) = char (P) and H_{p_m} is proportional to H_{p_1}, p_1 denoting the principal symbol of EP, so $p_1 = e_{1-m} p_m$, we can assume without loss of generality that $m = 1$, i.e. $P \in OPS^1$. We can also suppose that $u \in H^{s-1/2}$ on γ to start off with, and without loss of generality we can assume $u \in \mathscr{E}'(M)$.

Now let $\mathscr{M}$ be a set of pseudodifferential operators C with real valued symbols satisfying the following two properties:

(2.1) $$\mathscr{M} \subset \{C \in OPS^{s-1}_{1,0} : ES(C) \subset \Gamma\}$$

(2.2) $\mathscr{M}$ is a bounded subset of $OPS^s_{1,0}$.

Here Γ is a convenient small conic neighborhood of γ in $T^*(M)\backslash 0$. Write $P = A + iB$ with $A = A^*$, $B = B^*$. Then for $C \in \mathcal{M}$ we have

$$Im(Cf, Cu) = Re(BCu, Cu) + Im([C, A]u, Cu) + Re([C, B]u, Cu). \tag{2.3}$$

The plan of attack is to estimate this from below. Let us consider separately the three terms on the right in (2.3).

Since the principal symbol of B is ≥ 0 on Γ, the sharp Gårding inequality (which we prove in the following chapter) implies

$$Re(BCu, Cu) \geq -C_1\|Cu\|_{L^2}^2 - C_2 \tag{2.4}$$

with C_1, C_2 constants independent of $C \in \mathcal{M}$. If $c(x, \xi)$ is the symbol of C, then the principal symbol of $C^*[C, B]$ is $-icH_c b$, which is pure imaginary, if we denote by a, b the principal symbols of A, B. Hence

$$Re([C, B]u, Cu) \geq -C_3 \tag{2.5}$$

with C_3 independent of $C \in \mathcal{M}$. Furthermore a principal symbol of $C^*[C, A]$ is $icH_a c = (i/2)H_a c^2 = (i/2)\{a, c^2\}$. Hence

$$Im([C, A]u, Cu) \geq \tfrac{1}{2}(\{a, c^2\}(x, D)u, u) - C_4 \tag{2.6}$$

with C_4 indpendent of $C \in \mathcal{M}$.

Now (2.3) through (2.6) yield

$$\begin{aligned}\tfrac{1}{2}(\{a, c^2\}(x, D)u, u) - C_1\|Cu\|_{L^2}^2 &\leq C_5 + Im(Cf, Cu)\\ &\leq C_5 + \tfrac{1}{2}\|Cu\|_{L^2}^2 + \tfrac{1}{2}\|Cf\|_{L^2}^2\end{aligned}$$

or, since Cf is in a bounded subset of L^2 for $C \in \mathcal{M}$,

$$Re(e(x, D)u, u) \leq C_6 \tag{2.7}$$

where

$$e(x, \xi) = \{a, c^2\}(x, \xi) - (2C_1 + 1)c(x, \xi)^2. \tag{2.8}$$

The strategy we want to use is the following. By constructing C in a clever manner and using (2.7), we want to find an $r \in OPS^s_{1,0}$ with

$$Re(r(x, D)^2u, u) < \infty$$

and $r(x, \xi) \geq c|\xi|^s$ on a conic neighborhood of γ, which would yield $u \in H^s$ on γ.

The family $\mathcal{M}$ will consist of operators with symbols

$$c_{\lambda,\epsilon}(x, \xi) = ce^{\lambda a_0}(1 + \epsilon^2 a_1^2)^{-1/2} \tag{2.9}$$

where λ is a conveniently fixed real number and $0 < \epsilon \leq 1$. We make the following four requirements.

(2.10) $c(x, \xi)$ is homogeneous of degree s in ξ, $\operatorname{supp} c \subset \Gamma$, and $c(x, \xi) \geq 0$.

(2.11) $H_a c \geq 0$ on $\Gamma \backslash U$, where U is a small conic neighborhood of $\gamma(t_1)$, with strict inequality on $\gamma \backslash U$.

(2.12) $a_0 \in S^0, \qquad H_a a_0 = 1$ on a neighborhood of γ.

(2.13) $a_1 \in S^1, \qquad H_a a_1 = 0, \qquad a_1$ nonzero on γ.

Let us show how equations (2.10) through (2.13) can be arranged. We assume that, at $\gamma(t_0)$, H_a is not radial, since otherwise the theorem is trivial. Now we can construct a conic hypersurface Ω in $T^*(M) \backslash 0$ through $\gamma(t_0)$, transversal to H_a in a neighborhood of $\gamma(t_0)$. Let g be an arbitrary smooth positive function supported on a small conic neighborhood of γ, with $g(x, \xi)$ homogeneous of degree s in ξ, $|\xi| \geq 1$. Now solve the initial value problem

(2.14) $$H_a c = g; \qquad c = 0 \text{ on } \Omega.$$

Then multiply by a smooth function identically 1 in a neighborhood of $\gamma \backslash U$ to cut off c in a neighborhood of $\gamma(t_1)$, so that (2.10) and (2.11) hold. In a similar fashion a_0 and a_1 are constructed so that (2.12) and (2.13) hold.

With equations (2.10) through (2.13) arranged, we proceed. We have

(2.15) $$\begin{aligned} e_{\lambda,\epsilon}(x, \xi) &= \{a, c_{\lambda,\epsilon}^2\} - (2C_1 + 1)c_{\lambda,\epsilon}^2 \\ &= (\{a, c^2\} + (\lambda - 2C_1 - 1)c^2)e^{\lambda a_0}(1 + \epsilon^2 a_1^2)^{-1/2}. \end{aligned}$$

Now fix $\lambda > 2C_1 + 1$. Hence $(\{a, c^2\} + (\lambda - 2C_1 - 1)c^2)e^{\lambda a_0}$ is homogeneous of degree $2s$, is ≥ 0 on $\Gamma \backslash U$, and is > 0 on $\gamma \backslash U$. Thus we can find r, $q \in S^s$, $r, q \geq 0$, with $\operatorname{supp} q \subset U$ and

(2.16) $$r^2 \leq (\{a, c^2\} + (\lambda - 2C_1 - 1)c^2)e^{\lambda a_0} + q^2.$$

We now apply the sharp Gårding inequality to the difference of the two sides of (2.16), multiplied by $(1 + \epsilon^2 a_1^2)^{-1/2}$, using (2.7) with $e = e_{\lambda,\epsilon}$ given by (2.15). It follows that (letting $r_\epsilon = r(1 + \epsilon^2 a_1^2)^{-1/4}$, etc.)

$$\begin{aligned} \|r_\epsilon(x, D)u\|_{L^2}^2 = (r_\epsilon^2 u, u) &\leq Re(e_{\lambda,\epsilon}(x, D)u, u) + \|q_\epsilon(x, D)u\|_{L^2}^2 + C_7 \\ &\leq C_8 \end{aligned}$$

where C_7, C_8 are independent of $\epsilon > 0$, assuming $u \in H^{s-1/2}$ on γ. Letting $\epsilon \downarrow 0$, we get $\|r(x, D)u\|_{L^2}^2 \leq C_8$ so $u \in H^s$ on γ, as asserted. The proof is complete.

We remark that (2.14) is a tool from geometrical optics, known as the transport equation. In Chapter VIII, we shall see a great deal more of such tools from geometrical optics.

Exercise

2.1. Let Φ be the fundamental solution of the hyperbolic operator L considered in Chapter IV, Section 4, Exercise 4.1. Prove that $WF(\Phi)$ is equal to the set of null bicharacteristic strips over the upper half space which pass over the origin, plus the fiber over the origin Hint: Every null bicharacteristic strip passes into the lower half space, where $\Phi = 0$. Some miss the origin, where $L\Phi$ is singular; some do not.

§3. Local Solvability

Here we use Theorem 2.1 to derive some results on local solvability of a *PDE* $Pu = f$.

PROPOSITION 3.1. *Let $P \in OPS^m$ have real principal symbol. If no null bicharacteristic strip lies over a compact set K of Ω, then $u \in \mathscr{E}'(\Omega)$, $Pu \in C^\infty(\Omega)$,* imply $u \in C^\infty(\Omega)$.

Proof. Theorem 2.1 immediately implies $WF(u) = \emptyset$.

PROPOSITION 3.2. *Suppose $P \in OPS^m$ is properly supported and assume*

$$u \in \mathscr{E}'(K), \qquad Pu \in C^\infty \Rightarrow u \in C_0^\infty(K) \tag{3.1}$$

where K is some compact subset of Ω. Then, for any k, s, there exists $\ell \in \mathbf{R}$ such that

$$\|u\|_{H^k} \leq C\|Pu\|_{H^\ell} + C\|u\|_{H^s}, \qquad u \in C_0^\infty(K). \tag{3.2}$$

Proof. The space $F = \{u \in H_0^s(K) : Pu \in C^\infty(M)\}$ is a Frechet space, with the obvious topology. Here $H_0^s(K) = \{u \in H^s(\Omega) : \text{supp } u \subset K\}$. The inclusion $C_0^\infty(K) \xrightarrow{j} F$ is continuous and 1-1, and hypothesis (3.1) says it is onto. The conclusion follows from the open mapping theorem.

For the local solvability theorem we derive from this, we shall assume that $p(x, D)$ is a differential operator of order m, with real principal part. Define a closed, densely defined operator $P: H_0^k(K) \to H_0^\ell(K)$ by $\mathscr{D}(P) = \{u \in H_0^k(K) : p(x, D)u \in H_0^\ell(K)\}$, $Pu = p(x, D)u$. If (3.2) holds, P has closed range and finite dimensional kernel. Hence P^* has closed range of finite codimension. Since $\mathscr{D}(P^*) \subset H^{-\ell}(K)$, which is not a space of distributions on Ω (but rather a space of equivalence classes of such distributions) some care must be taken in interpreting this. However, any $u \in H^{-\ell}(K)$ defines a distribution on $\mathring{K}$, the interior of K, and $P^*u = p(x, D)^*u$ on $\mathring{K}$. If $p(x, D)$ has real principal part, $p(x, D)^*$ has the same principal part, hence the same set of characteristics and null bicharacteristic strips. Thus the roles of $p(x, D)$ and $p(x, D)^*$ can be reversed in the above argument. We sum up what we have so far.

PROPOSITION 3.3. *Let $p(x, D)$ be a differential operator with real principal part, K a compact subset of Ω, and suppose no null bicharacteristic strip lies entirely over K. Then there is a finite dimensional linear space $E \subset C_0^\infty(K)$ with the property that for any $f \in \mathscr{E}'(K)$, $f \in H^{-k}$, $f \perp E$, we can find $u \in H^{-\ell}(\Omega)$ satisfying the equation*

$$Pu = f$$

on $\mathring{K}$.

We are now ready to state and prove a local solvability theorem.

THEOREM 3.4. *Let $p(x, D)$ be a differential operator of order m, with real principal part, and assume that no null bicharacteristic strip lies over a single point. Then, for any $f \in \mathscr{D}'(\Omega)$, $P_0 \in \Omega$, there exists a neighborhood U of p_0 and $u \in \mathscr{D}'(U)$ such that*

$$p(x, D)u = f \text{ on } U.$$

Proof. First we note that there is a compact neighborhood K of p_0 over which no null bicharacteristic lies. In fact, if such strips γ_ν lie over $K_\nu = \{x \in \Omega: |x - p_0| \leq 1/\nu\}$, since by homogeneity of H_{p_m} we can suppose γ_ν passes through the cosphere bundle of Ω at some point over K_ν, we have a limiting null bicharacteristic strip γ lying over p_0, a contradiction.

Thus for each ν sufficiently large there is a finite dimensional linear space $E_\nu \subset C_0^\infty(K_\nu)$ such that $p(x, D)u = f$ can be solved, say on $K_{\nu+1}$, for all $f \perp E_\nu$. It remains only to show that $E_\nu = 0$ for sufficiently large ν. Now clearly $E_j \supset E_{j+1} \supset \cdots$, so the dimensions, being finite and decreasing must stabilize: $E_k = E_{k+1} = \cdots$ for a certain k. But since $f \in E_k$ implies $\operatorname{supp} f \subset K_\nu$ for all $\nu \geq k$, we have $f = 0$ since $f \in C^\infty$. Thus $E_k = 0$, and the proof is complete.

Note in particular that we can take U independent of f.

The argument we have given, via Propositions 3.1 and 3.2, has used only the fact that $WF(u)\backslash WF(p(x, D)u)$ is invariant under the flow generated by H_{p_m}. Actually, Theorem 2.1 yields the following stronger form of the inequality (3.2), when $p(x, D)$ satisfies the hypotheses of Proposition 3.1.

$$\|u\|_{H^{\ell+m-1}} \leq C\|Pu\|_{H^\ell} + C\|u\|_{H^s}, \qquad u \in C_0^\infty(K). \tag{3.3}$$

The local solvability result, Theorem 3.4, was first proved by Hörmander; see Hörmander [6]. In Hörmander [10] he introduced the sharp Gårding inequality for such a purpose. Our treatment here follows Hörmander [16]. For differential operators with complex coefficients in the principal part, results could be derived from Theorem 2.1, but they would be less complete than they are in the real case. Serious obstructions to local solvability in that case were first noticed by Lewy. See Chapter 6 of Hörmander [6]. Nirenberg and Trèves [1] have given necessary and sufficient conditions for solvability in the case of operators with analytic coefficients, whose principal symbol p_m is of principal type, i.e.,

$$Re\, p_m(x, \xi) \Rightarrow \nabla_\xi\, Re\, p_m(x, \xi) \neq 0.$$

Beals and Fefferman [1] have extended these results to operators with merely C^∞ coefficients.

Exercises

3.1. Prove that if $p(x, D)$ satisfies the conditions of Theorem 3.4, then each $p_0 \in \Omega$ has a neighborhood U on which the equation

$$p(x, D)u = f$$

can be solved for $u \in H^{m-1+s}$, given $u \in H^s(\Omega)$.

3.2. In the context of Exercise 3.1, prove that $p(x, D)u = f$ is locally solvable for smooth u, given $f \in C^\infty(\Omega)$.

3.3. Show that the operator $\partial/\partial\theta$ on the annulus $\Omega = \{z \in \mathbf{C} : 1 < |z| < 2\}$ has null bicharacteristic strips lying over compact sets. Check $\partial/\partial\theta$ against Proposition 3.3.

§4. Systems: An Exponential Decay Result

We begin by generalizing Theorem 2.1 to systems.

THEOREM 4.1. *Let $p(x, D)$ be a $k \times k$ system in OPS^m. Assume that $q(x, \xi) = \det p(x, \xi)$ has real principal part q_M, of order $M = km$. Let $p(x, D)u = f$, and let $\gamma : [t_0, t_1] \to T^*(\Omega)\backslash 0$ be a null bicharacteristic strip for q_M. If $f \in H^s$ on γ and if $u \in H^{s+m-1}$ at $\gamma(t_0)$, then $u \in H^{s+m-1}$ on γ.*

Proof. If ${}^{co}p(x, \xi)$ is the cofactor matrix of $p(x, \xi)$, then ${}^{co}p(x, D)p(x, D) = q_M(x, D) + r(x, D)$ where $r \in S^{M-1}$. Then $(q_M(x, D) + r(x, D))u = {}^{co}p(x, D)f \in H^{s-m(k-1)}$ on γ. Since the principal part of $q_M + r$ is scalar, Theorem 2.1 yields the desired conclusion.

Suppose now that $L = (\partial/\partial t) - G(x, D_x)$ is a symmetric first order $k \times k$ hyperbolic system on $\mathbf{R} \times \Omega$, where Ω is a compact manifold. Suppose in fact that

$$G(x, D_x) + G(x, D_x)^* = -F(x) \tag{4.1}$$

where $F(x)$ is a positive semidefinite self adjoint matrix function. We call L a dissipative symmetric hyperbolic operator. If u satisfies $Lu = 0$ and

$u(0) \in L^2(\Omega)$, then $u(t) \in L^2(\Omega)$ for all t, and

$$\frac{d}{dt}\|u(t)\|^2 = 2\,Re(G(x, D_x)u, u) = -2\int_\Omega F(x)u \cdot u\,dx \le 0.$$

We aim to show that, under certain conditions, $u(t)$ decays to zero at an exponential rate: $\|u(t)\| \le ce^{-\alpha t}$ as $t \uparrow \infty$.

Suppose $-F(x) \le -\eta < 0$ on an open subset U of Ω. We shall make the following assumption, with $q(x, \xi) = i^{-k} \det(i\tau - G(x, \xi))$.

(4.2) There is a number T such that every null bicharacteristic strip of q_M in $T^*(\mathbf{R} \times \Omega)\backslash 0$ passes through $T^*((0, T) \times U)\backslash 0$.

Assumption (4.2) together with Theorem 4.1 and the closed graph theorem, gives us the following inequality, with $\omega = [0, T] \times U$, $X = [0, T] \times \Omega$.

$$\|u\|^2_{L^2(X)} \le C\|Lu\|^2_{L^2(X)} + C\|u\|^2_{L^2(\omega)} + C\|u\|^2_{H^{-s}(X)}. \tag{4.3}$$

Thus if $A: L^2(X) \to L^2(X) \oplus L^2(\omega)$ is the closed linear operator given by $Au = \{Lu, u|_\omega\}$, $\mathscr{D}(A) = \{u \in L^2(X): Lu \in L^2(X)\}$, it follows that A has closed range. We make a further assumption:

(4.4) If T is taken sufficiently large, then $Lu = 0$ on X, $u = 0$ on ω implies $u \equiv 0$ on X.

This assumption is satisfied, for example, if Ω is an analytic manifold and L has analytic coefficients, by Holmgren's uniqueness theorem. A result of Rauch and Taylor [2] together with a dimension argument like that used in the proof of the local solvability theorem of the last section can be used to show that assumption (4.4) holds if G is elliptic and has the unique continuation property. For more on this, see Chapter XIV.

Granted assumption (4.4), $A^{-1}: R(A) \to L^2(X)$ is well defined and, by the closed graph theorem, continuous. This yields the inequality

$$\|u\|^2_{L^2(X)} \le C\|Lu\|^2_{L^2(X)} + C\|u\|^2_{L^2(\omega)}.$$

Now if $Lu = 0$, $u(0) \in L^2(\Omega)$, it follows that

$$\frac{d}{dt}\|u(t)\|^2_{L^2(\Omega)} = -2\int_\Omega Fu \cdot u\,dx \le -2\eta\|u(t)\|^2_{L^2(U)}.$$

And thus

$$\begin{aligned}\|u(T)\|^2_{L^2(\Omega)} &\le \|u(0)\|^2_{L^2(\Omega)} - 2\eta\|u\|^2_{L^2(\omega)}\\ &\le \|u(0)\|^2_{L^2(\Omega)} - C\|u\|^2_{L^2(X)}\\ &\le \|u(0)\|^2_{L^2(\Omega)} - CT^2\|u(T)\|^2_{L^2(\Omega)}.\end{aligned}$$

Therefore,

$$\|u(T)\|_{L^2(\Omega)}^2 \leq (1 + CT)^{-1}\|u(0)\|_{L^2(\Omega)}^2.$$

From this inequality, exponential decay is a very simple consequence.

This result was proved in Rauch and Taylor [2], and further results, for nondissipative equations in bounded domains, were given in Rauch and Taylor [3]. For these latter results such energy estimates did not seem effective, and the authors used Fourier integral operators. A much deeper connection, between propagation of singularities and local exponential decay of energy of solutions to hyperbolic equations in unbounded domains, was stressed by Lax and Phillips, in [1]; see also Majda and Taylor [2].

CHAPTER VII

The Sharp Gårding Inequality

In Chapter I was stated and in the last chapter used, a sharpened form of Gårding's inequality: if $p(x, \xi) \geq 0$, $p \in S^m_{1,0}$, then

$$(0.1) \qquad Re(p(x, D)u, u) \geq -C\|u\|^2_{H^{(m-1)/2}}, \qquad u \in C_0^\infty.$$

This result was first proved by Hörmander [10] in the scalar case, which is really all we have used. It was extended to the case where $p(x, \xi)$ is a positive matrix valued function by Lax and Nirenberg [1]. Friedrichs, using multiple symbols, and Vaillancourt [2], simplified their proof.

We follow Friedrichs in modifying $p(x, D)$ to make it positive by means of a multiple symbol. The treatment of such a multiple symbol will parallel the development in Chapter II, Section 3. We generalize (0.1) for $p(x, D) \in OPS^m_{\rho,\delta}$, $0 \leq \delta < \rho \leq 1$, following Kumano-Go [2]. A further generalization is given in Beals and Fefferman [2].

Recently, Hörmander, for use in semiglobal solvability, has strengthened the semiboundedness

$$(0.2) \qquad Re(p(x, D)u, u) \geq -C\|u\|^2_{L^2}$$

which by (0.1) holds for $p(x, D) \in OPS^1_{1,0}$, to $p(x, D) \in OPS^{6/5}_{1,0}$, provided $p(x, \xi) \geq 0$, and Fefferman and Phong [1] have gone further, producing the superb result that (0.2) is valid for $p(x, D) \in OPS^2_{1,0}$, provided $p(x, \xi) \geq 0$. There is also interest in proving (0.2) for $p(x, D) \in OPS^2$ with positive principal symbol, but whose total symbol is not necessarily ≥ 0. Under certain geometrical restrictions on the characteristic variety of $p_2(x, \xi)$ and certain hypotheses on $p_1(x, \xi)$, Hörmander [20] has some results of this nature, which we shall discuss in Chapter XIII.

§1. A Multiple Symbol

In Chapter II we considered symbols of the form $a(x, y, \xi)$, which defined operators $a(x, D, x)$, subsequently shown to be of the form $p(x, D)$ under appropriate circumstances. For the proof of the sharp Gårding inequality it is convenient to study operators of the form $b(D, x, D)$ and their associated multiple symbols.

DEFINITION 1.1. *We say* $b(\xi_2, x, \xi_1) \in S^{m_1,m_2}_{\rho,\delta_1,\delta_2}$ *if for K compact we have*

$$(1.1)\quad |D_x^\beta D_{\xi_2}^\gamma D_{\xi_1}^\alpha b(\xi_2, x, \xi_1)| \le C_{K,\alpha,\beta,\gamma}(1+|\xi_2|)^{m_2-\rho|\gamma|+\delta_2|\beta|}(1+|\xi_1|)^{m_1-\rho|\alpha|+\delta_1|\beta|}$$

for $x \in K$ *and* $\xi_j \in \mathbf{R}^n$.

For convenience we shall assume that $b(\xi_2, x, \xi_1) = 0$ for large x. Let

$$(1.2)\qquad \hat{b}(\xi_2, \eta, \xi_1) = \int b(\xi_2, x, \xi_1)e^{ix\cdot\eta}\,dx.$$

DEFINITION 1.2. *The operator* $B = b(D, x, D)$ *is defined by the formula*

$$(1.3)\quad Bu(x) = (2\pi)^{3n/2}\iiint b(\xi_2, y, \xi_1)e^{ix\cdot\xi_2+iy\cdot(\xi_2-\xi_1)}\hat{u}(\xi_1)\,d\xi_1\,dy\,d\xi_2.$$

Equivalent to (1.3) is the formula

$$\begin{aligned}(Bu)\hat{}(\xi_2) &= (2\pi)^{-n}\iint b(\xi_2, y, \xi_1)e^{iy\cdot(\xi_2-\xi_1)}\hat{u}(\xi_1)\,d\xi_1\,dy\\ &= (2\pi)^{-n}\int \hat{b}(\xi_2, \xi_2-\eta, \eta)\hat{u}(\eta)\,d\eta.\end{aligned}$$

Consequently, we can write

$$(1.4)\qquad b(D, x, D)u = (2\pi)^{-n}\int a(x, \xi_1)e^{ix\cdot\xi_1}\hat{u}(\xi_1)\,d\xi_1$$

where

$$(1.5)\qquad \begin{aligned}a(x, \xi) &= \int b(\xi_2, y, \xi)e^{i(y-x)\cdot(\xi_2-\xi)}\,dy\,d\xi_2\\ &= \int \hat{b}(\xi+\zeta, \zeta, \xi)e^{-ix\cdot\zeta}\,d\zeta.\end{aligned}$$

We analyze the amplitude $a(x, \xi)$, thus showing that $b(D, x, D)$ is a pseudo-differential operator.

PROPOSITION 1.1. *If* $b \in S^{m_1,m_2}_{\rho,\delta_1,\delta_2}$, *then* $a \in S^m_{\rho,\delta}$ *with* $m = m_1 + m_2$, $\delta = \delta_1 + \delta_2$, *provided* $0 \le \delta < \rho \le 1$. *Furthermore we have the asymptotic expansion*

$$(1.6)\qquad a(x, \xi) \sim \sum_{\alpha\ge 0}\frac{i^{|\alpha|}}{\alpha!}D_x^\alpha D_{\xi_2}^\alpha b(\xi_2, x, \xi)|_{\xi_2=\xi}.$$

Proof. By Taylor's formula we have

$$\hat{b}(\xi+\zeta, \zeta, \xi) = \sum_{|\alpha|<N}\frac{1}{\alpha!}\zeta^\alpha D_{\xi_2}^\alpha \hat{b}(\xi_2, \zeta, \xi)|_{\xi_2=\xi} + R_N(\zeta, \xi)$$

with

$$|R_N(\zeta, \xi)| \le C_N \sup_{|\gamma|=N,\,0\le t\le 1}|D_{\xi_2}^\alpha \hat{b}(\xi+t\zeta, \zeta, \xi)|\,|\zeta|^N.$$

Taking inverse Fourier transforms with respect to ζ yields

$$(1.7)\qquad a(x,\xi) = \sum_{|\alpha|<N} \frac{i^{|\alpha|}}{\alpha!} D_x^\alpha D_{\xi_2}^\alpha b(\xi_2, x, \xi)\big|_{\xi_2=\xi} + \tilde{R}_N(x,\xi)$$

where $\tilde{R}_N(x,\xi) = \int R_N(\zeta,\xi) e^{-ix\cdot\zeta}\, d\zeta$. The general term in the sum in (1.7) clearly belongs to $S_{\rho,\delta}^{m-(\rho-\delta)|\alpha|}$. To complete the proof, we apply Theorem 3.2 of Chapter II, Section 3. So it is only necessary to verify the following estimates:

$$(1.8)\qquad |\tilde{R}_N(x,\xi)| \le C_N(1+|\xi|)^{m+n-(\rho-\delta)N};$$

$$(1.9)\qquad |D_x^\beta D_\xi^\alpha a(x,\xi)| \le C_{\alpha\beta}(1+|\xi|)^{\mu(\alpha,\beta)}.$$

To prove (1.8), note that, if $b \in S_{\rho,\delta_1,\delta_2}^{m_1,m_2}$, then

$$(1.10)\quad |D_{\xi_2}^\gamma \hat{b}(\xi_2,\eta,\xi_1)| \le C_{\gamma,\nu}(1+|\xi_2|)^{m_2-\rho|\gamma|+\delta_2\nu}(1+|\xi_1|)^{m_1+\delta_1\nu}(1+|\eta|)^{-\nu}.$$

Therefore,

$$\begin{aligned} |R_N(\zeta,\xi)| &\le C_N \sup_{|\gamma|=N,\ 0\le t\le 1} |D_{\xi_2}^\gamma \hat{b}(\xi+t\zeta,\zeta,\xi)|\, |\zeta|^N \\ &\le C_{N,\nu} \sup_{0\le t\le 1} (1+|\xi+t\zeta|)^{m_2-\rho N+\delta_2\nu}(1+|\xi|)^{m_1+\delta_1\nu}(1+|\zeta|)^{N-\nu}. \end{aligned}$$

With $\nu = N$ we obtain a bound

$$|R_N(\zeta,\xi)| \le C(1+|\xi|)^{m-(\rho-\delta)N} \qquad \text{for} \qquad |\zeta| \le \tfrac{1}{2}|\xi|,$$

and if ν is large, we get a bound

$$\begin{aligned} |R_N(\zeta,\xi)| &\le C_M(1+|\zeta|)^{-M} \qquad \text{for} \qquad |\xi| \le 2|\zeta| \\ &\le C_M(1+\tfrac{1}{2}|\xi|)^{-M/2}(1+|\zeta|)^{-M/2}. \end{aligned}$$

From these estimates, (1.8) follows.

To prove (1.9), write

$$\begin{aligned} &D_x^\beta D_\xi^\alpha a(x,\xi) \\ &= \sum_{\alpha_1+\alpha_2=\alpha} C_{\alpha_1\alpha_2} \int (y-x)^{\alpha_1}(\xi_2-\xi)^\beta e^{i(y-x)\cdot(\xi_2-\xi)} D_\xi^{\alpha_2} b(\xi_2,y,\xi)\, dy\, d\xi_2. \end{aligned}$$

Thus we need a bound

$$(1.11)\qquad \left|\int y^\epsilon \xi_2^\gamma e^{i(y-x)\cdot(\xi_2-\xi)} D_\xi^{\alpha_2} b(\xi_2,y,\xi)\, dy\, d\xi_2\right| \le C(1+|\xi|)^\mu.$$

The left side is equal to (with $b_1 = y^\epsilon b$)

$$\begin{aligned} &\int \xi_2^\gamma e^{-ix\cdot(\xi_2-\xi)} D_{\xi_1}^{\alpha_2}\hat{b}_1(\xi_2,\xi_2-\xi,\xi_1)\big|_{\xi_1=\xi}\, d\xi_2 \\ &\qquad = \int (\xi+\zeta)^\gamma e^{-ix\cdot\zeta} D_{\xi_1}^{\alpha_2}\hat{b}_1(\zeta+\xi,\zeta,\xi_1)\big|_{\xi_1=\xi}\, d\zeta. \end{aligned}$$

The integrand in this last expression is bounded in absolute value by

$$C_\nu|\xi+\zeta|^{|\gamma|}(1+|\zeta+\xi|)^{m_2+\delta_2\nu}(1+|\xi|)^{m-|\alpha|\rho+\delta_1\nu}(1+|\zeta|)^{-\nu}$$

by (1.10). From this, (1.11) easily follows, and we obtain (1.9). This completes the proof of the proposition.

Exercises

1.1. If $a(\xi, x) \in S^{0,m}_{\rho,0,\delta}$, $0 \le \delta < \rho \le 1$, show that $a(D, x) \in OPS^m_{\rho,\delta}$ is the same operator obtained by the method of this section as defined in Chapter II, Section 3.

1.2. If $a(\xi, x)$ is as above and $p(x, \xi) \in S^\mu_{\rho,\delta}$, show that

$$a(D, x)p(x, D) = b(D, x, D)$$

where $b(\xi_2, x, \xi_1) = a(\xi_2, x)p(x, \xi_1)$.

1.3. Define more general classes of multiple symbols and associated pseudodifferential operators, such as $b(x, D, x, D, x, D)$, etc.

§2. Friedrichs' Symmetrization: Proof of the Sharp Gårding Inequality

Suppose $p(x, \xi) \in S^m_{\rho,\delta}$, $0 \le \delta < \rho \le 1$. Let $q(\xi) \ge 0$ be an even function, $q \in C_0^\infty(|\xi| \le 1)$, and suppose $\int q(\xi)^2\, d\xi = 1$. We define a multiple symbol

$$b(\xi_2, x, \xi_1) = \int F(\xi_2, \zeta)p(x, \zeta)F(\xi_1, \zeta)\, d\zeta \tag{2.1}$$

where

$$F(\xi, \zeta) = (1+|\xi|^2)^{-(n/4)\tau}q((1+|\xi|^2)^{-\tau/2}(\zeta-\xi)), \qquad \tau = \tfrac{1}{2}(\rho+\delta). \tag{2.2}$$

This is called the Friedrichs' symmetrization of $p(x, \xi)$. The reason for defining (2.1) is that, if $p(x, \xi) \ge 0$, then $B = b(D, x, D)$ will be a positive pseudodifferential operator, close to $p(x, D)$. In fact, the positivity of B is easy. If $p(x, \xi) \ge 0$, then, for any decent f,

$$\begin{aligned}\int f(\xi_1)b(\xi_2, x, \xi_1)\overline{f(\xi_2)}\, d\xi_2\, d\xi_1 \\ = \int \left\{\int f(\xi_1)F(\xi_1, \zeta)\, d\xi_1\right\} p(x, \zeta) \overline{\left\{\int f(\xi_2)F(\xi_2, \zeta)\, d\xi_2\right\}}\, d\zeta \\ \ge 0.\end{aligned}$$

The next goal is to prove that $b(D, x, D)$ is a pseudodifferential operator.

PROPOSITION 2.1. $b(\xi_2, x, \xi_1) \in S^{m,0}_{\tau,\delta,0}$.

Proof. An inductive argument shows that $D_\xi^\beta F(\xi, \zeta)$ has the form

(2.4) $D_\xi^\beta F(\xi, \zeta)$

$$= (1+|\xi|^2)^{-(\eta/4)\tau} \sum_{|\gamma| \le \beta,\ \gamma_1 \le \gamma} \psi_{\beta,\gamma,\gamma_1}(\xi)((1+|\xi|^2)^{-\tau/2}(\zeta - \xi))^{\gamma_1} D_\sigma^\gamma q(\sigma)$$

where

$$\sigma = (1+|\xi|^2)^{-\tau/2}(\zeta - \xi)$$

and

$$\psi_{\beta,\gamma,\gamma_1}(\xi) \in S_{1,0}^{-(|\beta|-(1-\tau)|\gamma-\gamma_1|)} \subset S_{1,0}^{-\tau|\beta|}. \tag{2.5}$$

Now applying Schwartz' inequality to (2.1), we get

(2.6) $|D_{\xi_2}^\gamma D_x^\beta D_{\xi_1}^\alpha b(\xi_2, x, \xi_1)|$

$$\le \left(\int |D_{\xi_2}^\gamma F(\xi_2, \zeta) D_x^\beta p(x, \zeta)|^2\, d\zeta\right)^{1/2} \left(\int |D_{\xi_1}^\alpha F(\xi_1, \zeta)|^2\, d\zeta\right)^{1/2}.$$

We can estimate the second factor, using (2.4) plus the fact that $|\sigma| \le 1$ on supp $q(\sigma)$, to obtain

$$\int |D_{\xi_1}^\alpha F(\xi_1, \zeta)|^2\, d\zeta \le C_\alpha (1+|\xi_1|^2)^{-\tau|\alpha|}, \tag{2.7}$$

and similarly we get an estimate for the first term

(2.8) $\int |D_{\xi_2}^\gamma F(\xi_2, \zeta) D_x^\beta p(x, \zeta)|^2\, d\zeta$

$$\le C_{\beta,\gamma}(1+|\xi_2|^2)^{-\tau|\gamma|} \int |q_{|\gamma|}(\sigma) p(x + \xi_2 + \sigma((1+|\xi_2|^2)^{\tau/2})|^2\, d\sigma$$

$$\le C'_{\beta,\gamma}(1+|\xi_2|^2)^{m+\delta|\beta|-\tau|\gamma|}$$

with $q_\ell(\sigma) = \max_{|\gamma| \le \ell} |D_\sigma^\gamma q(\sigma)|$. Applying (2.7) and (2.8) to (2.6), we obtain the proposition.

As a consequence of Proposition 1.1 we have $b(D, x, D) \in OPS_{\tau,\delta}^m$. We next sharpen these results and estimate the difference between $b(D, x, D)$ and $p(x, D)$.

THEOREM 2.2. *Let $b(D, x, D)$ be the Friedrichs' symmetrization of $p(x, D)$. Then $b(D, x, D) \in OPS_{\rho,\delta}^m$ and more precisely*

$$b(D, x, D) - p(x, D) \in OPS_{\rho,\delta}^{m-(\rho-\delta)}. \tag{2.9}$$

Proof. We know that $b(D, x, D) = a(x, D)$ is a pseudodifferential operator and we know the symptotic expansion of $a(x, \xi)$:

$$a(x, \xi) \sim \sum_{\alpha \ge 0} p_\alpha(x, \xi)$$

where

$$p_\alpha(x, \xi) = \frac{i^{|\alpha|}}{\alpha!} D_x^\alpha D_{\xi_2}^\alpha b(\xi_2, x, \xi)|_{\xi_2 = \xi}.$$

Previous estimates show that $p_\alpha(x, \xi) \in S^{m-(\rho-\delta)|\alpha|/2}_{\tau,\delta}$, but we aim to improve these estimates, and to show that, for all N sufficiently large,

$$p(x, \xi) - \sum_{|\alpha| \leq N} p_\alpha(x, \xi) \in S^{m-(\rho-\delta)}_{\rho,\delta}. \tag{2.10}$$

Clearly (2.10) implies (2.9).

Using formula (2.4), write

$$\begin{aligned} p_\alpha(x, \xi) \\ = \sum_{|\gamma| \leq |\alpha|,\ \gamma_1 \leq \gamma} \psi_{\alpha,\gamma,\gamma_1}(\xi) \int D_x^\alpha p(x, \xi + \sigma(1+|\xi|^2)^{\tau/2})\sigma^{\gamma_1}(D_\sigma^\gamma q(\sigma))q(\sigma)\, d\sigma. \end{aligned} \tag{2.11}$$

Then (2.5) and an elementary estimate of the integral shows that

$$p_\alpha(x, \xi) \in S^{m-(\tau-\delta)|\alpha|}_{\rho,\delta} \subset S^{m-(\rho-\delta)}_{\rho,\delta} \qquad \text{if} \qquad |\alpha| \geq 2. \tag{2.12}$$

Thus we only have to check $p_\alpha(x, \xi)$ for $|\alpha| = 0$ and 1.

Since $q(\sigma)$ is even and $\int q(\sigma)^2\, d\sigma = 1$, we have

$$\begin{aligned} p_0(x, \xi) &= \int p(x, \xi + \sigma(1+|\xi|^2)^{\tau/2})q(\sigma)^2 \\ &= p(x, \xi) + p'(x, \xi) \end{aligned} \tag{2.13}$$

where

$$p'(x, \xi) = 2(1+|\xi|^2)^\tau \sum_{|\gamma|=2} \int \frac{\sigma^\nu}{\gamma!} \int_0^1 (1-t) D_\xi^\gamma p(x, \xi + t\sigma(1+|\xi|^2)^{\tau/2})\, dt q(\sigma)^2\, d\sigma.$$

We have used Taylor's formula to order 2, expanding in powers of σ. Conveniently, the linear term integrates to zero, because q is even. Thus we obtain

$$p'(x, \xi) \in S^{m+2\tau-2\rho}_{\rho,\delta} = S^{m-(\rho-\delta)}_{\rho,\delta}. \tag{2.14}$$

To handle the case $|\alpha| = 1$, note that $\psi_{\alpha,\gamma,\gamma} \in S^{-1}_{1,0}$, so we shall get

$$p_\alpha(x, \xi) \in S^{m-(\rho-\delta)}_{\rho,\delta}, \qquad |\alpha| = 1, \tag{2.15}$$

provided we establish

$$\psi_{\alpha,\gamma,0}(\xi) \int D_x^\alpha p(x, \xi + \sigma(1+|\xi|^2)^{\tau/2}) D_\sigma^\gamma q(\sigma) \cdot q(\sigma)\, d\sigma \in S^{m-(\rho-\delta)}_{\rho,\delta}, \tag{2.16}$$

assuming $|\alpha| = |\gamma| = 1$. This is proved by writing

$$\begin{aligned} p(x, \xi + \sigma(1+|\xi|^2)^{\tau/2}) \\ = p(x, \xi) + (1+|\xi|^2)^{\tau/2} \sum_{|\alpha|=1} \sigma^\gamma \int_0^1 D_\xi^\gamma p(x, \xi + t\sigma(1+|\xi|^2)^{\tau/2})\, dt. \end{aligned} \tag{2.17}$$

Since $\int D_\sigma^\gamma q(\sigma) \cdot q(\sigma)\, d\sigma = 0$ (q is even), the first term in (2.17) gives no contribution in (2.16), and the desired estimates follows without difficulty.

Thus we have (2.16), and hence (2.15) which, together with equations (2.12) through (2.14), proves (2.10) and establishes the theorem.

Given Theorem 2.2, it is a simple matter to prove the main result of this chapter.

THEOREM 2.3. (Sharp Gårding inequality). *If* $p(x, \xi) \in S^m_{\rho,\delta}$, $0 \le \delta < \rho \le 1$, *and if* $p(x, \xi) \ge 0$, *then*

$$\text{(2.18)} \qquad Re(p(x, D)u, u) \ge -C\|u\|^2_{H^{(m-(\rho-\delta))/2}}, \qquad u \in C_0^\infty.$$

Proof. We have seen in (2.3) that the Friedrichs' symmetrization $B = b(D, x, D)$ of $p(x, D)$ is a positive operator provided $p(x, \xi) \ge 0$. Meanwhile, $b(D, x, D) - p(x, D) \in OPS^{m-(\rho-\delta)}_{\rho,\delta}$. From this, (2.18) is an immediate consequence.

A mysterious element in the symmetrization procedure used here is the necessity for so much care in (2.1) and (2.2). One is tempted to try the simpler formula

$$b_0(\xi_2, x, \xi_1) = \int q(\xi_2 - \zeta)p(x, \zeta)q(\xi_1 - \zeta)\, d\zeta.$$

Unfortunately, this does not yield a b_0 which belongs to one of our classes of symbols. For example, with $q(\xi) = Ce^{-|\xi|^2}$ and $p(x, \xi) = 1 + |\xi|^2$, one has

$$b_0 = C'e^{-|\xi_1-\xi_2|^2/2}(1 + \tfrac{1}{4}|\xi_1 + \xi_2|^2)$$

which is easily seen not to belong to $S^{2,0}_{\rho,0,0}$ for any $\rho > 0$.

Exercises

2.1. Let M be a compact manifold, $K = K(t) \in OPS^1_{1,0}$ a smooth one-parameter family of pseudodifferential operators on M, and suppose $\sigma_K(x, \xi) + \sigma_K(x, \xi)^* \ge 0$. Discuss existence and uniqueness of solutions to the initial value problem

$$\frac{\partial}{\partial t} u = -Ku + g$$

$$u(0) = f.$$

2.2. If $K \in OPS^1_{1,0}$ on M and $|\sigma_K(x, \xi)| \le C < \infty$, show that

$$K: L^2(M) \to L^2(M).$$

2.3. Let $p(x, \xi) \in S^0_{1,0}$, $p(x, \xi) \ge 0$, and $p(x_0, r\xi_0) > 1$. Let $q \in S^0_{1,0}$, $q \ge 0$, $q(x, \xi)$ supported on $\{(x, \xi): p(x, \xi) \ge 1/2\}$. Show that

$$Re(q(x, D)u, u) \le C_1 Re(p(x, D)u, u) + C_2\|u\|^2_{H^{-1/2}},$$

and

$$\|q(x, D)u\|^2_{L^2} \le C_1 Re(p(x, D)u, u) + C_2\|u\|^2_{H^{-1/2}}.$$

2.4. Let $q(\epsilon, x, \xi)$ satisfy the estimates

$$|D_\xi^\alpha D_x^\beta q| \le C_{\alpha\beta}(1 + |\xi|)^{1-|\alpha|}(1 + \epsilon|\xi|)$$

and

$$Re\, q \le -C_0\epsilon|\xi|^2.$$

Show that

$$Re(q(\epsilon, x, D)u, u) \le -C_1\epsilon\|u\|_{H^1}^2 + C_2\|u\|_{L^2}^2.$$

Let $K \in OPS^1$, and let $B \in OPS^2$. Suppose K_1 and B_2 are their principal symbols. We shall say $(\partial/\partial t) - K - \epsilon B$ is *strictly hyperbolic-parabolic* provided the eigenvalues $\mu_j(\epsilon, t, x, \xi)$ of $K_1(t, x, \xi) + \epsilon B_2(t, x, \xi)$ are distinct and satisfy the estimates, for $0 \le \epsilon \le 1$,

$$Re\, \mu_j(\varepsilon, t, x, \xi) \le -C'\varepsilon|\xi|^2 + C''$$
$$|\mu_j(\epsilon, t, x, \xi) - \mu_k(\epsilon, t, x, \xi)| \ge C|\xi|(1 + \epsilon|\xi|), \qquad j \ne k.$$

We say $K + \epsilon B$ is *uniformly dissipatible* if there is a family of self-adjoint operators $R(\epsilon, t, x, D_x)$, bounded in $OPS_{1,0}^0$, such that

$$\|Ru\|_{H^s} \ge C_s\|u\|_{H^s} - C_s'\|u\|_{H^{s-1}}$$

and

$$Re(R(K+\epsilon B)u, u) \le -C\epsilon\|u\|_{H^1}^2 + C'\|u\|_{L^2}^2.$$

2.5. Show that if $(\partial/\partial t) - K - \epsilon B$ is strictly hyperbolic-parabolic, then $K + \epsilon B$ is uniformly dissipatible. Hint: Construct the appropriate symmetrizers R and apply Exercise 2.4.

2.6. If $(\partial/\partial t) - K - \epsilon B$ is strictly hyperbolic-parabolic, discuss the existence, uniqueness, and convergence as $\epsilon \to 0$ of solutions to

$$\frac{\partial}{\partial t} u = Ku + \epsilon Bu + g,$$

$$u(0) = f.$$

CHAPTER VIII

Geometrical Optics and Fourier Integral Operators

The term *geometrical optics* refers generally to phenomena involving consideration of the bicharacteristic strips of a pseudodifferential operator. The basic tool for the study of this subject is the enlarged class of operators (more general than pseudodifferential)

$$Au = \int a(x, \xi)e^{i\varphi(x,\xi)}\hat{u}(\xi)\, d\xi \tag{0.1}$$

with amplitude a and phase function φ satisfying certain conditions discussed in the following pages of this chapter. Such operators were used by Lax [1] in the study of propagation of singularities and developed further by Hörmander in [13] and [15].

The operator (0.1) is essentially the solution operator to a hyperbolic $(\partial/\partial t)u = i\lambda(t, x, D_x)u$, and much of the theory can be developed from this point of view. Indeed, for the first two sections we make no use of the representation (0.1).

The basic result in the theory of geometrical optics is Egorov's theorem, which characterizes the action of conjugating a pseudodifferential operator by the solution operator to a hyperbolic equation. This theorem is stated and proved in Section 1. The proof here is perhaps novel in that no use is made of the representation (0.1). Section 2 shows how Egorov's theorem yields Hörmander's theorem on propagation of singularities.

In the third section, solution operators to hyperbolic equations are shown, at least locally, to be represented in the form (0.1). We also obtain a second proof of Egorov's theorem (this time, using (0.1)). The fourth section discusses a "complex geometrical optics" construction, producing a parametrix for an elliptic boundary value problem.

In the fifth section I discuss (0.1) apart from its arising from a hyperbolic equation, as an example of a Fourier integral operator, and then consider some circumstances under which one can be sure such an operator is indeed the solution operator to a hyperbolic equation. I then give a brief exposition of the more general types of Fourier integral operators discussed by Hörmander in [15] and by Duistermaat in [1]. Since this is a book about pseudodifferential operators, it is not my purpose to report on this general global theory of Fourier integral operators. Rather, the local theory discussed in Sections 2 through 4 will suffice for the work in this book. The material in Section 1 is global.

Finally, in Section 7 I discuss the asymptotic behavior of

$$p(x, D)(a(x, \xi)e^{i\varphi(x,\xi)}),$$

which plays a key role in the method of geometrical optics. Just to be different, I use Egorov's theorem, as proved in Section 1 without reference to (0.1), to reduce this to the study of $p(x, D)(a(x)e^{ix \cdot \xi})$ (with parameters). Section 8 discusses Egorov's theorem for $OPS^m_{1/2,1/2}$.

§1. Egorov's Theorem

We shall study in this section the solution operator to the first order hyperbolic equation

$$\frac{\partial}{\partial t} u = i\lambda(t, x, D_x)u. \tag{1.1}$$

We suppose $u(s)$ is given in $\mathscr{D}'(M)$, where M is a compact manifold, $\lambda(t, x, D_x)$ is a smooth one-parameter family of pseudodifferential operators on M, with symbol $\lambda(t, x, \xi) \sim \lambda_1(t, x, \xi) + \lambda_0(t, x, \xi) + \cdots$, where $\lambda_j(t, x, \xi)$ is homogeneous in ξ of degree j. We assume the principal symbol $\lambda_1(t, x, \xi)$ is real valued (scalar), which makes (1.1) hyperbolic. The other symbols λ_0, λ_{-1}, etc., can be $k \times k$ matrix valued symbols. The solvability of (1.1) was analyzed in Chapter IV of this volume.

Let $S(t, s)$ denote the solution operator to (1.1), taking $u(s)$ to $u(t)$. The analysis of many properties of $S(t, s)$, such as how $WF(u(t))$ is related to $WF(u(s))$, or the norm modulo compacts of the operator $S(t, s)$ on $L^2(M)$ or some other Sobolev space, can be deduced from Egorov's theorem, which studies the conjugation by $S(t, s)$ of a pseudodifferential operator $P_0 = p(x, D)$ on M. Thus, let

$$P(s, t) = S(t, s)p(x, D)S(s, t). \tag{1.2}$$

Our principal result of this section is the following.

EGOROV'S THEOREM. *Let $p(x, D) \in OPS^m_{\rho,1-\rho}$ and assume $\rho > 1/2$. Then $P(s_0, t_0) \in OPS^m_{\rho,1-\rho}$ for all $s_0, t_0 \in \mathbf{R}$, and the principal symbol of $P(s_0, t_0)$ (modulo $S^{m-(2\rho-1)}_{\rho,1-\rho}$) at a point $(x_0, \xi_0) \in T^*(M)$ is equal to $p(y_0, \eta_0)$ where (y_0, η_0) is obtained from (x_0, ξ_0) by following the flow generated by* *the (time dependent) vector field*

$$H_{\lambda_1(t,x,\xi)}$$

from time t_0 to time s_0 provided $p(x, \xi)$ has a scalar principal part.

In this theorem, H_{λ_1} is the Hamiltonian vector field on $T^*(M)$

$$H_{\lambda_1} = \sum_{j=1}^{n} \left(\frac{\partial \lambda_1}{\partial \xi_j} \frac{\partial}{\partial x_j} - \frac{\partial \lambda_1}{\partial x_j} \frac{\partial}{\partial \xi_j} \right),$$

which we have seen in action in Chapter VI.

Proof. Let's regard s_0 as fixed and $t = t_0$ as variable, and write

$$P(t) = P(s_0, t) = S(t, s_0)p(x, D)S(s_0, t).$$

We want to differentiate this identity with respect to t. Note that $(\partial/\partial t)S(t, s_0) = i\lambda(t, x, D)S(t, s_0)$ and $(\partial/\partial t)S(s_0, t) = -iS(s_0, t)\lambda(t, x, D)$. It follows that $P(t)$ solves the initial value problem

$$\frac{\partial}{\partial t} P(t) = i\lambda(t, x, D)P(t) - iP(t)\lambda(t, x, D), \tag{1.3}$$

$$P(s_0) = p(x, D). \tag{1.4}$$

What we shall do next is construct an approximate solution, $Q(t)$, to (1.3) and (1.4), with $Q(t) \in OPS^m_{\rho,1-\rho}$. We shall then show that $Q(t) - P(t)$ must be a smoothing operator, and this will complete the proof of the theorem.

So we are looking for $Q(t) = q(t, x, D) \in OPS^m_{\rho,1-\rho}$, solving the equations

$$Q'(t) = i[\lambda(t, x, D), Q(t)] + R(t), \tag{1.5}$$

$$Q(s_0) = p(x, D) \tag{1.6}$$

where $R(t)$ is a smooth family of smoothing operators on M. We do this by constructing the symbol $q(t, x, \xi)$, as follows. We look for $q(t, x, \xi)$ in the form

$$q(t, x, \xi) \sim q_0(t, x, \xi) + q_1(t, x, \xi) + \cdots$$

where $q_j(t, x, \xi) \in S^{m-j(2\rho-1)}_{\rho,1-\rho}$. In such a case, the symbol of $i[\lambda(t, x, D), Q(t)]$ is asymptotic to

$$\sum_{|\alpha| \geq 0} \frac{i^{-|\alpha|+1}}{\alpha!} (\lambda^{(\alpha)} q_{(\alpha)} - q^{(\alpha)} \lambda_{(\alpha)}) \tag{1.7}$$

$$= i(\tilde{\lambda}_0 q - q\tilde{\lambda}_0) + H_{\lambda_1} q + \{\tilde{\lambda}_0, q\} + \sum_{|\alpha| \geq 2} \frac{i^{-|\alpha|+1}}{\alpha!} (\lambda^{(\alpha)} q_{(\alpha)} - q^{(\alpha)} \lambda_{(\alpha)})$$

where $\tilde{\lambda}_0 = \lambda - \lambda_1 \sim \lambda_0 + \lambda_{-1} + \cdots$. Since $(\partial/\partial t)q(t, x, \xi)$ must be asymptotic to (1.7), this suggests defining $q_0(t, x, \xi)$ by the equation

$$(1.8)\quad \left(\frac{\partial}{\partial t} - H_{\lambda_1}\right) q_0(t, x, \xi) = i(\tilde{\lambda}_0 q_0 - q_0 \tilde{\lambda}_0), \qquad q_0(s_0, x, \xi) = p(x, \xi).$$

Clearly one can solve (1.8) for all t. Note that, if $p(x, \xi)$ is scalar, q_0 is scalar, and the right-hand side of (1.8) is 0, so in this case $q_0(t, x, \xi)$ is constant on the integral curves of $\partial/\partial t - H_{\lambda_1}$. Note that $q_0(t, x, \xi)$ is a locally-bounded family of symbols in $S^m_{\rho,1-\rho}$; $(\partial^j/\partial t^j)q_0(t, x, \xi)$ is locally-bounded in $S^{m+j(1-\rho)}_{\rho,1-\rho}$.

We determine the other $q_j(t, x, \xi)$ recursively as follows. If $Q_0(t) = q_0(t, x, D_x)$, then, using (1.8), we have

$$(1.9)\quad \sigma_{i[\lambda, Q_0(t)] - Q_0'(t)} \sim \{\tilde{\lambda}_0, q_0\} + \sum_{|\alpha| \geq 2} \frac{i^{-|\alpha|+1}}{\alpha!} (\lambda^{(\alpha)} q_{0(\alpha)} - q_0^{(\alpha)} \lambda_{(\alpha)})$$
$$= a_0(t, x, \xi) \in S^{m-(2\rho-1)}_{\rho,1-\rho}.$$

In view of this, we let $q_1(t, x, \xi)$ solve the equation

$$(1.10)\quad \left(\frac{\partial}{\partial t} - H_{\lambda_1}\right) q_1(t, x, \xi)$$
$$= i(\tilde{\lambda}_0 q_1 - q_1 \tilde{\lambda}_0) - a_0(t, x, \xi), \qquad q_1(s_0 x, \xi) = 0.$$

This yields $q_1(t, x, \xi) \in S^{m-(2\rho-1)}_{\rho,1-\rho}$ and

$$\frac{\partial^j}{\partial t^j} q_1(t, x, \xi) \in S^{m-(2\rho-1)+j(1-\rho)}_{\rho,1-\rho}.$$

We continue in this fashion, obtaining $q_j(t, x, \xi) \in S^{m-j(2\rho-1)}_{\rho,1-\rho}$, and then find $q(t, x, \xi) \in S^m_{\rho,1-\rho}$ (with $(\partial^j/\partial t^j)q \in S^{m+j(1-\rho)}_{\rho,1-\rho}$) such that

$$q(t, x, \xi) \sim \sum_{j=0}^{\infty} q_j(t, x, \xi).$$

It follows that $Q(t) = q(t, x, D_x)$ solves (1.5), (1.6).

Our last step is to prove that $P(t) - Q(t)$ is a smoothing operator. We do this using the energy estimates for solutions to hyperbolic equations, established in Chapter IV. That is to say,

$$S(\sigma, \tau): H^\gamma(M) \to H^\gamma(M) \qquad \text{for all } \sigma, \tau, \gamma \in \mathbf{R}.$$

With this in mind, let $A(t) = Q(t)S(t, s_0)$, and our assertion that $P(t) - Q(t)$ is smoothing is seen to be equivalent to

$$(1.11)\qquad A(t)u = S(t, s_0)p(x, D)u \qquad (\text{mod } C^\infty)$$

for all $u \in \mathscr{D}'(M)$. In fact, let $v(t) = A(t)u$. Then $v(s_0) = p(x, D)u$, while

$$\frac{\partial}{\partial t} v = Q'(t)S(t, s_0)u + Q(t)\frac{\partial}{\partial t} S(t, s_0)u$$
$$= i[\lambda, Q(t)]S(t, s_0)u + R(t)S(t, s_0)u + iQ(t)\lambda S(t, s_0)u$$
$$= i\lambda(t, x, D)Q(t)S(t, s_0)u + w(t)$$

with $w(t) = R(t)S(t, s_0)u \in C^\infty(\mathbf{R} \times M)$, so

$$\frac{\partial}{\partial t} v = i\lambda(t, x, D)v + w(t),$$
$$v(s_0) = p(x, D)u.$$

It follows that $v(t) - S(t, s_0)p(x, D)u = \tilde{v}(t)$ solves the initial value problem

$$\frac{\partial}{\partial t}\tilde{v} = i\lambda(t, x, D)\tilde{v} + w(t), \tag{1.12}$$
$$\tilde{v}(s_0) = 0$$

and the energy estimates of Chapter IV applied to (1.12) yields $\tilde{v} \in C^\infty$, which establishes (1.11). This completes the proof of Egorov's theorem.

Let us note that a special case of Egorov's theorem is the invariance of pseudodifferential operators (of type $(\rho, 1-\rho)$ with $\rho > 1/2$) under coordinate transformations $y = y(x)$. In fact, since the behavior of pseudodifferential operators under linear changes of coordinates is trivial, we may as well suppose that $y(x_0) = x_0$ and $(\partial y_j/\partial x_k)(x_0) = \delta_{jk}$. Hence, near x_0, we can connect the identity map to the map $x \to y(x)$ via a smooth one parameter family y_t of diffeomorphisms, $y_0(x) = x$, $y_1(x) = y(x)$, and if $a(t, x) = \sum a_j(t, x)(\partial/\partial x_j)$ is the vector field $a(t) = (d/dt)y_t$, it follows that $S(1, 0)p(x, D)S(0, 1)$ represents the pseudodifferential operator $p(x, D)$ in the y coordinate system, where $S(t, s)$ is the solution operator to the simple hyperbolic equation

$$\frac{\partial}{\partial t} u = \sum a_j(t, x)\frac{\partial}{\partial x_j} u.$$

We now apply Egorov's theorem to the determination of $WF(S(t, s)u)$ in terms of $WF(u)$. We shall work with the special case where $\lambda = \lambda(x, D)$ is independent of t, and write $e^{it\lambda} = S(t, 0)$. This doesn't make things any easier, but it does make the formulas more striking, in my opinion, and the analysis of $WF(e^{it\lambda}u)$ will suffice for the more general analysis of the next section. The reader could work out the more general case as an exercise. Note that, in this special case, Egorov's theorem says that

$$P(t) = e^{it\lambda}p(x, D)e^{-it\lambda}$$

belongs to $OPS^m_{\rho,1-\rho}(\rho > 1/2)$, given $p(x, D)$ in such a class, and its principal symbol $p(t, x, \xi)$, mod $S^{m-(2\rho-1)}_{\rho,1-\rho}$ is given by $p(C(t)(x, \xi))$ where the flow $C(t)$ is the one parameter family of canonical transformations $T^*(M)$ generated by the Hamiltonian flow H_{λ_1}.

We now analyze $WF(e^{it\lambda}u)$ in terms of $WF(u)$. The wave front set of u is the set of common characteristics of operators $p(x, D) \in OPS^0$ which smooth u out:

$$WF(u) = \bigcap_{p(x, D)u \in C^\infty,\ p(x, D) \in OPS^0} \{(x, \xi) \in T^*(M) - 0 \text{ where } p_0(x, \xi) = 0\} \tag{1.13}$$

This definition, due to Hörmander has been discussed in Chapter VI. In the first section of that chapter, in Proposition 3, there is a partial converse to (1.13): $p(x, D) \in C^\infty$ provided $p(x, \xi)$ has order $-\infty$ on a conic neighborhood of $WF(u)$. With this in mind, suppose $WF(u) = \Sigma$, a closed conic subset of $T^*(M)\backslash 0$. Let $p_j(x, D)$ be a family of operators in OPS^0, each of whose symbols vanish in a neighborhood of Σ, but such that

$$\Sigma = \bigcap_j \{(x, \xi) : p_j(x, \xi) = 0\}.$$

Clearly $p_j(x, D) \in C^\infty$ for each j. Using Egorov's theorem, we want to construct a family of pseudodifferential operators $q_j(x, D) \in OPS^0$ such that $q_j(x, D)e^{it\lambda}u \in C^\infty$, this family being rich enough to describe the wave front set of $e^{it\lambda}u$.

Indeed, let $q_j(x, D) = e^{it\lambda}p_j(x, D)e^{-it\lambda}$. Egorov's theorem implies that $q_j(x, D) \in OPS^0$ and gives the principal symbol of $q_j(x, D)$. Since $p_j(x, D)u \in C^\infty$, we have $e^{it\lambda}p_j(x, D)u \in C^\infty$, but then $e^{it\lambda}p_j(x, D) = q_j(x, D)e^{it\lambda}$ implies that $q_j(x, D)e^{it\lambda}u \in C^\infty$. From this it follows that $WF(e^{it\lambda}u)$ is contained in the intersection of the characteristics of the $q_j(x, D)$, which is precisely $C(t)\Sigma$, the image of Σ under the canonical transformation $C(t)$. In other words,

$$WF(e^{it\lambda}u) \subset C(t)WF(u).$$

However, our argument is reversible; $u = e^{-it\lambda}(e^{it\lambda}u)$, so $WF(u) \subset C(-t)WF(e^{it\lambda}u)$. Combining these two inclusions, we have

THEOREM 1.2. $WF(e^{it\lambda}u) = C(t)WF(u)$.

Exercises

1.1. Suppose $(x_0, \xi_0) \notin C(t)WF(u)$ for $0 \le t \le t_0$. Let $0 \le s < t_0$ and let

$$v = \int_0^s e^{i\lambda t}u\, dt.$$

Show that $(x_0, \xi_0) \notin WF(v)$.

1.2. Let $\lambda(x, D)$, $\mu(x, D) \in OPS^1$ have the same principal symbol, $\lambda_1(x, \xi)$, assumed to be real valued (scalar). Show that

$$P(t) = e^{-it\mu(x,D)} e^{it\lambda(x,D)} \in OPS^0$$

for all t, and compute the principal symbol. Hint: Use the equation $P'(t) = -i\mu(x, D)P(t) + iP(t)\lambda(x, D)$, and follow the proof of Egorov's theorem.

1.3. Extend Theorem 1.2 to the analysis of $WF(S(t, 0)u)$ in terms of $WF(u)$.

1.4. Suppose $\lambda_\theta(t, x, D)$ is a set of elements of OPS^1, $\theta \in \Omega$, which is a smooth family, in the sense that

$$\{D_\theta^\gamma \lambda_\theta(t, x, \xi) : \theta \in \Omega\} \text{ is } \textit{bounded} \text{ in } S^1, \text{ for each } \gamma.$$

Prove that for the solution operators $S_\theta(t, s)$ to the hyperbolic equation $(\partial/\partial t)u = i\lambda_\theta(t, x, D)u$, we have the following. If $p(x, D) \in OPS^m_{\rho,1-\rho}$, $\rho > 1/2$, and if

$$P_\theta(s, t) = S_\theta(t, s)p(x, D)S_\theta(s, t),$$

then $P_\theta(s, t)$ is a smooth family of elements of $OPS^m_{\rho,1-\rho}$ in the sense that, for each s, t, each γ,

$$\{D_\theta^\gamma \sigma_{P_\theta(s,t)}(x, \xi) : \theta \in \Omega\} \text{ is } \textit{bounded} \text{ in } S^{m+|\gamma|(1-\rho)}_{\rho,1-\rho}.$$

1.5. Show that $S(t, s)^* = q(t, s, x, D_x)S(s, t)$ with $q(t, x, x, D_x) \in OPS^0_{1,0}$.

1.6. If S is a closed conic hypersurface in $T^*(M)$, define

$$\mathcal{N}^{m,k}_{\rho,1-\rho}(S) = \{p(x, \xi) : |D_x^\beta D_\xi^\alpha p(x, \xi)| \\ \le C|\xi|^{m+|\beta|}(|\xi|^\rho + \text{dist}(\xi, S))^{-|\alpha|-|\beta|-k}\}.$$

Note that $\mathcal{N}^{m,k}_{\rho,1-\rho}(S) \subset S^{m-k\rho}_{\rho,1-\rho}$. Show that, for an elliptic Fourier integral operator $K = S(t, 0)$ such as considered in this section, we have, for $1/2 < \rho \le 1$,

$$K(OP\mathcal{N}^{m,k}_{\rho,1-\rho}(S))K^{-1} \subset OP\mathcal{N}^{m,k}_{\rho,1-\rho}(\tilde{S})$$

where $\tilde{S} = C(S)$ is the image of S under the canonical transformation as described in the statement of Egorov's theorem.

1.7. In the context of Egorov's theorem, suppose that $p(x, D) \in OPS^m_{\rho,0}$, $\rho > 1/2$, and furthermore that $D_x^\beta D_\xi^\alpha p(x, \xi) \in S^{m-(1-\rho)/2-\rho|\alpha|}_{\rho,0}$, $|\alpha| + |\beta| \ge 1$. Show that in this case the principal symbol is given as in Egorov's theorem, modulo $S^{m-(2\rho-1)-(1-\rho)/2}_{\rho,1-\rho}$ rather than merely modulo $S^{m-(2\rho-1)}_{\rho,1-\rho}$.

§2. Propagation of Singularities

In this section we examine propagation of singularities for solutions of a pseudodifferential equation $Pu = f$, reproving the theorem of Chapter

VI, Section 1. We suppose that P is a $k \times k$ matrix of pseduodifferential operator in OPS^m, with symbol

$$P(x, \xi) \sim P_m(x, \xi) + P_{m-1}(x, \xi) + \cdots$$

the $P_j(x, \xi)$ being homogeneous of degree j in ξ. Thus

$$q(x, \xi) = \det P_m(x, \xi)$$

is a scalar valued symbol, homogeneous of degree km in ξ. We make the hypothesis

(H) $\qquad q(x, \xi)$ is real valued.

If $q_1(x, \xi) = |\xi|^{1-km} q(x, \xi)$, then

$$H_{q_1} = \sum_j \left(\frac{\partial q_1}{\partial \xi_j} \frac{\partial}{\partial x_j} - \frac{\partial q_1}{\partial x_j} \frac{\partial}{\partial \xi_j} \right)$$

generates a one parameter group of canonical transformations $C(t)$ on $T^*(M)\backslash 0$. Our propagation of singularities result is the following.

THEOREM 2.1. *If $Pu = f$ and hypothesis (H) is satisfied, then $WF(u)$ is contained in $WF(f) \cup \{(x, \xi): q_1(x, \xi) = 0\}$, and $WF(u)\backslash WF(f)$ is invariant under $C(t)$, acting on $T^*(M)\backslash WF(f)$.*

Proof. If $Pu = f$ only holds on some open set Ω, we can cut u off via a function $\varphi \in C_0^\infty(\Omega)$ equal to 1 near some point $x_0 \in \Omega$, so without loss of generality we can assume $u \in \mathscr{E}'(\Omega)$. Thus we may suppose u is a distribution on some compact manifold M, so the results of Section 1 apply. If $A(x, D) \in OPS^{1-m}$ has principal symbol $A_{1-m}(x, \xi) = |\xi|^{1-km}\, {}^{co}P_m(x, \xi)$, where ${}^{co}P_m(x, \xi)$ is the cofactor matrix of $P_m(x, \xi)$, so ${}^{co}P_m(x, \xi) P_m(x, \xi) = q(x, \xi)I$, write

$$\lambda(x, D)u = A(x, D)Pu = A(x, D)f = g \tag{2.1}$$

where $\lambda(x, D) \in OPS^1$ has principal symbol $\lambda_1(x, \xi) = q_1(x, \xi)I$.

In order to apply Egorov's theorem, we introduce an extra variable, t, and let $v(t, x) = u(x)$; $v \in \mathscr{D}'(\mathbf{R} \times M)$. Thus v solves the equation

$$\frac{\partial}{\partial t} v = i\lambda(x, D)v - ig.$$

We obtain $v(t)$ from $v(0) = u$ by Duhamel's principle:

$$v(t) = e^{it\lambda}u - i \int_0^t e^{i(t-s)\lambda} g \, ds. \tag{2.2}$$

Now suppose $(x_0, \xi_0) \notin WF(u)$, so $(x_0, \xi_0) \notin WF(f)$, which implies that $(x_0, \xi_0) \notin WF(g)$. The theorem asserts that, if $C(t)(x_0, \xi_0) \notin WF(f)$ for

$t \in (t_0, t_1)$, then $C(t)(x_0, \xi_0) \notin WF(u)$ for t in this interval. This is a simple consequence of (2.2). In fact, Theorem 1.2 implies that $C(t)(x_0, \xi_0) \notin WF(e^{it\lambda}u)$, for any t, and a similar application of Egorov's theorem shows that, for $t \in (t_0, t_1)$, $C(t)(x_0, \xi_0)$ is not in the wave front set of $e^{it\lambda} \int_0^t e^{-is\lambda} g\, ds$ (see Exercise 1.1). Thus $C(t)(x_0, \xi_0) \notin WF(v(t))$ for $t \in (t_0, t_1)$, by (2.2); but since $v(t) = u$, this establishes the theorem.

It is worth pointing out here that, for many systems one encounters in practice, it is best not to take the determinant of the principal symbol. To take a trivial example, suppose P is a 2×2 system whose principal symbol $P_m(x, \xi) = p_m(x, \xi)I$, where $p_m(x, \xi)$ is scalar, real valued. If $Pu \in C^\infty$, $WF(u) \subset \{(x, \xi): p_m(x, \xi) = 0\} = \gamma(P)$. Theorem 2.1, as stated, asserts that $WF(u)$ is invariant under the flow on $T^*(M)$ generated H_{q_1} where $q_1(x, \xi) = |\xi|^{1-2m} P_m(x, \xi)^2$. But clearly all the characteristics of q_1 are double, so $H_{q_1} = 0$ on $\gamma(P)$, and hence the flow $C(t)$ is the identity on $\gamma(P)$. Thus Theorem 2.1 trivializes in this case. However, the obvious result of interest is that $WF(u)$ is invariant under the flow generated by $H_{\tilde{q}_1}$ with $\tilde{q}_1 = |\xi|^{1-m} p_m$. More generally we have the following.

THEOREM 2.2. *With the system $Pu = f$ considered in Theorem 1, assume the hypothesis:*

(H′) *There is a symbol $\tilde{P}(x, \xi) \in S^{1-m}$ such that $\tilde{P}(x, \xi)P_m(x, \xi) = \tilde{q}_1(x, \xi)I$, with $\tilde{q}_1(x, \xi) \in S^1$ real valued.*

Then $WF(u) \subset WF(f) \cup \{(x, \xi): \tilde{q}_1(x, \xi) = 0\}$ and $WF(u) \backslash WF(f)$ is invariant under the flow generated by $H_{\tilde{q}_1}$ acting on $T^*(M) \backslash WF(f)$.

The proof is the same as the proof of Theorem 2.1.

A more important example of this phenomenon is given by the equations of linear elasticity for an isotropic medium:

$$Lu = \frac{\partial^2}{\partial t^2} u - (\lambda + \mu) \operatorname{grad} \operatorname{div} u - \mu\, \Delta u = 0 \tag{2.3}$$

where $u = u(t, x)$ in a 3-vector field on $\mathbf{R} \times \mathbf{R}^3$. The quantities λ and μ, called the Lamé constants, are assumed positive. The principal symbol of L is $L_2(t, x, \tau, \xi) = -\tau^2 I + (\lambda + \mu)|\xi|^2 P_\xi + \mu|\xi|^2$ where P_ξ is the orthogonal projection of $\mathbf{R}^3$ onto the space spanned by ξ. For each $(\tau, \xi) \neq 0$, this is a symmetric matrix, with a simple eigenvalue $(\lambda + 2\mu)|\xi|^2 - \tau^2$ and a double eigenvalue $\mu|\xi|^2 - \tau^2$. The analysis of propagation of singularities of solutions to these equations of linear elasticity is a simple consequence of Theorem 2.2, and a special case of the following more general result, whose proof we leave to the reader.

COROLLARY 2.3. *Let $P_m(x, \xi)$ be a $k \times k$ self adjoint matrix, and suppose 0 is an eigenvalue of multiplicity exactly m_j on each connected component*

Γ_j of $\gamma(P) = \{(x, \xi): \det P_m(x, \xi) = 0\}$. Furthermore, suppose that there is an eigenvalue $\mu_j(x, \xi)$ of $P_m(x, \xi)$, of multiplicity exactly m_j on some conic neighborhood of Γ_j, $\mu_j(x, \xi) = 0$ on Γ_j. If $Pu = f$, then

$$WF(u) \subset \bigcup_j \Gamma_j \cup WF(f),$$

and $(WF(u) \cap \Gamma_j) \backslash WF(f)$ is invariant under the flow (on $\Gamma_j \backslash WF(f)$) generated by $H_{|\xi|^{1-m}\mu_j}$.

Exercises

2.1. Define the operator E on $\{f \in \mathcal{D}'(\mathbf{R} \times M): u = 0 \text{ for } t < 0\}$ to be the solution $u = Ef$ to the initial value problem

$$\frac{\partial}{\partial t} u = i\lambda(t, x, D_x)u + f, \qquad u(0) = 0. \tag{2.4}$$

Note that Ef is given by Duhamel's principle

$$Ef(t) = \int_0^t S(t, s)f(s)\, ds. \tag{2.5}$$

Show that $WF(Ef)$ is contained in the union of the null bicharacteristic strips of $(\partial/\partial t) - i\lambda(t, x, D_x)$, passing over $WF(f)$, going in the positive t-direction, plus $WF(f)$, assuming $WF(f) \cap \{|\xi| = 0\} = \emptyset$. Hint: Don't use (2.5). It may be useful to do the following exercises.

2.2. Assume $WF(u) \cap \{\xi = 0\} = \emptyset$. Show that there exists $q(t, x, D_{t,x}) \in OPS^1$ such that $q(t, x, \tau, \xi) = i\tau - i\lambda(t, x, \xi)$ off a small conic neighborhood of $\{\xi = 0\}$, and such that

$$q(t, x, D_{t,x})u - \left(\frac{\partial}{\partial t} - i\lambda(t, x, D)\right)u \in C^\infty.$$

2.3. If u solves (2.4) and $WF(f) \cap \{\xi = 0\} = \emptyset$, show that $WF(u) \cap \{\xi = 0\} = \emptyset$.

§3. The Geometrical Optics Construction

In this section we shall construct an integral formula giving, up to a smooth error, the solution $S(t, s)u(s)$ to the hyperbolic equation

$$\frac{\partial}{\partial t} u = i\lambda(t, x, D)u \tag{3.1}$$

first discussed in Section 1. For notational convenience, we set $s = 0$. We want to obtain $u(t)$ by an integral formula

$$u(t) = \int a(t, x, \xi) e^{i\varphi(t,x,\xi)} \hat{v}_0(\xi)\, d\xi \qquad (\mathrm{mod}\ C^\infty) \tag{3.2}$$

at least for t close to 0. This looks like the Fourier integral representation for a pseudodifferential operator, except for the replacement of $x \cdot \xi$ in the exponent by $\varphi(t, x, \xi)$, which we call a *phase function*. The requirements we place on a phase function φ are:

(3.3) φ is smooth, real valued, and homogeneous of degree 1 in ξ.

(3.4) The gradient $\nabla_x \varphi$ is nowhere vanishing on the conic support of a, for $\xi \neq 0$.

The symbol $a(t, x, \xi)$, which we call the amplitude, is assumed to be in a symbol class $S^m_{\rho,\delta}$, $\rho > 0$, $\delta < 1$. In fact, for, our particular problem of constructing a parametrix for (3.1), $a \in S^0$. There is no loss of generality in assuming $a(t, x, \xi) = 0$ for $|\xi| < 1$.

We first show that, under the hypotheses (3.3) and (3.4), the expression in (3.2) is a well-defined distribution, for $v_0 \in \mathscr{E}'$, using an integration by parts procedure similar to that used to show that pseudodifferential operators act on $\mathscr{E}'$. Our proof starts with the observation that $Le^{i\varphi} = ie^{i\varphi}$ where

$$L = |\nabla_x \varphi|^{-2}\, \nabla_x \varphi \cdot \nabla_x$$

which is a vector field whose coefficients are homogeneous in ξ of degree -1. If $M = L^t$ is the formal adjoint of L and if $\psi \in C_0^\infty$, we have

$$\left\langle \psi, \int a e^{i\varphi} \hat{v}_0(\xi)\, d\xi \right\rangle = \iint \psi(x) a(t, x, \xi) e^{i\varphi(t,x,\xi)} \hat{v}_0(\xi)\, d\xi\, dx \quad \text{(formally)} \tag{3.5}$$

$$= \iint M^k(\psi a) e^{i\varphi} \hat{v}_0(\xi)\, d\xi\, dx \quad \text{(formally)}$$

Now if $a \in S^m_{\rho,\delta}$ it is easy to see that $M^k(\psi a) \in S^{m-k\sigma}_{\rho,\delta}$ where $\sigma = \min(\rho, 1-\delta) > 0$. Since for any $v_0 \in \mathscr{E}'$, $\hat{v}_0(\xi)$ has at most polynomial growth, we see that, for k large enough, the last integral in (3.5) will be absolutely convergent. We take this formula as the definition of $\int a e^{i\varphi} \hat{v}_0(\xi)\, d\xi \in \mathscr{D}'$.

If we are to arrange that (3.2) provide a parametrix for (3.1), we need to compute

$$\lambda(t, x, D_x) u(t) = \int \lambda(t, x, D_x)(a e^{i\varphi}) \hat{v}_0(\xi)\, d\xi \tag{3.6}$$

and also rigorously establish the validity of (3.6). The crucial tool in this analysis is the asymptotic behavior of $\lambda(t, x, D_x)(a e^{i\varphi})$ as $|\xi| \to \infty$, known as the Fundamental Asymptotic Expansion lemma for pseudo-differential operators:

$$\lambda(t, x, D_x)(ae^{i\varphi}) = be^{i\varphi} \tag{3.7}$$

where $b \in S^{m+1}_{\rho,1-\rho}$ assuming $a \in S^m_{\rho,1-\rho}$ and $\rho > 1/2$. Furthermore,

$$b(t, x, \xi) = a(t, x, \xi)\lambda(t, x, \nabla_x\varphi), \qquad \text{mod } S^{m+1-(2\rho-1)}_{\rho,1-\rho}. \tag{3.8}$$

We establish the formulas (3.7) and (3.8) in Section 7. Granted these formulas, we have

$$\int \lambda(t, x, D_x)(ae^{i\varphi})\hat{v}_0(\xi)\, d\xi = \int be^{i\varphi}\hat{v}_0(\xi)\, d\xi,$$

and hence, if $u(t)$ is given by (3.2), we claim that

$$\lambda(t, x, D)u(t) = \int b(t, x, \xi)e^{i\varphi(t,x,\xi)}\hat{v}_0(\xi)\, d\xi. \tag{3.9}$$

In fact, if $v_0 \in C_0^\infty$, $\hat{v}_0$ is rapidly decreasing and the validity of (3.6) is apparent, since only absolutely integrable quantities are involved. Hence (3.9) is valid for $v_0 \in C_0^\infty$. But C_0^∞ is dense in $\mathscr{E}'$. As we have just seen, the operator in (3.2) is continuous from $\mathscr{E}'$ to $\mathscr{D}'$ if φ satisfies hypotheses (3.3) and (3.4); similarly for the operator on the right-hand side of (3.9). Thus, by continuity, (3.9) is valid for all $v_0 \in \mathscr{E}'$.

We now have enough machinery to construct, up to a smooth error, the solution $u(t)$ to (3.1) (for t small), given $u(0) = u_0$. Indeed, if $u(t)$ is given by (3.2), we see that

$$\left(\frac{\partial}{\partial t} - i\lambda(t, x, D_x)\right)u = \int c(t, x, \xi)e^{i\varphi(t,x,\xi)}\hat{v}_0(\xi)\, d\xi \tag{3.10}$$

where

$$c(t, x, \xi) = i\varphi_t a + a_t - ib \tag{3.11}$$

and

$$b = e^{-i\varphi}\lambda(t, x, D_x)(ae^{i\varphi}) \sim a\lambda(t, x, \nabla_x\varphi) + \cdots.$$

We suppose that

$$a(t, x, \xi) \sim \sum_{j \le 0} a_j(t, x, \xi)$$

with a_j homogeneous in ξ of degree j. We want to arrange that $c(t, x, \xi)$ is a symbol of order $-\infty$. In fact, $c(t, x, \xi)$ is asymptotic to a sum

$$\sum_{j \le 1} c_j(t, x, \xi),$$

the c_j being homogeneous of degree j in ξ:

(3.12) $$c_j(t, x, \xi) = i\varphi_t a_{j-1} + \frac{\partial}{\partial t} a_j - ib_j.$$

In particular,

(3.13) $$c_1(t, x, \xi) = i(\varphi_t - \lambda_1(t, x, \nabla_x \varphi))a_0,$$

and setting this term equal to zero yields the *eikonal equation*

(3.14) $$\varphi_t = \lambda_1(t, x, \nabla_x \varphi).$$

This is a first order nonlinear partial differential equation for φ, and it can be solved, for small t, if $\varphi(0, x, \xi)$ is prescribed. We suppose $\varphi(0, x, \xi)$ is homogeneous of degree 1 in ξ and satisfies requirement (3.4). Then, for $|t|$ small, $\varphi(t, x, \xi)$ will also satisfy requirements (3.3) and (3.4). A typical choice for $\varphi(0, x, \xi)$ will be $x \cdot \xi$. The next term to set equal to zero is

$$\begin{aligned} c_0(t, x, \xi) &= i(\varphi_t - \lambda_1(t, x, \nabla_x \varphi))a_{-1} + \frac{\partial}{\partial t} a_0 \\ &\quad - \sum_{j=1}^{n} \frac{\partial \lambda_1}{\partial \xi_j} \frac{\partial}{\partial x_j} a_0 - i\lambda_0 a_0 - \sum_{|\alpha|=2} \frac{1}{\alpha!} \lambda_1^{(\alpha)} \varphi_{(\alpha)} a_0 \end{aligned}$$

and in view of (3.14) this becomes

$$c_0(t, x, \xi) = \left(\frac{\partial}{\partial t} - \sum_{j=1}^{n} \frac{\partial \lambda_1}{\partial \xi_j} \frac{\partial}{\partial x_j}\right) a_0 - i\lambda_0 a_0 - \sum_{|\alpha|=2} \frac{1}{\alpha!} \lambda_1^{(\alpha)} \varphi_{(\alpha)} a_0.$$

Setting c_0 equal to zero thus yields the *first transport equation* for a_0:

(3.15) $$\left(\frac{\partial}{\partial t} - \sum_{j=1}^{n} \frac{\partial \lambda_1}{\partial \xi_j} \frac{\partial}{\partial x_j}\right) a_0 - \left(i\lambda_0 + \sum_{|\alpha|=2} \frac{1}{\alpha!} \lambda_1^{(\alpha)} \varphi_{(\alpha)}\right) a_0 = 0.$$

This is a linear partial differential equation of the first order for a_0, and since

$$X = \frac{\partial}{\partial t} - \sum \frac{\partial \lambda_1}{\partial \xi_j} \frac{\partial}{\partial x_j}$$

is a real vector field, it can be solved on the t-interval for which a solution to (3.14) exists, with $a_0(0, x, \xi)$ specified. A typical initial condition is $a_0(0, x, \xi) = 1$. Setting the other c_j equal to zero yields analogous transport equation for a_{-1}, a_{-2}, etc. In fact, as one can see from (3.12), for $j \le -1$,

(3.16) $$c_j(t, x, \xi) = Xa_j - i\left(\lambda_0 + \sum_{|\alpha|=2} \frac{1}{\alpha!} \lambda_1^{(\alpha)} \varphi_{(\alpha)}\right) a_j - d_j$$

where $d_j(t, x, \xi)$ is expressible in terms of $\varphi, a_0, a_{-1}, \dots, a_{j+1}$. Thus we can solve $Xa_j - i\lambda_0 a_j - d_j = 0$ specifying $a_j(0, x, \xi)$ to be homogeneous of degree j in ξ. A typical initial condition is $a_j(0, x, \xi) = 0, j \le -1$.

Such a determination of the $a_j(t, x, \xi)$ having been made, you can choose a smooth family $a(t, x, \xi) \in S^0$ asymptotic to this infinite sum, and it follows that $c(t, x, \xi)$ given by (3.11) will be a smooth family of symbols in $S^{-\infty}$ (thinking of t as a parameter). Thus if $u(t)$ is given by (3.2), it follows that $((\partial/\partial t) - i\lambda)u \in C^\infty(I \times M)$ where I is the t-interval on which the eikonal equation (3.14) is solved. This proves the following result:

THEOREM 3.1. *Suppose* φ *solves the eikonal equation* (3.14) *and suppose, for* $t \in I$, $\varphi(t, x, \xi)$ *satisfies the hypotheses* (3.3) *and* (3.4). *Suppose* $a \sim \sum_{j \le 0} a_j$ *and the* a_j *solve the transport equations,* (3.15) *and* (3.16). *Suppose*

$$u_0 = \int a(0, x, \xi) e^{i\varphi(0,x,\xi)} \hat{v}_0(\xi)\, d\xi$$

for some $v_0 \in \mathscr{E}'$. *Then* $u(t)$, *defined by* (3.2), *differs from the solution to*

$$\frac{\partial}{\partial t} u = i\lambda(t, x, D_x)u, \qquad u(0) = u_0, \qquad (t \in I)$$

by a smooth error. In particular, if $\varphi(0, x, \xi) = x \cdot \xi$ *and* $a(0, x, \xi) = 1$, *we have* $u(0) = v_0$.

Exercise

3.1. Extend the theorem to cover $a(0, x, \xi) \in S^m_{\rho,1-\rho}$ $\rho > 1/2$.

The last assertion of the theorem follows from the Fourier inversion formula. This is the most usual case, but it is sometimes useful to have the flexibility afforded by our more general statement. To illustrate this, we give another proof of Egorov's theorem. For notational simplicity, we assume $t = 0$, and consider

$$P(s) = S(0, s)p(x, D)S(s, 0). \tag{3.17}$$

For $|s|$ small we have

$$S(s, 0)v = \int a(s, x, \xi) e^{i\varphi(s,x,\xi)} \hat{v}(\xi)\, d\xi \tag{3.18}$$

with a, φ as described in Theorem 3.1, taking $\varphi(0, x, \xi) = x \cdot \xi$, $a(0, x, \xi) = 1$. Therefore,

$$p(x, D)S(s, 0)v = \int b(s, x, \xi) e^{i\varphi(s,x,\xi)} \hat{v}(\xi)\, d\xi \tag{3.19}$$

where $b(s, x, \xi) = e^{-i\varphi} p(x, D)(ae^{i\varphi}) \in S^m_{\rho,1-\rho}$ provided

$$p(x, D) \in OPS^m_{\rho,1-\rho} \qquad \rho > 1/2.$$

Now let $\tilde{b}(s, x, \xi) = b(s, x, \xi)$ and let $\psi(s, x, \xi) = \varphi(s, x, \xi)$; we think of s as being fixed. We have

$$p(x, D)S(s, 0)v = \int \tilde{b}(s, x, \xi) e^{i\psi(s,x,\xi)} \hat{v}(\xi)\, d\xi, \tag{3.20}$$

and Theorem (3.1) is directly applicable:

$$S(0, s)p(x, D)S(s, 0)v = \int \tilde{b}(0, x, \xi)e^{i\psi(0,x,\xi)}\hat{v}(\xi)\, d\xi \tag{3.21}$$

where $\tilde{b}(t, x, \xi) \in S^m_{\rho,1-\rho}$ solves certain transport equations and $\psi(t, x, \xi)$ solves the eikonal equation

$$\frac{\partial}{\partial t}\psi = \lambda_1(t, x, \nabla_x\psi), \tag{3.22}$$

with initial condition at $t = s$:

$$\psi(s, x, \xi) = \varphi(s, x, \xi). \tag{3.23}$$

But clearly the unique solution to (3.22), (3.23) is

$$\psi(t, x, \xi) = \varphi(t, x, \xi). \tag{3.24}$$

In particular, $\psi(0, x, \xi) = \varphi(0, x, \xi) = x \cdot \xi$, so

$$S(0, s)p(x, D)S(s, 0)v = \int \tilde{b}(0, x, \xi)e^{ix\cdot\xi}\hat{v}(\xi)\, d\xi \tag{3.25}$$

which shows that $P(s)$ is a pseudodifferential operator, for $|s|$ small. By the same argument, $P(s, t) = S(t, s)p(x, D)S(s, t)$ is seen to be a pseudodifferential operator for s close to t. But iterating the formula

$$S(t, s)p(x, D)S(s, t) = S(t, \sigma)S(\sigma, s)p(x, D)S(s, \sigma)S(\sigma, t) \tag{3.26}$$

shows that $P(s, t)$ is a pseudodifferential operator for all s, t. The principal symbol can be computed from the transport equations yielding $b(s, x, \xi)$ and $\tilde{b}(s, x, \xi)$. This method of computing the principal symbol of $P(s, t)$ is less straightforward than the method of Section 1, but instructive. The following argument was worked out by Mark Farris and David Yingst. Assume here that P is scalar.

The first transport equations for $a(t, x, \xi)$ and $\tilde{b}(t, x, \xi)$ are

$$\begin{gathered} Xa_0(t, x, \xi) - \left[i\lambda_0(t, x, \nabla_x\varphi) + \sum_{|\alpha|=2} \frac{1}{\alpha!}\lambda_1^{(\alpha)}\varphi_{(\alpha)}\right]a_0 = 0, \\ a_0(0, x, \xi) = 1 \end{gathered} \tag{3.27}$$

and

$$\begin{gathered} X\tilde{b}_0(t, x, \xi) - \left[i\lambda_0(t, x, \nabla_x\psi) + \sum_{|\alpha|=2} \frac{1}{\alpha!}\lambda_1^{(\alpha)}\psi_{(\alpha)}\right]\tilde{b}_0 = 0, \\ \tilde{b}_0(s, x, \xi) = a_0(s, x, \xi)p(x, \nabla\varphi), \end{gathered} \tag{3.28}$$

and recall that $\varphi \equiv \psi$. X is the vector field

$$\frac{\partial}{\partial t} - \sum_j \frac{\partial\lambda_1}{\partial\xi_j}(t, x, \nabla\varphi)\frac{\partial}{\partial x_j}.$$

Write

$$\tilde{b}_0 = a_0 d_0,$$

and derive an equation for $d_0(t, x, \xi)$. In fact, (3.28) yields

$$a_0 X d_0 + d_0 X a_0 = \left[i\lambda_0 + \sum_{|\alpha|=2} \frac{1}{\alpha!} \lambda_1^{(\alpha)} \varphi_{(\alpha)} \right] \tilde{b}_0,$$

and substituting in (3.27) gives $a_0 X d_0 = 0$, or

$$X d_0 = 0. \tag{3.29}$$

Thus in (3.25) the principal symbol is $\tilde{b}_0(0, x, \xi) = d_0(0, x, \xi)$, which is determined from (3.29) via the initial condition

$$d_0(s, x, \xi) = p(x, \nabla_x \varphi(s, x, \xi)). \tag{3.30}$$

So it remains to consider the initial value problem (3.29), (3.30) and evaluate $d_0(0, x, \xi)$. Consider the following maps

$$\begin{array}{ccc} (\nabla_\xi \varphi(t, x, \xi), \xi) & \xrightarrow{C(t)} & (x, \nabla_x \varphi(t, x, \xi)) \\ & {\scriptstyle B(t)} \searrow \quad \swarrow {\scriptstyle A(t)} & \\ & (x, \xi) & \end{array}$$

Set $(x(\sigma), \xi(\sigma)) = B(\sigma)(x_0, \xi_0)$. We claim that

$$\begin{aligned} \dot{x} &= -\nabla_\xi \lambda_1(\sigma, x(\sigma), \nabla_x \varphi(\sigma, x(\sigma), \xi_0)), \\ \dot{\xi} &= 0. \end{aligned} \tag{3.31}$$

Clearly $\dot{\xi} = 0$. Differentiating $x_0 = \nabla_\xi \varphi(\sigma, x(\sigma), \xi_0)$ yields

$$0 = \nabla_\xi \varphi_t(\sigma, x(\sigma), \xi_0) + \dot{x} \cdot \nabla_x \nabla_\xi \varphi(\sigma, x(\sigma), \xi_0),$$

while $\varphi_t = \lambda_1(t, x, \nabla_x \varphi)$ yields

$$\nabla_\xi \varphi_t = \nabla_\xi \nabla_x \varphi(t, x, \xi) \cdot \nabla_\xi \lambda_1(t, x, \nabla_x \varphi).$$

For small $|t|$, the matrix of second partials $\nabla_x \nabla_\xi \varphi$ is invertible, so the first formula of (3.31) follows. Consequently, $(0, x_0, \xi_0) \mapsto (\sigma, B(\sigma)(x_0, \xi_0))$ are the integral curves for X and hence

$$d_0(0, x_0, \xi_0) = d_0(x, B(s)(x_0, \xi_0)). \tag{3.32}$$

Combining this with (3.30), we see that

$$\begin{aligned} d_0(0, x_0, \xi_0) &= p(A(s)^{-1} B(s)(x_0, \xi_0)) \\ &= p(C(s)(x_0, \xi_0)), \end{aligned}$$

which completes the second proof of Egorov's theorem.

§4. Parametrix for Elliptic Evolution Equations

Now we construct a parametrix for the equation

$$(4.10)\qquad \frac{\partial}{\partial t} u = a(x, D_x)u, \qquad u(0) = v$$

where $a(x, \xi) \leq -c|\xi|$, $a \in S^1$. This will be in a spirit similar to Section 3, but it will lead to the appearance of complex valued phase functions. Thus, our parametrix will be of the form

$$(4.2)\qquad w(t) = \int b(t, x, \xi)e^{i\varphi(t,x,\xi)}\hat{v}(\xi)\, d\xi$$

where $b \in S^0$ and φ satisfies hypotheses (3.3) and (3.4) except that, instead of assuming φ is real valued, we will have $Im\varphi \geq 0$. It is easy to see that this minor change does not affect the proof given in Section 3 that the operator in (4.2) is well defined on $v \in \mathscr{E}'$.

In order to avoid complications, we shall restrict our attention to the case where $B = -a(x, D)^2$ is a *differential operator*, i.e., $-a(x, D_x)$ is the square root of the elliptic differential operator $-B$. (As we shall see later in Chapter XII, square roots of second order strongly elliptic operators are always pseudodifferential operators.) For a treatment of the more general case, we refer the reader to the work of Melin and Sjöstrand [1] on Fourier integral operators with complex phase.

Thus, a solution to (4.1) with $u(0) = v$ solves the boundary-value problem

$$(4.3)\qquad \left(\frac{\partial^2}{\partial t^2} + B\right)u = 0, \qquad u(0) = v,$$

and $(\partial^2/\partial t^2) + B$ is an elliptic differential operator on $\mathbf{R} \times X$. As we have seen in Chapter VI (4.3), the Dirichlet problem is a regular elliptic boundary value problem. It follows that any $w(t)$ on $I \times X$ satisfying

$$\left(\frac{\partial^2}{\partial t^2} + B\right)w \in C^\infty(I \times X), \qquad w(0) - v \in C^\infty$$

must differ from the exact solution to (4.1) by a smooth function on $I \times X$. We now obtain such a w in the form (4.2).

Applying $(\partial^2/\partial t^2) + B$ to (4.2), we get

$$(4.4)\qquad \left(\frac{\partial^2}{\partial t^2} + B\right)w = \int c(t, x, \xi)e^{i\varphi(t,x,\xi)}\hat{v}(\xi)\, d\xi$$

where $c(t, x, \xi) \in S^2$ is given by

$$c(t, x, \xi) = -\left[\varphi_t^2 + \sum_{j,k} b_{jk}(x)(D_j\varphi)(D_k\varphi)\right]b$$
$$+ 2i[\varphi_t b_t + \sum b_{jk}(x)(D_j\varphi)(D_k b)] + ib\left(\frac{\partial^2}{\partial t^2} + B - e\right)\varphi$$
$$+ \left(\frac{\partial^2}{\partial t^2} + B\right)b$$

where we suppose

$$B = \sum_{j,k} b_{jk}(x)D_jD_k + \sum_j d_j(x)D_j + e(x).$$

The ellipticity condition on B implies

$$\sum_{j,k} b_{jk}(x)\xi_j\bar{\xi}_k \geq c|\xi|^2.$$

Now, as in Section 3, we shall try to make $c(t, x, \xi)$ to have order $-\infty$. Thus we will suppose

$$b(t, x, \xi) \sim \sum_{j \leq 0} b_j(t, x, \xi),$$

with b_j homogeneous of degree j in ξ, separate out terms homogeneous of degree 2, 1, 0, . . . in $c(t, x, \xi)$, and attempt to make each of these terms vanish. The leading term is

$$c_2(t, x, \xi) = -\left[\varphi_t^2 + \sum_{j,k} b_{jk}(x)(D_j\varphi)(D_k\varphi)\right]b_0,$$

so setting it equal to zero is equivalent to solving the eikonal equation

$$\varphi_t^2 + \sum_{j,k} b_{jk}(x)(D_j\varphi)(D_k\varphi) = 0. \tag{4.5}$$

We would like to solve (4.5), prescribing $\varphi(0, x, \xi) = x \cdot \xi$. But note that the left-hand side of (4.5) is a positive quadratic form in $(\varphi_t, \nabla_x\varphi)$. Thus a real valued solution cannot possibly exist. Any solution must be complex valued. But then our basic local existence theorem for first order (non-linear) partial differential operators breaks down. However, we can obtain a formal power series in t,

$$\varphi \sim \sum_{j=0}^{\infty} \varphi_j(x, \xi)t^j$$

which formally solves (4.5), since the surface $t = 0$ is noncharacteristic. Given $\varphi(0, x, \xi) = x \cdot \xi$, there are two such expansions, but since we want $Im\varphi \geq 0$, we choose the one with $Im\varphi_1 > 0$. More specifically:

$$\frac{\partial}{\partial t}\varphi(0, x, \xi) = i\left[\sum_{j,k} b_{jk}(x)\xi_j\xi_k\right]^{1/2}. \tag{4.6}$$

Having obtained the formal series $\sum \varphi_j(x, \xi)t^j$, we use Borel's theorem to obtain a $\varphi(t, x, \xi)$ asymptotic to this sum, i.e.

$$\varphi(t, x, \xi) - \sum_{j=0}^{N} t^j\varphi_j(x, \xi) = 0(t^{N+1}), \qquad t \to 0.$$

This will be our phase function. It follows easily that

$$\varphi_t^2 + \sum_{j,k} b_{jk}(x)(D_j\varphi)(D_k\varphi) = 0(t^\infty), \qquad t \to 0. \tag{4.7}$$

Also (4.6) is valid, which yields

$$Im\varphi(t, x, \xi) \geq ct|\xi|, \qquad 0 \leq t \leq t_0. \tag{4.8}$$

We now consider the next term

$$\begin{aligned} c_1(t, x, \xi) = 2i&\left[\varphi_t\frac{\partial}{\partial t}b_0 + \sum b_{jk}(x)(D_j\varphi)\frac{\partial}{\partial x_k}b_0\right] \\ &+ ib_0\left(\frac{\partial^2}{\partial t^2} + B\right)\varphi - \left[\varphi_t^2 + \sum b_{jk}(x)(D_j\varphi)(D_k\varphi)\right]b_{-1} \end{aligned} \tag{4.9}$$

As with c_2, we shall only make this vanish to infinite order at $t = 0$. Since b_{-1} will be found to be smooth, the last term can hence be neglected, and what remains is a linear partial differential equation for b_0, our first transport equation. As with the eikonal equation, this cannot be set equal to zero and the resulting equation solved exactly because the vector field

$$\varphi_t\frac{\partial}{\partial t} + \sum b_{jk}(x)(D_j\varphi)\frac{\partial}{\partial x_j}$$

does not have real coefficients. However, it can be solved to infinite order at $t = 0$, with $b_0(0, x, \xi) = 1$ specified.

Requiring the remaining $c_j(t, x, \xi)$ to vanish to infinite order at $t = 0$, we can obtain recursively appropriate $b_j(t, x, \xi)$, $j \leq -1$, and we can require that $b_j(0, x, \xi) = 0, j \leq -1$. Having obtained all these b_j, we can find

$$b(t, x, \xi) \sim \sum_{j\leq 0} b_j(t, x, \xi),$$

$b(0, x, \xi) = 1$. This determines our parametrix $w(t)$, via (4.2). Clearly $w(0) = v$, $((\partial^2/\partial t^2) + B)w$ is given by (4.4) with φ satisfying (4.8) and

$$c(t, x, \xi) = 0(t^\infty) + 0(|\xi|^{-\infty}), \tag{4.10}$$

a similar estimate holding for all derivatives of $c(t, x, \xi)$. From (4.8) and (4.10) we want to deduce that $((\partial^2/\partial t^2) + B)w$ is smooth, in $\mathbf{R}^+ \times X$, up to the boundary $t = 0$.

We begin by showing that, whenever φ and c satisfy (4.8) and (4.10), then

$$\int c(t, x, \xi) e^{i\,\varphi(t,x,\xi)} \hat{v}(\xi)\, d\xi \tag{4.11}$$

is continuous on $\mathbf{R} \times X$, for any $\hat{v}(\xi)$ with polynomial growth, say $|\hat{v}(\xi)| \le C|\xi|^k$. Indeed, the integrand in (4.11) is bounded by

$$c_j(t^j + (1+|\xi|)^{-j})(1+|\xi|)^k e^{-tc_0|\xi|} \le c_j(1+|\xi|)^{k-j} + c_j' t^j e^{-tc_0(|\xi|+1)}(1+|\xi|)^k$$

But clearly

$$\sup_{0 \le t < \infty} t^j e^{-tc_0(|\xi|+1)} = c_j(1+|\xi|)^{-j},$$

so we have

$$|c(t, x, \xi) e^{i\varphi(t,x,\xi)} \hat{v}(\xi)| \le c_j(1+|\xi|)^{k-j}. \tag{4.12}$$

For j large enough, the right-hand side of (4.12) is absolutely integrable. It follows that (4.11) in continuous on $\mathbf{R}^+ \times X$.

More generally, any ℓ^{th} order (t, x) derivative of $((\partial^2/\partial t^2) + B)w(t)$ is of the form (4.11), with $c(t, x, \xi)$ replaced by $\tilde{c}(t, x, \xi)$ satisfying

$$|\tilde{c}(t, x, \xi)| \le c_j(t^j + (1+|\xi|)^{-j})(1 + |\xi|)^\ell$$

so the same argument shows that all derivatives of $((\partial^2/\partial t^2) + B)w$ are continuous on $\mathbf{R} \times X$, which is what we wanted to establish.

This completes our construction of the parametrix for (4.1). In Chapter XII we shall look at it in further detail and obtain some information about the spectrum of the elliptic operator B.

Exercises

4.1. If $\varphi(t, x, \xi)$ is the complex phase function constructed in this section, with $\varphi(0, x, \xi) = x \cdot \xi$, consider the function

$$\alpha(t, x, \xi) = e^{i\varphi(t,x,\xi) - ix\cdot\xi}.$$

Show that, for $0 \le t \le 1$,

(i) $\alpha(t, x, \xi)$ is a bounded set of symbols in $S^0_{1,0}$.

(ii) $D_t^k \alpha(t, x, \xi)$ is a bounded set of symbols in $S^k_{1,0}$.

(iii) $\alpha(t, x, \xi) \in S^{-\infty}_{1,0}$ for $t > 0$.

(iv) $t^\ell |\xi|^\ell \alpha(t, x, \xi)$ is a bounded set of symbols in $S^0_{1,0}$, $\ell = 0,1,2\ldots$

4.2. Using the parametrix constructed from the solution to (4.1), show that, if $p(x, D_x) \in OPS^0$, $p(x, \xi)$ is supported in a conic set disjoint from $WF(v) \subset T^*(X)$, then $p(x, D_x)u$ is smooth up the boundary $t = 0$. (This is a microlocal version of the regularity up to the boundary for solutions to regular elliptic boundary value problems.) Hint: Use the solution to Exercise 4.1.

§5. Fourier Integral Operators

The operators of the form

$$Au = \int a(x, \xi) e^{i\varphi(x,\xi)} \hat{u}(\xi) \, d\xi \tag{5.1}$$

with $a(x, \xi) \in S^m_{\rho,\delta}$, $0 < \rho \leq 1$, $0 \leq \delta < 1$, which we have dealt with in previous sections, are special cases of operators Hörmander calls Fourier integral operators. Without getting too far into the general theory of Fourier integral operators, in this section we do want to look at some more properties of the operators of type (5.1) and discuss how and for what reasons one generalizes this sort of operator. We assume φ satisfies hypotheses (3.3) and (3.4).

We first look at a general method of obtaining $WF(Au)$ in terms of WFu. By one characterization of wave front sets, to say $(x_0, \xi_0) \notin WF(Au)$ is equivalent to saying that for some $\chi \in C_0^\infty$, $\chi(x) = 1$ near x_0, $\langle \chi(x) e^{-ix \cdot \theta}, Au \rangle$ is rapidly decreasing as $\theta \to \infty$ on some conic neighborhood Γ of ξ_0. Now we have

$$\langle \chi(x) e^{-ix \cdot \theta}, Au \rangle = \iiint u(y) \chi(x) a(x, \xi) e^{i\varphi(x,\xi) - iy \cdot \xi - ix \cdot \theta} \, dy \, dx \, d\xi \tag{5.2}$$

the integral with respect to y being taken in the distribution sense, and the ξ integral being regarded as an oscillatory integral as in Section 4.

Before we proceed with the analysis of (5.2), let us make some preliminary observations that will simplify the analysis. Suppose $WF(u)$ is contained in a small conic neighborhood of (y_0, η_0). We may as well suppose that $u(y)$ is supported near y_0. Also, since Au defined by (5.1) would only be altered by a smooth function, we may as well suppose that $a(x, \xi)$ is supported for ξ in a small conic neighborhood of η_0, and that $a(x, \xi) = 0$ for $|\xi| < 1$. Finally, without loss of generality, we can suppose $u(y)$ is continuous. Indeed, any $u \in \mathscr{E}'$ can be smoothed out by applying some negative power of the Laplace operator, which would be compensated in (5.1) if $a(x, \xi)$ had its order increased.

With these hypotheses, we shall show that (5.2) is rapidly decreasing as $\theta \to \infty$ in a cone Γ with the property that, for $\theta \in \Gamma$, the phase function

$$\Phi = \varphi(x, \xi) - y \cdot \xi - x \cdot \theta$$

has no critical point as a function of x and ξ, i.e., assuming that for $\theta \in \Gamma$, $|\nabla_x \varphi - \theta| + |\xi|\,|\nabla_\xi \varphi - y|$ is bounded away from $0(|\xi|, |\theta| \geq 1)$. Let

$$L = [|\nabla_x\varphi - \theta|^2 + (|\xi| + |\theta|)^2 |\nabla_\xi \varphi - y|^2]^{-1}$$
$$\times [(\nabla_x \varphi - \theta)\cdot \nabla_x + (|\xi| + |\theta|)^2(\nabla_\xi\varphi - y)\cdot\nabla_\xi].$$

It follows that

$$Le^{i\Phi} = ie^{i\Phi}. \tag{5.3}$$

Note that the coefficient of ∇_x in L is homogeneous of degree -1 in (θ, ξ), and the coefficient of ∇_ξ is homogeneous of degree 0 in (θ, ξ). Also, by hypothesis, these coefficients are smooth for $\theta \in \Gamma$, $(x, \xi) \in \operatorname{supp} a$. We may integrate by parts, using (5.3), to obtain

$$\langle \chi(x)e^{-ix\cdot\theta}, Au\rangle = \iiint u(y)(L^t)^k(\chi(x)a(x,\xi))e^{i\Phi}\,dy\,dx\,d\xi.$$

The coefficient of ∇_x in L^t is homogeneous of degree -1 in (θ, ξ), and the coefficient of ∇_ξ in L^t is homogeneous of degree 0 in (θ, ξ); also L^t contains a zero order term, which is homogeneous of degree -1 in (θ, ξ). If $a(x, \xi) \in S^m_{\rho,\delta}$, it follows that $(L^t)^k(\chi(x)a(x,\xi))$ is a sum of symbols of the form

$$b(x, y, \xi, \theta) = \gamma(x, y, \xi, \theta) D_x^\beta D_\xi^\alpha a(x, \xi)$$

where $0 \leq |\alpha| + |\beta| \leq k$ and where $\gamma(x, y, \xi, \theta)$ is homogeneous of degree $-(k - |\alpha|)$ in (ξ, θ), and compactly supported in x. Therefore,

$$\begin{aligned} |b(x, y, \xi, \theta)| &\leq C(|\xi| + |\theta|)^{-k+|\alpha|}(1 + |\xi|)^{m-\rho|\alpha|+\delta|\beta|} \\ &\leq C(1+|\xi|)^{m+(1-\rho)|\alpha|+\delta|\beta|-k}(|\theta| \geq 1) \\ &\leq C(1+|\xi|)^{m-\sigma k} \end{aligned}$$

where $\alpha = \min(\rho, 1-\delta)$, assuming $\theta \in \Gamma$. More generally,

$$|D_x^\gamma b| \leq C(1 + |\xi|)^{m+\delta|\gamma|-\sigma k}. \tag{5.4}$$

Thus $\langle \chi(x)e^{-ix\cdot\theta}, Au\rangle$ is a sum of terms of the form

$$\iiint u(y)b(x, y, \xi, \theta)e^{i\varphi(x,\xi) - iy\cdot\xi - ix\cdot\theta}\,dy\,d\xi\,dx = \int c(x,\theta)e^{-ix\cdot\theta}\,dx$$

where $c(x, \theta) = \iint u(y)b(x, y, \xi, \theta)e^{i\varphi(x,\xi)-iy\cdot\xi}\,dy\,d\xi$. Now by (5.4), for each γ, $D_x^\gamma c(x, \theta)$ is a continuous compactly supported function of x, bounded by a constant independent of θ provided $|\gamma| < \sigma k - m - n$. Hence $\int c(x,\theta)e^{-ix\cdot\theta}\,dx$ is rapidly decreasing as $\theta \to \infty$, $\theta \in \Gamma$. Since any $u \in \mathscr{E}'$ may be decomposed into a finite sum of $u_j \in \mathscr{E}'$ with small wave front sets, the above argument establishes the following.

THEOREM 5.1. *If Au is given by* (5.1), *then* $WF(Au) \subset \{(x, \theta) : (\nabla_\xi\varphi, \xi) \in WF(u)$ *for some* $(x, \xi) \in \text{cone supp } a$ *and* $\nabla_x\varphi(x,\xi) = \theta\}$.

Thus, $WF(u)$ and $WF(Au)$ are related by the canonical relation

$$(\nabla_\xi \varphi, \xi) \mapsto (x, \nabla_x \varphi) \tag{5.5}$$

which is in general not necessarily single valued, nor is its domain necessarily all of T^*X. Of course, if $\varphi(x, \xi)$ is sufficiently close to $x \cdot \xi$ in the C^∞ topology, the relation (5.5) will indeed be a function on T^*X, close to the identity map, and hence a diffeomorphism. It is not too hard to show that such transformations of T^*X always preserve the symplectic form; hence the name canonical transformations. We shall not present the details here, but the reader may try to work it out as an exercise, or see Duistermaat [1].

In particular, if $Au = S(s, t)u$ is the solution operator to the hyperbolic equation $(\partial/\partial t)v = i\lambda(t, x, D_x)v$ considered in Sections 1 and 3,

$$S(t, 0)u = \int a(t, x, \xi)e^{i\varphi(t,x,\xi)}\hat{u}(\xi)\, d\xi$$

it follows that, for $|t|$ small, $\varphi(t, \cdot, \cdot)$ is close to $x \cdot \xi$ in the C^∞ topology, where φ solves the eikonal equation $\varphi_t = \lambda_1\, (t, x, \nabla_x \varphi)$ with initial condition $\varphi(0, x, \xi) = x \cdot \xi$. Consequently, the relation

$$(\nabla_\xi \varphi(t, x, \xi), \xi) \mapsto (x, \nabla_x \varphi(t, x, \xi)) \tag{5.6}$$

defines a bijective canonical transformation $C(t)$ on $T^*(X)$, for each t close to $t = 0$. It is not hard to see that $C(t)$ corresponds to the flow on T^*X generated by the (t-dependent) Hamiltonian field $H_{\lambda_1(t,x,\xi)}$. In fact, clearly $C(0)$ is the identity map. On the other hand, letting $C(t)(x_0, \xi_0) = (x(t), \xi(t))$, which is equivalent to saying

$$x_0 = \nabla_\xi \varphi(t, x(t), \xi_0), \qquad \xi(t) = \nabla_x \varphi(t, x(t), \xi_0),$$

an essentially straightforward application of the chain rule, which we leave to the reader, plus use of the eikonal equation, yields,

$$\begin{aligned} \dot{\xi}(t) &= \nabla_x \lambda_1(t, x(t), \xi(t)), \\ \dot{x}(t) &= -\nabla_\xi \lambda_1(t, x(t), \xi(t)). \end{aligned}$$

Thus this special case of Theorem 5.1 is contained in Exercise 1.3.

We next show that, if on a conic set $U \times \Gamma \subset T^*(U)$, $\varphi(x, \xi)$ is close enough to $x \cdot \xi$ in the C^2 topology, then the operator (5.1) is necessarily of the form

$$A = p(x, D)S(1, 0) \bmod OPS^{-\infty} \tag{5.7}$$

where $S(t, 0)$ is the solution operator to a hyperbolic equation as in Section 1 and $p(x, D) \in OPS^m_{\rho,\delta}$, assuming in (5.1) that $a(x, \xi) \in S^m_{\rho,\delta}$ with $0 \le \delta < \rho \le 1$, $\rho + \delta = 1$, and $a(x, \xi)$ is supported in a proper subcone of $U \times \Gamma$.

Indeed, let

$$\psi(t, x, \xi) = t\varphi(x, \xi) + (1 - t)(x \cdot \xi). \tag{5.8}$$

Now for $\varphi(x, \xi)$ close to $x \cdot \xi$ in the C^2 topology, we can specify $\lambda_1(t, x, \xi) \in S^1$, real valued, such that

$$\lambda_1(t, x, \xi + t(\nabla_x \varphi(x, \xi) - \xi)) = \varphi(x, \xi) - x \cdot \xi, \tag{5.9}$$

which implies

$$\psi_t = \lambda_1(t, x, \nabla_x \psi). \tag{5.10}$$

Now let $S(t, 0)$ be the solution operator to the hyperbolic equation

$$\frac{\partial}{\partial t} u = i\lambda_1(t, x, D_x)u \tag{5.11}$$

considered in Section 1. (Extend $\lambda_1(t, x, \xi)$ arbitrarily off $U \times \Gamma$, keeping it real.) The geometrical optics construction of Section 3 represents $S(t, 0)$ in the form

$$S(t, 0)u = \int b(t, x, \xi) e^{i\psi(t,x,\xi)} \hat{u}(\xi)\, d\xi \tag{5.12}$$

where ψ solves the eikonal equation (5.10) and $\psi(0, x, \xi) = x \cdot \xi$; hence ψ is given by (5.8); in particular, the representation (5.12) is valid for $0 \le t \le 1$. In particular,

$$S(1, 0)u = \int b(1, x, \xi) e^{i\varphi(x,\xi)} \hat{u}(\xi)\, d\xi. \tag{5.13}$$

Note that $b(1, x, \xi) \in S^0$ with principal symbol solving a transport equation of the form (3.15); in particular the principal symbol is nowhere vanishing.

Now if $p(x, D) \in OPS^m_{\rho,1-\rho}$, $1/2 < \rho \le 1$, it follows that

$$p(x, D)S(1, 0)u = \int q(x, \xi) e^{i\varphi(x,\xi)} \hat{u}(\xi)\, d\xi \tag{5.14}$$

where $q(x, \xi) = e^{-i\varphi} p(x, D)(b(1, x, \xi)e^{i\varphi}) \in S^m_{\rho,1-\rho}$, and in fact

$$q(x, \xi) = b(1, x, \xi)p(x, \nabla\varphi) \bmod S^{m-(2\rho-1)}_{\rho,1-\rho}. \tag{5.15}$$

If we want (5.7) to hold, we try to construct the symbol $p(x, \xi)$ by successive approximation:

$$p(x, \xi) \sim p_0(x, \xi) + p_1(x, \xi) + \cdots$$

with $p_j(x, \xi) \in S^{m-j(2\rho-1)}_{\rho,1-\rho}$. Thus, define $p_0(x, \xi)$ by

$$p_0(x, \nabla\varphi) = b(1, x, \xi)^{-1} a(x, \xi).$$

From (5.15) it follows that

$$p_0(x, D)S(1, 0)u = \int q_1(x, \xi)e^{i\varphi(x,\xi)}\hat{u}(\xi)\, d\xi$$

where $a(x, \xi) - q_1(x, \xi) = r_1(x, \xi) \in S^{m-(2\rho-1)}_{\rho,1-\rho}$. Consequently, let

$$p_1(x, \nabla\varphi) = b(1, x, \xi)^{-1}r_1(x, \xi)$$

define $p_1(x, \xi)$, so

$$(p_0(x, D)+p_1(x, D))S(1, 0)u = \int q_2(x, \xi)e^{i\varphi(x,\xi)}\hat{u}(\xi)\, d\xi$$

where $a(x, \xi) - q_2(x, \xi) = r_2(x, \xi) \in S^{m-2(2\rho-1)}_{\rho,1-\rho}$. Continue inductively in this fashion, obtaining $p_k(x, \nabla\varphi) = b(1, x, \xi)^{-1}r_k(x, \xi)$, so

$$(p_0(x, D)+\cdots+p_k(x, D))S(1, 0)u = \int q_{k+1}(x, \xi)e^{i\varphi(x,\xi)}\hat{u}(\xi)\, d\xi$$

with $a(x, \xi) - q_{k+1}(x, \xi) = r_{k+1}(x, \xi) \in S^{m-(k+1)(2\rho-1)}_{\rho,1-\rho}$. Now set $p(x, \xi) \sim p_0(x, \xi) + \cdots + p_k(x, \xi) + \cdots$. It follows that we have (5.7).

Suppose $a(x, \xi)$ is an elliptic symbol on $U_1 \times \Gamma_1 \subset U \times \Gamma$, i.e., $|a(x, \xi)| \geq C|\xi|^m$ for large $|\xi|$. It follows that $p(x, \xi)$ is an elliptic symbol on the appropriate subcone, and hence has a parametrix, which we will denote by $p(x, D)^{-1}$. Now $S(1, 0)$ is invertible; its inverse is $S(0, 1)$. It follows that, for $WF(u) \subset U_1 \times \Gamma_1$,

$$S(0, 1)p(x, D)^{-1}Au = u \bmod C^\infty \tag{5.16}$$

and, for $WF(v)$ similarly restricted,

$$AS(0, 1)p(x, D)^{-1}v = v \bmod C^\infty. \tag{5.17}$$

As a consequence, we call the Fourier integral operator A elliptic on $U_1 \times \Gamma_1$ provided its amplitude $a(x, \xi)$ is an elliptic symbol.

This result on representing an operator (5.1) in the form (5.7) can be generalized considerably. Now let A be given by (5.1) and suppose that, at a point (x_0, ξ_0), $\nabla_x \nabla_\xi \varphi(x_0, \xi_0)$ is an invertible matrix of second order derivatives, which implies that, near the point $(\nabla_\xi\varphi(x_0, \xi_0), \xi_0)$, the canonical relation (5.5), $(\nabla_\xi\varphi, \xi) \mapsto (x, \nabla_x\varphi)$, is a single valued canonical transformation. Suppose the amplitude $a(x, \xi)$ in (5.1) is supported in a small conic neighborhood of (x_0, ξ_0).

First we make a linear change of coordinates. Let

$$v(x) = u(Tx+y_0) = Fu(x) \tag{5.18}$$

where $T: \mathbf{R}^n \to \mathbf{R}^n$ is a linear isomorphism. It follows that

$$\hat{v}(\xi) = (\det T)^{-1}e^{iT^{-1}y_0\cdot\xi}\hat{u}(T^{-1*}\xi) \qquad \text{or}$$
$$\hat{u}(\xi) = (\det T)e^{-iy_0\cdot\xi}\hat{v}(T^*\xi).$$

Hence (5.1) becomes

$$Au = Bv = (\det T) \int a(x, \xi) e^{i\varphi(x,\xi) - iy_0 \cdot \xi} \hat{v}(T^* \xi)\, d\xi \tag{5.19}$$
$$= \int b(x, \xi) e^{i\psi(x,\xi)} \hat{v}(\xi)\, d\xi$$

where $b(x, \xi) = a(x, T^{*-1}\xi)$ and, with $V = T^{*-1}$, $z_0 = T^{-1}y_0$,

$$\psi(x, \xi) = \varphi(x, V\xi) - z_0 \cdot \xi. \tag{5.20}$$

We pick V such that the operator B given by (5.19) can be put in the form (5.7). To do this, note that we only have to find λ_1 such that the analogue of (5.9) holds:

$$\lambda_1(t, x, \xi + t(\nabla_x \psi(x, \xi) - \xi)) = \psi(x, \xi) - x \cdot \xi \tag{5.21}$$

and to solve (5.21) requires much less than that $\psi(x, \xi)$ be close to $x \cdot \xi$ in the C^2 topology. Indeed, we need only be sure that, for each $t \in [0, 1]$, the map $(x, \xi) \mapsto (x, \xi + t\nabla_x \psi(x, \xi) - t\xi)$ is a diffeomorphism near $(x_0, V^{-1}\xi_0)$, and this follows if $\nabla_\xi(\xi + t\nabla_x \psi(x, \xi) - t\xi)$ is invertible at this point, in particular if

$$(1-t)I + t\nabla_\xi \nabla_x \psi(x, \xi) = I \quad \text{at} \quad (x_0, V^{-1}\xi_0),$$

which is equivalent, by (5.20), to

$$V\nabla_\xi \nabla_x \varphi(x_0, \xi_0) = I.$$

Thus we have merely to take V equal to the inverse of the matrix of second partial derivatives $\nabla_\xi \nabla_x \varphi(x_0, \xi_0)$, and then, with $T = V^{-1*}$, F given by (5.18), we have

$$Au = p(x, D)S(1, 0)Fu \bmod C^\infty. \tag{5.22}$$

Consequently, any operator of the form (5.1) whose canonical relation is microlocally a canonical transformation, is microlocally equal to an operator of the form (5.22).

We shall briefly mention some more general classes of Fourier integral operators, referring to Hörmander [12], and Duistermaat [1] for further details. More generally than (5.1), one considers

$$Au(x) = \int a(x, y, \theta) e^{i\psi(x,y,\theta)} u(y)\, dy\, d\theta. \tag{5.23}$$

Here $x \in X$, $y \in Y$, and $\theta \in \mathbf{R}^N \backslash 0$. We consider $A: \mathscr{E}'(Y) \to \mathscr{D}'(X)$ under certain assumptions. We make the following assumption on the phase function ψ and the amplitude a.

(5.24) ψ is real valued, smooth, and homogeneous of degree 1 in θ.

(5.25) $\nabla_x\psi$ and $\nabla_\theta\psi$ are never both zero, on conic support of a.

(5.26) $\nabla_y\psi$ and $\nabla_\theta\psi$ are never both zero, on conic support of a.

(5.27) $a \in S^m_{\rho,1-\rho}(X \times Y; \mathbf{R}^N)$ with $\rho > 1/2$, and $a = 0$ for $|\theta| < 1$.

The integral (5.23) is not absolutely convergent, even when $u \in C_0^\infty(Y)$, unless a has large negative order, but an integration by parts procedure will make A a well defined operator. For example, with

$$L = (|\nabla_x\psi|^2 + |\theta|^2|\nabla_\theta\psi|^2)^{-1}(\nabla_x\psi \cdot \nabla_x + |\theta|^2\, \nabla_\theta\psi \cdot \nabla_\theta)$$

which has smooth coefficients on $X \times Y \times (\mathbf{R}^N\backslash 0)$ by (5.25), we have, formally,

$$Au(x) = \int (L^t)^k(a(x, y, \theta))u(y)e^{i\psi(x,y,\theta)}\, d\theta\, dy$$

which is absolutely convergent for k sufficiently large; and we take this to define A; we see that $A: C_0^\infty(Y) \to C^\infty(X)$. Similarly, hypothesis (5.26) implies that $A: \mathscr{E}'(Y) \to \mathscr{D}'(X)$.

In the case (5.1), $X = Y$ and $\psi(x, y, \xi) = \varphi(x, \xi) - y \cdot \xi$. In this case, hypothesis (5.26) is always satisfied, since $\nabla_y\psi = \xi$, and hypothesis (5.25) is equivalent to

(5.28) $\nabla_x\varphi$ and $\nabla_\xi\varphi - y$ never both zero.

This is equivalent to $\nabla_x\varphi \neq 0$, which is condition (3.4). The Fourier integral operators of the form (5.23) are more general than the ones like (5.1) in the following respects:

(i) X and Y need not be equal. In fact, they can have different dimensions.

(ii) Not only is ψ a more general sort of phase function than $\varphi(x, \xi) - y \cdot \xi$, but $\theta \in \mathbf{R}^N\backslash 0$, and N need not be equal to either dim X or dim Y.

An example of (i) is the map

$$u(t, x) = \int a(t, x, \xi)e^{i\varphi(t,x,\xi)}\hat{v}(\xi)\, d\xi$$

$$= \int a(t, x, \xi)e^{i\varphi(t,x,\xi) - iy \cdot \xi}v(y)\, dy\, d\xi$$

solving the wave equation

$$\frac{\partial}{\partial t}u = i\lambda(t, x, D_x)u; \qquad u(0, x) = v(x).$$

In this case, $v \mapsto u$ is a map from $\mathscr{E}'(X)$ to $\mathscr{D}'(I \times X)$, where I is an open interval in $\mathbf{R}$.

The possibility (ii) also arises in the Fourier integral representation of the solution operator $S(t, 0)$. For recall that the representation (3.2) only works for small t. This is due to the fact that the eikonal equation (3.14) does not in general have a globally defined solution. However, it is always possible to choose a partition $0 = s_0 < s_1 < \cdots < s_\ell = t$ so that each operator $S(s_{k+1}, s_k)$ has a representation of the form (3.2). Now use the formula

$$S(t, 0) = S(t, s_{\ell-1}) \cdots S(s_2, s_1)S(s_1, 0). \tag{5.29}$$

Substituting in the Fourier integral representations of each $S(s_{k+1}, s_k)$, after some manipulation, we find it is possible to represent $S(t, 0)$ as a Fourier integral operator of the form (5.23), with θ involving somewhat more than n variables. This is a special case of the general problem of multiplying two Fourier integral operators; for a nice exposition of this see Duistermaat [1, pp. 57–60].

In addition to the hypotheses (5.24) through (5.27), one imposes the following *nondegeneracy* condition on the phase function ψ:

(5.30) $\nabla_\theta \psi(x, y, \theta) = 0 \Rightarrow \nabla_{(x,y,\theta)} \dfrac{\partial \psi}{\partial \theta_j}(x, y, \theta)$ linearly independent for $j = 1, \ldots, N$, given $(x, y, \theta) \in$ conic support of $a(x, y, \theta)$.

In this case, the set

$$\Lambda_\psi = \{(x, y, \nabla_{x,y}\psi) : (x, y, \theta) \in \tilde{\Gamma}, \nabla_\theta \psi(x, y, \theta) = 0\}$$

is a submanifold of $T^*(X \times Y)$. Here $\tilde{\Gamma}$ is some conic neighborhood of the conic support of $a(x, y, \theta)$. In fact, Λ_ψ is a conic manifold of dimension equal to $\dim(X \times Y)$, and it is a Lagrangian manifold; the symplectic form on $T^*(X \times Y)$ annihilates pairs of vectors tangent to Λ_ψ. The set

$$\Lambda'_\psi = \{((x, \xi), (y, \eta)) \in T^*(X) \times T^*(Y) : (x, y, \xi, -\eta) \in \Lambda_\psi\}$$

generalizes the canonical relation (5.5).

The following important properties of Fourier integral operators are established in Hörmander [15]:

(i) *Equivalence of phase functions.* Suppose another phase function $\tilde{\psi}$ satisfies hypotheses (5.24) through (5.26) and (5.30), and suppose $\Lambda_\psi = \Lambda_{\tilde{\psi}}$. Then any operator of the form (5.23) can be written in the form

$$Au = \int \tilde{a}(x, y, \tilde{\theta}) e^{i\tilde{\psi}(x,y,\tilde{\theta})} u(y)\, dy\, d\tilde{\theta} \tag{5.31}$$

where $\tilde{a}(x, y, \tilde{\theta}) \in S^{\tilde{m}}_{\rho,1-\rho}(X \times Y; \mathbf{R}^{\tilde{N}})$. Here $\tilde{m} = m + (1/2)(N - \tilde{N})$. Furthermore, any conic Lagrangian submanifold of $T^*(X \times Y)$ can, at least locally, be represented in the form Λ_ψ for some nondegenerate phase function ψ.

(ii) *Principal symbol of a Fourier integral operator.* The representations (5.23) and (5.31) for the operator A involve completely different amplitudes, even defined on different sets. No one of them is satisfactory as a "principal symbol" of A. Hörmander associates with A a "principal symbol" which is a section of a certain line bundle $\Omega_{1/2} \otimes L$ over the Lagrangian manifold $\Lambda = \Lambda_\psi$. Here $\Omega_{1/2}$ is the bundle of "densities of order 1/2" and L is the so-called "Keller-Maslov line bundle."

(iii) *Composition of Fourier integral operators.* If B is a Fourier integral operator taking $\mathscr{D}'(X)$ to $\mathscr{D}'(Z)$, with associated Lagrangian manifold Λ_φ, then BA is a Fourier integral operator, granted certain geometrical assumptions on Λ'_ψ and Λ'_φ (see Duistermaat [1], p. 57). Furthermore, the principal symbol of BA can be obtained from the principal symbols of B and of A. When B and A do compose nicely, the Lagrangian manifold Λ associated to BA has the property that $\Lambda' = \Lambda'_\varphi \circ \Lambda'_\psi$, where Λ' is regarded as a relation between $T^*(Y)$ and $T^*(Z)$.

We might note that a special case of (i) is that any Fourier integral operator on X for which Λ'_ψ is the identity map on $T^*(X)$ can be represented as a pseudodifferential operator. One can deduce Egorov's theorem from properties (i) through (iii) as follows. The canonical relation associated with $S(0, t)$ is the inverse of the canonical relation associated with $S(t, 0)$. Thus, for any pseudodifferential operator $p(x, D) \in OPS^m_{\rho,1-\rho}$, $\rho > 1/2$, the composition $S(0, t)p(x, D)S(t, 0)$ is a Fourier integral operator whose associated canonical relation is the identity, hence is a pseudodifferential operator, and the formula alluded to in (iii) gives its principal symbol.

We now show directly that properties (i) and (iii) hold for the operators $S(t, s)$ which are solution operators to hyperbolic equations, on a compact manifold M. Let $S(t, s)$ be the solution operator to

$$\frac{\partial}{\partial t} u = i\lambda(t, x, D_x)u$$

and $T(t, s)$ the solution operator to

$$\frac{\partial}{\partial t} u = i\mu(t, x, D_x)u$$

where $\lambda, \mu \in OPS^1$ both have real principal symbol. We claim products of such operators are themselves solution operators to some hyperbolic equations. It suffices to show that $S(t, 0)T(0, t)$ is such a solution operator, for all t. Let

$$u(t) = S(t, 0)T(0, t)f.$$

Therefore,

$$\frac{\partial}{\partial t}u = i\lambda u - iS(t, 0)T(0, t)\mu f.$$

By Egorov's theorem,

$$S(t, 0)T(0, t)\mu = \tilde{\mu}(t, x, D_x)S(t, 0)T(0, t)$$

for a certain $\tilde{\mu} \in OPS^1$ whose principal symbol is obtained from that of μ by an appropriate canonical transformation (so in particular is real). Therefore,

$$\frac{\partial}{\partial t}u = i(\lambda + \tilde{\mu})u, \tag{5.32}$$

so

$$S(t, 0)T(0, t) = U(t, 0) \tag{5.33}$$

where $U(t, s)$ is the solution operator to (5.32).

Next, we show that if $S(1, 0)$ and $T(1, 0)$ are associated to the same canonical transformation, then

$$T(1, 0) = p(x, D)S(1, 0) \tag{5.34}$$

for some elliptic $p(x, D) \in OPS^0$. Indeed, $T(1, 0)S(0, 1)$ is of the form $U(1, 0)$ as above and its canonical transformation is the identity. It remains to show that such a solution operator is a pseudo-differential operator: $U(1, 0) = p(x, D)$. In order to prove $U(1, 0) \in OPS^0_{1,0}$, we use the following characterization of $OPS^0_{1,0}(M)$, pointed out to the author by H. O. Cordes.

LEMMA 5.2. *Let $A: \mathscr{D}'(M) \to \mathscr{D}'(M)$, M compact, and suppose that, given any $P_j \in OPS^1_{1,0}$, you have*

$$Ad(P_1) \cdots Ad(P_k)A: L^2(M) \to L^2(M). \tag{5.35}$$

Then $A \in OPS^0_{1,0}$. Here $Ad(P)A = [P, A] = PA - AP$.

Proof. From (5.35) it is easy to show that, in local coordinates,

$$(Ad\ x)^\alpha (Ad\ D)^\beta A: H^s \to H^{s+|\alpha|}. \tag{5.36}$$

In such local coordinates, define $a(x, \xi) = e^{-ix\cdot\xi}A(e^{ix\cdot\xi})$, so for $u \in \mathscr{E}'$, $Au = \int a(x, \xi)e^{ix\cdot\xi}\hat{u}(\xi)\, d\xi$. We need $a(x, \xi) \in S^0_{1,0}$. Now $D^\alpha_\xi D^\beta_x a(x, \xi) = e^{-ix\cdot\xi}B_{\alpha\beta}(e^{ix\cdot\xi})$ where $B_{\alpha\beta} = (Ad\ x)^\alpha(Ad\ D)^\beta A$, and (5.36) implies

$$\|B_{\alpha\beta}(e^{ix\cdot\xi})\|_{H^s} \le c\|e^{ix\cdot\xi}\varphi(x)\|_{H^{s-|\alpha|}} \tag{5.37}$$

where $\varphi \in C_0^\infty$ is a convenient cut-off (M is compact, recall). For $|\alpha| = 0$, $|\beta| = [n/2]+1$, $s = 0$, with Sobolev's imbedding theorem, (5.37) yields

$$|a(x, \xi)| \le C_0. \tag{5.38}$$

To estimate $D_\xi^\alpha D_x^\beta a(x, \xi)$, which is the symbol $b_{\alpha\beta}(x, \xi)$ of $B_{\alpha\beta}$, note that $b_{\alpha\beta}(x, \xi)(1+|\xi|^2)^{|\alpha|/2}$ is the symbol of $B_{\alpha\beta}(1-\Delta)^{|\alpha|/2} = E_{\alpha\beta}$, and using (5.35), it follows that $E_{\alpha\beta}$ satisfies the condition (5.35) hypothesized for A. So the reasoning leading to (5.38) applies; $E_{\alpha\beta}$ has a bounded symbol, or equivalently

$$|D_\xi^\alpha D_x^\beta a(x, \xi)| \leq c(1+|\xi|)^{-|\alpha|},$$

which proves the lemma.

We now show that the hypotheses of Lemma 5.2 apply to $U = U(1, 0)$ provided its canonical transformation is the identity. Indeed, for $P \in OPS^1$, $Ad\, PU = PU - UP = (P - UPU^{-1})U$, and by Egorov's theorem, $UPU^{-1} \in OPS^1$ has the same principal symbol as P, so $P - UPU^{-1} = A \in OPS^0$. More generally,

$$\begin{aligned} Ad\, P(AU) &= (PA - AUPU^{-1})U \\ &= A_1 U \end{aligned}$$

where again, by Egorov's theorem, $A_1 \in OPS^0$. Inductively, we obtain, for $P_j \in OPS^1$,

$$Ad\, P_1 \cdots Ad\, P_j(U) = A_j U, \qquad A_j \in OPS^0,$$

and since A_j and U both map $H^s(M)$ to itself, the hypotheses of Lemma 5.2 apply. Hence $U(1, 0) \in OPS^0_{1,0}$ if its canonical transformation is the identity.

By considering the conjugation action $U(1, 0)AU(0, 1)$, $A \in OPS^1$, we can determine the principal symbol of $U(1, 0) = P$, up to a multiplicative constant, by solving some transport equations, as follows. For $A \in OPS^1$, Egorov's theorem implies $UAU^{-1} \in OPS^1$ has the same principal symbol as A, so

$$UAU^{-1} - A = PAP^{-1} - A = B \in OPS^0, \tag{5.39}$$

and the proof of Egorov's theorem in Section 1 allows us to obtain the principal symbol b_0 of B via the solution to a certain transport equation, so $b_0 = \mathscr{L} a_1$. Rewriting (5.39) as $BP = PA - AP$, we have, for the principal symbol p_0 of P,

$$b_0 p_0 = -H_{a_1} p_0$$

or

$$-H_{a_1} \log p_0 = b_0 = \mathscr{L} a_1. \tag{5.40}$$

Choosing sufficiently many $a_1 \in S^1$, we see that (5.40) determines $\log p_0$ uniquely, up to an additive constant. Hence the principal symbol p_0 of $P = U(1, 0)$ is determined up to a multiplicative constant. We can get

$|p_0|^2$, the principal symbol of $U(1, 0)U(1, 0)^*$, by applying the method of proof of Egorov's theorem in Section 1 to $U(t, 0)U(t, 0)^*$ and solving an appropriate transport equation, but the phase of p_0 seems more difficult to obtain. On the other hand, facet (ii) above of Hörmander's machinery yields p_0 as a solution of a transport equation. The Maslov line bundle enters into the computation of the phase.

Finally, we consider operators of the form $Au(x) = \int K(x, y)a(y)\,dy$ with distribution kernels $K(x, y)$ of the form

$$p(x, y, D_{x,y})\delta(\varphi(x, y)) \tag{5.41}$$

where $\varphi \in C^\infty(X \times Y)$ has nonvanishing gradient on $\{\varphi = 0\}$ and $P = p(x, y, D_{x,y}) \in OPS^m(X \times Y)$. In case $P = I$, it is clear that A is of the form (5.23) with $\psi(x, y, \theta) = \theta\varphi(x, y)$. We make the following analogue of hypotheses (5.25), (5.26).

$$\text{On } \varphi(x, y) = 0, \qquad \nabla_x\varphi \neq 0 \text{ and } \nabla_y\varphi \neq 0. \tag{5.42}$$

A particular case would be $\varphi(x, y) = \sigma(x-y)$, so $\delta(\sigma(x-y))$ is the kernel of a convolution operator; then (5.42) is $\nabla_x\sigma \neq 0$ when $\sigma(x) = 0$. Note that the nondegeneracy hypothesis (5.30) is automatically satisfied.

LEMMA 5.3. *Assuming* (5.42), *we see the kernel K given by* (5.41) *has the form*

$$K(x, y) = \hat{q}_{x,y}(\varphi(x, y)) \bmod C^\infty \tag{5.43}$$

where $q_{x,y}(\lambda) \in S^m(\mathbf{R})$, *varying smoothly with x,y.*

Proof. Letting $z = (x, y)$, $\varphi(z) = \varphi(x, y)$, we must show that $p(z, D_z)\delta(\varphi(z)) = \hat{q}_z(\varphi(z))$. Now make a change of coordinates on $Z = X \times Y$ such that $z_1 = \varphi$ (locally); in this coordinate system, $p(z, D_z)$ becomes $\tilde{p}(z, D_z) \in OPS^m$, and K becomes

$$\begin{aligned}\tilde{p}(z, D_z)\delta(z_1) &= \int \tilde{p}(x, \xi)e^{iz\cdot\xi}\delta(\xi')\,d\xi \\ &= \int \tilde{q}(z, \xi_1)e^{iz_1\xi_1}\,d\xi_1\end{aligned}$$

where $\tilde{q}(x, \xi_1) = \tilde{p}(z, \xi_1, 0)$. Now this is of the form $\tilde{q}(z', z_1, D_{z_1})\delta(z_1) = \hat{q}_z(z_1)$ with $q_z(\lambda) \in S^m(\mathbf{R})$, which proves (5.43), locally, and one can patch the local results together.

Note that (5.43) is equivalent to

$$Au(x) = \int q(x, y, \lambda)e^{i\lambda\varphi(x,y)}u(y)\,d\lambda\,dx\,dy, \tag{5.44}$$

so such an operator generally is of the form (5.23), with $q(x, y, \lambda) \in S^m$. Conversely, any operator of the form (5.44) has the form (5.41). Note that the wave front set of the distribution $K \in \mathcal{D}'(X \times Y)$ is contained in the normal bundle $N^\perp$ to $\{\varphi(x, y) = 0\}$, from which it is not difficult to show that

$$WF(Au) \subset \{(x, \xi) \in T^*(X): (x, y, \xi, -\eta) \in N^\perp\{\varphi = 0\}, (y, \eta) \in WFu\},$$

so the canonical relation, relating $WF(u)$ to $WF(Au)$, is given by

$$(5.45) \qquad (y, -\lambda\nabla_y\varphi) \mapsto (x, \lambda\nabla_x\varphi); \qquad \varphi(x, y) = 0$$

and is a submanifold of $T^*(X) \times T^*(Y)$.

It is clear that an operator with kernel $p_1(x, D_x)K$ or $Kp_2(y, D_y)$ is of the form (5.41) if K is; indeed if $p_1(x, D_x)$ or $p_2(y, D_y)$ is regarded as a pseudodifferential operator on $X \times Y$, where its symbol is singular has empty intersection with $WF(K)$, granted hypothesis (5.42). We want to see when the converse is true, and hence obtain Egorov's theorem, or an appropriate generalization. We assume that X and Y have the same dimension.

First, consider the case when (5.45) is actually a canonical transformation, so the projection of Λ' onto T^*X and onto T^*Y is a local diffeomorphism. In this case we claim that any K of the form (5.41) must also have the form $p_1(x, D_x)\delta(\varphi(x, y))$. Indeed, we have written, in local coordinates on $Z = X \times Y$ such that $\varphi = z_1$,

$$(5.46) \qquad p(x, y, D_{x,y})\delta(\varphi) = \tilde{q}(z', z_1, D_{z_1})\delta(z_1)$$

where $\tilde{q}(z', z_1, \xi) = \tilde{p}(z, \xi_1, 0)$, $\tilde{p}$ representing p in the new coordinate system. Similarly, regarding $p_1(x, D_x)$ as a pseudodifferential operator on Z, at least near $WF\delta(\varphi)$, we have

$$(5.47) \qquad p_1(x, D_x)\delta(\varphi) = q_1(z', z_1, D_{z_1})\delta(z_1)$$

where $q_1(z', z_1, \xi_1) = \tilde{p}_1(z, \xi_1, 0)$, $\tilde{p}_1$ representing p_1 in the new coordinate system. By the transformation law for principal symbols, given $p(x, y, D_{x,y})$ and $p_1(x, D_x) \in OPS^m$, it is clear that the principal symbols of $\tilde{q}(z', z_1, D_{z_1})$ in (5.46) and of $q_1(z', z_1, D_{z_1})$ in (5.47) coincide on $WF\delta(z_1)$ if, and only if, the principal symbols of $p(x, y, D_{x,y})$ and $p_1(x, D_x)$ coincide on $WF\delta(\varphi)$, i.e., on the normal bundle of $\{\varphi = 0\}$, i.e.,

$$(5.48) \qquad p_{1m}(x, \nabla_x\varphi) = p_m(x, y, \nabla_{x,y}\varphi) \quad \text{on} \quad \varphi(x, y) = 0.$$

Now given $p_m(x, y, \xi, \eta)$, there exists a unique $p_{1m}(x, \xi)$ satisfying (5.48), provided (5.45) is a canonical transformation. It follows that the difference $p(x, y, D_{x,y})\delta(\varphi) - p_{1m}(x, D_x)\delta(\varphi) = r(x, y, D_{x,y})\delta(\varphi)$ with $r(x, y, D_{x,y}) \in OPS^{m-1}$ and one can continue inductively, finally obtaining $p_1(x, D_x)$ with $p(x, y, D_{x,y})\delta(\varphi) - p_1(x, D_x)\delta(\varphi) \in C^\infty$, assuming (5.45) is a canonical transformation, dim X = dim Y. Note that this reasoning applies in particular to kernels of the form $p_2(y, D_y)\delta(\varphi)$, and (5.48) becomes $p_{lm}(x, \nabla_x\varphi) = p_{2m}(y, \nabla_y\varphi)$ on $\varphi(x, y) = 0$, so we recover Egorov's theorem for operators of this sort.

A typical example of a kernel satisfying the above conditions is $t^{-1}\delta(t-|x-y|) = K_t(x, y)$, $t \neq 0$, which is, up to a constant factor, the fundamental solution to the wave equation if $X = Y = \mathbf{R}^3$. Generally, for small nonzero t, the solution operator to $(\partial^2/\partial t^2) - A(x, D_x)$ can be written in terms of such operators as satisfy (5.41), (5.42), if $A(x, D_x)$ is a second order elliptic differential operator, and also in many more general contexts. Indeed, such constructions preceeded the use of Fourier integral operators; see Hadamard [1].

The following is an example of a kernel satisfying (5.41), (5.42), whose canonical relation is not a canonical transformation. Consider the convolution kernel $\delta(\sigma(x-y))$ on $\mathbf{R}^2$, with

$$\sigma(x_1, x_2) = x_2 - \tfrac{1}{3}x_1^3. \tag{5.49}$$

It is not difficult to see (exercise) that the canonical relation $\Lambda' \subset T^*\mathbf{R}^2 \times T^*\mathbf{R}^2$ projects onto each factor as a Whitney fold (see Golubitsky and Guillemin [1] for material on folds and other singularities). This generates the following concept.

Definition *If $\Lambda' \subset T^*X \times T^*Y$ projects onto each factor as a fold, we call Λ' a folding canonical relation.* (dim X = dim Y.)

Thus an operator with kernel $\alpha(x, y)\delta(\sigma(x - y))$, $\alpha \in C^\infty$, σ given by (5.49), is a Fourier integral operator with a folding canonical relation.

For Fourier integral operators with folding canonical relations, there is a variant of Egorov's theorem, a little more involved than in the case of canonical transformations. Suppose A has kernel $\delta(\varphi(x, y))$, and let A' have kernel $\alpha(x, y)\delta(\varphi(x, y))$, $\alpha \in C^\infty(X \times Y)$. You can think of α as being in $C^\infty(T^*X \times T^*Y)$, and denote its restriction to Λ' by β. Since we are assuming $\pi_1 : \Lambda' \to T^*X$ is a fold, there is defined an involution $\mathscr{J}_1$ on Λ', leaving fixed the fold set.

Proposition 5.4. *Let $\zeta \in \Lambda'$ and suppose, with the terminology just introduced, $\beta - \mathscr{J}_1^*\beta$ vanishes to precisely first order on the fold set in Λ', near ζ. Then, microlocally near $\pi_2\zeta$, for any $P \in OPS^m$ on Y, we can write*

$$AP = \tilde{P}_1 A + \tilde{P}_2 A' \operatorname{mod} OPS^{-\infty} \tag{5.50}$$

for some $\tilde{P}_j \in OPS^m$ on X. More generally, given any operator B with kernel of the form (5.41), *we can write*

$$B = \tilde{P}_1 A + \tilde{P}_2 A' \operatorname{mod} OPS^{-\infty} \tag{5.51}$$

for some $\tilde{P}_j \in OPS^m$ on X.

Proof. Equations (5.46) and (5.47) still hold in this context, so the kernel of B is given by (5.46) and the kernel of $\tilde{P}_1 A$ is given by (5.47). Similarly, the kernel of $\tilde{P}_2 A'$ is given by

$$q_2(z', z_1, D_z)\delta(z_1)$$

where $q_2(z', z_1, \xi_1) = p_2^{\#}(z, \xi_1, 0)$, $p_2^{\#}$ representing $\tilde{P}_2(x, D_x)\alpha(x, y) = P_2^b(x, y, D_{x,y})$ in the appropriate new coordinate system where $\varphi = z_1$. To construct the principal symbols of $\tilde{P}_1$ and $\tilde{P}_2$, replace (5.48) by

$$\text{(5.52)} \quad \tilde{p}_{1m}(x, \nabla_x\varphi) + \tilde{p}_{2m}(x, \nabla_x\varphi)\alpha(x, y) = p_m(x, y, \nabla_{x,y}\varphi) \quad \text{on} \quad \varphi(x, y) = 0.$$

The hypothesis that (5.45) is a folding canonical relation and the hypothesis on $\beta = \alpha|_\Lambda$ implies that smooth homogeneous $\tilde{p}_{lm}$ and $\tilde{p}_{2m}$ can be found so that (5.52) holds. The same type of inductive argument used before allows one to construct the lower order terms in $\tilde{P}_1$ and $\tilde{P}_2$ so that (5.51) is true.

Fourier integral operators with folding canonical relations play a role in diffraction theory and are further developed in Melrose and Taylor [1]; see the last section in Chapter X.

We end this section by mentioning some further developments in the theory of Fourier integral operators. Duistermaat and Guillemin [1] and also Weinstein [1] have defined compositions of Fourier integral operators under more general geometrical conditions on Λ'_ψ and Λ'_ψ than those given in Duistermaat [1] and Hörmander [15]. Also, amplitudes in $S^m_{1/2,1/2}$ have been used. In particular $S(0, t)p(x, D)S(t, 0)$ is a pseudo-differential operator if $p(x, \xi) \in S^m_{1/2,1/2}$, as we show in Section 8, but the formula for its principal symbol breaks down. An even more general result is given in Beals [2]; a related result also appears in Boutet de Monvel [2]. Melin and Sjöstrand [1] have developed a theory of Fourier integral operator with complex phase function, which has been applied in Melin and Sjöstrand [2] and Boutet de Monvel and Sjöstrand [1]. They also handle amplitudes in $S^m_{1/2,1/2}$, in Melin and Sjöstrand [1]. Also, Fourier integral operators in which Airy functions as well as exponential functions appear arise in the treatment of the diffraction problem and appear in the papers of Melrose [1], [2] and Taylor [5], [6], and will be discussed in Chapter X.

Exercises

The following exercises do not depend on the discussion in the latter part of this section. In fact, they mainly depend on Section 1.

5.1. Let $p(x, \xi) \in S^m$ have real valued principal symbol $p_m(x, \xi)$. Suppose that $(\partial/\partial\xi_n)p_m(x, \xi) \neq 0$ in a conic neighborhood of (x_0, ξ_0). Show that there exist $q(x, D), r(x, D) \in OPS^{m-1}$, $q(x, D)$ elliptic, and $\lambda(x_n, x', D_{x'}) \in OPS^1$ (with $\lambda(x_n, x', \xi')$ real valued) such that

$$p(x, D)u = q(x, D)\left(\frac{\partial}{\partial x_n} - i\lambda(x_n, x', D_{x'})\right)u + r(x, D)u \bmod C^\infty$$

for all $u \in \mathscr{D}'(X)$ with wave front set in a small conic neighborhood Γ of (x_0, ξ_0).

5.2. Let $L = (\partial/\partial x_n) - i\lambda(x_n, x', D_{x'})$ be considered as an operator on distributions with wave front set in Γ. Let $S(t, s,)v = w(t)$ be the solution operator for the initial value problem

$$\frac{\partial}{\partial t} w = i\lambda(t, x', D_{x'})w,$$

$$w(s) = v \in \mathscr{D}'(X').$$

Let $v(x_n, x') = S(0, x_n)u(x_n, x')$. (Suppose $x_0 = (0, x_0')$.) Show that, for some $a(x_n, x', D_{x'}) \in OPS^{-1}$,

$$Lu = S(x_n, 0)\left(\frac{\partial}{\partial x_n} + a(x_n, x', D_{x'})\right)v(x_n, x') \bmod C^\infty$$

for all u with wave front set in Γ.

5.3. If v is defined as in Exercise 5.2, show that for $WF(u) \subset \Gamma$, $u \in H^\nu(X) \Leftrightarrow v \in H^\nu(X)$. Hint: First do it for $u \in L^2(X)$. Then note that, for $WF(u) \subset \Gamma$, elliptic operators on X' of order ν map $u \in H^\mu(X)$ to $H^{\mu-\nu}(X)$.

5.4. Conclude that, if $u \in H^{s+m-1}(X)$, $WF(u) \subset \Gamma$, and $p(x, D)u = f \in H^s(X)$, then

$$\frac{\partial}{\partial x_n} v = S(0, x_n)q(x, D)^{-1}f \bmod H^{s+m}(X)$$

and that $g = S(0, x_n)q(x, D)^{-1}f \in H^{s+m-1}(X)$.

For further details, and for the implications of these problems, see Nirenberg and Trèvcs [1].

§6. Operators with Singular Phase Functions

In this section we study operators of the form

$$Au = \int a(x, \xi)e^{i\varphi(x,\xi)}\hat{u}(\xi)\, d\xi \tag{6.1}$$

where $a \in S^m_{\rho,\delta}$ has support in some open cone Γ with smooth boundary, say Γ is given (locally) by $\gamma(x, \xi) > 0$, homogeneous of degree 0 in ξ, $\nabla_{x,\xi}\gamma \neq 0$ on $\partial\Gamma$. As usual we suppose φ is real valued, and homogeneous of degree 1 in ξ, but instead of supposing $\varphi(x, \xi) \in C^\infty(\bar{\Gamma}\backslash 0)$, we make the following weaker hypotheses:

$$\varphi(x, \xi) \in C^1(\bar{\Gamma}\backslash 0), \tag{6.2}$$

$$|D_x^\beta D_\xi^\alpha \varphi(x, \xi)| \leq c|\xi|^{1-|\alpha|}\gamma(x, \xi)^{(1+a)-|\alpha|-|\beta|}, \qquad \text{if } |\alpha| + |\beta| \geq 2 \tag{6.3}$$

where $0 \leq a < 1$. Typically, we shall have $a = 1/2$. We also assume, as usual,

$$|\nabla_x \varphi(x, \xi)| \geq c|\xi| \qquad \text{on} \qquad \Gamma. \tag{6.4}$$

We shall make sense out of these operators for $u \in \mathscr{E}'(\mathbf{R}^n)$ provided the amplitude satisfies the condition

$$a(x, \xi) \in S^m_{\rho,\delta} \text{ is supported on } \gamma(x, \xi) \geq c|\xi|^{-b} \tag{6.5}$$

where $c > 0$, $0 < b < 1$. Of course, if $b = 0$, there is no problem. As usual, we shall suppose $a(x, \xi)$ vanishes for $|\xi| \leq 1$.

As in the case of φ smooth, we use the identity $Le^{i\varphi} = e^{i\varphi}$ where $L = i^{-1}|\nabla_x\varphi|^{-2}\,\nabla_x\varphi \cdot \nabla_x$, and we manipulate the inner product $\langle Au, v\rangle$ formally, for $u \in \mathscr{E}'$, $v \in C_0^\infty$, until we get a well-defined convergent integral. Indeed, we have, formally,

$$\langle Au, v\rangle = \iint (L^t)^k(a(x, \xi)v(x))e^{i\varphi}\hat{u}(\xi)\, d\xi\, dx. \tag{6.6}$$

To estimate the integrand here, note that

$$\begin{aligned} L^t &= -i^{-1}|\nabla_x\varphi|^{-2}\,\nabla_x\varphi \cdot \nabla_x + i^{-1}\,\nabla_x \cdot (|\nabla_x\varphi|^{-2}\,\nabla_x\varphi) \\ &= \sum_{|\alpha|=1} a_\alpha(x, \xi)D_x^\alpha + b(x, \xi) \end{aligned}$$

where, by (6.2)–(6.4), we have $a_\alpha(x, \xi) \in C(\bar{\Gamma}\backslash 0)$, homogeneous of degree -1 in ξ, $b(x, \xi)$ homogeneous of degree -1 in ξ, and

$$\begin{aligned} |D_x^\beta D_\xi^\gamma a_\alpha(x, \xi)| &\leq c|\xi|^{-1-|\gamma|}\gamma(x, \xi)^{a-|\beta|-|\gamma|}, & |\beta| + |\gamma| \geq 1, \\ |D_x^\beta D_\xi^\gamma b(x, \xi)| &\leq c|\xi|^{-1-|\gamma|}\gamma(x, \xi)^{-1+a-|\beta|-|\gamma|}, & |\beta| + |\gamma| \geq 0. \end{aligned}$$

Consequently, we see that

$$(L^t)^k = \sum_{0 \leq |\sigma| \leq k} A_\sigma^{(k)}(x, \xi)D_x^\sigma \tag{6.7}$$

where $A_\sigma^{(k)}(x, \xi) \in C(\bar{\Gamma}\backslash 0)$ for $|\sigma| \leq k$, $A_\sigma^{(k)}(x, \xi)$ is homogeneous of degree $-k$ in ξ, and

$$|D_x^\beta D_\xi^\alpha A_\sigma^{(k)}(x, \xi)| \leq c|\xi|^{-k-|\alpha|}(1 + \gamma(x, \xi)^{a-(k-|\sigma|)-|\beta|-|\alpha|}). \tag{6.8}$$

Consequently, on the support of $a(x, \xi)$ we have

$$\text{(6.9)} \qquad \begin{aligned} |A_\sigma^{(k)}(x, \xi)| &\le C|\xi|^{-k}(1 + |\xi|^{(k-|\sigma|)b - ab}) \\ &\le C|\xi|^{-k(1-b) - |\sigma|b}. \end{aligned}$$

With this we can estimate

$$\int A_\sigma^{(k)}(x, \xi) D_x^\sigma(v(x)a(x, \xi)) e^{i\varphi} \hat{u}(\xi)\, d\xi\, dx$$

as follows. Since $a(x, \xi) \in S_{\rho,\delta}^m$ is supported on $\gamma(x, \xi) \ge C|\xi|^{-b}$, (6.9) implies

$$\begin{aligned} |A_\sigma^{(k)}(x, \xi) D_x^\sigma(v(x)a(x, \xi))| &\le C|\xi|^{-k(1-b) - |\sigma|b}|\xi|^{m + \delta|\sigma|} \\ &= C|\xi|^{m - k(1-b) + (\delta - b)|\sigma|}. \end{aligned}$$

If $1 > \delta \ge b$, which one can assume without loss of generality, this is bounded by

$$C|\xi|^{m - k(1-\delta)}.$$

Since $\hat{u}(\xi)$ has only polynomial growth for any given $u \in \mathscr{E}'$, this shows that the right-hand side of (6.6) is absolutely convergent, for k sufficiently large. Since (6.6) is a true identity for $a(x, \xi) \in S_{1,0}^{-\infty}$, satisfying (6.5), we see that also, given $\psi \in C_0^\infty(\mathbf{R}^n)$, $\psi(\xi) = 1$ for $|\xi| \le 1$,

$$\text{(6.10)} \qquad Au = \lim_{\varepsilon \to 0} \int \psi(\varepsilon\xi) a(x, \xi) e^{i\varphi(x,\xi)} \hat{u}(\xi)\, d\xi,$$

so Au is independent of the choice of k.

The above argument shows that differential operators can be brought under the integral sign:

$$\frac{\partial}{\partial x_j} Au = \int \left[i \frac{\partial \varphi}{\partial x_j} a + \frac{\partial a}{\partial x_j} \right] e^{i\varphi} \hat{u}(\xi)\, d\xi$$

etc., as is the case with ordinary Fourier integral operators.

The analysis of $WF(Au)$ in terms of $WF(u)$ also proceeds by a similar generalization of the argument of Section 5, and, under hypotheses (6.1) through (6.5), we have the same result:

$$\begin{aligned} WF(Au) \subset \{(x, \nabla_x \varphi(x, \xi)) \colon \text{for } y = \nabla_\xi \varphi(x, \xi), \quad (y, \xi) \in WF(u), \\ \text{and } (x, \xi) \in \text{conic support of } a\}. \end{aligned}$$

Thus, as before, the relation between $WF(u)$ and $WF(Au)$ is described by $(\nabla_\xi \varphi, \xi) \mapsto (x, \nabla_x \varphi)$. This subset of $T^*(X) \times T^*(X)$ is evidently the image of a C^a map applied to a closed conic subset of $T^*(X)\backslash 0$. Under certain natural conditions it may be the graph of a continuous, but not smooth, canonical transformation.

The type of operators studied in this section appear in the analysis of the diffraction problem. We shall use them in Chapter X, Section 2.

§7. The Fundamental Asymptotic Expansion Lemma

The purpose of this section is to analyze the behavior of

$$p(x, D)(a(x, \xi)e^{i\varphi(x,\xi)}) \tag{7.1}$$

assuming that $p(x, D) \in OPS^m_{\rho,1-\rho}(\rho > 1/2)$, $a(x, \xi) \in S^\mu_{\rho,1-\rho}$, and that $\varphi(x, \xi)$ is a real valued phase function satisfying hypotheses (3.3) and (3.4). We assume $p(x, D)$ is a properly supported operator. Furthermore, we assume $a(x, D)$ is properly supported, which involves no loss of generality since this can be achieved by adding some symbol in $S^{-\infty}$, and it is easy to verify that (7.1) is rapidly decreasing as $|\xi| \to \infty$ for $a \in S^{-\infty}$. The derivation we give of the behavior of (7.1) will be done in two stages. First we do it when $\varphi(x, \xi) = x \cdot \xi$; then we handle the general case using the results of Section 1.

For the case $\varphi = x \cdot \xi$, we have done all the work in the analysis of products of pseudodifferential operators, given in Chapter II, Section 4. The operator

$$b(x, D) = p(x, D)a(x, D)$$

belongs to $OPS^{m+\mu}_{\rho,1-\rho}$, and its symbol has the asymptotic expansion

$$b(x, \xi) \sim \sum_{\alpha \geq 0} \frac{1}{\alpha!} p^{(\alpha)}(x, \xi)a_{(\alpha)}(x, \xi).$$

However,

$$\begin{aligned} p(x, D)a(x, D)u &= p(x, D) \int a(x, \xi)e^{ix\cdot\xi}\hat{u}(\xi)\, d \\ &= \int p(x, D)(a(x, \xi)e^{ix\cdot\xi})\hat{u}(\xi)\, d\xi. \end{aligned}$$

It follows that

$$p(x, D)(a(x, \xi)e^{ix\cdot\xi}) = b(x, \xi)e^{ix\cdot\xi}. \tag{7.2}$$

We next consider the behavior of

$$p(x, D)(a(x, \lambda\xi_0)e^{i\lambda\varphi(x)}), \qquad |\xi_0| = 1, \tag{7.3}$$

as $\lambda \to \infty$, assuming $\nabla_x\varphi \neq 0$. Of course, if coordinates are so chosen that φ is one of them (locally), e.g. $\varphi(x) = x \cdot \xi_0$, then (7.3) is a special case of (7.2). On the other hand, we have seen in Section 1 that the invariance of $OPS^m_{\rho,1-\rho}$ under coordinate changes follows from Egorov's theorem. Thus if F denotes a diffeomorphism of $\mathbf{R}^n$ such that (at least locally, and one can suppose $a(x, \xi)$ has small support) $\varphi(F(x)) = x \cdot \xi_0$, and if $p_F(x, D)$

denotes the conjugated operator $p(x, D)$, we see that (letting $x = F(y)$)

$$\begin{aligned} p(x, D)(a(x, \lambda\xi_0)e^{i\lambda\varphi(x)}) &= p_F(y, D)(a_F(y, \lambda\xi_0)e^{i\lambda y \cdot \xi_0}) \\ &= \tilde{b}(y, \lambda\xi_0)e^{i\lambda y \cdot \xi_0} \\ &= \tilde{b}(F^{-1}(x), \lambda\xi_0)e^{i\lambda\varphi(x)} \end{aligned}$$

where $\tilde{b}(y, D) = p_F(y, D)a_F(y, D) \in OPS^{m+\mu}_{\rho,1-\rho}$. Thus we have

$$p(x, D)(a(x, \lambda\xi_0)e^{i\lambda\varphi(x)}) = b(x, \lambda\xi_0)e^{i\lambda\varphi(x)} \tag{7.4}$$

where $b(x, \xi) = \tilde{b}(F^{-1}(x), \xi) \in S^{m+\mu}_{\rho,1-\rho}$. Furthermore, using the formula for the principal symbol of $p_F(y, D)$ which is contained in Egorov's theorem, we see that

$$b(x, \lambda\xi_0) = a(x, \lambda\xi_0)p(x, \lambda\nabla_x\varphi), \qquad \text{mod } S^{m+\mu-(2\rho-1)}_{\rho,1-\rho}. \tag{7.5}$$

Now our goal, the analysis of (7.1), is simply (7.4) with parameters thrown in. Indeed, let

$$\theta = \frac{\xi}{|\xi|} \in S^{n-1}, \qquad \lambda = |\xi|,$$

and define the parametrized family $\varphi_\theta(x) = \varphi(x, \theta)$. Then $\varphi(x, \xi) = \lambda\varphi_\theta(x)$, so

$$p(x, D)(a(x, \xi)e^{i\varphi(x,\xi)}) = p(x, D)(a_\theta(x, \lambda\xi_0)e^{i\lambda\varphi_\theta(x)})$$

where $\xi_0 \in S^{n-1}$ is fixed and we arrange that $a_\theta(x, \lambda\xi_0) = a(x, \lambda\theta)$ (with $D^\gamma_\theta a_\theta$ bounded in $S^{\mu+|\gamma|(1-\rho)}_{\rho,1-\rho}$). If F_θ is a smooth family of diffeomorphisms such that $\varphi_\theta(F_\theta(x)) = x \cdot \xi_0$, it follows from Exercise 1.5 of Section 1 that the family $p_\theta(x, D) = p_{F_\theta}(x, D)$ of pseudodifferential operators obtained by conjugation by F_θ is a smooth family of pseudodifferential operators; more precisely, $D^\gamma_\theta p_\theta(x, \xi)$ is bounded in $S^{m+|\gamma|(1-\rho)}_{\rho,1-\rho}$, $\theta \in S^{n-1}$. Now we have, with $x = F_\theta(y)$,

$$\begin{aligned} p(x, D)(a(x, \xi)e^{i\varphi(x,\xi)}) &= p_\theta(y, D)(a_\theta(F_\theta(y), \lambda\xi_0)e^{i\lambda y \cdot \xi_0}) \\ &= p_\theta(y, D)(\tilde{a}_\theta(y, \lambda\xi_0)e^{i\lambda y \cdot \xi_0}) \\ &= \tilde{b}_\theta(y, \lambda\xi_0)e^{i\lambda y \cdot \xi_0} \\ &= \tilde{b}_\theta(F^{-1}(x), \lambda\xi_0)e^{i\varphi(x,\xi)} \end{aligned}$$

where $\tilde{b}_\theta(y, D) = p_\theta(y, D)\tilde{a}_\theta(y, D)$ has the property that $D^\gamma_\theta\tilde{b}_\theta(y, \xi)$ is bounded in $S^{m+\mu+|\gamma|(1-\rho)}_{\rho,1-\rho}$, $\theta \in S^{n-1}$. It follows that

$$p(x, D)(a(x, \xi)e^{i\varphi(x,\xi)}) = b(x, \xi)e^{i\varphi(x,\xi)} \tag{7.6}$$

where $b(x, \xi) = \tilde{b}_\theta(F^{-1}(x), \lambda\xi_0)$ is seen to belong to $S^{m+\mu}_{\rho,1-\rho}$. Furthermore the same argument which yields (7.5) implies that

$$b(x, \xi) = a(x, \xi)p(x, \nabla_x\varphi(x, \xi)), \qquad \text{mod } S^{m+\mu-(2\rho-1)}_{\rho,1-\rho}. \tag{7.7}$$

By solving the rest of the transport equations which arise in the proof of Egorov's theorem in Section 1, expanding the complete symbol of a product, etc., one could in principle obtain a complete asymptotic formula for the symbol $b(x, \xi)$ occurring in (7.6). However, this would be rather tedious. A better method is to use the complete asymptotic expansion of the symbol of the conjugated operator $p_F(x, D)$ which by Theorem 5.1 of Chapter II has the form

$$p_F(F(x), \xi) \sim \sum_{\alpha \geq 0} \frac{1}{\alpha!} \varphi_\alpha(x, \xi) p^{(\alpha)}(x, {}^tF'(x)\xi)$$

where $\varphi_\alpha(x, \xi)$ is a polynomial in ξ of degree $\leq (1/2)|\alpha|$ and $\varphi_0(x, \xi) = 1$. Plugging this into (7.6) yields for $b(x, \xi) = e^{i\varphi}p(x, D)e^{i\varphi}$,

$$b(x, \xi) \sim \sum_{\substack{\alpha \geq 0 \\ \beta \leq \alpha}} p^{(\alpha)}(x, \nabla\varphi) D_x^\beta a(x, \xi) \psi_{\alpha\beta}(x, \xi) \tag{7.8}$$

where $\psi_{\alpha\beta}(x, \xi)$ is a polynomial in ξ of degree $\leq (1/2)|\alpha - \beta|$ and $\psi_{00}(x, \xi) = 1$. In fact, the derivation of Melin and Sjöstrand [1] shows that, with $\rho(x, y, \xi) = \varphi(x, \xi) - \varphi(y, \xi) - (y - x) \cdot \nabla_x \varphi(x, \xi)$,

$$\psi_{\alpha\beta}(x, \xi) = \frac{1}{\alpha!} \binom{\alpha}{\beta} D_y^{\alpha - \beta} e^{i\rho(x, y, \xi)}\big|_{y = x}.$$

As a last remark, we would like to untangle some of the convoluted logic that may have confused the astute reader. Egorov's theorem was proved by energy estimates requiring us to work on a compact manifold. Then we established the invariance of $OPS^m_{\rho, 1-\rho}$ under coordinate changes in order to work on compact manifolds in the first place. The way to get around this is to work on the torus $\mathbf{T}^n$ to start with, where coordinate changes cause no problem. Also, our second proof of Egorov's theorem, given in Section 3, depends on the results of this appendix, and hence, if we must rely on Egorov's theorem to analyze (7.1), this proof can not be taken seriously. Of course, there are other proofs of (7.6), (7.7), given in Hörmander [7] and Melin and Sjöstrand [1], and if one uses them as a basis, the second proof is a valid one. At any rate it has a certain appeal (to the author) and perhaps further explains why the theorem is true.

§8. Egorov's Theorem for $OPS^m_{1/2,1/2}$

Egorov's theorem, stated and proved in Section 1, does not apply to $p(x, D) \in OPS^m_{1/2,1/2}$. Nevertheless, we show below that $S(t, s)p(x, D)S(s, t)$ is a member of $OPS^m_{1/2,1/2}$. However, one cannot give a simple explicit formula for a "principal symbol" of such a conjugated operator. We begin with a characterization of $OPS^m_{1/2,1/2}$ due to R. Beals [3]; compare with Lemma 5.2.

THEOREM 8.1. *Let $A:\mathcal{D}'(\mathbf{T}^n) \to \mathcal{D}'(\mathbf{T}^n)$, $\mathbf{T}^n$ the torus, and suppose that, given any $P_j \in OPS^1_{1,0}$, you have*

$$Ad(P_1)\cdots Ad(P_k)A:H^s(\mathbf{T}^n) \to H^{s-m-k/2}(\mathbf{T}^n). \tag{8.1}$$

Then $A \in OPS^m_{1/2,1/2}$.

Proof. Taking a partition of unity ψ_j on $\mathbf{T}^n$ and considering $\psi_j A \psi_k$, which we relabel A, we see that (8.1) applies to this modified operator, and we can regard A as mapping $\mathcal{D}'(\mathbf{R}^n)$ to $\mathcal{E}'(\mathbf{R}^n)$. It follows from (8.1) that

$$(Ad\ x)^\alpha(Ad\ D)^\beta A:H^s(\mathbf{R}^n) \to H^{s+|\alpha|/2-|\beta|/2-m}(\mathbf{R}^n). \tag{8.2}$$

Note that, if we define $a(x, \xi)$ by

$$a(x, \xi) = e^{-ix\cdot\xi}A(e^{ix\cdot\xi}),$$

then $Au = \int a(x, \xi)e^{ix\cdot\xi}\hat{u}(\xi)\,d\xi$. It remains to show that $|D^\beta_x D^\alpha_\xi a(x, \xi)| \le c_{\alpha\beta}(1 + |\xi|)^{|\beta|/2-|\alpha|/2+m}$, provided (8.2) holds. Note that if we replace A by $A(1-\Delta)^{-m/2}$ (on $\mathbf{T}^n$), then (8.1) and (8.2) hold with $m=0$ for $A(1-\Delta)^{-m/2}$, and if we show $A(1-\Delta)^{-m/2} \in OPS^0_{1/2,1/2}$, it follows that $A \in OPS^m_{1/2,1/2}$, so without loss of generality we can take $m = 0$. Note that, with $B_{\alpha\beta} = (Ad\ x)^\alpha(Ad\ D)^\beta A$,

$$B_{\alpha\beta}u = \int b_{\alpha\beta}(x, \xi)e^{ix\cdot\xi}\hat{u}(\xi)\,d\xi$$

where $b_{\alpha\beta}(x, \xi) = D^\beta_x D^\alpha_\xi a(x, \xi)$.

Let $V_t(u(x)) = t^{-n/2}u(t^{-1}x)$, and set $A_t = V_t^{-1}AV_t$. Then

$$B_{t\alpha\beta} = (Ad\ x)^\alpha(Ad\ D)^\beta A_t = t^{|\beta|-|\alpha|}V_t^{-1}\{(Ad\ x)^\alpha(Ad\ D)^\beta A\}\,V_t,$$

and it has "symbol" $b_{t\alpha\beta}(x, \xi) = e^{-ix\cdot\xi}B_{t\alpha\beta}(e^{ix\cdot\xi})$. Note that the symbol of A_t is $a(tx, t^{-1}\xi)$, and that $b_{t\alpha\beta}(x, \xi) = t^{|\beta|-|\alpha|}b_{\alpha\beta}(tx, t^{-1}\xi)$.

It is convenient to introduce a function $g \in \mathcal{S}$ with $g(0) = 1$ and $\hat{g}(\xi)$ supported in $|\xi| \le 1$. It is easy to verify that

$$a(x, \xi) = \int e^{i(x-y)\cdot(\eta-\xi)}a_0(x, y, \eta)\,d\eta\,dy \tag{8.3}$$

where

$$a_0(x, y, \eta) = e^{-ix\cdot\xi}A(g_y e^{ix\cdot\xi}) \tag{8.4}$$

with $g_y(x) = g(y-x)$. We have

$$a_0(x, y, \xi) = b_{0t}(t^{-1}x, t^{-1}y, t\xi) \tag{8.5}$$

where

$$(8.6)\qquad b_{0t}(x, y, \xi) = e^{-ix\cdot\xi}A_t(g_y e^{ix\cdot\xi}) = e^{-ix\cdot\xi}V_t^{-1}AV_t(g_y e^{ix\cdot\xi}).$$

Now, with $\langle\xi\rangle = (1+|\xi|^2)^{1/2}$, we have the bound $\|V_t g_y e^{ix\cdot\xi}\|_{H^s} \le C\langle t^{-1}\xi\rangle^s$, and plugging this into (8.6) yields

$$\|D_x^\beta D_y^\gamma D_\xi^\alpha b_{0t}(\cdot, y, \xi)\|_{L^2} \le C_{\alpha\beta\gamma} t^{|\beta|-|\alpha|}\langle t^{-1}\xi\rangle^{|\beta|/2-|\alpha|/2}.$$

Replace ξ by $t\xi$ and set $t = \langle\xi\rangle^{-1/2}$ to get

$$(8.7)\qquad \|D_x^\beta D_y^\gamma D_\xi^\alpha b_{0t}(\cdot, y, t\xi)\|_{L^2} \le C_{\alpha\beta\gamma}, \qquad t = \langle\xi\rangle^{-1/2}.$$

Applying the Sobolev imbedding theorem, with β replaced by $\beta+\beta'$, $|\beta'| > n/2$ in (8.7), yields

$$|D_x^\beta D_y^\gamma D_\xi^\alpha b_{0t}(x, y, t\xi)| \le C_{\alpha\beta\gamma}, \qquad t = \langle\xi\rangle^{-1/2},$$

so, by (8.5),

$$(8.8)\qquad |D_x^\beta D_y^\gamma D_\xi^\alpha a_0(x, y, \xi)| \le C_{\alpha\beta\gamma}\langle\xi\rangle^{|\beta|/2-|\alpha|/2}.$$

It remains to show that (8.8) implies the correct estimates on $D_x^\beta D_\xi^\alpha a(x, \xi)$, given by (8.3). This is a special case of the following proposition, which is of interest in its own right. More general results are given in Beals and Fefferman [2] and Beals [2].

PROPOSITION 8.2. *Let $b(x, y, \eta)$ satisfy the estimates*

$$(8.9)\qquad |D_y^\gamma D_\eta^\alpha b(x, y, \eta)| \le C_{\alpha\gamma}\langle\eta\rangle^{m-|\alpha|/2+|\gamma|/2}.$$

Then

$$(8.10)\qquad a(x, \xi) = \int e^{i(x-y)\cdot(\eta-\xi)} b(x, y, \eta)\, dy\, d\eta$$

satisfies the estimate

$$(8.11)\qquad |a(x, \xi)| \le C\langle\xi\rangle^m.$$

More generally, if

$$(8.12)\qquad |D_x^\beta D_y^\gamma D_\eta^\alpha b(x, y, \eta)| \le C_{\alpha\beta\gamma}\langle\eta\rangle^{m-|\alpha|/2+|\beta|/2+|\gamma|/2},$$

then

$$(8.13)\qquad |D_x^\beta D_\xi^\alpha a(x, \xi)| \le C\langle\xi\rangle^{m+|\beta|/2-|\alpha|/2}.$$

Proof. First note that, upon integration by parts, (8.10) yields

$$D_x^\beta D_\xi^\alpha a(x, \xi) = \int e^{i(x-y)\cdot(\eta-\xi)}(D_y + D_x)^\beta D_\eta^\alpha b(x, y, \eta)\, dy\, d\eta.$$

Consequently, if we deduce (8.11) from (8.9) we shall automatically have (8.13) from (8.12).

To prove (8.11), set

$$R = \langle\xi\rangle + \langle\eta\rangle, \qquad r = \langle\xi\rangle^{-1} + \langle\eta\rangle^{-1},$$
$$S = \langle\xi\rangle^{1/2} + \langle\eta\rangle^{1/2}, \qquad s = \langle\xi\rangle^{-1/2} + \langle\eta\rangle^{-1/2},$$
$$h = e^{i(x-y)\cdot(\eta-\xi)},$$
$$g_0 = 1 + \langle\xi\rangle|x-y|^2 + \langle\xi\rangle^{-1}|\xi-\eta|^2,$$
$$g = 1 + R|x-y|^2 + r|\xi-\eta|^2.$$

Let L be the differential operator

$$Lu = g^{-1}(1 - R\Delta_\eta - r\Delta_y)u.$$

Then $Lh = h$ and (8.10) implies

(8.14) $$a(x, \xi) = \int hb_N(x, y, \xi, \eta)\, dy\, d\eta$$

where

$$b_N(x, y, \xi, \eta) = (L^t)^N b.$$

Hypothesis (8.9) implies

$$|(R^{1/2}D_\eta)^\alpha b| + |(r^{1/2}D_y)^\alpha b| \le C_\alpha (Rr)^{|\alpha|/2}\langle\eta\rangle^m.$$

Hence

(8.15) $$|b_N| \le Cg^{-N}(Rr)^N\langle\eta\rangle^m.$$

We estimate a by writing (8.14) as a sum of integrations over three subregions, $a = a_1 + a_2 + a_3$, and using (8.15) in each region. These regions are:

Region (1). $s^{-1}|x-y| + S^{-1}|\xi-\eta| \le c_0$,

(8.16) Region (2). $s^{-1}|x-y| + S^{-1}|\xi-\eta| \ge c_0$, but

(8.17) $$r^{1/2}|\xi-\eta| + r^{-1/2}|x-y| \le c_1 R^\delta,$$

Region (3). Complement of (1) $\cup$ (2).

In Region (1), $g \sim g_0$, $\langle\xi\rangle \sim \langle\eta\rangle$, and, $Rr \sim 1$, so (8.15) becomes

(8.18) $$|b_N| \le Cg_0^{-N}\langle\xi\rangle^m, \qquad \text{in Region (1).}$$

Choose $N > n$ and integrate (8.18) with respect to y and η to obtain

(8.19) $$|a_1| \le C\langle\xi\rangle^m.$$

To handle Region (2), choose c_1 sufficiently small that, by (8.17), $\langle\xi\rangle \sim \langle\eta\rangle$. (8.16) implies $g_0 \sim R|x-y|^2 + R^{-1}|\eta-\xi|^2 \ge C_2$, so (8.15) implies

$$|b_N| \le Cg_0^{-N}\langle\xi\rangle^m, \qquad \text{in Region (2)}$$

which as before yields

(8.20) $$|a_2| \le C\langle\xi\rangle^m.$$

Finally, in Region (3),

$$\begin{aligned} g &> R|x-y|^2 + r|\xi-\eta|^2 \\ &= (Rr)R^{2\delta}[R^{-2\delta}r^{-1}|x-y|^2 + R^{-2\delta}R^{-1}|\xi-\eta|^2] \\ &\ge c_1(Rr)R^{2\delta} \ge c_2(Rr)^{1+\delta}R^{\delta}. \end{aligned}$$

If $\epsilon > 0$ is small enough, $g \ge c(Rr)R^{\epsilon}g^{\epsilon}$. Also, $\langle\eta\rangle^m \le CR^M\langle\xi\rangle^m$. Hence (8.15) yields

(8.21) $$|b_N| \le Cg_0^{-N\epsilon}R^{-N\epsilon+M}\langle\xi\rangle^m.$$

Take $N > n/\epsilon$ and $N \ge M/\epsilon$ and integrate (8.21) to get

(8.22) $$|a_3| \le C\langle\xi\rangle^m$$

(8.19), (8.20), and (8.22) prove (8.11) and hence the proposition.

Proposition 8.2 completes the proof of Theorem 8.1. It also shows that, if $P \in OPS^m_{1/2,1/2}$, $Q \in OPS^{\mu}_{1/2,1/2}$, then

(8.23) $$P^* \in OPS^m_{1/2,1/2} \quad \text{and}$$

(8.24) $$PQ \in OPS^{m+\mu}_{1/2,1/2}.$$

Indeed, the proofs of Theorems 4.1 through 4.3 of Chapter II generalize, with Theorem 3.8 of Chapter II replaced by Proposition 8.2. Generally, one does not have a covenient formula for the symbols of P^* and PQ, analogous to (4.2) and (4.5) of Chapter II. In fact, commutators of elements of $OPS^0_{1/2,1/2}(\mathbf{T}^n)$ are known not to be necessarily compact, so any symbol calculus would have to reflect this essential noncommutativity. Symbol calculi for some important subclasses of $S^m_{1/2,1/2}$ are given in Boutet de Monvel [2] and Dynin [1].

We are now ready for the main result of this section. If $S(t,s)$ is the solution operator to a hyperbolic equation $(\partial/\partial t)u = i\lambda(t,x,D_x)u$ on $\mathbf{R}\times\mathbf{T}^n$, we establish the following.

THEOREM 8.3. *If $p(x,D) \in OPS^m_{1/2,1/2}(\mathbf{T}^n)$, then*

$$S(t,s)p(x,D)S(s,t) = P(s,t) \in OPS^m_{1/2,1/2}.$$

Proof. We verify that $A = P(s,t)$ satisfies (8.1). Note that, with $U = S(s,t)$,

$$\begin{aligned} Ad(P_1)(U^{-1}p(x,D)U) &= P_1U^{-1}p(x,D)U - U^{-1}p(x,D)UP_1 \\ &= U^{-1}[UP_1U^{-1}p(x,D) - p(x,D)UP_1U^{-1}]U \\ &= U^{-1}Ad(UP_1U^{-1})p(x,D)U, \end{aligned}$$

and inductively we obtain

$$
\begin{aligned}
(8.25)\quad & Ad(P_1)\cdots Ad(P_k)(U^{-1}p(x,D)U) \\
&= U^{-1}\,Ad(UP_1U^{-1})\cdots Ad(UP_kU^{-1})p(x,D)U.
\end{aligned}
$$

Now by Egorov's theorem, each UP_jU^{-1} belongs to OPS^1. Hence Theorem 4.4 of Chapter II yields $d(UP_1U^{-1})\cdots Ad(UP_kU^{-1})p(x,D)$ in $OPS^{m-k/2}_{1/2,1/2}$. Since $U: H^s \to H^s$ for all s, we can apply the Calderon-Vaillancourt theorem of Chapter XIII to conclude that the right side of (8.25) maps $H^s(\mathbf{T}^n)$ to $H^{s-m-k/2}(\mathbf{T}^n)$, verifying hypothesis (8.1). This proves the theorem.

CHAPTER IX

Reflection of Singularities

In this chapter we shall examine reflection of singularities of solutions to hyperbolic systems in regions with boundary,

$$Lu = f \quad \text{on} \quad \mathbf{R} \times \Omega, \tag{0.1}$$

$$B_j u|_{\mathbf{R} \times \partial\Omega} = g_j \tag{0.2}$$

where $u = u(t, x)$, and L is a hyperbolic operator, of order m, so initial data

$$D_t^j u(0, x) = f_j, \qquad j = 0, \ldots, m-1 \tag{0.3}$$

are also supposed given. In order to understand the singularities of such solutions, it suffices, as will be seen, to handle the case where u solves (0.1), (0.2), with $f = 0$, $g_j \in \mathscr{E}'(\mathbf{R} \times \partial\Omega)$, and

$$u(t, x) = 0 \quad \text{for} \quad t \ll 0. \tag{0.4}$$

We analyze (0.1), (0.2), (0.4) by constructing an approximate solution, i.e. function v such that

$$Lv \in C^\infty(\mathbf{R} \times \bar{\Omega}), \quad B_j v|_{\mathbf{R} \times \partial\Omega} - g_j \in C^\infty(\mathbf{R} \times \partial\Omega), \quad v \in C^\infty \text{ for } t \ll 0. \tag{0.5}$$

In order to insure that v differs from u by a smooth function, we need to have the following condition on the hyperbolic system:

(0.6) If $Lw \in C^\infty(\mathbf{R} \times \bar{\Omega})$, $B_j w|_{\mathbf{R} \times \partial\Omega} \in C^\infty(\mathbf{R} \times \partial\Omega)$, and $w = 0$ for $t \ll 0$, then $w \in C^\infty(\mathbf{R} \times \bar{\Omega})$.

Condition (0.6) generally follows from the well posedness of (0.1) through (0.3). We consider here a particular class of hyperbolic systems for which (0.6) can be established very cleanly, with a minimum of technical problems. For other results on systems satisfying (0.6), the reader is referred to Rauch and Massey [1].

The class we consider here consists of second order hyperbolic systems of the form

$$\left(\frac{\partial^2}{\partial t^2} - A(x, D_x)\right) u = f, \tag{0.7}$$

(0.8) $$B_j(x, D_x)u|_{\partial\Omega} = g_j,$$

(0.9) $$u(0, x) = a(x),\ u_t(0, x) = b(x).$$

We assume A is a second order elliptic operator, Ω is bounded, that the boundary conditions B_j are coercive, and that the operator with domain $\mathcal{D}(A) = \{u \in L^2(\Omega): Au \in L^2(\Omega),\ B_j u|_{\partial\Omega} = 0\}$ is a self adjoint, negative semi-definite operator. The unique solvability of (0.7) – (0.9), with $f \in C(\mathbf{R}, \mathcal{D}(A^{1/2}))$, $g = 0$, $a \in \mathcal{D}(A^{1/2})$, $b \in L^2(\Omega)$ is an immediate consequence of the spectral theorem, assuming $f = 0$ for $t \ll 0$. If $f = 0$ for $t \leq 0$, we can write the solution as

(0.10) $$u = \cos t\sqrt{-A}a + \frac{\sin t\sqrt{-A}}{\sqrt{-A}}b + \frac{1}{\sqrt{-A}}\int_0^t \sin(t-s)\sqrt{-A}f(s)\,ds.$$

THEOREM 0.1. *Let u satisfy* (0.7), (0.8) *with $f \in C^\infty(\mathbf{R}\times\bar{\Omega})$, $g_j \in C^\infty(\mathbf{R}\times\partial\Omega)$, $f, g_j = 0$ for $t \leq 0$, and suppose $u = 0$ for $t \leq 0$. Then $u \in C^\infty(\mathbf{R}\times\bar{\Omega})$.*

Proof. Subtracting off a smooth function satisfying (0.8), we can suppose without loss of generally that $g_j = 0$. Also, by solving for the formal power series, which is possible since $\mathbf{R}\times\partial\Omega$ is assumed to be noncharacteristic, we can find a smooth function satisfying (0.7) to infinite order at the boundary, and subtracting this off, we may suppose that f vanishes to infinite order on $\mathbf{R}\times\partial\Omega$. This makes $f \in C^\infty(\mathbf{R}, \mathcal{D}(A^j))$, all j. Now $D_t^\ell u$ satisfies

$$\left(\frac{\partial^2}{\partial t^2} - A\right)(D_t^\ell u) = D_t^\ell f, \qquad B_j(D_t^\ell u)|_{\partial\Omega} = 0,$$

$D_t^\ell u = 0$ for $t \leq 0$, so (0.10) implies $D_t^\ell u \in C(\mathbf{R}, \mathcal{D}(A^j))$, for all j, ℓ, which yields $u \in C^\infty(\mathbf{R}\times\bar{\Omega})$, as asserted.

When we construct an approximate solution v satisfying (0.5), we may as well suppose that $WF(g_j)$ is contained in a small conic neighborhood $\mathfrak{U}$ of some point $(z_0, \zeta_0) \in T^*(\mathbf{R}\times\partial\Omega) - 0$. In this chapter we consider only the case where all null bicharacteristics of L which pass over $\mathfrak{U}$ hit $\mathbf{R}\times\partial\Omega$ *transversally*. In the following chapter we shall discuss some cases when such rays hit the boundary tangentially, the case of grazing rays.

In order to construct an approximate solution v to (0.5), we first change variables, so let y denote some normal variable to $\partial\Omega$ so $y = 0$ defines $\partial\Omega$, and relabel the x and t variables as x, thus identifying a neighborhood of $\mathbf{R}\times\partial\Omega$ in $\mathbf{R}\times\bar{\Omega}$ with $[0, 1) \times X$. Then (0.1) becomes, under the hypothesis that $\partial\Omega$ is noncharacteristic for L,

(0.11) $$\left(\frac{\partial^m}{\partial y^m} + \sum_{j=0}^{m-1} P_{m-j}(y, x, D_x)\frac{\partial^j}{\partial y^j}\right)u = f,$$

and we suppose (0.2) goes over into certain boundary conditions for $u = u(y, x)$ at the boundary $y = 0$:

$$\tilde{B}_j u(0, x) = h_j. \tag{0.12}$$

Next we convert this to a first order system for m times as many unknowns, by the standard device. Namely, set

$$w_j = D_y^{j-1}\Lambda^{m-j}u, \qquad j = 1, \ldots, m,$$

so (0.11) becomes

$$\frac{\partial}{\partial y} w = \begin{pmatrix} 0 & \Lambda & 0 & \cdots & 0 \\ & 0 & \Lambda & & \cdot \\ & & 0 & & \cdot \\ & & & \ddots & \cdot \\ & & & \cdot & \Lambda \\ b_1 & b_2 & & \cdots & b_m \end{pmatrix} w + \begin{pmatrix} 0 \\ \cdot \\ \cdot \\ \cdot \\ 0 \\ f \end{pmatrix} = Gw + \tilde{f} \in C^\infty \tag{0.13}$$

where $b_j = -P_{m-j+1}(y, x, D_x)\Lambda^{j-m}$. Here $\Lambda = \Lambda(x, D_x)$ is some elliptic first order operator, say $\Lambda(x, \xi) = |\xi|$. The boundary condition (0.12) becomes a boundary condition for w, say

$$\beta(x, D_x)w(0, x) = h. \tag{0.14}$$

Here we assume $WF(h)$ is contained in a small conic neighborhood $\mathfrak{U}$ of some point $(x_0, \xi_0) \in T^*(x) - 0$.

We suppose that j null bicharacteristics of L pass over (x_0, ξ_0), so there are j real solutions $\eta_1, \ldots, \eta_j$ of $q(0, x_0, \eta, \xi_0) = \det(i\eta - G_1(0, x_0, \xi_0)) = 0$, where $G_1(y, x, \xi)$ is the principal symbol of G. Since we are assuming that all these bicharacteristics intersect the boundary $y = 0$ transversally, we have $\dot{y} = \partial q/\partial n \neq 0$ at $(0, x_0, \eta_\nu, \xi_0)$, so all these real zeros $\eta_1, \ldots, \eta_j$ are simple. It follows that, for (x, ξ) in a small conic neighborhood of (x_0, ξ_0) and for $y \geq 0$ small, $G_1(y, x, \xi)$ has j simple pure imaginary eigenvalues $i\lambda_1(y, x, \xi), \ldots, i\lambda_j(y, x, \xi)$. Here $\lambda_\nu(0, x_0, \xi_0) = \eta_\nu$.

Suppose that on the bicharacteristics associated with $\lambda_1, \ldots, \lambda_\ell$, t is increasing as y increases, while on those associated with $\lambda_{\ell+1}, \ldots, \lambda_j$, t is decreasing as y increases. We want to find conditions on the system (0.13), (0.14) which will allow us to construct an approximate solution w (with $\tilde{f}$ smooth) which is singular only along the first ℓ bicharacteristics. We need only make the construction for y small, since in the interior of $\mathbf{R} \times \Omega$, we can use the results of Chapter VIII.

§1. Decoupling First Order Systems

Our first task in analyzing the system (0.13) will be to put it in a very simple form via a change of dependent variable. Since it will turn out that $WF(u(y))$ will be contained in a small conic neighborhood of (x_0, ξ_0), for y small, we can work locally in $T^*(X)\backslash 0$. We can construct a smooth invertible matrix $U(y, x, \xi)$, homogeneous of degree zero in ξ, such that, on a conic neighborhood $\mathfrak{U}_1$ of (x_0, ξ_0), for $0 \le y \le y_1$,

$$UG_1U^{-1} = \tilde{G}_1 = \begin{pmatrix} i\lambda_i & & & & \\ & \ddots & & & \\ & & i\lambda_j & & \\ & & & E_+ & \\ & & & & E_- \end{pmatrix} \tag{1.1}$$

where E_+ is a square matrix whose eigenvalues have positive real part, and E_- one whose eigenvalues have negative real part. Extend $\tilde{G}_1$, keeping the form (1.1). Now let

$$w_1 = U(y, x, D)w.$$

If $WF(w)$ is contained in $\mathfrak{U}_1$, (0.13) is equivalent to

$$\frac{\partial}{\partial y} w_1 = \tilde{G}w_1 + F \tag{1.2}$$

where $\tilde{G} = UGU^{-1} + U_yU^{-1}$ and $F = U\tilde{f}$. The principal symbol of $\tilde{G}$ is $\tilde{G}_1$, given by (1.1). Thus we have

$$\frac{\partial}{\partial y} w_1 = \begin{pmatrix} i\lambda_1 & & & & \\ & \ddots & & & \\ & & i\lambda_j & & \\ & & & E_+ & \\ & & & & E_- \end{pmatrix} w_1 + Bw_1 \pmod{C^\infty} \tag{1.3}$$

where $B \in OPS^0$. This equation is decoupled in the principal part, but the B produces zero order coupling. We now want to decouple (1.3) *completely*, modulo a smooth remainder.

We consider a slightly more general problem. Let v solve the system

$$\frac{\partial}{\partial y} v = \begin{pmatrix} F & \\ & E \end{pmatrix} v + Av \tag{1.4}$$

where $G = \begin{pmatrix} F & \\ & E \end{pmatrix}$ has symbol homogeneous of degree 1 in ξ and $A \in OPS^0$. The assumption we make on the symbols $F(y, x, \xi)$ and $E(y, x, \xi)$ is that these two matrices, or order $\mu \times \mu$ and $\nu \times \nu$, respectively, have disjoint eigenvalues, for each $(y, x, \xi) \in [0, y_0) \times T^*(X) - 0$.

We first try for a modest goal, to decouple terms of order zero. Let $w^{(1)} = (1 + K_1)v$ with $K_1 \in OPS^{-1}$ to be determined. We have

$$\begin{aligned}\frac{\partial}{\partial y} w^{(1)} &= (1 + K_1)G(1 + K_1)^{-1}w^{(1)} + (1 + K_1)A(1 + K)^{-1}w^{(1)} + \cdots \\ &= Gw^{(1)} + (K_1 G - GK_1 + A)w^{(1)} + \cdots\end{aligned}$$

where the remainder involves terms of order at most -1 operating on $w^{(1)}$. We would like to be able to pick K_1 such that the off diagonal terms of $K_1 G - GK_1 + A$ vanish. We demand that K_1 be of the form

$$\begin{pmatrix} 0 & K_{12} \\ K_{21} & 0 \end{pmatrix}.$$

If

$$A = \begin{pmatrix} A_{11} & A_{12} \\ A_{21} & A_{22} \end{pmatrix},$$

we are led to require that

$$K_{12}E - FK_{12} = -A_{12}, \tag{1.5}$$

$$K_{21}F - EK_{21} = -A_{21}. \tag{1.6}$$

That (1.5) and (1.6) have unique solutions is a consequence of the following lemma.

LEMMA 1.1. *Let $F \in M_{\nu \times \nu}$, the set of $\nu \times \nu$ matrices, and $E \in M_{\mu \times \mu}$. Define $\varphi : M_{\nu \times \mu} \to M_{\nu \times \mu}$ by*

$$\varphi(T) = TF - ET.$$

Then φ is bijective, if E and F have disjoint spectra.

Proof. In fact, if $\{f_j\}$ are the eigenvalues of F and $\{e_k\}$ those of E, it is easily seen that the eigenvalues of φ are $\{f_j - e_k\}$.

So we obtain unique solutions K_{12} and K_{21} to (1.5) and (1.6). With such a choice of the symbol of K_1, we have

$$\frac{\partial}{\partial y} w^{(1)} = Gw^{(1)} + \begin{pmatrix} A_1 & \\ & A_2 \end{pmatrix} w^{(1)} + Bw^{(1)}$$

with $B \in OPS^{-1}$. To decouple the part of order -1, we try $w^{(2)} = (1 + K_2)w^{(1)}$ with $K_2 \in OPS^{-2}$. We get

$$\frac{\partial}{\partial y} w^{(2)} = GW^{(2)} + \begin{pmatrix} A_1 & \\ & A_2 \end{pmatrix} w^{(2)} + (K_2 G - GK_2 + B)w^{(2)} + \cdots,$$

so we want to choose K_2 so that, on the symbol level, the off diagonal terms of $K_2G - GK_2 + B$ vanish. This is the problem solved above, so we are in good shape.

From here we continue, defining $w^{(j)} = (1 + K_j)w^{(j-1)}$ with $K_j \in OPS^{-j}$ decoupling further out along the line. Letting $w = (1 + K)v$ with $K \in OPS^{-1}$, $1 + K \sim \cdots (1 + K_2)(1 + K_1)$, we have

$$\frac{\partial}{\partial y} w = \begin{pmatrix} E & \\ & F \end{pmatrix} w + \begin{pmatrix} \alpha & \\ & \beta \end{pmatrix} w \pmod{C^\infty}$$

with $\alpha, \beta \in OPS^0$. The system (1.4) is now completely decoupled.

Applying such a decoupling procedure inductively to (1.3), we can write $\tilde{w} = (1 + K)w_1$ with $K \in OPS^{-1}$, so that (1.3) becomes

$$\text{(1.7)} \quad \frac{\partial}{\partial y} \tilde{w} = \begin{pmatrix} i\lambda_1 & & & & \\ & \ddots & & & \\ & & i\lambda_j & & \\ & & & E_+ & \\ & & & & E_- \end{pmatrix} \tilde{w} + \begin{pmatrix} a_1 & & & & \\ & \ddots & & & \\ & & a_j & & \\ & & & \alpha & \\ & & & & \beta \end{pmatrix} \tilde{w} \pmod{C^\infty},$$

which effects the decoupling of (0.13), into equations for $w_1, \ldots, w_j$, w_+, w_-, the components of $\tilde{w}$. Of course, these components of $\tilde{w}$ are related to each other via the boundary condition, derived from (0.14), which is

$$\text{(1.8)} \quad \beta(x, D_x)U^{-1}(1 + K)^{-1}\tilde{w}(0, x) = h.$$

The equations for $w_1, \ldots, w_j$ are hyperbolic, and the construction of parametrices for these components was done in Chapter VIII. The equations for w_+ and w_- are backward and forward elliptic evolution equations, respectively, and these equations will be discussed in the next section.

§2. Elliptic Evolution Equations

In this section we shall consider solutions to first order systems

$$\text{(2.1)} \quad \begin{aligned} \frac{\partial}{\partial y} u &= E(y, x, D_x)u, \\ u(0) &= f \end{aligned}$$

where E is a $k \times k$ matrix of first order pseudodifferential operators, $E(y) \in OPS^1$, whose principal symbol $E_1(y, x, \xi)$ satisfies the condition

$$\text{(2.2)} \quad Re(\text{spectrum of } E_1(y, x, \xi)) \leq -c_0|\xi| < 0, \qquad |\xi| \neq 0.$$

In such a case, we call (2.1) an elliptic evolution equation. The regularity of solutions to such an equation is a special case of the results of Chapter

V; indeed the details of the analysis of (2.1) are a little easier than the more general case discussed there. We shall briefly sketch this analysis, in order to make this section somewhat self-contained, though the reader is referred to Chapter V for a fuller treatment. Then we shall look a little more closely at the structure of the solution operator to (2.1.).

First, there exists a "symmetrizer" for (2.1), i.e., a family $R(y)$ of positive definite elliptic operators in OPS^0 such that the principal symbol of $R(y)E(y) + E(y)^*R(y)$ is $\leq -c_1|\xi|$, given $0 < c_1 < c_0$. This construction is given in Chapter V. Consequently, any sufficiently smooth solution to (2.1) satisfies the energy inequality

$$\begin{aligned} \frac{d}{dy}(R(y)u, u) &= ((RE + E^*R)u, u) + (R'(y)u, u) \\ &\leq -c\|u\|^2_{H^{1/2}(M)} + c'\|u\|^2_{L^2} \\ &\leq c''(R(y)u, u) \end{aligned} \tag{2.3}$$

which by Gronwall's inequality and the positive definiteness of $R(y)$ yields the a priori inequality:

$$\|u(y)\|^2_{L^2} \leq c(y)\|f\|^2_{L^2}, \qquad y \geq 0. \tag{2.4}$$

Similarly one obtains the a priori energy inequality

$$\|u(y)\|^2_{H^s(M)} \leq c_s(y)\|f\|^2_{H^s}, \qquad y \geq 0, \tag{2.5}$$

and by standard methods (e.g. Chapter IV) this leads to the existence of a unique $u \in C([0, T], H^s(M))$, given any $f \in H^s(M)$. We get sharper results on smoothness of u by improving the a priori inequality (2.5).

Denoting by $\Lambda^{1/2} \in OPS^{1/2}$ any positive self-adjoint elliptic operator on M with scalar symbol, we have

$$\begin{aligned} \frac{d}{dy}(R(y)\Lambda^{1/2}u, \Lambda^{1/2}u) &= ((RE+E^*R)\Lambda^{1/2}u, \Lambda^{1/2}u) + (R[\Lambda^{1/2}, E]u, \Lambda^{1/2}u) \\ &\quad + (R\Lambda^{1/2}u, [\Lambda^{1/2}, E]u) + (R'\Lambda^{1/2}u, \Lambda^{1/2}u) \\ &\leq -c\|u\|^2_{H^1(M)} + c'\|u\|^2_{H^{1/2}(M)} \end{aligned}$$

Adding $c\|u\|^2_{H^1(M)}$ to each side and using Poincare's inequality yields, with new c and c',

$$c\|u\|^2_{H^1(M)} + \frac{d}{dy}(R(y)\Lambda^{1/2}u, \Lambda^{1/2}u) \leq c'\|u\|^2_{L^2(M)},$$

and if we integrate with respect to y and recall the positive definiteness of $R(y)$, we get

$$c\int_0^T \|u(y)\|^2_{H^1(M)}\,dy + \|u(T)\|^2_{H^{1/2}(M)} \leq c\|u\|^2_{L^2(\Omega)} + c\|f\|^2_{H^{1/2}(M)}$$

where $\Omega = [0, T] \times M$. Since

$$\int_0^T \|u_y\|_{L^2}^2\, dy = \int_0^T \|Eu\|_{L^2}^2\, dy \le c \int_0^T \|u\|_{H^1}^2\, dy$$

and $\|u(y)\|_{L^2}^2$ is estimated by (2.4), this yields

$$\|u\|_{H^1(\Omega)}^2 \le c\|f\|_{H^{1/2}(M)}^2. \tag{2.6}$$

The a priori inequality (2.6) is first established for sufficiently smooth u, but then the Friedrichs mollifier method, as used in Chapter V, shows that, given $f \in H^{1/2}(M)$, the unique solution u to (2.1) belongs to $H^1(\Omega)$. Similarly one obtains the a priori inequalities

$$\|u\|_{H^k(\Omega)}^2 \le c\|f\|_{H^{k-1/2}(M)}^2, \qquad k = 1, 2, 3, \ldots, \tag{2.7}$$

and the regularity result that the map $f \mapsto u$ takes $H^{k-1/2}(M)$ to $H^k(\Omega)$. Interpolation shows that $f \mapsto u$ takes $H^s(M)$ to $H^{s+1/2}(\Omega)$, $s \ge 1/2$. Indeed, this conclusion is valid for any $s \in \mathbf{R}$, but for $s < 1/2$ it is perhaps easier to obtain such a conclusion from a parametrix for the solution operator to (2.1).

If we denote the solution operator to (2.1) by $S(y)$, so $u(y) = S(y)f$, we aim to show that $S(y)$ is a bounded family of operators in $OPS_{1,0}^0$, with the further property that

$$y^k D_y^j S(y) \text{ bounded in } OPS_{1,0}^{j-k}, \qquad 0 \le y \le T. \tag{2.8}$$

In fact, we shall prove a stronger result, which will be useful in Chapter XI, Section 5, expressing $S(y)$ as the product of such a family of operators and the Poisson semigroup.

If U is a coordinate patch in M and $f \in \mathscr{E}'(U)$, define

$$P^y f = \int e^{-y|\xi|} e^{ix\cdot\xi} \hat{f}(\xi)\, d\xi.$$

This is the familiar Poisson integral. Our result is the following.

PROPOSITION 2.1. *For $f \in \mathscr{E}'(U)$, we can write*

$$S(y)f = A(y, x, D_x)P^{c_1 y} f, \qquad \text{mod } C^\infty(\bar{\Omega}) \tag{2.9}$$

where c_1 is fixed, $0 < c_1 < c_0$, c_0 being as in (2.2) and

$$D_y^j A(y, x, D_x) \text{ bounded in } OPS_{1,0}^j. \tag{2.10}$$

Proof. We shall prove this by constructing a parametrix for (2.1) of the form (2.9). Thus we look for an approximate solution to (2.1), say $v(y)$, of the form

$$v(y) = \int A(y, x, \xi) e^{-c_1 y|\xi|} e^{ix\cdot\xi} \hat{f}(\xi)\, d\xi \tag{2.11}$$

where we shall pick $A(0, x, \xi) = 1$, $D_y^j A(y, x, \xi) \in S^j_{1,0}$, bounded for $0 \le y \le T$. We shall obtain $A(y, x, \xi)$ as asymptotic to a sum

$$A(y, x, \xi) \sim \sum_{j=0}^{\infty} A_j(y, x, \xi) \tag{2.12}$$

where $A_0(0, x, \xi) = 1$, other $A_j(0, x, \xi) = 0$, and $D_y^k A_j(y, x, \xi)$ are bounded in $S^{k-j}_{1,0}$ for $0 \le y \le T$. We describe how to recursively obtain the $A_j(y, x, \xi)$.

Applying $(\partial/\partial y) - E(y)$ to (2.11) and using the fundamental asymptotic expansion lemma, we get

$$\left(\frac{\partial}{\partial y} - E\right)v = \int \{-c_1|\xi|A + A_y - B\}e^{-c_1 y|\xi|}e^{ix\cdot\xi}\hat{f}(\xi)\,d\xi \tag{2.13}$$

where B is defined by $E(Ae^{ix\cdot\xi}) = Be^{ix\cdot\xi}$. Thus

$$B \sim \sum_{\alpha\ge 0} \frac{1}{\alpha!} E^{(\alpha)} A_{(\alpha)}$$

in the sense that, granted $A(y, x, \xi)$ to have the asserted properties, the difference between B and

$$\sum_{0\le|\alpha|\le N} \frac{1}{\alpha!} E^{(\alpha)} A_{(\alpha)},$$

say $D_N(y, x, \xi)$, has the property that $D_y^j D_N(y, x, \xi)$ is bounded in $S^{-N+j}_{1,0}$, $0 \le y \le T$. If we pull out the terms of order one in (2.13) and set their sum equal to zero, we obtain the equation

$$\frac{\partial}{\partial y} A_0 = (E_1 + c_1|\xi|)A_0. \tag{2.14}$$

If we solve this equation, with initial data $A_0(0, x, \xi) = I$, then the fact that $E_1 + c_1|\xi|$ has spectrum in the left-half plane yields that $D_y^j A_0$ is bounded in $S^j_{1,0}$, $0 \le y \le T$. Setting the terms of order 0 in (2.13) equal to zero yields

$$\frac{\partial}{\partial y} A_1 - (E_1 + c_1|\xi|)A_1 = \sum_{\alpha=1} E_0^{(\alpha)} D_x^\alpha A_0 + E_0 A_0, \tag{2.15}$$

and we can solve this, with initial condition $A_1(0, x, \xi) = 0$, obtaining $D_y^j A_1$ bounded in $S^{-1+j}_{1,0}$, $0 \le y \le T$. Similarly, the symbols $A_2, A_3, \ldots$ are obtained, and then $A(y, x, \xi)$ is obtained so that (2.12) is satisfied. It follows immediately that

$$\left(\frac{\partial}{\partial y} - E\right)v \in C^\infty([0, T] \times M),$$

$$v(0) = f.$$

Consequently $v - u \in C^\infty([0, T] \times M)$, so the proposition is proved.

Exercise

2.1. Compare the construction of this section with the complex geometrical optics construction of Chapter VIII, Section 4.

§3. Reflection of Singularities

We can now describe when a solution to (0.13), (0.14), i.e.

$$\frac{\partial}{\partial y} w = Gw \ (\text{mod } C^\infty), \tag{3.1}$$

$$\beta(x, D_x)w(0, x) = h \ (\text{mod } C^\infty) \tag{3.2}$$

can be constructed for $y \geq 0$ small, given that $WF(h)$ is contained in a small conic neighborhood of some point $(x_0, \xi_0) \in T^*(X) - 0$, with the following property:

(3.3) w is smooth along the null bicharacteristics associated with the roots $\lambda_{\ell+1}, \ldots, \lambda_j$, i.e. along the null bicharacteristics such that t is decreasing as y increases.

In fact, as seen in Sections 1 and 2, if $\tilde{w} = (1 + K)Uw = (w_1, \ldots, w_j, w_+, w_-)$, we can construct such a solution to (3.1), (3.2), provided we can specify $w_1, \ldots, w_\ell$ and w_- at $y = 0$ such that

$$\beta(x, D)U^{-1}(1 + K)^{-1}(w_1(0), \ldots, w_\ell(0), 0, \ldots, 0, 0, w_-(0)) = h \ (\text{mod } C^\infty). \tag{3.4}$$

We immediately have the following result.

THEOREM 3.1. *Suppose the equation* (3.4) *is a square system of* ℓ + *rank* E_- *equations in* ℓ + *rank* E_- *unknowns, which is elliptic, at* (x_0, ξ_0). *Then, for* $WF(h)$ *in a small conic neighborhood of* (x_0, ξ_0), *there exists a solution* w *to* (3.1)–(3.3), *for* $0 \leq y \leq y_0$, *small. We also have:*

(3.5) *On* $0 < y < y_0$, $WF(w)$ *is contained in the union of the null bicharacteristics associated with* $\lambda_1, \ldots, \lambda_j$ (i.e. *such that* t *is increasing as* y *increases) which pass over* $WF(h)$ *at* $y = 0$.

(3.6) *If* $\psi(x, \xi)$ *has order* $-\infty$ *on a conic neighborhood of* $WF(h)$, *then* $\psi(x, D_x)w \in C^\infty([0, y_1] \times X)$ *for* y_1 *sufficiently small.*

Note that, since K has order -1, the factor $(1 + K)^{-1}$ in (3.4) does not affect the ellipticity. So the condition is simply that the principal symbol of $\beta U^{-1}J$ be elliptic, where $J(a_1, \ldots, a_\ell, a_-) = (a_1, \ldots, a_\ell, 0, \ldots, 0, 0, a_-)$.

We give another condition on the boundary value problem (3.1), (3.2), which implies the ellipticity of (3.4). In the formulation of this condition, it is convenient not to lump the t variable together with the x variables, so we write the problem as

$$\frac{\partial}{\partial y} w = G(y, x, t, D_x, D_t)w, \tag{3.7}$$

$$\beta(x, t, D_x, D_t)w(0, x, t) = h. \tag{3.8}$$

As usual, $G \in OPS^1$, $\beta \in OPS^0$. Assume $G_1(y, x, t, \xi, \tau)$ and $\beta_0(x, t, \xi, \tau)$ are *analytic* in (ξ, τ), near $(x_0, t_0, \xi_0, \tau_0)$, which, of course, can be arranged if (3.7), (3.8) arise from converting a high order system of differential equations into a first order system of pseudodifferential equations. Now consider the matrix

$$M(\eta) = G(x_0, t_0, \xi_0, \tau_0 + i\eta)$$

with $\eta < 0$ small. In such a case, the decomposition (1.1) remains valid, and the eigenvalues of E_+(respectively, E_-) have positive (respectively, negative) real parts, for η small, but the λ_ν are not real. In fact, along a null bicharacteristic strip associated with λ_ν we have, at $\eta = 0$,

$$\dot{t} = -\frac{\partial \lambda_\nu}{\partial \tau} = -\frac{1}{i}\frac{\partial \lambda_\nu}{\partial \eta}, \tag{3.9}$$

so if $\dot{t} < 0$ we must have $Re\, i\lambda_\nu > 0$ for small $\eta < 0$, while if $\dot{t} > 0$, we have $Re\, i\lambda_\nu < 0$ for small $\eta < 0$.

In particular, for small $\eta < 0$, we see that all the eigenvalues of $M(\eta)$ have nonzero real part. Let $V(\eta)$ denote the linear span of all the generalized eigenvectors of $M(\eta)$ corresponding to eigenvalues with negative real part.

PROPOSITION 3.2. *The hypothesis*

(3.10) *$V(\eta)$ is bounded away from the kernel of $\beta_0(x_0, t_0, \xi_0, \tau_0)$, as $\eta \uparrow 0$ implies the ellipticity of* (3.4), *assuming h takes values in $\mathbf{C}^\mu$ with $\mu = \ell + \text{rank}\, E_-$.*

Proof. If E^I denotes the span of the eigenvectors of G_1 corresponding to $i\lambda_1, \dots, i\lambda_\ell$, E^{II} that corresponding to $i\lambda_{\ell+1}, \dots, i\lambda_j$, E^{III} that corresponding to E_+ and E^{IV} that corresponding to E_-, we see that ellipticity of (3.4) is equivalent to $\mu = \ell + \text{rank}\, E_-$ and

$$E^{II} \oplus E^{IV} \cap \ker \beta_0 = \{0\}.$$

As the conclusion from (3.9) shows that $V(\eta)$ approaches $E^{II} \oplus E^{IV}$ as $\eta \uparrow 0$, we have proved the proposition.

For first order systems of differential equations, hypothesis (3.10) was studied by Kreiss [1], where it was shown that such a condition, uniformly valid on $T^*(X) - 0$, leads to well posedness of the initial-boundary value problem; for higher order equation, see Sakamoto [1]; for smoothness results in this context analogous to theorem 0.1 in the introduction to this chapter, see Rauch and Massey [1].

We now consider the special case of a second order wave equation

$$\left(\frac{\partial^2}{\partial t^2} - A_2(t, x, y, D_x, D_y)\right)u = 0 \tag{3.11}$$

with boundary condition either Dirichlet:

$$u|_{y=0} = f(x, t) \tag{3.12}$$

or Neuman, or more generally, oblique derivative:

$$\left(\frac{\partial}{\partial y} + X\right)u|_{y=0} = f(x, t) \tag{3.13}$$

where $X = \sum a_j(\partial/\partial x_j)$, a real vector field. We suppose

$$A_2 = \sum_{|\alpha| \le 2} a_\alpha(t, x, y)D^\alpha_{x,y},$$

$D_{x_j} = \partial/\partial x_j$, and

$$\sum_{|\alpha| = 2} a_\alpha(t, x, y)\zeta^\alpha \ge c|\zeta|^2$$

for real ζ. Making a change of coordinates, we can suppose that, at $y = 0$, $(t, x) = (t_0, x_0)$, we have

$$A_2(t_0, x_0, 0, D_x, D_y) = \Delta_x + \frac{\partial^2}{\partial y^2}. \tag{3.14}$$

We want to convert this to a first order system and check ellipticity of (3.4) in these two special cases.

Note that, if Λ has symbol $(|\xi|^2 + \tau^2)^{1/2}$ and

$$w = \begin{pmatrix} \Lambda u \\ \dfrac{\partial}{\partial y} u \end{pmatrix},$$

we get

$$G_1(0, x_0, t_0, \xi, \tau) = \begin{pmatrix} 0 & (|\xi|^2 + \tau^2)^{1/2} \\ (|\xi|^2 - \tau^2)(|\xi|^2 + \tau^2)^{-1/2} & 0 \end{pmatrix},$$

and its eigenvalues are $\pm\sqrt{|\xi|^2 - \tau^2}$. Thus $T^*_{(x_0,t_0)}(X) - 0$ is naturally divided into two regions.

Region 1. $|\xi|^2 < \tau^2$. Here $G_1(0, x_0, t_0, \xi, \tau)$ has two pure imaginary roots, $\pm i\sqrt{\tau^2 - |\xi|^2}$, and two rays pass over (x_0, t_0, ξ, τ), one with $\dot{t} < 0$ for $\dot{y} > 0$, and the other with $\dot{t} > 0$ for $\dot{y} > 0$. Thus $\ell = 1$, rank $E_- = 0$, and either (3.12) or (3.13) yields $\mu = 1 = \ell + \text{rank } E_-$. This region is called the *hyperbolic region.*

Region 2. $|\xi|^2 > \tau^2$. Here $G_1(0, x_0, t_0, \xi, \tau)$ has two real roots, $\pm\sqrt{|\xi|^2 - \tau^2}$, so there are no rays passing over (x_0, ξ_0, ξ, τ). $\ell = j = 0$, rank $E_+ = \text{rank } E_- = 1$. Here also, either (3.12) or (3.13) yields $\mu = 1 = \ell + \text{rank } E_-$. This region is called the elliptic region.

These two regions are separated by the hypersurface $|\xi|^2 = \tau^2$; over a point here passes one ray, tangent to the boundary, which we are not treating in this chapter. These glancing rays will be discussed in the next chapter.

The Dirichlet boundary condition (3.12) converts into the form (3.8) with

$$\beta_0(x_0, t_0, \xi, \tau) \cdot \begin{pmatrix} a \\ b \end{pmatrix} = a,$$

assuming $h = \Lambda f$. Since we have, at (x_0, t_0, ξ, τ),

$$U^{-1}\begin{pmatrix} 1 \\ 0 \end{pmatrix} = \kappa \begin{pmatrix} \sqrt{|\xi|^2 - \tau^2} \\ \sqrt{|\xi|^2 + \tau^2} \end{pmatrix}$$

when κ is a nonzero scalar. Consequently,

$$\beta_0 U^{-1}\begin{pmatrix} 1 \\ 0 \end{pmatrix} = \kappa\sqrt{|\xi|^2 - \tau^2} \neq 0 \text{ in Region 1,}$$

and similarly one gets ellipticity in Region 2, so the conclusion of Theorem 3.1 holds for the system obtained from the equation (3.11), with the Dirichlet boundary condition. Similarly, the boundary condition (3.13) converts into the form (3.8) with

$$\beta_0(x_0, t_0, \xi, \tau)\begin{pmatrix} a \\ b \end{pmatrix} = b + [i \textstyle\sum a_j \xi_j (|\xi|^2 + \tau^2)^{-1/2}]a$$

so in this case

$$\beta_0 U^{-1}\begin{pmatrix} 1 \\ 0 \end{pmatrix} = i\kappa[\sqrt{\tau^2 - |\xi|^2} + \textstyle\sum a_j \xi_j] \neq 0$$

which yields ellipticity in Region 1 provided $a_j \equiv 0$, and one similarly gets ellipticity in Region 2. Thus the conclusion of Theorem 3.1 also holds for the system obtained from (3.11), with boundary condition given by (3.13), provided $a_j \equiv 0$ in the appropriate coordinates.

The following two examples of operators defined by either (3.11), (3.12) or by (3.11), (3.13) occur naturally. Let

$$A_2(x, D_x) = \sum a_{jk}(x) \frac{\partial^2}{\partial x_j \, \partial x_k},$$

and set

$$Q(u, v) = -\int_\Omega \sum a_{jk}(x) \frac{\partial u}{\partial x_j} \frac{\partial v}{\partial x_k} \, dx.$$

An integration by parts shows that, if $A = \sum (\partial/\partial x_j) a_{jk}(x)(\partial/\partial x_k)$,

$$(Au, v) - Q(u, v) = \int_{\partial\Omega} (Yu)\bar{v} \, dS(x)$$

where Y is a real vector field on $\partial\Omega$ which is a nonvanishing scalar multiple of a vector field of the type (3.13). Note that in a coordinate system such that (3.14) holds, one has $a_j \equiv 0$ in this case.

Thus if we let $A = \sum (\partial/\partial x_j) a_{jk}(x)(\partial/\partial x_k)$ and specify that u in the domain of A should satisfy either the Dirichlet boundary condition $u|_{\partial\Omega} = 0$ or the boundary condition $Yu|_{\partial\Omega} = 0$, which is called the Neumann boundary condition, it follows that A is a formally self-adjoint, negative operator. Now each of these conditions is a regular elliptic boundary condition, by the results of Chapter V, Section 3. Consequently the theorem of the introduction to this chapter applies to each of these cases, so any approximate solution (modulo a C^∞ error) we construct to such boundary value problems for the wave equation $(\partial^2/\partial t^2) - A$ must differ from an exact solution by a smooth error. Consequently we are in a position to analyze the singularities of solutions to these two boundary value problems, whenever no singularities along grazing rays are involved.

In fact, suppose Ω is an open subset of $\mathbf{R}^n$ with smooth boundary, and let u be a solution to the wave equation

$$\left(\frac{\partial^2}{\partial t^2} - A\right) u = 0, \qquad x \in \Omega \tag{3.15}$$

with A as above, satisfying either the Dirichlet boundary condition

$$u|_{\mathbf{R} \times \partial\Omega} = 0 \tag{3.16}$$

or the Neuman boundary condition,

$$Yu|_{\mathbf{R} \times \partial\Omega} = 0, \tag{3.17}$$

and suppose u satisfies the initial conditions

$$u(0, x) = f(x) \in \mathscr{E}'(\Omega), \qquad \frac{\partial}{\partial t} u(0, x) = g(x) \in \mathscr{E}'(\Omega). \tag{3.18}$$

We shall make an analysis of the singularities of the solution u, assuming the following *nongrazing* hypothesis:

(3.19) Each null bicharacteristic strip of $(\partial^2/\partial t^2) - A$ passing over $WF(f) \cup WF(g)$ at $t = 0$, only hits $\mathbf{R} \times \partial\Omega$ transversally, hence lies over a point ζ_0 in the hyperbolic region of $T^*(\mathbf{R} \times \partial\Omega)$. If such a ray is continued by reflection, i.e. if one continues along the other null bicharacteristic strip of $(\partial^2/\partial t^2) - A$ lying over ζ_0, this ray also only hits transversally, and, iterating this process, one obtains rays that are bounded away from grazing rays, for $t \in (T_0, T_1)$, $T_0 < 0 < T_1$.

THEOREM 3.3. *Let u solve the wave equation* (3.15), *with boundary condition given by either* (3.16) *or* (3.17), *and initial condition* (3.18), *and suppose the nongrazing hypothesis* (3.19) *is satisfied. Then u is only singular along the union $\mathfrak{U}$ of the rays passing over $WF(f) \cup WF(g)$ at $t = 0$. If $\mathcal{O}$ is an open subset of $\mathbf{R} \times \partial\Omega$ over which none of these rays pass, then u is C^∞ near $\mathcal{O}$, up to the boundary.*

Proof. For $t \in (-\varepsilon, \varepsilon)$, the solution vanishes near $\partial\Omega$, and the result follows from the propagation of singularities results of Chapter VIII. It is a simple matter to write u in this interval as a finite sum of solutions whose wave front sets have small supports, so without loss of generality we can suppose $\mathfrak{U}$ is a tiny conic neighborhood of a single ray γ. Propagation of singularities results interior to $\mathbf{R} \times \Omega$ show that u is C^∞ except near γ, up to time $t_0 - \varepsilon$, where γ hits $\partial\Omega$ at time t_0, so applying a smooth cut-off one obtains a solution mod C^∞ which is only singular near γ, for $t < t_0 - \varepsilon$, and is C^∞ near $\partial\Omega$. Continue this solution for $t < t_0 + \varepsilon$, on a neighborhood of $\mathbf{R} \times \bar{\Omega}$, disregarding the boundary condition, so the continued solution will be smooth except near γ, but will fail to satisfy the appropriate boundary condition. Rather we have

$$u|_{\partial\Omega} = \alpha, \qquad Yu|_{\partial\Omega} = \beta$$

where $WF(\alpha)$, $WF(\beta)$ are contained in a small conic neighborhood U of the point $\zeta_0 \in T^*(\mathbf{R} \times \partial\Omega)$ over which γ passes. Call the continued solution $\tilde{u}$.

Now write a collar neighborhood of $\mathbf{R} \times \partial\Omega$ in $\mathbf{R} \times \Omega$, in the form $[0, Y] \times X$, $X = \mathbf{R} \times \partial\Omega$, and apply Theorem 3.1 to construct a solution (mod C^∞) to the boundary value problem

$$\left(\frac{\partial^2}{\partial t^2} - A\right)v = 0,$$

and either $v|_{\partial\Omega} = -\alpha$ or $Yv|_{\partial\Omega} = -\beta$, such that v is smooth, up to $\partial\Omega$, for $t < t_0 - \varepsilon$, and v is only singular along the rays passing over $WF(\alpha)$,

$WF(\beta)$ with $t > 0$ for $\dot{y} > 0$, which is in a small conic neighborhood of the *reflected* ray γ. Then $u = \tilde{u} + v$ gives a solution mod C^∞ to our boundary value problem, with singularities appropriately specified, for $t < t_0 + \varepsilon$. By the nongrazing hypothesis, the reflected ray γ will remain in Ω for some interval of time, bounded away from zero, and here again the propagation of singularities results on open domains applies, until γ perhaps hits $\partial\Omega$ a second time, when the above method can be applied.

The nongrazing hypothesis guarantees that, after iterating the above construction a finite number of times, one gets a solution mod C^∞ to the boundary value problem, with an appropriate description of its singularities and the theorem of the introduction to this chapter shows that such an approximate solution differs from the exact solution by a smooth error. The proof is complete.

Another interesting example where one can analyze the singularities along nongrazing rays by the techniques of this chapter is provided by the transmission problem for the wave equation with different sound speeds in two adjacent regions, Ω_1 and Ω_2, separated by a smooth surface S. For example, one can take $(D_t^2 - \Delta)u = 0$ on Ω_1, and

$$\left(\frac{1}{c^2} D_t^2 - \Delta + a\frac{\partial}{\partial t}\right)u = 0$$

on Ω_2, while the values of u and its normal derivatives match up on the two sides of S. This can be converted into a second order system by identifying Ω_1 with Ω_2 smoothly in a neighborhood of S. Theorem 3.1 can be shown to apply to the first order system one obtain from this.

Note that, for the transmission problem, over any nongrazing $\rho_0 \in T^*(\mathbf{R} \times S)\backslash 0$ there pass either 4, 2, or 0 rays. If only two rays pass over p_0, they must be incident and reflected rays for the slow speed region (Ω_1 if $c > 1$). This is the case of total internal reflection. If four rays pass over p_0, all nongrazing, then one can check that if u is smooth along one ray coming in from negative time and belongs to H^s along the other ray, but not to $H^{s+\varepsilon}$, then along the two rays over p_0 going into positive time, u belongs to H^s but not to $H^{s+\varepsilon}$. An interesting contrast occurs if we take $c \equiv 1$. Then the transmission problem is merely the wave equation with a discontinuous "friction" term $a(\partial/\partial t)$. In this case, if $u \in H^s$ along one incident ray and smooth along the other, then $u \in H^s$ along the transmitted ray, but $u \in H^{s+1}$ along the reflected ray, i.e. reflected waves are relatively weak.

The reflection of singularity results just established can be put together with other limiting arguments to construct examples of phenomena involving propagation of singularities when the nongrazing hypothesis is violated. We give a couple of examples of this.

Let Ω be a smooth strictly convex bounded region in $\mathbf{R}^m$. We will construct a solution to

$$\left(\frac{\partial^2}{\partial t^2} - \Delta\right)u = 0, \qquad u|_{\partial\Omega} = 0$$

which is smooth on the interior of $\mathbf{R} \times \Omega$ but is singular at a curve running along the boundary $\mathbf{R} \times \partial\Omega$. To begin this construction, pick points $x_j \in \Omega$ tending to a point $p_0 \in \partial\Omega$, and rays issuing from x_j, pointing parallel to a given tangent to $\partial\Omega$ at p_0. We suppose these rays γ_j travel in straight lines at unit speed and bounce off $\partial\Omega$ according to the usual rule: angle of incidence equals angle of reflection. Exhaust Ω by open sets $\Omega_1 \Subset \Omega_2 \Subset \cdots$ with the property that γ_j lies in $\Omega\backslash\Omega_j$ for $t \in [0, 1]$. Define functions $\varphi_j \in L^2_{\text{comp}}(\Omega)$ with wave front set consisting of just one ray lying over x_j such that $\varphi_j \notin H^1(\Omega)$, i.e. φ_j has infinite energy, and the solution v_j to the mixed problem

$$\left(\frac{\partial^2}{\partial t^2} - \Delta\right)v_j = 0, \qquad v_j|_{\partial\Omega} = 0, \tag{3.20}$$

$$v_j(0, x) = \varphi_j, \qquad \frac{\partial}{\partial t}v_j(0, x) = 0 \tag{3.21}$$

has a singularity just along γ_j in the interior of $\mathbf{R} \times \Omega$, and is smooth up to the boundary except at points where γ_j hits $\mathbf{R} \times \partial\Omega$. That this can be done for small t follows from the results of Chapter VIII, and that it will hold for all t follows from the reflection of singularities results we have just established. We also suppose supp φ_j belongs to a small neighborhood $\mathcal{Q}_j$ of x_j.

Now we want to smooth v_j out just slightly and multiply by a small constant, say

$$u_j(t) = c_j \int \rho_j(\tau)v_j(t - \tau)\, d\tau,$$

where $\rho_j \in C_0^\infty(\mathbf{R})$ has small support, $\int\rho_j = 1$, and c_j is chosen so that

$$\sum_{|\alpha| \le j} \int_0^1 \int_{\Omega_j} |D^\alpha u_j|^2\, dx\, dt \le 2^{-j}, \tag{3.22}$$

$$\int_{\mathcal{Q}_j} |\nabla u_j(x, 0)|^2\, dx \ge 10^j\left(1 + \sum_{k<j} \int_\Omega |\nabla u_k(x, 0)|^2\right), \tag{3.23}$$

$$\|u_j(x, 0)\|_{L^2(\Omega)} \le 2^{-j}. \tag{3.24}$$

Let

$$u = \sum_{j=1}^\infty u_j.$$

By (3.22) and (3.24), $u \in C^\infty((0, 1) \times \Omega)$ and solves the initial value problem (3.20) there, in a weak sense. It follows that u continues to solve such a mixed problem for all t, and, by reflection of singularities, u must be smooth in the interior of $\mathbf{R} \times \Omega$. However, $u(0)$ has infinite energy, by (3.23), so u cannot be smooth on $\mathbf{R} \times \bar{\Omega}$. In fact, one sees that u is smooth on the complement of the curve γ_0 on $\mathbf{R} \times \partial\Omega$ which is the limit of the broken rays γ_j. Such a curve γ_0 is known as a "gliding ray."

Our second example involves a solution to the transmission problem

$$\left(\frac{1}{c^2}\frac{\partial^2}{\partial t^2} - \Delta\right)u = 0 \text{ in } \Omega_1,$$

$$\left(\frac{\partial^2}{\partial t^2} - \Delta\right)u = 0 \text{ in } \mathbf{R}^2 \backslash \Omega_1,$$

where $\Omega_1 \subset \mathbf{R}^2$ is a bounded strictly convex region. We suppose u and ∇u are continuous across $\partial\Omega_1$. We suppose $c > 1$, i.e., sound speed greater than 1 in Ω_1. Then, as mentioned previously, a ray coming in from $\mathbf{R}^2\backslash\Omega_1$ and hitting $\partial\Omega_1$ too obliquely produces a singularity along the reflected ray, but no singularity going into Ω_1. There is a critical angle below which this phenomenon of total internal reflection in $\mathbf{R}^2\backslash\Omega_1$ holds. What we now do is construct an example of a solution to the transmission problem with a singularity along a single ray γ_0, for $t < t_0$, which hits $\partial\Omega$, at $t = t_0$, at this critical angle.

A ray hitting $\partial\Omega$ at an angle just greater than this critical angle produces reflected and refracted rays as indicated in Figure IX.1. If u is in L^2 along the ray γ_1 but not in H^1, then u has the same property along all the rays indicated in Figure IX.1. Therefore, if we consider a sequence of such rays, approaching a critical ray γ_0, then a simple modification of the construction of the previous example will produce an example of a solution u, with a singularity only along the ray γ_0 for $t < t_0$, which spews off a two-dimensional sheet of singularities for $t > t_0$.

One important problem to which Theorem 3.1 does not apply is the following boundary value problem from linear elasticity:

$$(3.25) \quad Lu = \frac{\partial^2}{\partial t^2} u - (\lambda + \mu) \operatorname{grad} \operatorname{div} u - \mu \Delta u = 0, \qquad x \in \Omega \subset \mathbf{R}^3$$

with boundary condition

$$(3.26) \qquad \sum_i n_i \sigma_{ij} = 0 \quad \text{on} \quad \partial\Omega$$

where

$$\sigma_{ij} = \lambda(\operatorname{div} u)\delta_{ij} + \mu\left(\frac{\partial u_i}{\partial x_j} + \frac{\partial u_j}{\partial x_i}\right)$$

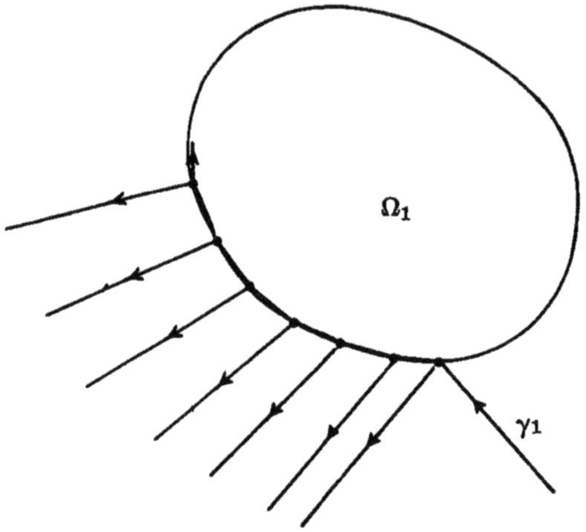

Figure IX.1

is the stress tensor. Here u takes values in $\mathbf{R}^3$. n_i represents the components of the unit normal vector to $\partial\Omega$. The constants λ and μ are assumed to be positive. The principal symbol of L is $L_2(t, x, \tau, \xi) = -\tau^2 I + (\lambda + \mu)|\xi|^2 P_\xi + \mu|\xi|^2$ where P_ξ is the orthogonal projection of $\mathbf{R}^3$ onto the space spanned by ξ. For each $(\tau, \xi) \neq 0$, this is a symmetric matrix, with a simple eigenvalue $(\lambda + 2\mu)|\xi|^2 - \tau^2$ and a double eigenvalue $\mu|\xi|^2 - \tau^2$. The analysis of propagation of singularities away from the boundary for this case was discussed at the end of Section 2, Chapter VIII.

The operator $A = (\lambda + \mu)$ grad div $+ \mu\Delta$, with boundary condition given by (3.26), is an elliptic self-adjoint negative operator, and the boundary condition is coercive, so the problem (3.25), (3.26), with initial conditions $u(0, x) = f \in \mathscr{D}(A^{1/2})$, $u_t(0, x) \in L^2(\Omega)$, is well posed, and the theorem in the introduction applies.

Since the operator L has two sound speeds, $T^*(\mathbf{R} \times \partial\Omega)$ is divided into three regions. It turns out that, over the "elliptic" region (over which no rays pass), the boundary operator (3.4) which one gets after converting (3.25), (3.26) to a first order system of the form (3.1), (3.2), is *not* elliptic. Indeed, there is a matrix factor α such that the principal part of $\alpha\beta U^{-1}J$ is a real valued scalar, with simple characteristics. In fact, the rays for $\alpha\beta U^{-1}J$ will travel along $\mathbf{R} \times \partial\Omega$ at a speed slightly slower than the slower sound speed of L. From the results of Chapter VIII, one can construct a solution mod C^∞ to

$$\alpha\beta U^{-1}J(v_-) = h \tag{3.27}$$

and analyze its singularities. Such singularities propagate along $\mathbf{R} \times \partial\Omega$ at this third slow sound speed. Returning to constructing the parametrix for (3.1), (3.2) in this case, with $u_-(0) = v_-$, one gets an approximate solution the boundary value problem (3.25), with an inhomogeneous boundary condition

$$\sum_i n_i \sigma_{ij} = G \in \mathscr{E}'(\mathbf{R} \times \partial\Omega) \tag{3.28}$$

under the nongrazing hypothesis, and the solution has singularities which run along the boundary at the slow speed, called the Rayleigh sound speed. These waves are called Rayleigh waves, and were first studied for flat boundaries in Rayleigh [1]. For the details of the analysis in the presence of curved boundaries, see Taylor [10]. A straightforward generalization of the analysis leading to Theorem 3.1 yields the following result, which contains the analysis of Rayleigh waves.

THEOREM 3.4. *In the context of Theorem* 3.1, *suppose the system* (3.4) *is a square system of* ℓ + rank E_- *equations in* ℓ + ranks E_- *unknowns. Suppose that, given* $h \in \mathscr{E}'(X)$, *we can solve the system* (3.4), *mod* C^∞, *for* $w_1(0), \dots, w_\ell(0), w_-(0)$. *If we can deduce that* $WFw_\nu(0) \subset \Gamma_\nu$, $WFw_-(0) \subset \Gamma_-$ *where* $\Gamma_\nu(\nu = 1, \dots, \ell)$ *and* Γ_- *are closed conic subsets of* $T^*(X)$ *obtained from* $WF(h)$ *by some process, it follows that u is smooth except along these rays passing over*

$$\bigcup_\nu \Gamma_\nu,$$

with $t > 0$ *for* $\dot{y} > 0$. *Furthermore,* $w_-(y)$ *is smooth up to the boundary* $y = 0$ *except at points* $x \in X$ *such that* $(x, \xi) \in \Gamma_-$ *for some* ξ.

Exercises

3.1. Consider the following boundary value problem for the electromagnetic field on a region Ω in $\mathbf{R}^3$ bounded by a perfect conductor:

$$\left(\frac{\partial^2}{\partial t^2} - \Delta\right)E = 0, \qquad \left(\frac{\partial^2}{\partial t^2} - \Delta\right)B = 0,$$

$$\nu \times E\big|_{\partial\Omega} = 0, \qquad \operatorname{div} E\big|_{\partial\Omega} = 0; \qquad \nu \cdot B\big|_{\partial\Omega} = 0, \qquad \nu \times \operatorname{curl} B\big|_{\partial\Omega} = 0.$$

Show that Δ on $\binom{E}{B}$, a 6 vector, with these boundary conditions, is a negative self-adjoint operator with coercive boundary conditions. Show that in this case the system (3.4) is elliptic, and analyze reflection of singularities of such electromagnetic waves, granted the nongrazing hypothesis.

3.2. Check out ellipticity of (3.4) for the transmission problem and work out the details of the discussion of the transmission problem given in this section.

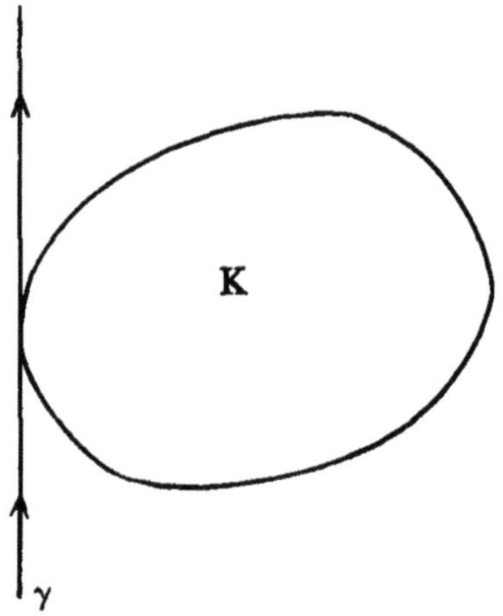

Figure IX.2

3.3 Construct an example of a solution to the transmission problem which is singular only along a critical ray γ_0 for $t < t_0$, but for $t > t_0$, instead of spewing off a 2-dimensional set of singularities, is singular only along the ray obtained from γ_0 by reflection.

3.4. For the boundary value problem (3.25), (3.26) of linear elasticity, verify the assertion made about making $\alpha\beta U^{-1}J$ have real scalar principal part with simple characteristics, so Theorem 3.4 applies.

3.5. Consider a smooth bounded strictly convex obstacle $K \subset \mathbf{R}^n$. Using the method of this section, construct a solution to

$$\left(\frac{\partial^2}{\partial t^2} - \Delta\right)u = 0,$$

on $\mathbf{R} \times (\mathbf{R}^n - K)$, $u|_{\partial K} = 0$, which is smooth except along the grazing ray γ (see Figure IX.2) hitting ∂K at $t = t_0$, and smooth up to the boundary except at the point of contact of γ with $\mathbf{R} \times \partial K$. In the next chapter, we shall see this is the only type of situation that can exist, granted that u is smooth except along γ for $t < t_0$.

3.6. Consider a smooth compact $K \subset \mathbf{R}^2$, as in Figure IX.3, which is convex from p_2 to p_1, and concave from p_1 to p_2, going clockwise. Suppose the curvature of ∂K vanishes to first order at p_1 and p_2. Construct an example of a solution to

$$\left(\frac{\partial^2}{\partial t^2} - \Delta\right)u = 0$$

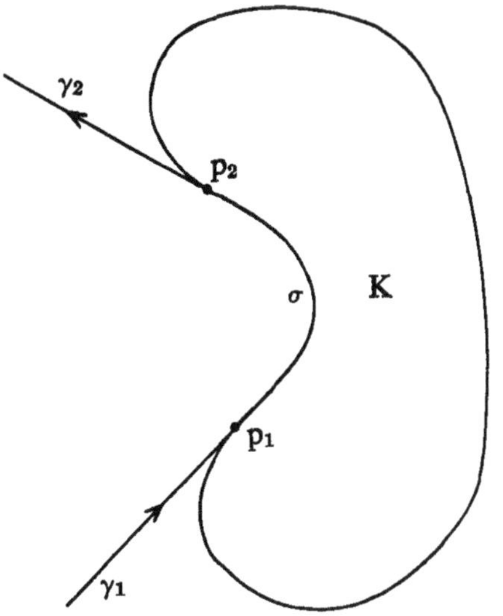

Figure IX.3

on $\mathbf{R} \times (\mathbf{R}^2 \backslash K)$, $u|_{\partial K} = 0$, which is smooth except along the ray γ_1 for $t < t_0$, whose singularity runs along the segment σ of ∂K from p_1 to p_2, at unit speed, from $t = t_0$ to $t = t_1 = t_0 +$ (length of σ), and whose singularity then runs along the ray γ_2, where γ_1 hits p_1 tangentially at $t = t_0$ and γ_2 leaves p_2 tangentially at $t = t_1$.

3.7. Construct an obstacle K as in Figure IX.1, but with the curvature of ∂K vanishing to infinite order at p_1, with the property that there is a broken light ray which hits σ and bounces and hits σ again, an infinite number of times, each time with angle of incidence equal to angle of reflection, at points tending to p_1.

3.8. Let Γ be any concave curve in $\mathbf{R}^2$ with a light ray which reflects infinitely often on Γ, at points accumulating to $p_1 \in \bar{\Gamma}$. Prove that the curvature of Γ must vanish to infinite order at p_1.

3.9. Let $\tilde{K}$ be a domain in $\mathbf{R}^3 = \{(x, y, z)\}$ such that, for $z = 0$, one has the set K of Exercise 3.7. Suppose that, for $z > 0$, the boundary is independent of z on intervals $z \in (2^{-j}, 2^{-j+1})$, and, for each $z > 0$ the boundary, a curve in $\mathbf{R}^2$, will be flat at p_1, but only to finite order. Show that there is a solution to

$$\left(\frac{\partial^2}{\partial t^2} - \Delta\right) u = 0$$

on $\mathbf{R}^3 \backslash \tilde{K}$, $u|_{\partial K} = 0$, with a singularity along a ray $\tilde{\gamma}_1$ hitting $(p_1, 0) \in \partial\tilde{K}$ tangentially, for $t < 0$, which produces a singularity for $t > 0$ along a broken light ray reflecting at $\partial\tilde{K}$ an infinite number of times, at points accumulating at $(p_1, 0)$ and times accumulating at $t = 0$, as well as a solution with singularity along $\tilde{\gamma}_1$ for $t < 0$, which produces a singularity hugging $\partial\tilde{K}$ between $(p_1, 0)$ and $(p_2, 0) \in \partial\tilde{K} \subset \mathbf{R}^3$. Thus a linear combination of such solutions splits the beam along $\tilde{\gamma}_1$ (for $t < 0$) into two rays.

CHAPTER X

Grazing Rays and Diffraction

In the last chapter we studied reflection of singularities for boundary value problems for hyperbolic equations $Lu = f$ on $\mathbf{R} \times \Omega$, assuming the rays involved hit $\mathbf{R} \times \partial\Omega$ transversally. In this chapter we consider the case when rays can hit $\mathbf{R} \times \partial\Omega$ tangentially, but we suppose that, if such a ray is tangent to $\mathbf{R} \times \partial\Omega$ at $t = t_0$, then the contact with $\mathbf{R} \times \partial\Omega$ is of exactly second order, and that the ray is in the interior of $\mathbf{R} \times \Omega$ for $t \in (t_0 - \varepsilon, t_0)$ and $t \in (t_0, t_0 + \varepsilon)$. Thus, the boundary is assumed to be *convex* with respect to these grazing rays. The operators needed to construct parametrices in this case are a little more complicated than the Fourier integral operators we have dealt with previously. Such operators, called Fourier-Airy integral operators, will be introduced in Section 1 and studied in Section 2.

We shall devote this chapter to the study of the wave equation

$$\left(\frac{\partial^2}{\partial t^2} - \Delta\right)u = 0$$

on the exterior of a smooth, strictly convex obstacle $K \subset \mathbf{R}^n$, of positive curvature, as most of the analytical difficulties already appear in this case. From the material of Chapter IX, we see that it suffices to construct solutions mod C^∞ to

$$\left(\frac{\partial^2}{\partial t^2} - \Delta\right)u = 0 \text{ on } \mathbf{R} \times \Omega, \qquad \Omega = \mathbf{R}^n \backslash K,$$

$$Bu|_{\partial\Omega} = f \in \mathscr{E}'(\mathbf{R} \times \partial K),$$

$$u = 0 \qquad \text{for } t \ll 0,$$

where $Bu|_{\partial\Omega} = f$ is the boundary value problem, assuming $WF(f)$ is contained in a small conic neighborhood of a point $\zeta_0 \in T^*(\mathbf{R} \times \partial K)$ over which a grazing ray passes. We shall first treat the Dirichlet problem, and in Section 5 treat the Neumann problem and several other boundary value problems. We conclude this chapter with a study of the Kirchhoff approximation, a tool encountered in classical scattering theory, to analyze the solution u_s to the reduced wave equation

$$(\Delta + \lambda^2)u_s = 0 \text{ on } \mathbf{R}^3\backslash K$$

with boundary condition

$$u_s|_{\partial K} = -e^{-i\lambda x \cdot \omega}$$

and satisfying the "outgoing radiation condition"

$$\frac{\partial u_s}{\partial r} - i\lambda u_s = o(|x|^{-1}), \qquad u_s = O(|x|^{-1}) \qquad \text{as } |x| \to \infty.$$

The Kirchhoff approximation says

$$\left.\frac{\partial u_s}{\partial \nu}\right|_{\partial K} \approx -i\lambda|\nu \cdot \omega|e^{-i\lambda x \cdot \omega},$$

and we provide a justification for this approximation, showing that

$$\left.\frac{\partial u_s}{\partial \nu}\right|_{\partial K} = K(x, \lambda)e^{-i\lambda x \cdot \omega}$$

with

$$|K(x, \lambda) + i\lambda|\nu \cdot \omega|\,| \le c_\varepsilon \lambda^{3/4+\varepsilon}(1 + \lambda^{1/6}|\nu \cdot \omega|)^{-9/2}.$$

Some applications for classical scattering are indicated.

The analysis in Sections 1 through 4 was worked out independently by Melrose [2] and Taylor [5], the latter author using previous work of Ludwig [2].

§1. The Ansatz

In this section we shall write down the ansatz for the solution to the Dirichlet problem for the wave equation on the exterior of a smooth strictly convex obstacle $K \subset \mathbf{R}^n$,

$$\left(\frac{\partial^2}{\partial t^2} - \Delta\right)u = 0, \tag{1.1}$$

$$u|_{\partial K} = f \in \mathscr{E}'(\mathbf{R} \times \partial K), \tag{1.2}$$

$$u = 0 \qquad \text{for} \qquad t \ll 0. \tag{1.3}$$

We assume $WF(f)$ is contained in a small conic neighborhood of a point $(z_0, \zeta_0) \in T^*(\mathbf{R} \times \partial K)$ over which a grazing ray passes.

The key to the ansatz we describe here lies in considering solutions to the wave equation with caustics. A caustic is a family of rays with a smooth convex envelope, as indicated in Figure X.1. Though we did not consider such an approach in Chapter VIII, solutions to the wave equation which

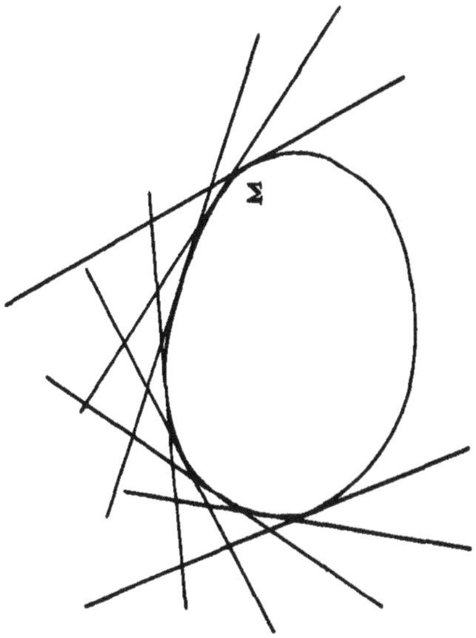

Figure X.1

are singular along such rays can be written in the form of superposition of solutions of the type

$$[g(t, x, \lambda)Ai(\lambda^{2/3}\rho(x)) + ih(t, x, \lambda)\lambda^{-1/3}Ai'(\lambda^{2/3}\rho(x))]e^{i\lambda\theta(x)} \tag{1.4}$$

where g and h are a pair of amplitudes, ρ and θ are a pair of phase functions, and $Ai(s)$ is the Airy function, given by

$$Ai(s) = \frac{1}{2\pi}\int_{-\infty}^{\infty} e^{i(ts+t^3/3)}\,dt. \tag{1.5}$$

Properties of $Ai(s)$ will be developed in the exercises. For material on this approach to caustics, see Ludwig [1], Duistermaat [2], and for a very nice treatment, Guillemin and Schaefer [1]. Solutions obtained as superpositions of solutions of the type (1.4) are smooth on one side of the envelope $\sum$ of the rays, called the "shadow side," and singular along the rays which form this envelope, which fill up a neighborhood of the other side of the envelope, called the "illuminated side."

The ansatz for our grazing ray problem is motivated by making a caustic decomposition of rays near the grazing ray passing over ζ_0, into families of rays with convex caustics as envelopes, these envelopes being surfaces roughly parallel to ∂K. In particular, grazing rays to ∂K determine a family of rays whose envelope is precisely ∂K. In our case, superpositions of solutions of the form (1.4) would not be appropriate, since they involve

Figure X.2

solutions with singularities that go in and hit the boundary and then go off and our condition (1.3) requires constructing a solution whose singularities simply leave the boundary, so lie only on half of each ray of Figure 1, as indicated in Figure X.2. The ansatz will be of the form

$$u(t, x) = \iint \left[g \frac{A(|\xi|^{-1/3}\rho)}{A(-|\xi|^{-1/3}\eta)} + ih|\xi|^{-1/3} \frac{A'(|\xi|^{-1/3}\rho)}{A(-|\xi|^{-1/3}\eta)} \right] e^{i\theta} \hat{F}(\xi, \eta)\, d\xi\, d\eta \tag{1.6}$$

where $g(t, x, \xi, \eta)$ and $h(t, x, \xi, \eta)$ are a pair of amplitudes and $\rho(t, x, \xi, \eta)$ and $\theta(t, x, \xi, \eta)$ are a pair of phase functions. $A(s)$ is one of two different Airy functions, $A_+(s)$ or $A_-(s)$, defined by

$$A_+(s) = Ai(-e^{2\pi i/3}s), \qquad A_-(s) = Ai(-e^{-2\pi i/3}s), \tag{1.7}$$

$(\xi, \eta) \in \mathbf{R}^n$, which we regard as a straightened out version of the boundary $\mathbf{R} \times \partial K$, locally. $F \in \mathscr{E}'(\mathbf{R}^n)$ is a distribution which will be related to the unknown $f \in \mathscr{E}'(\mathbf{R} \times \partial K)$ of (1.2) in Section 4.

The Airy function $Ai(s)$ defined by (1.5) is easily seen to satisfy the Airy equation $Ai''(s) - sAi(s) = 0$, which implies that $Ai(s)$ is an entire holomorphic function of s, so (1.7) makes sense. The following asymptotic formula for $Ai(z)$, valid for $-\pi < \arg z < \pi$, may be found in Erdelyi [1]:

$$\begin{aligned} &Ai(z) = \Phi(z)e^{-2z^{3/2}/3}, \qquad \text{with} \\ &\Phi(z) \sim z^{-1/4}\left(\frac{1}{2\sqrt{\pi}} + a_1 z^{-3/2} + \cdots\right) \qquad \text{as } |z| \to \infty. \end{aligned} \tag{1.8}$$

Thus,

$$\begin{aligned} A(s) &= \Phi(e^{-\pi i/3}s)e^{-2is^{3/2}/3}, \qquad s > 0, \\ A(s) &= \Phi(-e^{2\pi i/3}s)e^{2(-s)^{3/2}/3}, \qquad s < 0. \end{aligned} \tag{1.9}$$

Consequently, $A(s)$ blows up as $s \to -\infty$, which should not be alarming since one should not expect the approximate solution to be defined inside K, which is the only place this behavior will lead to a blow up of (1.6). It is known (see Olver [2]) that the only zeros of $Ai(z)$ occur for z real and negative, and similarly for $Ai'(z)$. Consequently, neither $A(s)$ nor $A'(s)$ has any real zeros.

We shall look for real valued functions ρ and θ that are homogeneous of degree 1 in (ξ, η), and we shall look for amplitudes g, h of the form

$$g \sim \sum_{j \geq 0} g_j(t, x, \xi, \eta), \qquad h \sim \sum_{j \geq 0} h_j(t, x, \xi, \eta)$$

where g_j and h_j are homogeneous of degree $-j$ in (ξ, η). We want to derive equations for these quantities, analogous to the eikonal and transport equations of geometrical optics. In order to do this, we shall apply $(\partial^2/\partial t^2) - \Delta$ to (1.6), express the result in a similar form, using the Airy equation $A''(s) + sA(s) = 0$, and group terms of like orders together, and set them successively equal to zero. Indeed, if we apply $(\partial^2/\partial t^2) - \Delta$ under the integral in (1.6), the integrand becomes

$$\frac{1}{A(-|\xi|^{-1/3}\eta)} \hat{F}(\xi, \eta)$$

times

$$\begin{aligned}
(1.10) \quad & -Ae^{i\theta}\Big(|\nabla\theta|^2 g + \frac{\rho}{|\xi|}|\nabla\rho|^2 g - \theta_t^2 g - \frac{\rho}{|\xi|}\rho_t^2 g + 2h\frac{\rho}{|\xi|}\nabla\theta\cdot\nabla\rho \\
&\qquad - 2h\frac{\rho}{|\xi|}\theta_t\rho_t\Big) \\
& -i|\xi|^{-1/3}A'e^{i\theta}\Big(|\nabla\theta|^2 h + \frac{\rho}{|\xi|}|\nabla\rho|^2 h - \theta_t^2 h - \frac{\rho}{|\xi|}\rho_t^2 h \\
&\qquad + 2g\,\nabla\theta\cdot\nabla\rho - 2g\theta_t\rho_t\Big) \\
& + iAe^{i\theta}\Big(2\nabla g\cdot\nabla\theta + g\,\Delta\theta + 2\frac{\rho}{|\xi|}\nabla h\cdot\nabla\rho - \frac{\rho}{|\xi|}(\Delta\rho)h - |\xi|^{-1}|\nabla\rho|^2 \\
&\qquad - 2g_t\theta_t - g\theta_{tt} - 2\frac{\rho}{|\xi|}h_t\rho_t + \frac{\rho}{|\xi|}\rho_{tt}h + |\xi|^{-1}\rho_t^2\Big) \\
& + |\xi|^{-1/3}A'e^{i\theta}(2\nabla g\cdot\nabla\rho + g\,\Delta\rho + 2\nabla h\cdot\nabla\theta - h\,\Delta\theta - 2g_t\rho_t \\
&\qquad - g\rho_{tt} - 2h_t\theta_t - h\theta_{tt}) \\
& + Ae^{i\theta}(\Delta g - g_{tt}) + i|\xi|^{-1/3}A'e^{i\theta}(\Delta h - h_{tt})
\end{aligned}$$

Setting the first two terms equal to zero, we obtain a pair of equations for θ and ρ which we shall call eikonal equations:

$$\theta_t^2 + \frac{\rho}{|\xi|}\rho_t^2 - |\nabla\theta|^2 - \frac{\rho}{|\xi|}|\nabla\rho|^2 = 0, \tag{1.11}$$
$$\theta_t\rho_t - \nabla\theta\cdot\nabla\rho = 0.$$

Since our particular equation has t-independent coefficients and the domain is t-independent, we can choose $\theta = t|\xi| + \psi(x, \xi, \eta)$ and ρ, g, h independent of t. Then the eikonal equations become

$$|\nabla\psi|^2 + \frac{\rho}{|\xi|}|\nabla\rho|^2 = |\xi|^2, \tag{1.12}$$
$$\nabla\psi\cdot\nabla\rho = 0.$$

We also obtain transport equations for g and h by setting the other terms equal to 0. If we suppose g and h are independent of t, these become

$$2\nabla\psi\cdot\nabla g + 2\frac{\rho}{|\xi|}\nabla\rho\cdot\nabla h + (\Delta\psi)g + \frac{\rho}{|\xi|}(\Delta\rho)h + \frac{1}{|\xi|}|\nabla\rho|^2 h = -\Delta g, \tag{1.13}$$
$$2\nabla\rho\cdot\nabla g + 2\nabla\psi\cdot\nabla h + (\Delta\rho)g + (\Delta\psi)h = -\Delta h$$

where (1.13) is satisfied in the formal sense that terms of like order of homogeneity in (ξ, η) are separated out and set equal.

In Section 3, we shall discuss the solvability of the eikonal and transport equations, (1.12) and (1.13), and a number of their properties. First, in Section 2, we shall examine the Fourier-Airy integral operators of the form (1.6).

Exercises

1.1. Define $Ai(s)$ to be the Fourier transform of $e^{(i/3)t^3}$, which is formally equivalent to (1.5); certainly $Ai(s)$ exists as a tempered distribution. Show that $Ai(s)$ solves the differential equation

$$Ai''(s) - sAi(s) = 0. \tag{1.14}$$

Hint: Fourier transform an obvious first order differential equation satisfied by $e^{(i/3)t^3}$.

1.2. From (1.14) deduce that $Ai(s)$ is smooth, in fact an entire analytic function of its argument.

1.3. Let $\mathscr{L}$ be any contour that begins at a point at infinity in the sector $-(1/2)\pi \leq \arg v \leq -(1/6)\pi$ and ends at infinity in the sector $(1/6)\pi \leq \arg v \leq (1/2)\pi$. Show that

$$Ai(z) = \frac{1}{2\pi i}\int_{\mathscr{L}} e^{(1/3)v^3 - zv}\,dv. \tag{1.15}$$

Hint: Deform the contour in

$$(1.16)\qquad Ai(x) = \frac{1}{2\pi i}\int_{-\infty i}^{\infty i} e^{(1/3)v^3 - xv}\, dv \qquad (x \in \mathbf{R}^+).$$

Note that the right side of (1.15) defines an entire analytic function of z.

1.4. Show that, for $|\arg z| < \pi$,

$$(1.17)\qquad Ai(z) = \frac{1}{2\pi} e^{-(2/3)z^{3/2}} \int_0^\infty e^{-tz^{1/2}} \cos\left(\tfrac{1}{3}t^{3/2}\right) t^{-1/2}\, dt.$$

Hint: For $z = x \in \mathbf{R}^+$, use (1.16) and set $v = x^{1/2} + it^{1/2}$ on the upper half of the deformed path and $x^{1/2} - it^{1/2}$ on the lower half. Then extend (1.17) by analytic continuation.

1.5. Let $q(t)$ be a bounded function on $\mathbf{R}^+$, smooth for $t > 0$, such that, near $t = 0$,

$$(1.18)\qquad q(t) \sim \sum_{j=0}^{\infty} a_j t^{(j+\lambda-\mu)/\mu}$$

given $\mu > 0$, $Re\,\lambda > 0$. Assume $q(t)$ well behaved as $t \to \infty$. Let

$$I(z) = \int_0^\infty e^{-zt} q(t)\, dt.$$

Show that

$$(1.19)\qquad I(z) \sim \sum_{j=0}^{\infty} \Gamma\left(\frac{j+\lambda}{\mu}\right) a_j z^{-(j+\lambda)/\mu},$$

uniformly as $|z| \to \infty$ in $|\arg z| \le (1/2)\pi - \delta$ ($\delta > 0$ fixed).

1.6. Show that, for $|z| \to \infty$ in $|\arg z| \le \pi - \delta$,

$$Ai(z) \sim \frac{1}{2\sqrt{\pi}} z^{-1/4} e^{-(2/3)z^{3/2}} \sum_{j=0}^{\infty} a_j z^{-(3/2)j}$$

with $a_0 = 1$. Hint: Use (1.17) and Exercise 1.5.

1.7. Let $A_\pm(s) = Ai(-e^{\pm(2/3)\pi i}s)$, as in (1.7). Show that

$$(1.20)\qquad A'_+(s)A_-(s) - A'_-(s)A_+(s) = \text{const.}$$

Hint: Differentiate. Use the asymptotic expansion to calculate the constant and show it is nonzero.

1.8. Show that $\overline{A_+(z)} = A_-(\bar{z})$ and hence that $z \in \mathbf{C}$ is a zero of A_+ if, and only if, $\bar{z}$ is a zero of A_-. Similarly $z \in \mathbf{C}$ is a zero of A'_+ if, and only if, $\bar{z}$ is a zero of A'_-. Using this, deduce from (1.20) that $A_\pm$ and $A'_\pm$ have no real zeros. Hint: $Ai(z)$ is real valued for $z \in \mathbf{R}$.

1.9. Show that with the transformation $\xi = (2/3)s^{3/2}$ and $W = s^{-1/2}Ai(s)$, equation (1.14) becomes the Bessel equation

$$\frac{d^2W}{d\xi^2} + \frac{1}{\xi}\frac{dW}{d\xi} - \left(1 + \frac{1}{9\xi^2}\right)W = 0.$$

Show that

$$Ai(z) = \frac{1}{\pi}\left(\frac{z}{3}\right)^{1/2} K_{1/3}\left(\frac{2z^{2/3}}{3}\right)$$

for $|\arg z| < (2/3)\pi$, where K_ν is the modified Bessel function. Hint: Examine the asymptotic behavior to determine the coefficients. We remark that one integral formula for $K_\nu(z)$ is

$$K_\nu(z) = \frac{\sqrt{\pi}}{\Gamma(\nu+(1/2))}((1/2)z)^\nu \int_1^\infty e^{-zt}(t^2-1)^{\nu-1/2}\,dt$$

if $Re\ \nu > -1/2$, $|\arg z| < (1/2)\pi$.

§2. Fourier-Airy Integral Operators

In this section we give a general discussion of operators of the form

$$A(F) = \iint\left[g\frac{A(|\xi|^{-1/3}\rho)}{A(-|\xi|^{-1/3}\eta)} + ih|\xi|^{-1/3}\frac{A'(|\xi|^{-1/3}\rho)}{A(-|\xi|^{-1/3}\eta)}\right]e^{i\theta}\hat{F}(\xi,\eta)\,d\xi\,d\eta. \tag{2.1}$$

Here $g = g(x, \xi, \eta)$, $h = h(x, \xi, \eta)$, $\rho = \rho(x, \xi, \eta)$, $\theta = t|\xi| + \psi(x, \xi, \eta)$, are defined and smooth for x in some small neighborhood $\bar{U}$ in $\mathbf{R}^n\backslash K$ of a boundary point $x_0 \in \partial K$, and $(\xi, \eta) \in \Gamma = \{|\eta| \le c_0|\xi|\}$. We make the following assumptions:

$$g, h \in S^m_{1,0}(\bar{U} \times \Gamma); \tag{2.2}$$

(2.3) ρ, ψ are real valued, homogeneous of degree 1 in (ξ, η);

$$\nabla_x\psi(x, \xi, \eta) \neq 0 \text{ on } \bar{U} \times (\Gamma\backslash 0); \tag{2.4}$$

$$\rho(x, \xi, \eta) = -\eta \text{ for } x \in \partial K, \quad \eta > 0; \quad \rho > 0 \text{ for } x \in \partial K, \quad \eta < 0; \tag{2.5}$$

$$|\xi|^{-1}\frac{\partial}{\partial\nu}\rho(x, \xi, \eta) \ge a_0 > 0 \text{ for } x \in \partial K; \tag{2.6}$$

$$F \in \mathscr{E}'(\mathbf{R}^n), \qquad WF(F) \subset \{|\eta| \le \tfrac{1}{2}c_0|\xi|\}. \tag{2.7}$$

Our first task is to give $A(F)$ a meaning for $F \in \mathscr{E}'$. For $F \in C_0^\infty(\mathbf{R}^n)$, the integral (2.1) is absolutely convergent, but even this fact requires an argument. Indeed, we have the following.

LEMMA 2.1. *We have the estimate*

$$\left| D_x^\beta \frac{A(|\xi|^{-1/3}\rho)}{A(-|\xi|^{-1/3}\eta)} \right| \le c(1+|\xi|)^{1/6+|\beta|}$$

for $|\xi| \le 1$, $(\xi, \eta) \in \Gamma$. *A similar estimate holds for*

$$|\xi|^{-1/3} \frac{A'(|\xi|^{-1/3}\rho)}{A(-|\xi|^{-1/3}\eta)}.$$

Proof. We have

$$\frac{A(|\xi|^{-1/3}\rho)}{A(-|\xi|^{-1/3}\eta)} \tag{2.8}$$

$$= \frac{\Phi(e^{-\pi i/3}|\xi|^{-1/3}\rho)}{\Phi(-e^{-\pi i/3}|\xi|^{-1/3}\eta)} e^{-2|\xi|/3\left(\left(e^{-\pi i/3}\frac{\rho}{|\xi|}\right)^{3/2} - \left(-e^{-\pi i/3}\frac{\eta}{|\xi|}\right)^{3/2}\right)}$$

where $\Phi(z)$ is the holomorphic function on $-\pi < \arg z < \pi$ given by (1.8), and the argument in $e^{-(2/3)z^{3/2}}$ is defined on the cut plane $-\pi < \arg z < \pi$. From the asymptotic expansion of Φ we see that

$$\left|\frac{\Phi}{\Phi}\right| \le c(1+|\xi|)^{1/6}.$$

Now use the hypotheses (2.5) and (2.6), which yield

$$\rho(x, \xi, \eta) \ge a_1|\xi|y + \rho(0, x', \xi, \eta) \tag{2.9}$$

on a sufficiently small neighborhood of ∂K, if coordinates are chosen, $x = (y, x')$, so $\partial K = \{y = 0\}$. This shows that the exponential factor in (2.8) is bounded, which yields the desired bound for $\beta = 0$. A similar argument handles $|\xi|^{-1/3}(A'/A)$. For general β, use the chain rule and the Airy equation $A''(s) + sA(s) = 0$ to get the general result from the $\beta = 0$ case.

From Lemma 2.1 it follows immediately that, if $F \in C_0^\infty(\mathbf{R}^n)$, then $A(F) \in C^\infty(\bar{U})$. In order to define $A(F)$ for $F \in \mathscr{E}'(\mathbf{R}^n)$, satisfying (2.7), we reduce (2.1) to Fourier integral operators with singular phase functions, studied in Chapter VIII, Section 6.

In order to do this, we want to break

$$\frac{A(|\xi|^{-1/3}\rho)}{A(-|\xi|^{-1/3}\eta)}$$

into sums of products of symbols and exponentials. Formula (2.8) by itself will not do, because these terms will fail to be smooth at $\eta = 0$ and at $\rho = 0$ (which do not coincide unless $x \in \partial K$). To take care of this, we use a partition of unity. Let $p_1(s)$, $p_2(s)$, $p_3(s)$ be smooth functions, ≥ 0, supported respectively on $s \geq 1$, $-2 \leq s \leq 2$, and $s \leq -1$, with $p_1 + p_2 + p_3 = 1$. Let $q_2(s) = 1 - p_1(s) = p_2(s) + p_3(s)$. Now write (with $\omega = e^{-\pi i/3}$)

$$
\begin{aligned}
(2.10)\quad & \frac{A(|\xi|^{-1/3}\rho)}{A(-|\xi|^{-1/3}\eta)} \\
&= p_1(-|\xi|^{-1/3}\eta)p_1(|\xi|^{-1/3}\rho)\,\frac{\Phi(\omega|\xi|^{-1/3}\rho)}{\Phi(-\omega|\xi|^{-1/3}\eta)}\,e^{-\frac{2i}{3}|\xi|\left(\left(\frac{\rho}{|\xi|}\right)^{3/2}-\left(\frac{-\eta}{|\xi|}\right)^{3/2}\right)} \\
&\quad + \frac{q_2(-|\xi|^{-1/3}\eta)}{A(-|\xi|^{-1/3}\eta)}\,p_1(|\xi|^{-1/3}\rho)\Phi(\omega|\xi|^{-1/3}\rho)e^{-\frac{2i}{3}|\xi|\left(\frac{\rho}{|\xi|}\right)^{3/2}} \\
&\quad + q_2(|\xi|^{-1/3}\rho)\frac{A(|\xi|^{-1/3}\rho)}{A(-|\xi|^{-1/3}\eta)} \\
&= T_1 + T_2 + T_3.
\end{aligned}
$$

We claim that the factors other than the exponentials in T_1 and T_2 are symbols, and also that T_3 is a symbol. More precisely, we shall have bounded families of symbols in (x', ξ, η), with y as a parameter, where y derivatives are also bounded families, of symbols of higher order. To keep track of this, we use the following symbol classes.

DEFINITION. *We say $p(y, x', \xi, \eta) \in S^m_{\rho,\delta,\nu}(\bar{U} \times \Gamma)$ if, on $\bar{U} \times \Gamma$, we have*

$$|D_y^k D_{x'}^\beta D_{\xi,\eta}^\alpha p(y, x', \xi, \eta)| \leq C(1 + |\xi|)^{m - \rho|\alpha| + \delta|\beta| + \nu k}.$$

Some of our factors don't depend on y, and they are easy to estimate. To shorten notation, we replace $\Phi(\omega s)$ by $\Phi(s)$.

LEMMA 2.2. *We have the inclusions, on Γ,*

$$\frac{p_1(-|\xi|^{-1/3}\eta)}{\Phi(-|\xi|^{-1/3}\eta)} \in S^{1/6}_{1/3,0}, \tag{2.11}$$

$$\frac{q_2(-|\xi|^{-1/3}\eta)}{A(-|\xi|^{-1/3}\eta)} \in S^0_{1/3,0}. \tag{2.12}$$

Proof. Note that

$$\frac{p_1(-s)}{\Phi(-s)} \in S^{1/4}_{1,0}(\mathbf{R})$$

and

$$\frac{q_2(-s)}{A(-s)} \in S^0_{1,0}(\mathbf{R}).$$

In general, if $r(s) \in S^m_{1,0}(\mathbf{R})$, $m \geq 0$, then $r(|\xi|^{-1/3}\eta) \in S^{2m/3}_{1/3,0}(\Gamma)$, as is a simple consequence of the chain rule.

More generally, $r(|\xi|^{-1/3}\rho) \in S^{2m/3}_{1/3,2/3,2/3}$, which yields the following:

LEMMA 2.3. *We have the inclusion*

$$p_1(|\xi|^{-1/3}\rho)\Phi(|\xi|^{-1/3}\rho) \in S^0_{1/3,2/3,2/3}(\bar{U} \times \Gamma). \tag{2.13}$$

It remains to analyze $T_3(y, x', \xi, \eta)$, and we claim that

$$T_3 \in S^{1/6}_{1/3,2/3,1}(\bar{U} \times \Gamma). \tag{2.14}$$

We start our analysis of T_3 with the following.

LEMMA 2.4. *We have*

$$p_2(|\xi|^{-1/3}\rho)\frac{A(|\xi|^{-1/3}\rho)}{A(-|\xi|^{-1/3}\eta)} \in S^{1/6}_{1/3,2/3,2/3}(\bar{U} \times \Gamma).$$

Proof. This term (call it $\tilde{T}_3$) is supported on $-2|\xi|^{1/3} \leq \rho \leq 2|\xi|^{1/3}$. On this set, we have $\eta \geq -2|\xi|^{1/3} + 0(|\xi|^{-\infty})$, so, for $|\xi| \geq 1$, we have

$$\tilde{T}_3 = \frac{q_2(-\frac{1}{2}|\xi|^{-1/3}\eta)}{A(-|\xi|^{-1/3}\eta)}[p_2(|\xi|^{-1/3}\rho)A(|\xi|^{-1/3}\rho)].$$

Since

$$\frac{q_2(-\frac{1}{2}s)}{A(-s)} \in S^{1/4}_{1,0}(\mathbf{R}),$$

we have the first factor in $S^{1/6}_{1/3,0}(\Gamma)$. Since $p_2(s)A(s) \in C_0^\infty(\mathbf{R}) \subset S^0_{1,0}(\mathbf{R})$, the second factor belongs to $S^0_{1/3,2/3,2/3}$, completing the proof.

More to the point is the following estimate.

LEMMA 2.5. *We have*

$$p_3(|\xi|^{-1/3}\rho)\frac{A(|\xi|^{-1/3}\rho)}{A(-|\xi|^{-1/3}\eta)} \in S^{1/6}_{1/3,2/3,1}(\bar{U} \times \Gamma).$$

Proof. This term (call it $\tilde{\tilde{T}}_3$) is supported on $-\rho \geq |\xi|^{1/3}$. Thus $\eta \geq |\xi|^{1/3}$ on supp $\tilde{\tilde{T}}_3$. Now write

$$\begin{aligned}\tilde{\tilde{T}}_3 &= p_3(|\xi|^{-1/3}\rho)\frac{\Phi(|\xi|^{-1/3}\rho)}{\Phi(-|\xi|^{-1/3}\eta)}e^{-(2/3)((|\xi|^{-1/3}\eta)^{3/2}-(-|\xi|^{-1/3}\rho^{3/2})}\\ &= p_3(\zeta)\frac{\Phi(\zeta)}{\Phi(\zeta_0)}e^{-(2/3)(\zeta_0^{3/2}-\zeta^{3/2})}\end{aligned}$$

where we have set $\zeta = -|\xi|^{-1/3}\rho$, $\zeta_0 = |\xi|^{-1/3}\eta$. Note that $\zeta_0 \geq \zeta + ay|\xi|^{2/3}$

on supp $\tilde{\tilde{T}}_3$. This implies that $\zeta_0^{3/2} - \zeta^{3/2} \geq by^{3/2}|\xi|$ on supp $\tilde{\tilde{T}}_3$, so

$$|e^{-(2/3)(\zeta_0^{3/2}-\zeta^{3/2})}| \leq e^{-(2/3)\,by^{3/2}|\xi|}.$$

Now we could arrange that $p_3(\zeta) = q_3(\zeta)^2$, with q_3 smooth. Then the term

$$q_3(\zeta)\,\frac{\Phi(\zeta)}{\Phi(\zeta_0)}$$

is easily handled; it belongs to $S^{1/6}_{1/3,2/3,2/3}$. We are left with $q_3(\zeta)e^{-(2/3)(\zeta_0^{3/2}-\zeta^{3/2})} = T'_3$. We have

$$(2.15)\quad D_y^j D_{x'}^\beta D_{\xi,\eta}^\alpha T'_3 = \sum_{\substack{j_1+\cdots+j_\nu=j\\ \beta_1+\cdots+\beta_\nu=\beta\\ \alpha_1+\cdots+\alpha_\nu=\alpha}}$$

$$\times\; cq_3(\zeta)D_y^{j_1}D_{x'}^{\beta_1}D_{\xi,\eta}^{\alpha_1}(\zeta_0^{3/2}-\zeta^{3/2})\cdots D_y^{j_\nu}D_{x'}^{\beta_\nu}D_{\xi,\eta}^{\alpha_\nu}(\zeta_0^{3/2}-\zeta^{3/2})e^{-(2/3)(\zeta_0^{3/2}-\zeta^{3/2})}$$

$+$ similar terms.

If $j_s \geq 1$, we use the estimate, on supp T'_3

$$(2.16)\qquad |D_y^{j_s}D_{x'}^{\beta_s}D_{\xi,\eta}^{\alpha_s}(\zeta_0^{3/2}-\zeta^{3/2})| \leq c|\xi|^{1-|\alpha_s|}\left(1+\left|\frac{\rho}{|\xi|}\right|^{3/2-j_s-|\beta_s|-|\alpha_s|}\right),$$

while for $j_s = 0$, since $\zeta(0, x', \xi, \eta) = \zeta_0$, we have

$$(2.17)\quad |D_{x'}^{\beta_s}D_{\xi,\eta}^{\alpha_s}(\zeta_0^{3/2}-\zeta^{3/2})| \leq cy|\xi|^{1-|\alpha_s|}\left(1+\left|\frac{\rho}{|\xi|}\right|^{1/2-j_s-|\beta_s|-|\alpha_s|}\right).$$

Now, on supp T'_3,

$$\left|\frac{\rho}{|\xi|}\right| \geq |\xi|^{-2/3},$$

so (2.16) is dominated by

$$(2.18)\qquad c|\xi|^{1-|\alpha_s|} + c|\xi|^{(2/3)(j_s+|\beta_s|)-|\alpha_s|/3} \leq c|\xi|^{j_s+2|\beta_s|/3-|\alpha_s|/3}$$

and (2.17), multiplied by $e^{-cy^{3/2}|\xi|}$, is dominated by

$$(2.19)\qquad c|\xi|^{1/3-|\alpha_s|}(1+|\xi|^{1/3+(2/3)(|\beta_s|+|\alpha_s|)}) \leq c|\xi|^{-|\alpha_s|/3+2|\beta_s|/3}.$$

Consequently, we dominate (2.15) by $c|\xi|^{j+2|\beta|/3-|\alpha|/3}$, which shows that $T'_3 \in S^0_{1/3,2/3,1}$, and completes the proof of Lemma 2.5.

In view of Lemmas 2.4 and 2.5, we have (2.14), which we restate as:

COROLLARY 2.6. *We have* $T_3 \in S^{1/6}_{1/3,2/3,1}(\bar{U}\times\Gamma)$. *More generally,*

$$(2.20)\qquad y^j T_3 \in S^{1/6-2j/3}_{1/3,2/3,1}.$$

The latter result, (2.20), is proved in the same fashion as in the analysis of T_3, noting that

$$y^j e^{-(2/3)\text{ by }^{3/2}|\xi|} \leq c|\xi|^{-2j/3}$$

and

$$y^j q_2(-\tfrac{1}{2}|\xi|^{-1/3}\eta)^{1/2} p_2(|\xi|^{-1/3}\rho)^{1/2} \in S^{-2j/3}_{1/3,2/3,2/3}.$$

Consequently, T_3 has a lot in common with the amplitude of a Poisson integral.

We can now analyze (2.1) in terms of ordinary Fourier integral operators, and Fourier integral operators with singular phase functions discussed in Section 6 of Chapter VIII. Indeed,

$$\text{(2.21)}\quad \iint g \frac{A(|\xi|^{-1/3}\rho)}{A(-|\xi|^{-1/3}\eta)} e^{i\theta}\hat{F}(\xi,\eta)\, d\xi\, d\eta$$
$$= \iint g a_1(x,\xi,\eta) e^{i\theta-(2/3)i|\xi|\left(\left(\frac{\rho}{|\xi|}\right)^{3/2}-\left(\frac{-\eta}{|\xi|}\right)^{3/2}\right)}\hat{F}(\xi,\eta)\, d\xi\, d\eta$$
$$+ \iint g a_2(x,\xi,\eta) e^{i\theta-(2/3)i|\xi|\left(\frac{\rho}{|\xi|}\right)^{3/2}}\hat{F}(\xi,\eta)\, d\xi\, d\eta$$
$$+ \iint g T_3(x,\xi,\eta) e^{i\theta}\hat{F}(\xi,\eta)\, d\xi\, d\eta$$

where

$$a_1(x,\xi,\eta) = p_1(-|\xi|^{-1/3}\eta)p_1(|\xi|^{-1/3}\rho)\frac{\Phi(|\xi|^{-1/3}\rho)}{\Phi(-|\xi|^{-1/3}\eta)}$$

and

$$a_2(x,\xi,\eta) = \frac{q_2(-|\xi|^{-1/3}\eta)}{A(-|\xi|^{-1/3}\eta)} p_1(|\xi|^{-1/3}\rho)\Phi(|\xi|^{-1/3}\rho).$$

As we have seen in Lemmas 2.2 and 2.3, we have

$$\text{(2.22)}\quad a_1(x,\xi,\eta) \in S^{1/6}_{1/3,2/3,2/3}, \quad \text{supported on } \{-\eta \geq |\xi|^{1/3}\} \cap \{\rho \geq |\xi|^{1/3}\}$$

and

$$\text{(2.23)}\quad a_2(x,\xi,\eta) \in S^{0}_{1/3,2/3,2/3}, \quad \text{supported on } \rho \geq |\xi|^{1/3}.$$

Consequently, the first two terms of (2.21) can be handled by the techniques of Section 6, Chapter VIII, on Fourier integral operators with singular phase functions, while the third term is an ordinary Fourier integral operator. The term in (2.1) involving $A'(|\xi|^{-1/3}\rho)$ is handled in the same fashion. Finally, one can read off $WF(Au)$ in terms of $WF(u)$, given the material of Chapter VIII.

For example, we see that, for

$$\begin{aligned}A_1'(F) &= \iint gT_1 e^{i\theta}\hat{F}\, d\xi\, d\eta \\ &= \iint p_1(-|\xi|^{-1/3}\eta)p_1(|\xi|^{-1/3}\rho)\frac{\Phi(|\xi|^{-1/3}\rho)}{\Phi(-|\xi|^{-1/3}\eta)} \\ &\quad\times e^{i\left(\theta-(2/3)|\xi|\left(\left(\frac{\rho}{|\xi|}\right)^{3/2}-\left(\frac{-\eta}{|\xi|}\right)^{3/2}\right)\right)}\hat{F}\, d\xi\, d\eta, \\ &= \iint p_1(-|\xi|^{-1/3}\eta)p_1(|\xi|^{-1/3}\rho)\frac{\Phi(|\xi|^{-1/3}\rho)}{\Phi(-|\xi|^{-1/3}\eta)} e^{i\varphi}\hat{F}\, d\xi\, d\eta,\end{aligned}$$

$WF(A_1'F) \subset \{(z, \theta) \in T^*(\mathbf{R} \times \Omega) : (\nabla_\zeta \varphi, \zeta) \in WF(F)$ and $\nabla_z \varphi = -\theta$, for some $(z, \zeta) \in$ conic supp of $p_1 p_1(\Phi/\Phi)\}$. Here we have set $\zeta = (\xi, \eta)$.

§3. The Eikonal and Transport Equations

Our first task in this section is to consider solvability of the eikonal equations for ρ and ψ:

$$|\nabla\psi|^2 + \frac{\rho}{|\xi|}|\nabla\rho|^2 = |\xi|^2, \tag{3.1}$$

$$\nabla\psi \cdot \nabla\rho = 0 \tag{3.2}$$

where $\rho = \rho(x, \xi, \eta)$, $\psi = \psi(x, \xi, \eta)$. We shall show that, given any convex surface S, which will be chosen to vary with (ξ, η), we can find a solution to (3.1), (3.2) with $\rho = 0$ on S, as long as ψ satisfies the obvious condition dictated by these equations, namely that $\nabla\psi$ be tangent to S and $|\nabla\psi| = |\xi|$ on S. Actually, an exact solution to (3.1), (3.2) will only exist on the "illuminated" side of S, the side filled out by the tangent lines to S, and we shall have to make a careful continuation of these functions onto the other side of S, in order to solve (3.1), (3.2) to infinite order on ∂K. The crucial fact about the solutions ρ and θ which permits such a continuation is that they are C^∞ up to the boundary S (see Proposition 3.1).

For $\eta = 0$, S will coincide with ∂K, so on ∂K, $\eta = 0$ will be equivalent to $\rho = 0$. In fact, as we shall see, it can be arranged that $-\eta$ and ρ match up to infinite order on ∂K,

$$|\xi|^{-1}\rho = -|\xi|^{-1}\eta + 0\left(\left(\frac{|\eta|}{|\xi|}\right)^\infty\right), \qquad x \in \partial K, \tag{3.3}$$

which will be very convenient, as will be seen in the next section. We also want $\theta|_{\mathbf{R}\times\partial K}$ to be the phase function of a Fourier integral operator with a locally bijective associated canonical transformation, i.e., if z parametrizes an open set in $\mathbf{R} \times \partial K$, $\zeta = (\xi, \eta)$, we want the $n \times n$ matrix of second order derivatives $\nabla^2_{z\zeta}\theta$ to be nonsingular, at least for z close to z_0, $\eta < c_0|\xi|$. As shown in the next section, these facts will guarantee that, for u given by (1.6), $f = u|_{\mathbf{R} \times \partial K} = JF$ where J is a Fourier integral operator with locally bijective associated canonical transformation.

We begin with the following existence theorem, in which (ξ, η) is dropped, say $\tilde{\rho} = \tilde{\rho}(x)$, $\tilde{\psi} = \tilde{\psi}(x)$.

PROPOSITION 3.1. *Let S be a smooth strictly convex surface in $\mathbf{R}^n$, and suppose $\tilde{\psi}_0$ is given on S, as a solution of the "surface eikonal equation"*

$$|\nabla_{\tan}\tilde{\psi}_0| = 1 \quad \text{on} \quad S \tag{3.4}$$

(S is not assumed closed). Then there is a neighborhood U of S in $\mathbf{R}^n$ and smooth functions $\tilde{\psi}$ and $\tilde{\rho}$ defined on the illuminated side of S intersected with U, smooth up to the boundary S, such that $\tilde{\rho} = 0$ on S, $\tilde{\psi}|_S = \tilde{\psi}_0$, and $\tilde{\rho}$, $\tilde{\psi}$ satisfy the eikonal equations

$$|\nabla\tilde{\psi}|^2 + \tilde{\rho}|\nabla\tilde{\rho}|^2 = 1, \tag{3.5}$$

$$\nabla\tilde{\psi} \cdot \nabla\tilde{\rho} = 0. \tag{3.6}$$

Furthermore, $\nabla\tilde{\rho} \neq 0$ on S; in fact $\partial\tilde{\rho}/\partial\nu > 0$ where ν is the unit normal to S pointing into the illuminated region.

Proof. Consider the functions $\varphi^{\pm} = \tilde{\psi} \pm (2/3)\tilde{\rho}^{3/2}$. The system (3.5), (3.6) yields

$$|\nabla\varphi^{\pm}|^2 = 1 \tag{3.7}$$

which is the usual eikonal equation of geometrical optics, while $\rho = 0$ on S, $\tilde{\psi}|_s = \tilde{\psi}_0$ yield

$$\varphi^{\pm}|_s = \tilde{\psi}_0. \tag{3.8}$$

Meanwhile, (3.4) implies $(\partial/\partial\nu)\varphi^{\pm} = 0$, so with the boundary data (3.8) we see that S is characteristic for (3.7). Nevertheless, the ray method of solving the eikonal equation does lead to two functions $\varphi^{\pm}$ defined on the illuminated side U^+ of S. Indeed, if one defines a section of $T^*(\mathbf{R}^n)$ over S to be $\nabla_{\tan}\tilde{\psi}_0$ on vectors tangent to S and to annihilate vectors normal to S (so $(\partial/\partial\nu)\psi_0 = 0$ on S), then the flow out of this section by the Hamiltonian vector field $H_{|\xi|^2}$ defines a Lagrangian manifold L which, under the natural projection $T^*(\mathbf{R}^n) \xrightarrow{\pi} \mathbf{R}^n$, is a two sheeted covering of the illuminated region U^+, with a fold over S. Thus it is the graph of two sections of the cotangent bundle over U^+, say $v^{\pm}$, smooth away from S, and in $C^{1/2}$ near S, with $dv^{\pm} = 0$ on U^+. Thus there exist $\varphi^{\pm}$ with $d\varphi^{\pm} = v^{\pm}$ on U^+, $\varphi^{\pm}|_S = \tilde{\psi}_0$. $\varphi^{\pm}$ are smooth away from S, and in $C^{3/2}$ near S. Pick the signs so that $\varphi^+ > \varphi^-$ near S, which is possible since the strict convexity of S shows that $|\varphi^+ - \varphi^-| \geq c|\tau|^{3/2}$, where τ denotes distance from S.

Having thus constructed $\varphi^\pm$, we define $\tilde{\psi}$ and $\tilde{\rho}$ by

$$\tilde{\psi} = \tfrac{1}{2}(\varphi^+ + \varphi^-), \tag{3.9}$$

$$\tilde{\rho} = [\tfrac{3}{4}(\varphi^+ - \varphi^-)]^{2/3}. \tag{3.10}$$

It is clear that $\tilde{\psi}$ and $\tilde{\rho}$ are smooth in U^+, at least away from S. We shall show that they are smooth up to S. First, we check that they satisfy (3.5), (3.6). In fact, $\nabla\tilde{\psi} = (1/2)(\nabla\varphi^+ + \nabla\varphi^-)$,

$$\nabla\tilde{\rho} = \tfrac{1}{2}\tilde{\rho}^{-1/2}(\nabla\varphi^+ - \nabla\varphi^-), \qquad \text{so } \nabla\tilde{\psi}\cdot\nabla\tilde{\rho} = \tfrac{1}{4}\tilde{\rho}^{-1/2}(|\nabla\varphi^+|^2 - |\nabla\varphi^-|^2) = 0,$$

establishing (3.6), while

$$\begin{aligned} |\nabla\tilde{\psi}|^2 + \tilde{\rho}|\nabla\tilde{\rho}|^2 &= \tfrac{1}{4}(|\nabla\varphi^+|^2 + 2\nabla\varphi^+\cdot\nabla\varphi^- + |\nabla\varphi^-|^2) \\ &\quad + \tfrac{1}{4}(|\nabla\varphi^+|^2 - 2\nabla\varphi^+\cdot\nabla\varphi^- + |\nabla\varphi^-|^2) = 1, \end{aligned}$$

establishing (3.5). Now we show $\tilde{\psi}$ and $\tilde{\rho}$ are smooth up to S.

In fact, the 1-forms $v^\pm$ on U^+ pull back to a smooth 1-form on L, namely $\tilde{v} = \kappa^*\alpha$ where $\alpha = \sum \xi_j\, dx_j$ is the canonical 1-form on $T^*(\mathbf{R}^n)$ and $\kappa: L \to T^*(\mathbf{R}^n)$. Clearly $d\tilde{v} = 0$, since L is Lagrangian.

$$\begin{array}{ccc} L & \xrightarrow{\kappa} & T^*(\mathbf{R}^n) \\ & {\scriptstyle \pi_L}\searrow \quad \mathbf{R}^n \quad \swarrow{\scriptstyle \pi} & \end{array}$$

Consequently, there is a smooth function Φ on L such that $d\Phi = \tilde{v}$, $\Phi \circ v^\pm = \tilde{\psi}_0$ on S, so $\varphi^\pm = \Phi \circ v^\pm$. Since the projection $\pi_L: L \to \mathbf{R}^n$ is a simple fold along S, it follows that coordinates can be chosen on a neighborhood of $\pi_L^{-1}S$ in L, and on a neighborhood of S in $\mathbf{R}^N$, such that π_L is the map $(y, z) \mapsto (y^2, z)$; $z \in \mathbf{R}^{n-1}$, and $y > 0$ defines U^+. In these coordinates, $v^\pm(\tau, z) = (\pm\tau^{1/2}, z)$. The map $(y, z) \overset{j}{\mapsto} (-y, z)$ is an involution on L commuting with π_L, and if $\Psi = \Phi \circ j \in C^\infty(L)$, we see that

$$\begin{aligned} \tilde{\psi} &= \tfrac{1}{2}(\Phi + \Psi) \circ v^\pm, \\ \tilde{\rho} &= [\tfrac{3}{4}(\Phi - \Psi) \circ v^\pm]^{2/3} = \sigma^{1/3}. \end{aligned}$$

In the (y, z) coordinates on L, $\Phi + \Psi$ and $(\Phi - \Psi)^2$ are both smooth functions that are even functions of y. Thus $\tilde{\psi}(\tau, z)$ and $\sigma(\tau, z)$ are smooth functions of $(\tau^{1/2}, z)$ that are even functions of $\tau^{1/2}$. It easily follows that $\tilde{\psi}(\tau, z)$ and $\sigma(\tau, z)$ are smooth functions of (τ, z), thus smooth in U^+, up to the boundary S. Since $c_1|\tau|^{3/2} \leq |\varphi^+(\tau, z) - \varphi^-(\tau, z)| \leq c_2|\tau|^{3/2}$, by the strict convexity of S, we see that $\sigma(\tau, z)$ vanishes to exactly third order on $S = \{\tau = 0\}$, which implies that $\sigma(\tau, z) = \tau^3\sigma_0(\tau, z)$, $\sigma_0(0, z) \neq 0$. Thus $\rho(\tau, z) = \tau\sigma_0(\tau, z)^{1/3}$ is smooth, and the proof is complete.

Remarks. Since $c_1|y|^3 \le |\Phi - \Psi| \le c_2|y|^3$ on L, we see that $[(3/4)(\Phi - \Psi)]^{1/3} = \gamma$ is smooth on L. Clearly $(\gamma \circ v^{\pm})^2 = \tilde{\rho}$. Thus $\pm\sqrt{\tilde{\rho}} \circ \pi_L = \gamma$ is smooth on the two sheeted covering L of U^+, taking a positive value on one sheet and a negative value on the other sheet. Consequently γ can stand for the y variable in defining the fold $\pi_L : L \to U^+$. With slight abuse of terminology, we say

$$y = \pm\sqrt{\tilde{\rho}}. \tag{3.11}$$

Thus with $\tau = y^2$, $\tau = \tilde{\rho}$. The fold π_L is described by $\tau = y^2$. We can arrange that all the z_j have gradient orthogonal to $\nabla\tilde{\rho}$.

We now want to study linearizations of the equations (3.1), (3.2) which will include the transport equations. To do this, we need to study the vector fields $(\nabla\tilde{\psi} \pm \tilde{\rho}^{1/2}\,\nabla\tilde{\rho}) \cdot \nabla$ on U^+, lifted to L.

PROPOSITION 3.2. *The double-valued vector field* $(\nabla\tilde{\psi} \pm \tilde{\rho}^{1/2}\,\nabla\tilde{\rho}) \cdot \nabla$ *lifts to a smooth vector field on L.*

Proof. In the (τ, z) coordinates on U^+, this is a smooth scalar multiple of the vector field

$$X = \sum a_j(\tau, z)\frac{\partial}{\partial z_j} \pm \tau^{1/2}\frac{\partial}{\partial \tau}, \qquad a_j \in C^\infty.$$

We may neglect the scalar multiple, and examine how X lifts to L, in its (y, z) coordinates. Clearly $\partial/\partial z_j$ lifts to $\partial/\partial z_j$. Meanwhile, $\pm\tau^{1/2}(\partial/\partial\tau)$ lifts to $(1/2)(\partial/\partial y)$. This completes the proof.

Remark. In fact, this vector field on L coincides with $H_{|\xi|^2}$.

PROPOSITION 3.3. *Consider the system for* $\tilde{g}, \tilde{h}$,

$$2\nabla\tilde{\psi} \cdot \nabla\tilde{g} + 2\tilde{\rho}\,\nabla\tilde{\rho} \cdot \nabla\tilde{h} + |\nabla\tilde{\rho}|^2\tilde{h} + A\tilde{g} + \tilde{\rho}B\tilde{h} = F_1, \tag{3.12}$$

$$2\nabla\tilde{\rho} \cdot \nabla\tilde{g} + 2\nabla\tilde{\psi} \cdot \nabla\tilde{h} + B\tilde{g} + A\tilde{h} = F_2 \tag{3.13}$$

where $\tilde{\rho}$, $\tilde{\psi}$ are as in Proposition 3.1, A, B, F_1, F_2 are given, smooth in a neighborhood of S. Then there exist unique solutions $\tilde{g}$, $\tilde{h}$ to (3.12), (3.13), in U^+, smooth up to S, with

$$\tilde{g}|_S = g_0 \tag{3.14}$$

where $g_0 \in C^\infty(S)$ is given. If g_0 satisfies the surface transport equation

$$(2\nabla\tilde{\psi} \cdot \nabla + A)g_0 = F_1 - |\nabla\tilde{\rho}|^2 h_0 \text{ (on } S), \tag{3.15}$$

then $\tilde{h} = h_0$ on S.

Proof. In analogy with the proof of Proposition 3.1, let $a^{\pm} = \tilde{g} \pm \tilde{\rho}^{1/2}\tilde{h}$. Then (3.12), (3.13) yield, with consistent choice of $+$ or $-$,

$$\{2(\nabla\tilde{\psi} \pm \tilde{\rho}^{1/2}\,\nabla\tilde{\rho}) \cdot \nabla + (A \pm \tilde{\rho}^{1/2}B)\}a^{\pm} = F_1 \pm \tilde{\rho}^{1/2}F_2. \tag{3.16}$$

We see that $a^{\pm} \circ \pi_L = a$, where a, defined on L, satisfies

$$\{2Y + (\tilde{A} + y\tilde{B})\}a = \tilde{F}_1 + y\tilde{F}_2. \tag{3.17}$$

Here $\tilde{A} = A \circ \pi_L$, etc., and Y is the *smooth* vector field on L to which $(\nabla\tilde{\psi} \pm \tilde{\rho}^{1/2}\,\nabla\tilde{\rho}) \cdot \nabla$ lifts, according to Proposition 3.2. We want $a = g_0 \circ \pi_2$ on $\pi_L^{-1}(S) = S_L$. Since Y is transverse to S_L, there is a unique solution to (3.17) with a specified on S_L.

Next set $b = a \circ j$ where j is the involution on L used before, and let

$$\tilde{g} = \tfrac{1}{2}(a + b) \circ v^{\pm} = \tfrac{1}{2}(a^{+} + a^{-}),$$
$$\tilde{h} = [\tfrac{1}{2}y^{-1}(a - b)] \circ v^{\pm} = \tfrac{1}{2}\tilde{\rho}^{-1/2}(a^{+} - a^{-}).$$

since $a + b$ and $y^{-1}(a - b)$ are both smooth even functions on L, it follows that $\tilde{g}$ and $\tilde{h}$ are smooth on U^{+}, up to S. Checking that (3.12) and (3.13) are satisfied in U^{+} follows along the same lines as the analogous argument in the proof of Proposition 3.1. Finally, comparing (3.12) and (3.15) yields $|\nabla\tilde{\rho}|^2(\tilde{h} - h_0) = 0$ on S, where $\tilde{\rho} = 0$. Since $|\nabla\tilde{\rho}| \neq 0$ on S, this implies $\tilde{h} = h_0$ on S. The proof is complete.

In our application of Proposition 3.1 to obtaining the functions ρ, ψ with the properties mentioned in the first two paragraphs, we let $\sigma = |\xi|^{-1}\xi$, $\omega = |\xi|^{-1}(\xi, \eta)$, $\alpha = |\xi|^{-1}\eta$, and prescribe $\tilde{\rho}(x, \omega) = |\xi|^{-1}\rho(x, \xi, \eta) = 0$ on S_ω, where $S_\omega = \partial K$ for $\eta = 0$. Furthermore we prescribe $\tilde{\psi}_0(x, \omega) = |\xi|^{-1}\psi_0(x, \xi, \eta)$ on S_ω, satisfying the surface eikonal equation on S_ω. To obtain such $\tilde{\psi}_0(x, \omega)$, we can take a smooth hypersurface C_ω of S_ω, and prescribe $\tilde{\psi}_0|_{C_\omega}$ to be any function whose gradient restricted to C_ω is <1, then solve the surface eikonal equation to obtain $\tilde{\psi}_0(x, \omega)$ on a neighborhood of C_ω in S_ω. In particular, we may specify $\nabla\tilde{\psi}_0$ on C_ω to be any unit field, tangent to S_ω but not to C_ω, whose projection on T^*C_ω is closed. However, in order to satisfy (3.3), some restrictions must be made. We cannot let S_ω be an arbitrary smooth family of convex surfaces with $S_\omega = \partial K$ for $\alpha = 0$, and we cannot let $\tilde{\psi}_0|_{C_\omega}$ be arbitrary, except when $\alpha = 0$. Nevertheless, as we shall see in the proof of the next proposition, the functions $(\partial^j/\partial\alpha^j)\,\tilde{\psi}_0$ can be specified on $C_{(\sigma,0)}$, and this will provide enough freedom to guarantee that $\nabla^2_{z\zeta}\theta$ is nonsingular.

PROPOSITION 3.4. *Let $\tilde{\psi}_0(x, \sigma)$ be given on ∂K as solutions of the surface eikonal equation $|\nabla_{\tan}\tilde{\psi}_0|^2 = 1$, for $\alpha = 0$, where $\omega = |\xi|^{-1}(\xi, \eta) = (\sigma, \alpha)$. Then there exist surfaces S_ω, with $S_\omega = \partial K$ for $\alpha = 0$, and smooth solutions $\tilde{\psi} = \tilde{\psi}(x, \omega)$, $\tilde{\rho} = \tilde{\rho}(x, \omega)$ to (3.5), (3.6), on the illuminated side of S_ω, including the boundary S_ω, with $\tilde{\rho}(x, \omega) = 0$ on S_ω, and $\tilde{\psi}(x, \sigma, 0)|_{\partial K} = \tilde{\psi}_0(x, \sigma)$, such that*

$$\tilde{\rho}(x, \omega) = -\alpha + O(\alpha^{\infty}), \qquad x \in \partial K. \tag{3.18}$$

Proof. We first construct a formal power series in α

$$\hat{\rho}(x, \sigma, \alpha) \sim \sum_{j=0}^{\infty} \hat{\rho}_j(x, \sigma) \frac{\alpha^j}{j!}, \tag{3.19}$$

$$\hat{\psi}(x, \sigma, \alpha) \sim \sum_{j=0}^{\infty} \hat{\psi}_j(x, \sigma) \frac{\alpha^j}{j!}, \tag{3.20}$$

which formally solves (3.5), (3.6), in the ring of formal power series in α, with coefficients smooth functions of (x, σ), and then we construct exact solutions $\tilde{\rho}(x, \omega)$, $\tilde{\psi}(x, \omega)$ which agree with $\hat{\rho}$, $\hat{\psi}$ to infinite order at $\alpha = 0$. We require that $\hat{\rho}(x, \sigma, \alpha) = -\alpha$ for $x \in \partial K$, i.e. $\hat{\rho}_j(x, \sigma) = 0$ for $j \neq 1$, $x \in \partial K$, and $\hat{\rho}_1(x, \sigma) = -1$. $\hat{\rho}_0$ and $\hat{\psi}_0$ are given by Proposition 3.1.

Substituting (3.19) and (3.20) into (3.5), (3.6), and equating each coefficient of α^j to zero, we see that $\hat{\rho}_j$, $\hat{\psi}_j$ must satisfy

$$2\nabla\hat{\psi}_0 \cdot \nabla\hat{\psi}_j + 2\hat{\rho}_0 \nabla\hat{\rho}_0 \cdot \nabla\hat{\rho}_j + |\nabla\hat{\rho}_0|^2 \hat{\rho}_j = F_{1j}, \tag{3.21}$$

$$\nabla\hat{\rho}_0 \cdot \nabla\hat{\psi}_j + \nabla\hat{\psi}_0 \cdot \nabla\hat{\rho}_j = F_{2j} \tag{3.22}$$

where F_{1j}, F_{2j} depend on $\hat{\rho}_0, \dots, \hat{\rho}_{j-1}$ and $\hat{\psi}_0, \dots, \hat{\psi}_{j-1}$, $F_{11} = F_{21} = 0$. Proposition 3.3 applies to this system, with $\tilde{\psi} = \hat{\psi}_0$, $\tilde{\rho} = \hat{\rho}_0$, $\tilde{g} = \hat{\psi}_j$, $\tilde{h} = \hat{\rho}_j$, and $A = B = 0$. Consequently, the formal power series (3.19), (3.20) have the desired properties provided $\hat{\psi}_j(x, \sigma)$ are specified on $S_{(\sigma,0)} = \partial K$ to satisfy the surface transport equations

$$2\nabla\hat{\psi}_0 \cdot \nabla\hat{\psi}_1 = |\nabla\hat{\rho}_0|^2 \text{ on } \partial K, \tag{3.23}$$

$$2\nabla\hat{\psi}_0 \cdot \nabla\hat{\psi}_j = F_{1j}, \qquad j \geq 2, \text{ on } \partial K. \tag{3.24}$$

Take arbitrary smooth extensions of $\hat{\psi}_j$, $\hat{\rho}_j$ to the shadow side of ∂K.

Now such formal power series as (3.19), (3.20) cannot be guaranteed to converge, but Borel's theorem guarantees that there exist smooth functions $\hat{\rho}(x, \sigma, \alpha)$, $\hat{\psi}(x, \sigma, \alpha)$ such that $(\partial^j/\partial\alpha^j)\hat{\rho}(x, \sigma, 0) = \hat{\rho}_j(x, \sigma)$ and $(\partial^j/\partial\alpha^j)\hat{\psi}(x, \sigma, 0) = \hat{\psi}_j(x, \sigma)$. Define the surface $S_\omega = S_{(\sigma,\alpha)}$ to be the zero set of $\hat{\rho}(x, \sigma, \alpha)$. We construct $\tilde{\rho}(x, \omega)$ and $\tilde{\psi}(x, \omega)$ as follows.

Let Σ_σ be some hypersurface or $\mathbf{R}^n$ which intersects ∂K transversally. Suppose $\nabla\hat{\psi}(x, \sigma, 0)$ is transversal to $\Sigma_\sigma \cap \partial K$. Then Σ_σ intersects S_ω transversally, for α small. Let $C_\omega = \Sigma_\sigma \cap S_\omega$. Let $\hat{\psi}_{00}(x, \omega) = \hat{\psi}(x, \omega)$ on C_ω, and define $\tilde{\psi}_0(x, \omega)$ on S_ω to solve the surface eikonal equation

$$|\nabla_{\tan}\tilde{\psi}_0(x, \omega)| = 1 \text{ on } S_\omega, \tag{3.25}$$

$$\tilde{\psi}_0(x, \omega)|_{C_\omega} = \tilde{\psi}_{00}(x, \omega) \text{ on } C_\omega. \tag{3.26}$$

The hypothesis on Σ_σ implies the part of $\nabla\tilde{\psi}_{00}$ tangent to C_ω has norm <1, so a smooth solution exists, on a neighborhood of C_ω. Applying Proposition 3.1, we construct $\tilde{\psi}$, $\tilde{\rho}$ on the illuminated side of S_ω, satisfying

(3.5), (3.6), with $\tilde{\rho}(x, \omega) = 0$ on S_ω and

$$\tilde{\psi}(x, \omega) = \tilde{\psi}_0(x, \omega) \text{ on } S_\omega.$$

It remains only to show that $(\partial^j/\partial\alpha^j)\tilde{\psi}(x, \sigma, 0) = (\partial^j/\partial\alpha^j)\hat{\psi}(x, \sigma, 0)$ and $(\partial^j/\partial\alpha^j)\tilde{\rho}(x, \sigma, 0) = (\partial^j/\partial\alpha^j)\hat{\rho}(x, \sigma, 0)$, for $x \in U^+$. For $j = 0$ this is clear.

Indeed, the same argument that we started this proof with shows that $\tilde{\psi}_j(x, \sigma) = (\partial^j/\partial\alpha^j)\tilde{\psi}(x, \sigma, 0)$ and $\tilde{\rho}_j(x, \sigma) = (\partial^j/\partial\alpha^j)\tilde{\rho}(x, \sigma, 0)$ satisfy the equations

$$2\nabla\tilde{\psi}_0 \cdot \nabla\tilde{\psi}_j + 2\tilde{\rho}_0 \nabla\tilde{\rho}_0 \cdot \nabla\tilde{\rho}_j + |\nabla\tilde{\rho}_0|^2\tilde{\rho}_j = F'_{1j}, \tag{3.27}$$

$$\nabla\tilde{\rho}_0 \cdot \nabla\tilde{\psi}_j + \nabla\tilde{\psi}_0 \cdot \nabla\tilde{\rho}_j = F'_{2j} \tag{3.28}$$

where F'_{kj} depend on $\tilde{\rho}_0, \ldots, \tilde{\rho}_{j-1}, \tilde{\psi}_0, \ldots, \tilde{\psi}_{j-1}$ in the same fashion as F_{kj} depend on $\hat{\rho}_0, \ldots, \hat{\psi}_{j-1}$. $F'_{11} = F'_{21} = 0$. Supposing it established that $\tilde{\rho}_\nu = \hat{\rho}_\nu$, $\tilde{\psi}_\nu = \hat{\psi}_\nu$ for $x \in U^+$, $\nu \le j - 1$, we see that $F'_{kj} = F_{kj}$ on U^+.

Let $x(\alpha)$ be a smooth curve in $\mathbf{R}^n$, with $x(\alpha) \in S_{\sigma,\alpha}$. Differentiating $\hat{\rho}(x(\alpha), \sigma, \alpha) = 0$ and $\tilde{\rho}(x(\alpha), \sigma, \alpha) = 0$ and setting $\alpha = 0$ yields

$$\hat{\rho}_1 + \nabla\hat{\rho}_0 \cdot \frac{\partial x}{\partial\alpha} = 0 \text{ on } \partial K,$$

$$\tilde{\rho}_1 + \nabla\tilde{\rho}_0 \cdot \frac{\partial x}{\partial\alpha} = 0 \text{ on } \partial K.$$

Since $\hat{\rho}_0 = \tilde{\rho}_0$ on U^+, we have $\nabla\hat{\rho}_0 = \nabla\tilde{\rho}_0$ on U^+, so we deduce that

$$\tilde{\rho}_1(x, \sigma) = \hat{\rho}_1(x, \sigma) = -1 \text{ on } \partial K. \tag{3.29}$$

From (3.27) it follows that

$$2\nabla\tilde{\psi}_0 \cdot \nabla\tilde{\psi}_1 = |\nabla\tilde{\rho}_0|^2 \text{ on } \partial K$$

which in view of the identities $\tilde{\psi}_0 = \hat{\psi}_0$, $\tilde{\rho}_0 = \hat{\rho}_0$ in U^+, coincides with the equation (3.23). Thus it will follow that $\tilde{\psi}_1 = \hat{\psi}_1$ on ∂K, if we know this identity on $C_{\sigma,0}$. This follows from differentiating the identity $\tilde{\psi}(x, \sigma, \alpha) = \hat{\psi}(x, \sigma, \alpha)$ on $C_{\sigma,\alpha}$. Since (3.21), (3.22) coincide with (3.27), (3.28) for $j = 1$, and $\tilde{\rho}_1 = \hat{\rho}_1$ on ∂K, $\tilde{\psi}_1 = \hat{\psi}_1$ on ∂K, it follows that $\tilde{\rho}_1 = \hat{\rho}_1$, $\tilde{\psi}_1 = \hat{\psi}_1$ on U^+. Arguing in a similar fashion, we deduce that $\tilde{\rho}_j = \hat{\rho}_j$, $\tilde{\psi}_j = \hat{\psi}_j$ on U^+, and the proof is complete.

We now obtain the phase functions, ρ, ψ as indicated in the paragraph preceeding Proposition 3.4, with that assertion satisfied, and with $|\xi|^{-1}\nabla\psi(x, \xi, 0)$ a prescribed unit vector field (not tangent to C) on $C_{\sigma,0} = C$, some hypersurface of ∂K. We prescribe the vector field so that, as σ varies, $|\xi|^{-1}\nabla_x\psi(x, \xi, 0)$ varies over an open subset of the tangent space to ∂K, say at $x = x_0 \in C$, so that $\nabla^2_{x\xi}\psi(x, \xi, 0)$ is an invertible matrix.

Proposition 3.4 guarantees that

$$|\xi|^{-1}|\nabla_x\psi(x, \xi, \eta)| \le 1 - c\frac{|\eta|}{|\xi|}$$

for small $\eta < 0$, which implies that $\nabla^2_{(x,t),(\xi,\eta)}\theta(t, x, \xi, \eta)$ is invertible, for x near x_0, $|\eta| \le C_0|\xi|$. This completes our construction of ρ, ψ, for $\eta \le 0$. We now describe how to obtain suitable smooth extensions to $\eta > 0$.

We shall arrange that (3.1), (3.2) be satisfied to infinite order on ∂K, for $\alpha \ge 0$. The first thing to do is specify ρ, θ and all their normal derivatives on ∂K. Let $\rho = -\eta$ on ∂K, for $\eta \ge 0$. Take $\psi|_{\partial K}$ to be any smooth extension of the ψ previously obtained for $\alpha \le 0$. Since ρ is independent of x on ∂K for $\alpha > 0$, $\nabla\rho$ is normal to ∂K, so $(\partial/\partial\nu)\psi = 0$ on ∂K for $\alpha > 0$. (Note that, for α close to 0, $(\partial/\partial\nu)\rho \ne 0$ on ∂K, by Proposition 3.1.). To specify $\partial\rho/\partial\nu$ on ∂K, we use the eikonal equation (3.1), obtaining

$$\rho_\nu^2 = \frac{1}{\alpha}(|\nabla\psi|^2 - |\xi|^2) \text{ on } \partial K. \tag{3.30}$$

This is well defined for $\alpha \ne 0$ since the quantities on the right side have been specified, but it looks suspicious at $\alpha = 0$. However, for $\alpha \le 0$ we know that ρ, θ are smooth, and $\rho_\nu \ne 0$, so (3.30) defines a smooth positive ρ_ν.

To obtain $\psi_{\nu\nu}$ and $\rho_{\nu\nu}$ on ∂K, differentiate the eikonal equations (3.1), (3.2). Denoting the commutator $[(\partial/\partial\nu), \nabla]$ by L, a first order differential operator, you get

$$\nabla\rho\cdot\nabla\psi_\nu + L\psi\cdot\nabla\rho + \nabla\psi\cdot\nabla\rho_\nu + \nabla\psi\cdot L\rho = 0, \tag{3.31}$$

$$\frac{2\rho}{|\xi|}\nabla\rho\cdot\nabla\rho_\nu + \frac{2\rho}{|\xi|}L\rho\cdot\nabla\rho + \frac{\rho_\nu}{|\xi|}|\nabla\rho|^2 + 2\nabla\psi_\nu\cdot\nabla\psi + L\psi\cdot\nabla\psi = 0. \tag{3.32}$$

(3.31) expresses $\rho_\nu\psi_{\nu\nu}$ as a known quantity on ∂K (since $\nabla\psi\cdot\nabla\rho_\nu$ on ∂K is a tangential derivative of ρ_ν), so $\psi_{\nu\nu}$ is determined, on ∂K. Next, (3.32) expresses

$$-\frac{2\rho}{|\xi|}\rho_\nu\rho_{\nu\nu}$$

as a known quantity on ∂K, so $\rho_{\nu\nu}$ is determined (trouble at $\rho = 0$, i.e., at $\alpha = 0$, is avoided as before).

Further differentiation of the eikonal equations leads to determination of all higher normal derivatives of θ and ρ on ∂K. Whitney's extension theorem (see Malgrange [2] Chapter 1) allows us to extend these quantities smoothly into the whole region $x \in \mathbf{R}^n \backslash K$, $|\alpha| \le c$. We can further require

$$\rho > -\eta, \qquad x \in \mathbf{R}^n \backslash K, \qquad \eta > 0 \tag{3.33}$$

which is easily arranged, since $|\xi|^{-1}\rho_\nu \ge a_0 > 0$ near ∂K.

The transport equations (1.13) can also be treated as a consequence of Proposition 3.3. Using an argument analogous to the proof of Proposition 3.4, we can arrange that

$$h(x, \xi, \eta) = 0\left(\left(\frac{|\eta|}{|\xi|}\right)^\infty\right), \qquad x \in \partial K. \tag{3.34}$$

Indeed, writing down a formal power series

$$\hat{g}(x, \xi, \eta) = \sum \hat{g}_j \frac{\alpha^j}{j!},$$

$$\hat{h}(x, \xi, \eta) = \sum \hat{h}_j \frac{\alpha^j}{j!}$$

we obtain equations for $\hat{g}_j$, $\hat{h}_j$ analogous to (3.21), (3.22). Use Borel's theorem to specify $\hat{g}(x, \xi, \eta)$, $\hat{h}(x, \xi, \eta)$. Then solve the appropriate surface transport equation on $S_{\sigma,\alpha}$ for g, taking value $\hat{g}$ on $C_{\sigma,\alpha}$, and then apply Proposition 3.3 to obtain g, h. An argument similar to the one for ρ, ψ, enables one to get smooth continuations of $g(x, \xi, \eta)$, $h(x, \xi, \eta)$ for $\eta > 0$, $x \in \mathbf{R}^n \backslash K$, which solve the transport equations to infinite order on ∂K; $h(x, \xi, \eta) = 0$ on ∂K for $\eta > 0$. By Proposition 3.4, you can specify $g(x, \xi, 0)$ on the hypersurface C of ∂K. Make sure that

$$g(x, \xi, 0) \ne 0 \text{ on } C, \qquad |\xi| \ne 0. \tag{3.35}$$

This completes our discussion of the eikonal and transport equations.

The main idea of the proof of Propositions 3.1 and 3.3 is due to Keller, and details were carried out by Ludwig [1], who also proved Proposition 3.4 in Ludwig [2].

§4. Justification and Analysis of the Parametrix

Our next order of business is to show that $F \in \mathscr{E}'(\mathbf{R}^n)$ can be chosen so that u, defined by (1.6), satisfies $u|_{\partial K} = f$, mod C^∞, given $f \in \mathscr{E}'(\mathbf{R} \times \partial K)$ with wave front set in a small conic neighborhood of ζ_0. We claim that $u|_{\mathbf{R} \times \partial K}$ is obtained by applying a Fourier integral operator to F. In order to simplify the expression (1.6) for $(t, x) \in \mathbf{R} \times \partial K$, write, for $\eta \leq 0$

$$(4.1)\quad \frac{A(|\xi|^{-1/3}\rho)}{A(-|\xi|^{-1/3}\eta)} = \frac{\Phi(-e^{2\pi i/3}|\xi|^{-1/3}\rho)}{\Phi(e^{2\pi i/3}|\xi|^{-1/3}\eta)} e^{(2/3)\pi i|\xi|\left[\left(\frac{\rho}{|\xi|}\right)^{3/2} - \left(\frac{-\eta}{|\xi|}\right)^{3/2}\right]}$$
$$= B(x, \xi, \eta)e^{i\gamma(x,\xi,\eta)}.$$

Set $B(x, \xi, \eta) = 1$ for $\eta > 0$, $\gamma(x, \xi, \eta) = 0$ for $\eta > 0$ so (4.1) is valid for $|\eta| \leq c_0|\xi|$, $x \in \partial K$. Recall from Section 3 that

$$(4.2)\quad |\xi|^{-1}\rho = -|\xi|^{-1}\eta + 0\left(\left(\frac{|\eta|}{|\xi|}\right)^{\infty}\right), \qquad x \in \partial K.$$

Thus it is clear that $\gamma(x, \xi, \eta)$ is a C^∞ function, homogeneous of degree 1 in (ξ, η), for $|\xi| > 0$, $|\eta| \leq c_0|\xi|$. We also have the following.

LEMMA 4.1. *We have $B(x, \xi, \eta) \in S^0_{1,0}(\partial K, \mathbf{R}^n)$, on $|\eta| < c_0|\xi|$.*

Proof. It is clear that $B(x, \xi, \eta)$ is continuous on this region, and there exist constants a_0, a_1 such that

$$0 < a_0 \leq |B(x, \xi, \eta)| \leq a_1, \qquad x \in \partial\Omega, \qquad |\eta| \leq c_0|\xi|$$

provided c_0 is small enough. Thus it suffices to show that

$$\begin{aligned} F(x, \xi, \eta) &= \log B(x, \xi, \eta) \\ &= \log \Phi(-e^{2\pi i/3}|\xi|^{-1/3}\rho) - \log \Phi(e^{2\pi i/3}|\xi|^{-1/3}\eta) \qquad (\text{for } \eta \leq 0) \\ &= \varphi(|\xi|^{-1/3}\rho) - \varphi(-|\xi|^{-1/3}\eta) \text{ belongs to } S^0_{1,0}. \end{aligned}$$

Note that $\varphi^{(j)}(x) \leq c|x|^{-j}$ for $|x| \geq 1$, $\varphi^{(j)}(x) \leq c + c|x|^{3/2-j}$, for $|x| \leq 1$. In fact, we shall only need

$$(4.3)\quad |\varphi^{(j)}(x)| \leq c_j|x|^{-j}, \qquad j \geq 1$$

to estimate the derivatives of $F(x, \xi, \eta)$. We have

$$(4.4)\quad D^j_\eta D^\alpha_\xi D^\beta_x F(x, \xi, \eta) = \sum_{\substack{j_1+\cdots+j_\nu=j \\ \alpha_1+\cdots+\alpha_\nu=\alpha \\ \beta_1+\cdots+\beta_\nu=\beta}} c\{(D^{j_1}_\eta D^{\alpha_1}_\xi D^{\beta_1}_x(|\xi|^{-1/3}\rho)) \cdots$$
$$(D^{j_\nu}_\eta D^{\alpha_\nu}_\xi D^{\beta_\nu}_x(|\xi|^{-1/3}\rho))\varphi^{(\nu)}(|\xi|^{-1/3}\rho) - (D^{j_1}_\eta D^{\alpha_1}_\xi D^{\beta_1}_x(-|\xi|^{-1/3}\eta)) \cdots$$
$$(D^{j_\nu}_\eta D^{\alpha_\nu}_\xi D^{\beta_\nu}_x(-|\xi|^{-1/3}\eta))\varphi^{(\nu)}(-|\xi|^{-1/3}\eta)\}$$

Write $\rho = -\eta + \sigma$, so estimate (4.2) says

$$|\xi|^{-1}\sigma = 0\left(\left(\frac{|\eta|}{|\xi|}\right)^{\infty}\right), \qquad x \in \partial K.$$

Replacing each term

$$D^{j_\mu}_\eta D^{\alpha_\mu}_\xi D^{\beta_\mu}_x(|\xi|^{-1/3}\rho)$$

in (4.4) by

$$D_\eta^{j_\mu} D_\xi^{\alpha_\mu} D_x^{\beta_\mu}(-|\xi|^{-1/3}\eta) + D_\eta^{j_\mu} D_\xi^{\alpha_\mu} D_x^{\beta_\mu}(|\xi|^{-1/3}\sigma)$$

and expanding out, we see that (4.4) is equal to a sum of terms of the following types, upon rearranging the order of some products:

$$\text{Type (1).}\quad D_\eta^{j_1} D_\xi^{\alpha_1} D_x^{\beta_1}(|\xi|^{-1/3}\sigma) \cdots D_\eta^{j_\mu} D_\xi^{\alpha_\mu} D_x^{\beta_\mu}(|\xi|^{-1/3}\sigma) D_\eta^{j_{\mu+1}} D_\xi^{\alpha_{\mu+1}} \times (-|\xi|^{-1/3}\eta) \cdots D_\eta^{j_\nu} D_\xi^{\alpha_\nu}(-|\xi|^{-1/3}\eta) \cdot \varphi^{(\nu)}(|\xi|^{-1/3}\rho).$$

$$\text{Type (2).}\quad D_\eta^{j_1} D_\xi^{\alpha_1}(-|\xi|^{-1/3}\eta) \cdots D_\eta^{j_\nu} D_\xi^{\alpha_\nu}(-|\xi|^{-1/3}\eta)[\varphi^{(\nu)}(|\xi|^{-1/3}\rho) - \varphi^{(\nu)}(-|\xi|^{-1/3}\eta)].$$

(Type (2) occurs only if $|\beta| = 0$). Note that $1 \le \nu \le j + |\alpha| + |\beta|$. Applying (4.2) and (4.3), we estimate a type (1) term by

$$c_N|\xi|^{2\nu/3 - j - |\alpha|}\left(\frac{|\eta|}{|\xi|}\right)^N (|\xi|^{-1/3}|\eta|)^{-\nu} \le c_N|\xi|^{-N+\nu-j-|\alpha|}|\eta|^{N-\nu} \quad (\text{choose } N > \nu)$$
$$\le c_N'|\xi|^{-j-|\alpha|} \text{ on } |\eta| \le c_0|\xi|.$$

Similarly we estimate a Type (2) term by

$$C|\xi|^{2\nu/3-|\alpha|-j} \int_{-|\xi|^{-1/3}\eta}^{|\xi|^{-1/3}\rho} |\varphi^{(\nu+1)}(\tau)|\, d\tau$$
$$\le c_N|\xi|^{2\nu/3-|\alpha|-j}(|\xi|^{-1/3}|\eta|)^{-\nu-1}|\xi|^{2/3}\left(\frac{|\eta|}{|\xi|}\right)^N$$
$$\le c_N|\xi|^{-N+\nu+1-|\alpha|-j}|\eta|^{N-\nu-1} \quad (\text{choose } N > \nu - 1)$$
$$\le c_N'|\xi|^{-|\alpha|-j} \text{ on } |\eta| \le c_0|\xi|.$$

Thus $F \in S_{1,0}^0$, which proves the lemma.

With this lemma applied to (4.1), we can deduce from (1.6) that

$$(4.5)\quad u(t, x)|_{\mathbf{R}\times\partial K} = \iint \left[g + ih|\xi|^{-1/3}\frac{A'(|\xi|^{-1/3}\rho)}{A(|\xi|^{-1/3}\rho)}\right] Be^{i(\theta+\gamma)}\hat{F}(\xi, \eta)\, d\xi\, d\eta$$
$$= J(F).$$

We can simplify this even further by writing

$$(4.6)\qquad \frac{A'(|\xi|^{-1/3}\rho)}{A'(-|\xi|^{-1/3}\eta)} = Ce^{i\gamma}.$$

A simple analogue of Lemma 2.1 shows that $C(x, \xi, \eta) \in S_{1,0}^0$ for $x \in \partial K$; $C = 1$ for $\eta \ge 0$. Thus

$$\frac{A'(|\xi|^{-1/3}\rho)}{A(|\xi|^{-1/3}\rho)} = \frac{C}{B}\frac{A'(-|\xi|^{-1/3}\eta)}{A(-|\xi|^{-1/3}\eta)} = \frac{C}{B}|\xi|^{1/3}q(\xi, \eta)$$

where

$$q(\xi, \eta) = |\xi|^{-1/3} \frac{A'}{A}(-|\xi|^{-1/3}\eta). \tag{4.7}$$

Since we easily see that $(A'/A)(\lambda) \in S^{1/2}_{1,0}(\mathbf{R})$, it easily follows that $q \in S^{0}_{1/3,0}$, and more precisely we have

$$|D^j_\eta D^\alpha_\xi q(\xi, \eta)| \leq c|\xi|^{-|\alpha|}(|\xi|^{1/3} + |\eta|)^{-j}, \qquad |\eta| \leq c_0|\xi|. \tag{4.8}$$

Indeed, (4.8) is a simple consequence of the chain rule, which we leave to the reader, remarking that this result and some of its consequences are investigated in Chapter XI, Section 4. Now recall from Section 3 that

$$h = 0\left(\left(\frac{|\eta|}{|\xi|}\right)^{\infty}\right), \qquad x \in \partial K \tag{4.9}$$

from which (4.8) easily yields

$$h(x, \xi, \eta)q(\xi, \eta) = \tilde{h}(x, \xi, \eta) \in S^0_{1,0}, \qquad x \in \partial K,$$

and also $|\tilde{h}|$ is small on $|\eta| \leq c_0|\xi|$ if c_0 is small enough. Now rewrite (4.5) as

$$u(t, x)|_{\mathbf{R}\times\partial K} = \iint [gB + i\tilde{h}C]e^{i(\theta+\gamma)}\hat{F}(\xi, \eta)\, d\xi\, d\eta = J(F). \tag{4.10}$$

Since as we saw in Section 3 we can take $|g| \geq c_1 > 0$ on $\eta = 0$, and θ is the generating function of a bijective canonical transformation from $\{|\eta| < c_0|\xi|\}$ to a small conic neighborhood of the characteristic variety, we have the following.

THEOREM 4.2. *The operator J is an elliptic Fourier integral operator taking distributions with wave front set in $|\eta| < c_0|\xi|$ to distributions with wave front set in a small conic neighborhood of the characteristic variety.*

The image of $|\eta| < c_0|\xi|$ under the canonical transformation $\mathscr{J}$ associated with J will contain $WF(f)$ if $WF(f)$ is contained in a sufficiently small conic neighborhood of a point $\zeta_0 \in T^*(\mathbf{R} \times \partial K)$ over which a grazing ray passes, as assumed at the beginning of this section. Now as shown in Chapter VIII, there is a Fourier integral operator with canonical transformation associated with $\mathscr{J}^{-1}$, call it J^{-1}, with the property that JJ^{-1} and $J^{-1}J$ each differ from the identity operator by a smoothing operator, when operating on distributions with wave front sets contained in a small conic neighborhood of the characteristic variety or $\eta = 0$, respectively. Consequently, in order to guarantee (1.2), mod C^∞, we need only take

$$F = J^{-1}(f). \tag{4.11}$$

Having shown that the parametrix (1.6) reproduces the correct boundary condition, we now show that $(\partial^2/\partial t^2) - \Delta)u \in C^\infty(\mathbf{R} \times \bar{\Omega})$. Indeed, the results of Section 3 on the eikonal and transport equations show that

$$\left(\frac{\partial^2}{\partial t^2} - \Delta\right)u = \iint \left[a \frac{A(|\xi|^{-1/3}\rho)}{A(-|\xi|^{-1/3}\eta)} + b|\xi|^{-1/3} \frac{A'(|\xi|^{-1/3}\rho)}{A(-|\xi|^{-1/3}\eta)}\right] e^{i\theta}\hat{F}(\xi, \eta)\, d\xi\, d\eta \tag{4.12}$$

and the amplitudes have asymptotic expansions $a \sim \sum a_j$, $b \sim \sum b_j$ with the properties

$$a_j = 0 \text{ and } b_j = 0 \qquad \text{for } \eta \le 0$$

$$|a_j(x, \xi, \eta)|,\ |b_j(x, \xi, \eta)| \le c_{jv}|\xi|^{-j}y^v, \text{ any } v, \qquad \text{for } \eta > 0$$

where y denotes the distance to ∂K. Then, in the region $\eta \ge 0$, the first term in the integrand (4.12) is estimated by

$$G = \left| a \frac{A(|\xi|^{-1/3}\rho)}{A(-|\xi|^{-1/3}\eta)} \right| \le c_v(|\xi|^{-v} + y^v)|e^{-(2/3)|\xi|(\alpha^{3/2} - (\alpha - a_0 y)^{3/2})}||\xi|.$$

We divide the estimate of G into parts, noting that the estimate for $\eta \le 0$ is trivial:

(i) $|\xi|^{-2/3}\eta \le 1$. Since $a \in S^0_{1,0}$ vanishes for $\eta < 0$, we see that

$$|a(x, \xi, \eta)| \le c_v\left(\frac{|\eta|}{|\xi|}\right)^v \le c_v|\xi|^{-v/3},$$

in this region, so G is rapidly decreasing.

(ii) $|\xi|^{-2/3}\eta \ge 1$. Consider two cases.

(a) $a_0 y \ge \alpha$. Here

$$\begin{aligned} G &\le c|\xi|^{-v+1} + cy^v|\xi|e^{-(2/3)|\xi|\alpha^{3/2}}, \\ &\le c|\xi|^{-v+1} + cy^v|\xi|e^{-(2/3)|\xi|^{1/2}(|\xi|^{-2/3}\eta)^{3/2}}, \\ &\le c|\xi|^{-v+1} + cy^v|\xi|e^{-(2/3)|\xi|^{1/2}}. \end{aligned}$$

(b) $0 \le a_0 y \le \alpha$. Here

$$G \le c|\xi|^{-v+1} + c|\xi| \sup_{0 \le y \le \alpha/a_0} y^v e^{-(2/3)|\xi|(\alpha^{3/2} - (\alpha - a_0 y)^{3/2}}.$$

We estimate the supremum of

$$q(y) = y^v e^{-(2/3)|\xi|(\alpha^{3/2} - (\alpha - a_0 y)^{3/2})}$$

as follows. Its critical points satisfy

$$y^2(\alpha - a_0 y) = \frac{v^2}{a_0^2}|\xi|^{-2},$$

which has two positive roots if $|\xi|^{-1/3}\eta$ is large. It is easily seen that $q'(y) > 0$ if $y = 0$ or $y = (\alpha/a_0)$. Hence if q assumes its maximum at one of its critical points in $[0, (\alpha/a_0)]$, it must be the smaller one,

$$y \approx \frac{\nu}{a_0}\alpha^{-1/2}|\xi|^{-1} = \frac{\nu}{a_0}(|\xi|^{-2/3}\eta)^{-1/2}|\xi|^{-5/6}.$$

In this case we get $G \le c_\nu|\xi|^{-5\nu/6+1}$. The case when q assumes its maximum at the end point $y = (\alpha/a_0)$ has been taken care of in (a).

Since ν can be taken arbitrarily large, it follows that the first term in the integrand of (4.12) is rapidly decreasing. A similar analysis holds for the second term, so (since $\hat{F}$ has only polynomial growth) the integral is absolutely convergent. Thus $((\partial^2/\partial t^2) - \Delta)u$ is continuous on $\mathbf{R} \times \bar{\Omega}$. Applying (x, t) derivatives to (4.12), we see that the same results hold, so $((\partial^2/\partial t^2) - \Delta)u \in C^\infty(\mathbf{R} \times \bar{\Omega})$, as desired.

The final step in showing that u given by (1.6) differs from the exact solution by an element of $C^\infty(\mathbf{R} \times \bar{\Omega})$ is to show that such a u is C^∞, up to ∂K, for $t < t_0$, where $f \in \mathscr{E}'(\mathbf{R} \times \partial K)$ is supported on $t \ge t_0$. As this result follows the analysis of the singularities of u given by (1.6), we turn to that study.

As shown in Section 2, given the results on Fourier integral operators with singular phase functions, from Chapter VIII, Section 6, we can read off $WF(u)$ in terms of $WF(F)$, if u is given by (1.6). Of course, $WF(F) = \mathscr{J}^{-1}WF(f)$. Comparing these two relates $WF(u)$ to $WF(f)$ via a canonical relation, which we can analyze as follows.

If $WF(f)$ is contained in a small conic neighborhood of a point $\zeta_0 \in T^*(\mathbf{R} \times \partial K)$ over which a grazing ray γ_0 passes, but in fact $WF(f)$ has empty intersection with the characteristic variety $\mathscr{J}(\{\eta = 0\})$, the ansatz (1.6) reduces to the ordinary geometric optics ansatz. Thus $WF(u)$ inside $\mathbf{R} \times \Omega$ consists of the single rays passing over points of $WF(f)$ in the hyperbolic region. Depending on whether $A = A_+$ or $A = A_-$ is chosen, these rays will either go in the positive or negative t direction as they go into $\mathbf{R} \times \Omega$ from $\mathbf{R} \times \partial\Omega$, as one can see from replacing $A_\pm$ by their asymptotic expansions in this context. Pick the Airy function A which yields up only rays passing forward into time. This establishes the canonical relation, relating $WF(u)$ to $WF(f)$, for $WF(f) \cap \mathscr{J}(\{\eta=0\}) = \emptyset$. The results of Section 2, and of Chapter VIII, Section 6, show that the general canonical relation between $WF(f)$ and $WF(u)$, valid when $WF(f)$ does intersect the characteristic variety, is contained in the *closure* of this previous canonical relation. Since the bicharacteristic flow of $T^*(\mathbf{R} \times \partial K)$ is continuous (just not smooth), we can deduce that in general, $WF(u)$ consists of the forward going rays passing over elements of $WF(f)$ in the hyperbolic region of $T^*(\mathbf{R} \times \partial K)$, plus the forward going halves of the grazing rays passing over elements of $WF(f)$ in the characteristic variety. If one regards the operators in Section 2 as a one parameter family of operators on $\mathbf{R} \times \partial K$, y being the parameter, we also get appropriate smoothness up to the boundary. This completes the discussion of regularity of the solutions.

§5. The Neumann Operator

In this section we want to analyze the normal derivative $\partial u/\partial \nu$ on $\mathbf{R} \times \partial K$ of the solution u to the outgoing Dirichlet problem for the wave equation, (1.1)–(1.3). Using this analysis we shall discuss constructing parametrices for the Neumann boundary problem and for the electric field on the exterior of a perfect conductor.

Assuming u is given by (1.6), we have

$$(5.1) \quad \frac{\partial}{\partial \nu} u\Big|_{\partial K} = \iint [\rho_\nu g - \theta_\nu h + ih_\nu] C e^{i(\theta + \gamma)} q(\xi, \eta) \hat{F}(\xi, \eta)\, d\xi\, d\eta$$

$$+ \iint [i\theta_\nu g + i|\xi|^{-1} \rho\rho_\nu h + g_\nu] B e^{i(\theta + \gamma)} \hat{F}(\xi, \eta)\, d\xi\, d\eta,$$

where $q(\xi, \eta)$ is given by (4.7). It is convenient to set $q(\xi, \eta)\hat{F}(\xi, \eta) = \widehat{QF}(\xi, \eta)$, $Q = q(D) \in OPS^0_{1/3,0}$, and write (5.1) as

$$(5.2) \qquad \frac{\partial}{\partial \nu} u\Big|_{\partial K} = K_1 QF + K_2 F$$

where K_1 and K_2 are Fourier integral operators, each of whose associated canonical transformation is $\mathcal{J}$, that associated with J.

As shown in Section 3, ρ_ν is nonvanishing on ∂K, for $|\eta| < c_0|\xi|$, while θ_ν vanishes on ∂K at $\eta \geq 0$. Thus we see from (5.1) that $J^{-1}K_1 = A_1$ is an elliptic operator in $OPS^1_{1,0}$ while $J^{-1}K_2 = A_2 \in OPS^1_{1,0}$ has principal symbol which vanishes at $\eta = 0$. Thus

$$(5.3) \qquad A_2(t, x, \xi, \eta) = \eta\alpha_0 + \beta_0$$

with $\alpha_0, \beta_0 \in S^0_{1,0}$. Now define the Neumann operator N as

$$Nf = \frac{\partial u}{\partial \nu}\Big|_{\mathbf{R} \times \partial K}.$$

We see from (5.2) that $Nf = (K_1 Q + K_2)J^{-1}f$, and the above argument yields

$$(5.4) \qquad N = J(A_1 Q + A_2)J^{-1}.$$

We thus have an expression for the Neumann operator. Note that we cannot appeal to Egorov's theorem to get a simplification of this operator, since $A_1Q + A_2 \in OPS^1_{1/3,0}$ is being conjugated by a Fourier integral operator, and Egorov's theorem does not apply in this context. Actually, making use of additional structure of Q, we shall prove that $N \in OPS^1_{1/3,2/3}$.

In order to discuss the Neumann boundary condition, it is convenient to rewrite (5.4) as

$$N = J(A_1 + A_2Q^{-1})QJ^{-1}. \tag{5.5}$$

Now $Q^{-1} \in OPS^{1/3}_{1/3,0}$, but from (5.3) it easily follows that $A_2Q^{-1} \in OPS^1_{1/3,0}$ and its symbol is small on $|\eta| < c_0|\xi|$ for c_0 small. Thus $A_1 + A_2Q^{-1} \in OPS^1_{1/3,0}$ is elliptic, near $\eta = 0$, so $(A_1 + A_2Q^{-1})^{-1} \in OPS^{-1}_{1/3,0}$ exists. Consequently, N is hypoelliptic, and an inverse mod smoothing operators is

$$N^{-1} = JQ^{-1}(A_1 + A_2Q^{-1})^{-1}J^{-1}. \tag{5.6}$$

Now it is an easy matter to construct a parametrix for a solution to the Neumann boundary problem on the exterior of the convex obstacle K, of the form

$$\left(\frac{\partial^2}{\partial t^2} - \Delta\right)u = 0 \quad \text{on} \quad \mathbf{R} \times (\mathbf{R}^n \backslash K),$$

$$\frac{\partial}{\partial \nu} u\big|_{\mathbf{R}\times\partial K} = g \in \mathscr{E}'(\mathbf{R} \times \partial K),$$

$$u = 0 \quad \text{for } t \ll 0.$$

Simply take u to be given by (1.6), solving (1.1) through (1.3), with

$$f = N^{-1}g.$$

In a similar fashion we can treat the boundary value problem for the electromagnetic field on the exterior of a smooth, convex, perfectly conducting obstacle $K \subset \mathbf{R}^3$. For example, the electric field satisfies the equations

$$\left(\frac{\partial^2}{\partial t^2} - \Delta\right)E = 0 \quad \text{on} \quad \mathbf{R}^3\backslash K,$$

$$\operatorname{div} E = 0 \text{ and } \nu \times E = 0 \quad \text{on} \quad \partial K \times \mathbf{R}$$

as mentioned in Chapter IX. So to analyze the diffraction of electric waves, in this case, it suffices to construct a parametrix for

$$\left(\frac{\partial^2}{\partial t^2} - \Delta\right)E = 0 \quad \text{on} \quad \mathbf{R}^3\backslash K, \tag{5.7}$$

$$\text{div } E|_{\mathbf{R}\times\partial K} = G_1 \in \mathscr{E}'(\mathbf{R}\times\partial K), \tag{5.8}$$

$$\nu \times E|_{\mathbf{R}\times\partial K} = G_2 \in \mathscr{E}'(\mathbf{R}\times\partial K), \tag{5.9}$$

$$E = 0 \qquad \text{for} \qquad t \ll 0. \tag{5.10}$$

Here G_2 takes values in the tangent bundle to ∂K. Picking a (local) basis of tangent vectors to ∂K, we can think of G_2 as taking values in $\mathbf{R}^2$. G_1 takes values in $\mathbf{R}^1$.

We will assume $WF(G_j)$ are contained in a small conic neighborhood of a point $\zeta_0 \in T^*(\mathbf{R}\times\partial K)$ over which a grazing ray passes, the other cases being simpler to handle (see Exercise 3.1 of Chapter IX). So define E by (1.6),

$$E = \iint\left[g\frac{A(|\xi|^{-1/3}\rho)}{A(-|\xi|^{-1/3}\eta)} + ih|\xi|^{-1/3}\frac{A'(|\xi|^{-1/3}\rho)}{A(-|\xi|^{-1/3}\eta)}\right]e^{i\theta}\hat{F}(\xi,\eta)\,d\xi\,d\eta \tag{5.11}$$

where now F takes values in $\mathbf{R}^3$, and as usual the problem is to solve for F in terms of G_1 and G_2. In analogy with (5.1), we have

$$\begin{aligned}(5.12)\quad \text{div } E|_{\mathbf{R}\times\partial K} &= \iint[g\,\nabla\rho - h\,\nabla\theta + i\,\nabla h]\cdot Ce^{i(\theta+\gamma)}\widehat{QF}(\xi,\eta)\,d\xi\,d\eta \\ &\quad + \iint[ig\,\nabla\theta + i|\xi|^{-1}\rho h\,\nabla\rho + \nabla g]\cdot Be^{i(\theta+\gamma)}\hat{F}(\xi,\eta)\,d\xi\,d\eta \\ &= L_1QF + L_2F\end{aligned}$$

where L_j are Fourier integral operators of order one. Note that the principal amplitudes of L_1 and L_2 at $\eta=0$ are $g\,\nabla\rho\cdot$ and $ig\,\nabla\theta\cdot$ respectively, and $\nabla\rho\cdot\nabla\theta = 0$ by (1.12). Meanwhile, (4.10) yields

$$\begin{aligned}(5.13)\quad \nu\times E|_{\mathbf{R}\times\partial K} &= \iint[gB + i\tilde{h}C]\nu\times e^{i(\theta+\gamma)}\hat{F}(\xi,\eta)\,d\xi\,d\eta, \\ &= MF,\end{aligned}$$

and the principal amplitude of M at $\eta = 0$ is $g\nu\times$. Now to solve the system

$$\begin{aligned}(L_1Q + L_2)F &= G_1 \qquad (\text{mod } C^\infty), \\ MF &= G_2 \qquad (\text{mod } C^\infty),\end{aligned} \tag{5.14}$$

it is convenient to split F into components tangent and normal to ∂K; indeed, set

$$F = \pi_1F_1 + \pi_2F_2$$

where F_1 takes values in $\mathbf{R}^1$, F_2 in $\mathbf{R}^2$, and $\pi_1:\mathbf{R}^1 \to$ normal bundle to ∂K, $\pi_2:\mathbf{R}^2 \to$ tangent bundle to ∂K, isomorphically, at least near sing supp G_j. We can rewrite (5.14) as

$$(5.15)\qquad \begin{aligned}(L_1Q\pi_1 + L_2\pi_1)F_1 + (L_1Q\pi_2 + L_2\pi_2)F_2 &= G_1,\\ M\pi_1F_1 + M\pi_2F_2 &= G_2.\end{aligned}$$

Note that the principal amplitudes of $L_2\pi_1$, $L_1\pi_2$, and $M\pi_1$ at $\eta = 0$ are, respectively, $ig\,\nabla\theta\cdot\pi_1$, $g\,\nabla\rho\cdot\pi_2$, and $gv\times\pi_1$, all of which vanish. Now multiplying (5.15) by J^{-1} yields

$$(5.16)\qquad \begin{aligned}(\tilde{A}_1Q\pi_1 + \tilde{A}_2\pi_1)F_1 + (\tilde{A}_1Q\pi_2 + \tilde{A}_2\pi_2)F_2 &= J^{-1}G_1 = \tilde{G}_1,\\ \tilde{B}\pi_1F_1 + \tilde{B}\pi_2F_2 &= J^{-1}G_2 = \tilde{G}_2\end{aligned}$$

where $\tilde{A}_1, \tilde{A}_2, \tilde{B} \in OPS^1_{1,0}$, and the principal symbols of $\tilde{A}_2\pi_1$, $\tilde{A}_1\pi_2$, and $\tilde{B}\pi_1$ all vanish at $\eta = 0$. As for the principal symbols of $\tilde{A}_1\pi_1$, $\tilde{A}_2\pi_2$, and $\tilde{B}\pi_2$, we see that

$$(5.17)\qquad \begin{aligned}\sigma_{\tilde{A}_1\pi_1} &= \alpha_1\,\nabla\rho\cdot\pi_1 = \tilde{\alpha}_1 &&\text{at } \eta = 0,\\ \sigma_{\tilde{A}_2\pi_2} &= \alpha_2\,\nabla\theta\cdot\pi_2 &&\text{at } \eta = 0,\end{aligned}$$

$$(5.18)\qquad \sigma_{\tilde{B}\pi_2} = \alpha_3 v\times\pi_2 \qquad \text{at } \eta = 0,$$

where α_j, $\tilde{\alpha}_1$ are nonvanishing scalars. Note that $\sigma_{B\pi_2}$ is an *injective* map of $\mathbf{R}^2$ into $\mathbf{R}^3$. To analyze (5.16), it is convenient to multiply both sides by Q^{-1}, and set $F_2 = Q\Phi_2$. We get

$$(5.19)\qquad \begin{aligned}&(Q^{-1}\tilde{A}_1Q\pi_1 + Q^{-1}\tilde{A}_2\pi_1)F_1\\ &\qquad + (Q^{-1}\tilde{A}_1Q\pi_2Q + Q^{-1}\tilde{A}_2\pi_2Q)\Phi_2 = Q^{-1}\tilde{G}_1 = \tilde{\tilde{G}}_1,\\ &Q^{-1}\tilde{B}\pi_1F_1 + Q^{-1}\tilde{B}\pi_2Q\Phi_2 = Q^{-1}\tilde{G}_2 = \tilde{\tilde{G}}_2.\end{aligned}$$

In order to simplify (5.19), it is convenient to note the following.

LEMMA 5.1. *If $P \in OPS^m_{1,0}$, then $P - Q^{-1}PQ \in OPS^{m-1/3}_{1/3,0}$.*

Proof. If $P = p(x, D)$, the symbol of $Q^{-1}PQ - P$ is asymptotic to

$$\sum_{\alpha>0} \frac{1}{\alpha!}(D^\alpha_{\xi,\eta}\tilde{q}(\xi,\eta))(D^\alpha_x p(x,\xi,\eta))\tilde{q}(\xi,\eta)^{-1}$$

where $\tilde{q}(\xi,\eta) = q(\xi,\eta)^{-1} = |\xi|^{1/3}\tilde{r}(|\xi|^{-1/3}\eta)$, with

$$\tilde{r}(s) = r(s)^{-1} = \frac{A(s)}{A'(s)}.$$

An elementary calculation verifies that this belongs to $S^{m-1/3}_{1/3,0}$.

Consequently, if R_j denotes the sum of an element of $OPS^1_{1/3,0}$ with symbol small for $|\eta| < c_0|\xi|$ and an element of $OPS^{2/3}_{1/3,0}$, and we see that

(5.19) is of the form

$$(5.20)\qquad \begin{aligned} (\tilde{A}_1\pi_1 + R_1)F_1 + (\tilde{A}_2\pi_2 + R_2)\Phi_2 &= \tilde{\tilde{G}}_1, \\ R_3F_1 + (\tilde{B}\pi_2 + R_4)\Phi_2 &= \tilde{\tilde{G}}_2. \end{aligned}$$

the ellipticity of (5.20), near $\eta = 0$, would follow from the ellipticity of

$$(5.21)\qquad \begin{pmatrix} \tilde{A}_1\pi_1 & \tilde{A}_2\pi_2 \\ 0 & \tilde{B}\pi_2 \end{pmatrix}.$$

But clearly if $\binom{\sigma}{\omega}$ belongs to the kernel of the symbol of (5.21) at $\eta = 0$ (5.17) and (5.18) imply $\sigma = 0$ and $\omega = 0$, so the ellipticity of (5.21) for $|\eta| < c_0|\xi|$ is easily established. This implies the solvability mod C^∞ of (5.14), with $WF(F) \subset \mathscr{J}^{-1}(WF(G_1) \cup WF(G_2))$. The construction of the parametrix for (5.7) through (5.10) is complete.

Having examined applications to the Neumann and electromagnetic boundary value problems, we return to a further analysis of the Neumann operator $N = J(A_1Q + A_2)J^{-1}$. To avoid some subscripts, let us replace A_1 by A, A_2 by B, so

$$(5.22)\qquad N = J(AQ + B)J^{-1}.$$

It is convenient to decompose $AQ + B$ into a sum of two terms, one supported very near $\eta = 0$, say on $|\eta| \le c|\xi|^a$, and the other supported on $|\eta| \ge (1/2)c|\xi|^a$. In order to do this, we look a little more closely at the operator $Q \in OPS^0_{1/3,0}$.

First, we note that

$$\sigma_Q = |\xi|^{-1/3}\frac{A'}{A}(-|\xi|^{-1/3}\eta)$$

belongs to $\mathscr{N}^0_{1/3}$, where the symbol class $\mathscr{N}^m_\rho$ is defined to be the set of symbols $p(x, \xi, \eta)$ such that

$$(5.23)\qquad |D_x^\beta D_\eta^j D_\xi^\alpha p(x, \xi, \eta)| \le c|\xi|^{m-|\alpha|}(|\xi|^\rho + |\eta|)^{-j}$$

for $|\eta| < |\xi|$, $p \in S^m_{1,0}$ on any open conic set not intersecting $\{\eta = 0\}$. A systematic study of $OP\mathscr{N}^m_\rho$ and the larger algebra $OP\mathscr{M}^m_\rho$ is made in the next chapter, in Section 4, where it is established that $Q \in OP\mathscr{N}^0_{1/3}$, (which the reader could now take as an exercise), so we shall be brief here.

Pick $\varphi_2 \in C_0^\infty(\mathbf{R})$ with $\varphi_2(s) = 1$ for $|s| \le 1$, and let $\varphi_1(s) = 1 - \varphi_2(s)$. Pick a with $1/2 < a < 1$, and let $\varphi_j(\xi, \eta) = \varphi_j(|\xi|^{-a}\eta)$. One easily verifies that $\varphi_j(\xi, \eta) \in \mathscr{N}^0_a$. Furthermore, as shown in Chapter XI, Propositions 4.14 and 4.15, we have

$$(5.24)\qquad Q_2 = \varphi_1(D)Q \in OP\mathscr{N}^0_a$$

and

$$Q_2 = \varphi_2(D)Q \in OP\mathcal{N}_{1/3}^{-(1-a)/2}. \tag{5.25}$$

Now we have $AQ + B = (AQ_1 + B) + AQ_2$ where the first term belongs to $OP\mathcal{N}_a^1 \subset OPS_{a,0}^1$ and the second term belongs to $OP\mathcal{N}_{1/3}^{-(1-a)/2+1}$. Egorov's theorem from C apter VIII applies to the first term, yielding

$$N_1 = J(AQ_1 + B)J^{-1} \in OPS_{a,1-a}^1 (\tfrac{1}{2} < a < 1). \tag{5.26}$$

Meanwhile, as will be shown in Chapter XI, a weak form of Egorov's theorem holds for $OP\mathcal{M}_\rho^m$, which contains $OP\mathcal{N}_\rho^m$. Namely, for any elliptic Fourier integral operator K, $0 < \rho \leq 1/2$, we have

$$K(OP\mathcal{M}_\rho^m)K^{-1} \subset OPS_{\rho,1-\rho}^{m+\varepsilon} \tag{5.27}$$

where $\varepsilon > 0$ can be taken arbitrarily small. It follows that

$$N_2 = J(AQ_2)J^{-1} \in OPS_{1/3,2/3}^{1/2+a/2+\varepsilon}, \tag{5.28}$$

and furthermore $KN_2K^{-1} \in OPS_{1/3,2/3}^{1/2+a/2+\varepsilon}$ for any elliptic Fourier integral operator K.

Combining (5.26) and (5.28) easily yields

$$N \in OPS_{1/3,2/3}^1. \tag{5.29}$$

We do not want to emphasize this result, since $OPS_{1/3,2/3}^1$ does not contain very much structure. Nevertheless, as we shall see in the next section, the methods used to obtain this will be useful in obtaining more interesting results.

We also note that the symbol $q_1(\xi, \eta)$ of $Q_1 = \varphi_1(D)Q$ has the property that

$$D_\eta^j D_\xi^\alpha q_1(\xi, \eta) \in \mathcal{N}_a^{-|\alpha|-aj-(1-a)/2} \subset S_{a,0}^{-(1-a)/2-a(j+|\alpha|)}$$

as is easily verified. This allows a better estimate on the symbol of N_1 than one merely gets from $AQ_1 + B \in OPS_{a,1-a}^1$. In fact, as Exercise 1.8 of Chapter VIII shows, we have

$$\sigma_{N_1}(t, x, \zeta) = \sigma_A \sigma_{Q_1} + \sigma_B \bmod S_{a,1-a}^{2-2a-(1-a)/2} \tag{5.30}$$

where the right-hand side of (5.30) is evaluated at the image of (t, x, ζ) under the canonical transformation $\mathcal{J}^{-1}$. This is an improvement over having identity modulo $S_{a,1-a}^{2-2a}$.

It is actually a little more convenient to write down the principal symbol of $N_1' = J(AQ + B)\varphi_1(D)J^{-1}$; note that $N_1 = N_1' + N_1''$ with $N_1'' = JB\varphi_2(D)J^{-1}$. Since the principal symbol of B vanishes on $\eta = 0$, it is easy to see that $B\varphi_2(D) \in OP\mathcal{N}_a^a \subset OPS_{a,1-a}^a$. Thus $N_1'' \in OPS_{a,1-a}^a$ has relatively low order. Now we can evaluate N_1' by substituting into (1.6) and using the fact that, on supp $\varphi_1(\xi, \eta)$, the Airy function $A(-|\xi|^{-1/3}\eta)$ can be replaced by its asymptotic expansion. This yields

$$(5.31)\quad N_1'(JF) = J(AQ + B)\varphi_1(D)J^{-1}(JF)$$

$$= \iint \left[g \frac{A(|\xi|^{-1/3}\rho)}{A(-|\xi|^{-1/3}\eta)} + ih|\xi|^{-1/3} \times \frac{A'(|\xi|^{-1/3}\rho)}{A(-|\xi|^{-1/3}\eta)} \right] (i\theta_v) e^{i\theta} \widehat{\varphi_1(D)F}\, d\xi\, d\eta$$

$$+ \iint \left[|\xi|^{-1/3}\rho_v g \frac{A'(|\xi|^{-1/3}\rho)}{A(-|\xi|^{-1/3}\eta)} + i(|\xi|^{-1/3}\rho_v) h|\xi|^{-1/3} \times \frac{A''(|\xi|^{-1/3}\rho)}{A(-|\xi|^{-1/3}\eta)} \right] e^{i\theta} \widehat{\varphi_1(D)F}\, d\xi\, d\eta + \cdots$$

$$= \iint \left[g \frac{A(|\xi|^{-1/3}\rho)}{A(-|\xi|^{-1/3}\eta)} + ih|\xi|^{-1/3} \frac{A'(|\xi|^{-1/3}\rho)}{A(-|\xi|^{-1/3}\eta)} \right] \times (i\theta_v + i(|\xi|^{-1}\rho)^{1/2}\rho_v)\varphi_1(|\xi|^{-a}\eta) \cdot e^{i\theta}\hat{F}\, d\xi\, d\eta + \cdots$$

where the dots indicate lower order terms. Indeed, one obtains

$$(5.32)\quad N_1'(JF) = \iint \left\{ (i\theta_v + i(|\xi|^{-1}\rho)^{1/2}\rho_v) \left[g + ih|\xi|^{-1/3} \frac{A'(|\xi|^{-1/3}\rho)}{A(|\xi|^{-1/3}\rho)} \right] \times B\varphi_1(|\xi|^{-a}\eta) + R \right\} e^{i(\theta+\gamma)}\hat{F}\, d\xi\, d\eta$$

where, with certain $a_j \in S^1_{1,0}$,

$$(5.33)\qquad R = \sum_{j=1}^{4} a_j(-|\xi|^{-1/3}\rho)^{-3j/2}\varphi_1(|\xi|^{-a}\eta) \bmod \mathcal{N}^0_a.$$

One sees that $(|\xi|^{-1/3}\rho)^{-3j/2}\varphi_1(|\xi|^{-a}\eta) \in \mathcal{N}_a^{-(3j/2)(a-1/3)}$. Hence $R \in \mathcal{N}_a^{(3/2)(1-a)} \subset \mathcal{N}_a^{3/4-\varepsilon}$, if a is slightly bigger than 1/2. It follows that the symbol of N_1' is

$$(5.34)\qquad n_1'(t, x, \zeta) = i(\theta_v + (|\xi|^{-1}\rho)^{1/2}\rho_v)\sigma_{\varphi_1} \bmod S^{(3/2)(1-a)}_{a,1-a}$$

where the right side of (5.34) is evaluated at (t, x, ξ, η) with $(y, \xi, \eta) = \mathcal{J}^{-1}(t, x, \zeta)$. Note that, even though $\theta_v + (|\xi|^{-1}\rho)^{1/2}\rho_v$ is not smooth at $\eta = 0$, since ρ vanishes there, $(\theta_v + (|\xi|^{-1}\rho)^{1/2}\rho_v)\varphi_1(|\xi|^{-a}\eta) \in S^1_{a,0}$.

Now, as we have seen in our treatment of the eikonal equation, $\varphi^+ = \theta + (2/3)|\xi|^{-1/2}\rho^{3/2}$ solves the ordinary eikonal equation of geometrical optics, exactly in the hyperbolic region, and to infinite order in the elliptic region, and $(\partial/\partial v)\varphi^+ = \theta_v + (|\xi|^{-1}\rho)^{1/2}\rho_v$.

Combining (5.28) with $N_1'' \in OPS^a_{a,1-a}$ and (5.31), and taking a slightly larger than 1/2, we obtain the following conclusion about N, which refines (5.29).

PROPOSITION 5.1. *The Neumann operator N is a pseudodifferential operator with symbol*

$$\sigma_N(t, x, \zeta) = i(\theta_\nu + (|\xi|^{-1}\rho)^{1/2}\rho_\nu)\sigma_{\varphi_1} \bmod S^{3/4+\varepsilon}_{1/3,2/3} \tag{5.35}$$

$$= i\frac{\partial\varphi^+}{\partial\nu}\sigma_{\varphi_1} \bmod S^{3/4+\varepsilon}_{1/3,2/3}$$

where $i(\partial\varphi^+/\partial\nu)\sigma_{\varphi_1} \in S^1_{a,1-a}$, and the right-hand side of (5.35) *is evaluated at (t, x, ξ, η) with $(y, \xi, \eta) = \mathscr{J}^{-1}(t, x, \zeta)$.*

We emphasize that Proposition 5.1 by itself does not give a terribly incisive analysis of the operator N. Indeed, the very fact that we were able to analyze the part of $AQ + B$ near $\eta = 0$ in such a crude fashion makes $N^{-1} = JQ^{-1}(A + BQ^{-1})^{-1}J^{-1}$ not amenable to such treatment. Proposition 5.1 is useful in so far as it provides a tool for the analysis of the Kirchhoff approximation in the next section. We might add that, in treating scattering problems for the Neumann boundary condition, we are very fortunate in often being able to reduce our study to $(AQ + B)^{-1}\Psi$, where the symbol of Ψ is relatively small near $\eta = 0$.

Exercises

5.1. If $P \in OPS^1$, show that $[P, N] \in OPS^{1+1/3+\varepsilon}_{1/3,2/3}$. Consider higher cummutators.

5.2. Using the fact that the principal symbol of B in (5.22) vanishes to infinite order at $\eta = 0$, rewrite the Neumann operator as

$$N = J(A'Q + B')J^{-1}$$

with $A' \in OPS^1_{1,0}$ elliptic and $B' \in OPS^0_{1,0}$.

§6. The Kirchhoff Approximation

Suppose K is a smooth-bounded strictly convex subset of $\mathbf{R}^3$, and we consider a plane wave solution to the wave equation on $\mathbf{R}^3 \backslash K$, $v = v(t, x, \omega)$, defined by

$$\left(\frac{\partial^2}{\partial t^2} - \Delta\right)v = 0 \quad \text{on} \quad \mathbf{R}^3\backslash K,$$

$$v|_{\mathbf{R}\times\partial K} = 0, \tag{6.1}$$

$$v = \delta(t - x\cdot\omega) \quad \text{for } t \ll 0.$$

We can write $v = \delta(t - x\cdot\omega) + w$ where the "scattered" wave $w = w(t, x, \omega)$ satisfies the equations

$$\left(\frac{\partial^2}{\partial t^2} - \Delta\right) w = 0 \qquad \text{on} \qquad \mathbf{R}^3 \backslash K,$$

$$(6.2) \qquad w|_{\mathbf{R}\times\partial K} = -\delta(t - x\cdot\omega),$$

$$w = 0 \qquad \text{for } t \ll 0.$$

The last condition is called the "outgoing" condition on w.

A fundamental result of scattering theory is that w decays exponentially on any compact subset $\mathcal{O}$ of $\mathbf{R}^3$, as $t \to \infty$. See Lax and Phillips [1], Morawetz and Ludwig [1], or Majda and Taylor [2]. Moreover, for t large w is smooth on $\mathcal{O}$ and all its derivatives decay exponentially. Granted this result, we can consider

$$(6.3) \qquad u_s(\lambda, x, \omega) = \int_{-\infty}^{\infty} e^{-i\lambda t} w(t, x, \omega)\, dt.$$

It easily follows that u_s satisfies the following reduced wave equation:

$$(\Delta + \lambda^2)u_s = 0 \qquad \text{on} \qquad \mathbf{R}^3\backslash K,$$

$$(6.4) \qquad u_s|_{\partial K} = -e^{i\lambda x\cdot\omega},$$

$$\frac{\partial u_s}{\partial r} - i\lambda u_s = o(|x|^{-1}), \qquad u_s = 0(|x|^{-1}), \qquad |x| \to \infty.$$

The last condition is known as the Sommerfeld radiation condition. It turns out that solutions to (6.4) are unique, but since u_s is well defined by (6.3) we needn't pursue this.

Now the exact solution of (6.4) can be recovered from knowledge of $u_s|_{\partial K}$ and

$$\left.\frac{\partial u_s}{\partial \nu}\right|_{\partial K}.$$

In fact, Green's formula yields

$$(6.5) \qquad u_s = \int_{\partial K} \left[u_s(y) \frac{\partial}{\partial \nu} G_\lambda(x - y) - \frac{\partial u_s}{\partial \nu} G_\lambda(x - y)\right] dS(y)$$

where

$$G_\lambda(x) = \frac{e^{i\lambda|x|}}{|x|}.$$

Out of this one can obtain a host of formulas from classical scattering theory, but it is necessary to know

$$\left.\frac{\partial u_s}{\partial \nu}\right|_{\partial K}.$$

What has commonly been used in calculations is the Kirchhoff approximation:

$$\left.\frac{\partial u_s}{\partial \nu}\right|_{\partial K} \approx -i\lambda|\nu \cdot \omega|e^{-i\lambda x \cdot \omega} \tag{6.6}$$

where $\nu = \nu(x)$ is the unit outward normal on ∂K. In this section we shall use the analysis of the Neumann operator from the previous section to examine the validity of (6.6), and estimate the error.

From (6.3) and the definition of w we see that

$$\left.\frac{\partial u_s}{\partial \nu}\right|_{\partial K} = \int_{-\infty}^{\infty} e^{-i\lambda t} N\gamma(t, x)\, dt \tag{6.7}$$

where $\gamma \in \mathscr{E}'(\mathbf{R} \times \partial K)$ is given by $\gamma = \delta(t - x \cdot \omega)|_{\mathbf{R} \times \partial K}$. Another convenient representation of

$$\left.\frac{\partial u_s}{\partial \nu}\right|_{\partial K}$$

is given by the following.

LEMMA 6.1. $N(e^{-i\lambda(x \cdot \omega - t)}) = e^{i\lambda t}(\partial u_s/\partial \nu)|_{\partial K}$.

Proof. We first remark that, due to the exponential decay result, N is well defined on $\mathscr{S}'(\mathbf{R}) \otimes \mathscr{D}'(\partial K)$, so $N(e^{-i\lambda(x \cdot \omega - t)})$ is well defined. Also, N commutes with the translation operator T_τ given by $(T_\tau f)(t, x) = f(t + \tau, x)$. Now

$$\begin{aligned} e^{i\lambda t} \left.\frac{\partial u_s}{\partial \nu}\right|_{\partial K} &= \int_{-\infty}^{\infty} e^{-i\lambda s}(N\gamma)(s + t, x)\, ds = \int_{-\infty}^{\infty} e^{-i\lambda s} T_s(N\gamma)(t, x)\, ds \\ &= N\left(\int_{-\infty}^{\infty} e^{-i\lambda s} T_s\gamma(t, x)\, ds\right) \qquad (\text{by } T_s N = N T_s) \\ &= N\left(\int_{-\infty}^{\infty} e^{-i\lambda s}\delta(x \cdot \omega - t - s)\, ds\right) = N(e^{-i\lambda(x \cdot \omega - t)}). \end{aligned}$$

This completes the proof.

Since, as already mentioned, the distribution kernel of N has the form $n(t - t', x, x')$, and is C^∞ for $(t, x) \neq (t', x')$ and exponentially decreasing, with all its derivatives, as $|t - t'| \to \infty$, we can localize Lemma 6.1.

COROLLARY 6.2. *Pick $\varphi \in C_0^\infty(\mathbf{R})$ with $\varphi(t) = 1$ for $|t| \leq 2$. Then, for $|t| \leq 1$,*

$$\left.\frac{\partial u_s}{\partial \nu}\right|_{\partial K} = e^{-i\lambda t} N(\varphi(t) e^{-i\lambda(x \cdot \omega - t)}) + 0(\lambda^{-\infty}). \tag{6.8}$$

Proof. It suffices to observe that $N((1-\varphi)e^{-i\lambda(x\cdot\omega-t)}) = 0(\lambda^{-\infty})$ for $|t| \leq 1$, which is immediate.

Consequently, if we write

$$N(\varphi(t)e^{-i\lambda(x\cdot\omega-t)}) = K(t, x, \lambda)e^{-i\lambda(x\cdot\omega-t)}, \tag{6.9}$$

we see that, if $\Psi \in C_0^\infty(-1, 1)$ is picked with $\int\Psi(t)\,dt = 1$, we have

$$\left.\frac{\partial u_s}{\partial \nu}\right|_{\partial K} = K(x, \lambda)e^{-i\lambda x\cdot\omega} + 0(\lambda^{-\infty}) \tag{6.10}$$

where

$$K(x, \lambda) = \int_{-1}^{1} \Psi(t)K(t, x, \lambda)\,dt. \tag{6.11}$$

Now, as shown in the previous section, for any elliptic Fourier integral operator L, we can write (with $1/2 < a < 1$)

$$LNL^{-1} = LN_1L^{-1} + LN_2L^{-1} = P_1 + P_2 \tag{6.12}$$

where $P_1 \in OPS^1_{a,1-a}$, $P_2 \in OPS^{1/2+a/2+\varepsilon}_{1/3,2/3}$. In particular, choose $Lu(t, x) = u(\chi(t, x))$ where $\chi(t, x)$ is a diffeomorphism of $\mathbf{R} \times \partial K$ chosen so that, in a convenient coordinate system on $\mathbf{R} \times \partial K$, $L(e^{-i\lambda(x\cdot\omega-t)}) = e^{i\lambda\alpha(t,x)}$ with $\alpha(t, x)$ a *linear* function of (t, x). We see that

$$LN(\varphi(t)e^{-i\lambda(x\cdot\omega-t)}) = LNL^{-1}(\varphi e^{i\lambda\alpha}) = P_1(\varphi e^{i\lambda\alpha}) + P_2(e^{i\lambda\alpha}).$$

Now $P_1(\varphi e^{i\lambda\alpha}) = \tilde{K}_1(t, x, \lambda)e^{i\lambda\alpha}$ where $\tilde{K}_1 \in S^1_{a,1-a}$ has a complete asymptotic expansion which one can in principle compute, since one can compute the full symbol of P_1. Meanwhile $P_2(\varphi e^{i\lambda\alpha}) = \tilde{K}_2(t, x, \lambda)e^{i\lambda\alpha}$ with $\tilde{K} \in S^{1/2+a/2+\varepsilon}_{1/3,2/3}$. Consequently,

$$N(\varphi(t)e^{-i\lambda(x\cdot\omega-t)}) = [K_1(t, x, \lambda) + K_2(t, x, \lambda)]e^{-i\lambda(x\cdot\omega-t)}$$

where $K_j(t, x, \lambda) = \tilde{K}_j((\chi^{-1}(t, x), \lambda)$. Thus we have

$$K(t, x, \lambda) = K_1(t, x, \lambda) + K_2(t, x, \lambda) \tag{6.13}$$

where $K_1 \in S^1_{a,1-a}$ and its full symbol can in principal be calculated by Egorov's theorem, while $K_2(t, x, \lambda) \in S^{1/2+a/2+\varepsilon}_{1/2,2/3}$. In particular, we have

$$\left.\frac{\partial u_s}{\partial \nu}\right|_{\partial K} = K_1(x, \lambda)e^{-i\lambda x\cdot\omega} + 0(\lambda^{1/2+a/2+\varepsilon}) \tag{6.14}$$

where

$$K_1(x, \lambda) = \int_{-1}^{1} \Psi(t)K_1(t, x, \lambda)\,dt \in S^1_{a,1-a}.$$

To analyze $K_1(t, x, \lambda)e^{-i\lambda\psi} = N_1(\varphi e^{-i\lambda\psi})$, where $\psi = t - x\cdot\omega|_{\mathbf{R}\times\partial K}$, we use the reasoning leading up to Proposition 5.1; if we pick a slightly larger than 1/2, we see that $N_1(\varphi e^{-i\lambda\psi}) = L^{-1}LJ(AQ_1 + B)J^{-1}L^{-1}(\varphi e^{-i\lambda\alpha})$ and $M_1 = LJ(AQ_1 + B)(LJ)^{-1}$ has symbol

$$(6.15)\quad \begin{aligned} m_1(t, x, \zeta) &= i(\theta_\nu + (|\xi|^{-1}\rho)^{1/2}\rho_\nu)\sigma_{\varphi_1}((\chi^* \circ \mathscr{J})^{-1}(t, x, \zeta)) \bmod S^{3/4+\varepsilon}_{1/3,2/3} \\ &= i\frac{\partial\varphi^+}{\partial\nu}\sigma_{\varphi_1}((\chi^* \circ \mathscr{J})^{-1}(t, x, \zeta)) \bmod S^{3/4+\varepsilon}_{1/3,2/3}. \end{aligned}$$

(In fact, we can say this of the symbol $m(t, x, \zeta)$ of LNL^{-1}.) Meanwhile, off the shadow boundary (where ordinary geometrical optics specifies $N(\varphi e^{-i\lambda\psi})$) one easily sees that

$$(6.16)\qquad \frac{\partial}{\partial\nu}\varphi^+ = -\lambda|\nu\cdot\omega|$$

the left side of (6.16) being evaluated at (x, t, ξ, η), where (y, ξ, η) is the image in $T^*(\mathbf{R}^n)$ of $(x, t, \lambda\nabla_{x,t}\psi) \in T^*(\mathbf{R}\times\partial K)$, under the canonical transformation $\mathscr{J}$. Since both sides of (6.16) are homogeneous and continuous, the formula extends by continuity, to yield

$$(6.17)\qquad \theta_\nu + (|\xi|^{-1}\rho)^{1/2}\rho_\nu = -\lambda|\nu\cdot\omega|,$$

the left side being evaluated at the appropriate spot. Consequently, we have

$$(6.18)\qquad K(t, x, \lambda) = -i\lambda|\nu\cdot\omega|\varphi_1 + 0(\lambda^{3/4+\varepsilon}).$$

Note that $\varphi_1 = 1$ for $|\eta| \geq c|\xi|^a$. Now the conic manifold $\Sigma = \{(t, x; s\nabla_{t,x}\psi) \in T^*(\mathbf{R}\times\partial K): s > 0\}$ intersects the characteristic variety $\mathscr{J}\{(\eta = 0\})$ only over the shadow boundary $\nu\cdot\omega = 0$. Furthermore, Σ belongs to the "hyperbolic region" of $T^*(\mathbf{R}\times\partial K)$, over which two rays pass, except over $\nu\cdot\omega = 0$, and in fact Σ and $\mathscr{J}(\{\eta = 0\})$ is tangent to each other at their intersection, having precisely second order contact there. It follows that, in (6.18), $\varphi_1 = 1$ for $|\nu\cdot\omega|^2 \geq c\lambda^{a-1}$. Thus

$$(6.19)\qquad |\nu\cdot\omega|\varphi_1 = |\nu\cdot\omega| + 0(\lambda^{-(1-a)/2}).$$

From (6.18) and (6.19) we have the following conclusion:

THEOREM 6.3. *The correction to the Kirchhoff approximation is given by*

$$\left.\frac{\partial u_s}{\partial\nu}\right|_{\partial K} = K(x, \lambda)e^{-i\lambda x\cdot\omega} + 0(\lambda^{-\infty})$$

where

$$(6.20)\qquad |K(x, \lambda) + i\lambda|\nu\cdot\omega|\,| \leq c\lambda^{3/4+\varepsilon}.$$

We now aim at a refinement of this result, estimating $(\nu \cdot \omega)^k(K(x, \lambda) + i\lambda|\nu \cdot \omega|)$, with the idea in mind that, outside a fixed neighborhood of $\nu \cdot \omega = 0$, the error (6.20) is $0(1)$. Thus we want an estimate on the "boundary layer" outside which the Kirchhoff approximation does a better job.

We begin with an estimate on

$$(\nu \cdot \omega)^k K_2(t, x, \lambda)e^{-i\lambda\psi} = R_k(t, x, \lambda) = (\nu \cdot \omega)^k JAQ_2J^{-1}(\varphi e^{-i\lambda\psi}).$$

Recall that $A \in OPS^1$, $Q_2 \in OP\mathcal{N}_{1/3}^{a/2-1/2}$. Pick $b > 1/2$ and let $\Phi_b \in OPS_{b,0}^0$ have symbol $\varphi_b(\xi, \eta) = \rho(|\xi|^{-b}\eta)$, given $\rho \in C_0^\infty(\mathbf{R})$, $\rho(s) = 1$ for $|s| \le 1$. Thus $Q_2 = Q_2\Phi_b \bmod OPS^{-\infty}$. Also, set $Tu = J^{-1}(\nu \cdot \omega)(Ju)$; $T \in OPS^0$. Thus,

$$\begin{aligned} (6.21)\quad R_k(t, x, \lambda) &= JT^kAQ_2\Phi_bJ^{-1}(\varphi e^{-i\lambda\psi}) \\ &= JAQ_2T^k\Phi_bJ^{-1}(\varphi e^{-i\lambda\psi}) + J([T^k, AQ_2]\Phi_b)J^{-1}(e^{-i\lambda\psi}) \\ &= R_k'(t, x, \lambda) + R_k''(t, x, \lambda). \end{aligned}$$

In order to analyze $R_k'(t, x, \lambda) = JAQ_2T^k\Phi_bJ^{-1}(\varphi e^{-i\lambda\psi})$, pick $P_b \in OPS_{b,1-b}^0$ such that

$$e^{i\lambda\psi}P_b(\varphi e^{-i\lambda\psi}) = \varphi \bmod S^{-\infty}$$

and such that the symbol of P_b has order $-\infty$ outside the set

$$\Sigma_b = \{(t, x, \zeta) \in T^*(\mathbf{R} \times \partial K) : \text{dist}\,((t, x, \zeta), \Sigma) \le c|\zeta|^b\}$$

where

$$\Sigma = \{(t, x, s\nabla_{t,x}\psi) : s > 0\}.$$

With $\tilde{P}_b = J^{-1}P_bJ$, we have

$$(6.22)\qquad \begin{aligned} R_k'(t, x, \lambda) &= J(AQ_2)(T^k\Phi_b\tilde{P}_b)J^{-1}(\varphi e^{-i\lambda\psi}) \bmod 0(\lambda^{-\infty}) \\ &= L^{-1}B_1B_2(\varphi e^{-i\lambda\alpha}) \end{aligned}$$

where $B_1 = (LJ)(AQ_2)(LJ)^{-1} \in OPS_{1/3,2/3}^{1/2+a/2+\varepsilon}$. We now look at $B_2 = (LJ)(T^k\Phi_b\tilde{P}_b)(LJ)^{-1}$. It will suffice to analyze $T^k\Phi_b\tilde{P}_b$ and apply Egorov's theorem. To look at this, consider the sets $\mathcal{J}^{-1}(\{\nu \cdot \omega = 0\}) = S_1$, $\{\eta = 0\} = S_2$, and $\mathcal{J}^{-1}(\Sigma) = S_3$, where $\mathcal{J}$ is the canonical transformation associated with the Fourier integral operator J. Since, as already mentioned, it is clear that Σ intersects the characteristic variety $\mathcal{J}(S_2)$ precisely over the shadow boundary $\nu \cdot \omega = 0$, we have $S_2 \cap S_3 \subset S_1$. Furthermore, S_2 and S_3 are tangent on their intersection, having precisely second order contact there. It follows that the symbol of $\Phi_b\tilde{P}_b$ has order $-\infty$ outside a neighborhood $\mathfrak{U}_b$ of S_1 of the form

$$(6.23)\qquad \mathfrak{U}_b = \{(t, x, \zeta) : \text{dist}\,((t, x, \zeta), S_1) \le c|\zeta|^{(1+b)/2}\}.$$

Since the symbol of T belongs to S^0 and its principal part vanishes on S_1, it is easy to infer that $T^k\Phi_b\tilde{P}_b \in OPS_{b,1-b}^{-(k/2)(1-b)}$, if $(k/2)(1-b) \le 1$. Consequently $B_2 \in OPS_{b,1-b}^{-(k/2)(1-b)}$, provided $b > 1/2$ and $(k/2)(1-b) \le 1$. If, in fact, $b > 2/3$, so $1 - b < 1/3$, we deduce that $B_1B_2 \in OPS_{1/3,2/3}^{1/2+a/2-(k/2)(1-b)}$. Thus, (6.22) yields

$$R_k'(t, x, \lambda) = 0(\lambda^{1/2+a/2-k/6+\varepsilon}) \qquad (k \le 6). \tag{6.24}$$

Next we analyze $R_k''(t, x, \lambda) = J([T^k, AQ_2]\Phi_b\tilde{P}_b)J^{-1}(\varphi e^{-i\lambda\psi})$. It's convenient to write, with $(Ad\ T) \cdot B = [T, B]$,

$$[T^k, AQ_2] = \sum_{\ell=1}^{k} \binom{k}{\ell}[(Ad\ T)^\ell \cdot (AQ_2)]T^{k-\ell}.$$

Now, for $\ell \ge 1$,

$$(Ad\ T)^\ell(AQ_2) \in OP\mathcal{N}_{1/3}^{1/2+a/2-2/3-(\ell-1)/3}.$$

Meanwhile, by the previous argument, we have

$$T^{k-\ell}\Phi_b\tilde{P}_b \in OPS_{b,1-b}^{-(k-\ell)(1-b)\cdot 1/2}.$$

Thus $R_k''(t, x, \lambda)$ is a sum of terms of the form (for $1 \le \ell \le k$)

$$\begin{aligned}&L^{-1}(LJ)OP\mathcal{N}_{1/3}^{1/2+a/2-2/3-(\ell-1)/3}(LJ)^{-1} \times \\ &\quad (LJ)OPS_{b,1-b}^{-k(1-b)/2+\ell(1-b)/2}(LJ)^{-1}(\varphi e^{-i\lambda\alpha}) \subset OPS_{1/3,2/3}^{1/2+a/2-(k+2)/6+\varepsilon}(\varphi e^{i\lambda\alpha})\end{aligned}$$

(taking $b > 2/3$ but close). Thus we have $R_k''(t, x, \lambda) = 0(\lambda^{1/2+a/2-(k+2)/6+\varepsilon})$ which, combined with (6.24), yields

$$(\nu \cdot \omega)^k K_2(t, x, \lambda) = 0(\lambda^{1/2+a/2-k/6+\varepsilon}), \qquad k \le 5. \tag{6.25}$$

In order to get good estimates on $(\nu \cdot \omega)^k K_1(t, x, \lambda)$, it will be necessary to get a refinement of the description of the remainder term in the complete symbol of conjugates of $AQ_1 + B$ by Fourier integral operators, further improving (5.30) and (6.15).

In fact, given a smooth conic hypersurface S of $T^*(\mathbf{R}^n)$, let

$$\tilde{\mathcal{N}}_{a,1-a}^{m,k}(S) = \{p(x, \zeta): |D_x^\beta D_\zeta^\alpha p(x, \zeta)| \le c|\zeta|^{m+|\beta|}(|\zeta|^a + \text{dist}(\zeta, S))^{-|\alpha|-|\beta|+k}\}$$

(for $\mathbf{R} \times \partial K$, replace x by (t, x), of course). The following results are easily established.

(6.26) $\quad \sigma_Q = q(\xi, \eta) \in \tilde{\mathcal{N}}_{1/3,2/3}^{-1/2,1/2}(S_2), \qquad$ with $S_2 = \{\eta = 0\}$.

(6.27) $\quad$ If $p \in \tilde{\mathcal{N}}_{1/3,2/3}^{m,k}(S_2)$, $\varphi_1 = \varphi_1(|\xi|^{-a}\eta)$ the previous cut-off, $a > 1/3$, then $p\varphi_1 \in \tilde{\mathcal{N}}_{a,1-a}^{m,k}(S_2)$.

(6.28) $\quad$ If $p \in \tilde{\mathcal{N}}_{a,1-a}^{m,k}(S)$, then $D_\zeta^\alpha p \in \tilde{\mathcal{N}}_{a,1-a}^{m,k-|\alpha|}(S)$.

In particular, the assertion

$$D^{\alpha}_{\xi,\eta} q_1 \in \mathcal{N}_a^{-(1-a)/2 - a|\alpha|} \subset S_{a,1-a}^{-(1-a)/2 - a|\alpha|}$$

for $|\alpha| \geq 1$, used to derive (5.30), can be sharpened to

$$q_1 \in \tilde{\mathcal{N}}_{a,1-a}^{-1/2,1/2}(S_2) \tag{6.29}$$

which implies

$$D^{\alpha}_{\xi,\eta} q_1 \in \tilde{\mathcal{N}}_{a,1-a}^{-1/2,1/2-|\alpha|}(S_2). \tag{6.30}$$

Note that $\tilde{\mathcal{N}}_{a,1-a}^{m,k}(S) \subset S_{a,1-a}^{m-a|k|}$ if $k < 0$, so $\tilde{\mathcal{N}}_{a,1-a}^{-1/2,1/2-|\alpha|} \subset S_{a,1-a}^{-(1-a)/2-a|\alpha|}$ if $|\alpha| \geq 1$.

Furthermore, the proof of Egorov's theorem (see Chapter VIII, Exercise 1.6) shows that, for an elliptic Fourier integral operator K with associated canonical transformation $\mathscr{K}$,

$$K\, OP\tilde{\mathcal{N}}_{a,1-a}^{m,k}(S) K^{-1} = OP\tilde{\mathcal{N}}_{a,1-a}^{m,k}(\mathscr{K}(S)), \qquad \tfrac{1}{2} < a \leq 1. \tag{6.31}$$

Also, the complete symbol of such an operator KPK^{-1} differs from $\sigma_K \circ \mathscr{K}^{-1}$ by an element of $\tilde{\mathcal{N}}_{a,1-a}^{m+1,k-2}(\mathscr{K}(S))$. In particular, we see that

$$K(AQ_1 + B)K^{-1} \in OP\tilde{\mathcal{N}}_{a,1-a}^{1/2,1/2}(\mathscr{K}(S_2)), \tag{6.32}$$

noting that since the principal symbol of B vanishes on $\eta = 0$ we have $B \in \tilde{\mathcal{N}}_{a,1-a}^{0,1}(S_2) \subset \tilde{\mathcal{N}}_{a,1-a}^{1/2,1/2}(S_2)$. Also, the complete symbol of $K(AQ_1 + B)K^{-1}$ differs from $\sigma_{(AQ_1+B)} \circ \mathscr{K}^{-1}$ by an element of $\tilde{\mathcal{N}}_{a,1-a}^{3/2,-3/2}(\mathscr{K}(S_2))$.

As in Section 5, it is more convenient to do computations on $(AQ + B)\varphi_1(D) = (AQ_1 + B) - B\varphi_2(D)$. There we noted that $B\varphi_2(D) \in OP\mathcal{N}_a^a \subset OPS_{a,1-a}^a$. Here we note that $T^k B\varphi_2(D)\tilde{P}_a \in OPS_{a,1-a}^{a-(k(1-a)\wedge 1)}$, which follows from the fact that the symbol of $B\varphi_2(D)\tilde{P}_a$ has order $-\infty$ outside a set $\mathfrak{U}_a$ as described in (6.23). Consequently, we see that

$$(\nu \cdot \omega)^k J B\varphi_2(D) J^{-1}(\varphi e^{-i\lambda\psi}) = 0(\lambda^{a-(k(1-a)\wedge 1)}). \tag{6.33}$$

Of course, $(AQ + B)\varphi_1(D)$ enjoys the same nice properties as $AQ_1 + B$. Namely, with $K = LJ$, $K(AQ + B)\varphi_1(D)K^{-1} \in OP\tilde{\mathcal{N}}_{a,1-a}^{1/2,1/2}(\mathscr{K}(S_2))$ and its complete symbol differs from

$$L(t, x, \zeta) = \sigma_{(AQ+B)\varphi_1(D)} \circ \mathscr{K}^{-1}(t, x, \zeta)$$

by an element $M(t, x, \zeta) \in \tilde{\mathcal{N}}_{a,1-a}^{3/2,-3/2}(\mathscr{K}(S_2))$. Let $M = M(t, x, D)$.

First we estimate

$$(\nu \cdot \omega)^k L^{-1} M(t, x, D)(\varphi e^{i\lambda\alpha}) = (\nu \cdot \omega)^k \tilde{M}(t, x, D)(\varphi e^{-i\lambda\psi}),$$

with

$$\tilde{M}(t, x, D) = L^{-1} M(t, x, D) L \in OP\tilde{\mathcal{N}}_{a,1-a}^{3/2,-3/2}(\mathscr{J}(S_2)).$$

With $P_a \in OPS^0_{a,1-a}$ as before, so $e^{i\lambda\psi}P_a(\varphi e^{-i\lambda\psi}) = \varphi \bmod S^{-\infty}$, this is equal to $(v \cdot \omega)^k \tilde{M} P_a(\varphi e^{-i\lambda\psi})$. Since the symbol of P_a is concentrated near $\mathscr{J}(S_3)$, the symbol of $\tilde{M}$ relatively small away from $\mathscr{J}(S_2)$, and $\mathscr{J}(S_2) \cap J(S_3) \subset \mathscr{J}(S_1)$, where $v \cdot \omega = 0$, these two sets $\mathscr{J}(S_2)$ and $\mathscr{J}(S_3)$ having exactly second order contact at their intersection, it is easy to see that

$$(6.34) \qquad (v \cdot \omega)^k \tilde{M} P_a \in OPS^{(3/2)(1-a)(1-k/3)}_{a,1-a}, \qquad 0 \le k \le 3.$$

Consequently, for a slightly bigger than 1/2,

$$(6.35) \quad (v \cdot \omega)^k L^{-1} M(t, x, D)(\varphi e^{i\lambda\alpha}) = 0(\lambda^{3/4 - k/4 + \varepsilon}), \qquad 0 \le k \le 3.$$

Now an examination of (5.32) shows that

$$(6.36) \quad L(t, x, \zeta) = i(\theta_v + (|\xi|^{-1}\rho)^{1/2}\rho_v)\sigma_{\varphi_1} \bmod \tilde{\mathcal{N}}^{3/2,-3/2}_{a,1-a}(\mathscr{K}(S_2))$$

where the right-hand side of (6.36) is evaluated at (t, x, ξ, η); $(y, \xi, \eta) = \mathscr{K}(t, x, \zeta)$. Thus, using (6.35), we have

$$(6.37) \quad (v \cdot \omega)^k J(AQ + B)\varphi_1(D) J^{-1}(\varphi e^{-i\lambda\psi})$$
$$= -i\lambda(v \cdot \omega)^k |v \cdot \omega| \varphi_1 + 0(\lambda^{3/4 - k/4 + \varepsilon}), \qquad k \le 3.$$

Since φ_1 on the right side of (6.37) is equal to 1 for $|v \cdot \omega|^2 \ge c\lambda^{a-1}$, we see that

$$(v \cdot \omega)^k |v \cdot \omega| \varphi_1 = (v \cdot \omega)^k |v \cdot \omega| + 0(\lambda^{-(1/2)(k+1)(1-a)}).$$

Thus (6.37) yields

$$(6.38) \quad (v \cdot \omega)^k J(AQ + B)\varphi_1(D) J^{-1}(\varphi e^{-i\lambda\psi})$$
$$= -i\lambda(v \cdot \omega)^k |v \cdot \omega| + 0(\lambda^{3/4 - k/4 + \varepsilon}), \qquad k \le 3.$$

Finally, combining (6.25), (6.33), and (6.38), we sharpen up Theorem 6.3 to our final result.

THEOREM 6.4. *The correction to the Kirchhoff approximation of Theorem* 6.1 *satisfies*

$$|K(x, \lambda) + i\lambda|v \cdot \omega|\,| \le c\lambda^{3/4+\varepsilon}(1 + \lambda^{1/6}|v \cdot \omega|)^{-9/2}.$$

§7. References to Further Work

Results in this chapter have been used to solve some basic problems in the classical theory of scattering of waves in $\mathbf{R}^3$ by an obstacle $K \subset \mathbf{R}^3$, smooth, compact, and strictly convex. In particular, problems arising in the study of the scattering amplitude, defined below, have been treated.

For Dirichlet boundary conditions, the scattering operator and scattering amplitude are defined as follows. Let $w = w(t, x, \omega)$ and $u_s = u_s(\lambda, x, \omega)$

be given by (6.2) and (6.3). The formula (6.4) for u_s implies that, as $|x| \to \infty$, $u_s(x) \sim (e^{i\lambda r}/r)a_s$, $r = |x|$, where a_s, given by

$$a_s(\theta, \omega, \lambda) = \lim_{r \to \infty} re^{-i\lambda r}u_s(\lambda, r\theta, \omega) \tag{7.1}$$

is called the scattering amplitude. The scattering operator is an operator defined on $\mathscr{E}'(S^2 \times \mathbf{R})$ to $\mathscr{D}'(S^2 \times \mathbf{R})$ with kernel $\hat{a}_s(\theta, \omega, s-t)$, where

$$\hat{a}_s(\theta, \omega, t) = \int_{-\infty}^{\infty} a_s(\theta, \omega, \lambda)e^{i\lambda t}\, d\lambda. \tag{7.2}$$

Using formula (6.5), one gets the integral representations

$$a_s(\theta, \omega, \lambda) = \int_{\partial K} e^{-i\lambda\theta \cdot y}\left[i\lambda(\nu \cdot \theta)u_s(\lambda, y, \omega) + \frac{\partial}{\partial \nu} u_s(\lambda, y, \omega)\right] dS(y), \tag{7.3}$$

$$\hat{a}_s(\theta, \omega, t) = \int_{\partial K} \left(\frac{\partial}{\partial \nu} - (\nu \cdot \theta)\frac{\partial}{\partial t}\right) w(t + y \cdot \theta, y, \omega)\, dS(y). \tag{7.4}$$

One wants to analyze the asymptotic behavior of a_s as $\lambda \to \infty$ and to analyze the singularities of $\hat{a}_s$. To locate the singularities of $\hat{a}_s$, use (7.4) and recall that $(\partial/\partial\nu)w = Nw$, N being pseudolocal, so $WF((N - (\nu \cdot \theta)(\partial/\partial t))w) \subset WF(\delta)$. Consequently, it is easy to show that $\hat{a}_s$ is only singular for those values t_0 for which the hypersurface $t = y \cdot \omega$ intersects the surface $t = t_0 + y \cdot \theta$, nontransversally. For $\theta = \omega$ this happens precisely at $t_0 = 0$, where the two hypersurfaces coincide exactly; this "big" singularity in $\hat{a}_s(\theta, \theta, t)$ leads to the "diffraction peak" in $a_s(\theta, \theta, \lambda)$. For $\theta \neq \omega$, fixed, there are two such values of t, given by the min and max, respectively, of $y \cdot (\theta - \omega)$, $y \in \partial K$. At these two points $y_1(\theta, \omega)$ and $y_2(\theta, \omega)$ the normals to ∂K are parallel (resp. antiparallel) to $\theta + \omega$. One can show that in fact there is only a singularity at

$$t = \min_{y \in \partial K} y \cdot (\theta - \omega).$$

The method of stationary phase can be applied to asymptotically evaluate (7.3) for $\theta \neq \omega$, yielding

$$a_s(\theta, \omega, \lambda) = k(y_1)^{-1/2}e^{-i\lambda y_1 \cdot (\theta - \omega)}\frac{\nu \cdot \theta}{|\theta - \omega|} + 0(\lambda^{-1}) \tag{7.5}$$

with indeed a complete asymptotic expansion. See Majda [1], Majda and Taylor [2]. As for the case $\theta = \omega$, use of the Kirchhoff approximation, Theorem 6.3, gives

$$\begin{aligned} a_s(\theta, \theta, \lambda) &= 2i\lambda \int_{\partial K_+} \nu(x) \cdot \theta\, dS(x) + 0(\lambda^{3/4+\varepsilon}) \\ &= 2i\lambda \text{ Area } (\mathscr{S}(\partial K, \theta)) + 0(\lambda^{3/4+\varepsilon}) \end{aligned} \tag{7.6}$$

where $\partial K^+ = \{x \in \partial K : \nu \cdot \omega \leq 0\}$ is the *illuminated side of* ∂K, $\mathscr{S}(\partial K, \theta)$ is the "shadow projection" of ∂K in the direction θ. Such an estimate was first obtained by Majda and Taylor [1], before uniform estimates on the error of the Kirchhoff approximation were proved. A better error estimate could be deduced from Theorem 6.4, but in fact Melrose has shown that

$$a_s(\theta, \theta, \lambda) = 2i\lambda \text{ Area } (\mathscr{S}(\partial K, \theta)) + a_1 \lambda^{1/3} + \cdots \tag{7.7}$$

with decreasing powers of $\lambda^{1/3}$, and $\log \lambda$ terms. More recently, Melrose and Taylor [1] have made a uniform analysis of $a_s(\theta, \omega, \lambda)$, for ω near θ. This is obtained by analyzing the scattering operator $\hat{a}_s$ given by (7.4), as

$$\hat{A} = T(N+P)U$$

where N is the Neumann operator, $P \in OPS^1$, and T and U are defined by

$$TF(\theta, t) = \int_{\partial K} F(y, t + y \cdot \theta)\, dS(y); \qquad T: \mathscr{E}'(\mathbf{R} \times \partial K) \to \mathscr{E}'(\mathbf{R} \times S^2),$$

$$Uf(y, t) = \int_{S^2} f(\omega, t - y \cdot \omega)\, d\omega; \qquad U: \mathscr{E}'(\mathbf{R} \times S^2) \to \mathscr{E}'(\mathbf{R} \times \partial K).$$

T and U are both Fourier integral operators with folding canonical relations, briefly mentioned at the end of Section 5, Chapter VIII. Putting such operators into a standard form is an essential tool in analyzing the scattering amplitude. As is shown in Melrose and Taylor [1], there exist elliptic Fourier integral operators J and K and classical pseudo-differential operators P_1, P_2 such that a Fourier integral operator U with folding canonical relation microlocally has the form

$$U = J(P_1 \mathscr{A}i + P_2 \mathscr{A}i')K \tag{7.8}$$

where the "canonical" Fourier integral operators with folding canonical relations are defined by

$$(\mathscr{A}iu)\hat{\ }(\xi, \eta) = Ai(\xi_1^{-1/3}\eta)\hat{u}(\xi, \eta),$$
$$(\mathscr{A}i'u)\hat{\ }(\xi, \eta) = Ai'(\xi_1^{-1/3}\eta)\hat{u}(\xi, \eta).$$

In Melrose and Taylor [1] is also derived a complete asymptotic expansion for the corrected Kirchhoff approximation, improving Theorems 6.3 and 6.4 to

$$K(x, \lambda) \sim K^c(x, \lambda) + \sum_{j,k,\ell \geq 0} \varkappa_{jk\ell}(x, \lambda)\Psi_{jk}^{(\ell)}(\lambda^{1/3} Y_\omega(x)) \tag{7.9}$$

with $K^c \in S^0$, $\kappa_{jk\ell} \in S^{2/3 - j/3 - k - 2\ell/3 + a(\ell)}$, $a(0) = a(1) = 0$, $a(\ell) = 1$ for $\ell \geq 2$. Here $\Psi_{jk}(\lambda) \in S^{1-2j}$ so $\Psi_{jk}(\lambda^{1/3} Y_\omega) \in S_{2/3,\, 1/3}^{(1/3)(1-2j)+}$. The function $Y_\omega(x)$ vanishes to first order on the "shadow boundary" $\{\nu \cdot \omega = 0\}$.

Other boundary value problems are amenable to the treatment of this chapter, see Taylor [6], Taylor, Farris, and Yingst [1]. Yingst [1] has analyzed the error in the Kirchhoff approximation for the wave equation with Neumann boundary conditions.

If P is a compactly supported pseudodifferential operator on $\mathbf{R}^3\backslash K$, K convex as above, $e^{it\sqrt{-\Delta}}Pe^{-it\sqrt{-\Delta}}$ certainly preserves wave front sets. M. Farris has analyzed this as a pseudodifferential operator, extending Egorov's theorem to this case. Of course, the canonical transformation involved is not smooth, and the conjugated operator is not of classical type.

Propagation of singularities of solutions to boundary value problems in Ω, when $\partial\Omega$ is not assumed to be convex with respect to the bicharacteristics of P in Ω, has been analyzed by Anderson and Melrose [1] for $\partial\Omega$ bicharacteristically *concave*, and by Melrose and Sjöstrand [1] for general smooth $\partial\Omega$. Also, Melrose [6] has a theory of transformations of boundary value problems which obtains parametrices for both the grazing ray problems considered in this chapter and for the case of bicharacteristically concave $\partial\Omega$. An essential tool in this analysis is the solution in Melrose [3] of the problem of equivalence of glancing hypersurfaces. Two hypersurfaces $F = \{f = 0\}$ and $G = \{g = 0\}$ in $T^*\Omega$, intersecting transversally, are said to be glancing at $(x_0, \xi_0) \in F \cap G$ if

$$\{f, g\}(x_0, \xi_0) = 0,$$

$$\{f, \{f, g\}\}(x_0, \xi_0) \neq 0, \qquad \text{and} \qquad \{g, \{g, f\}\}(x_0, \xi_0) \neq 0.$$

It is shown that there is a homogeneous smooth canonical transformation locally sending (F, G) to a standard example. This equivalence, with certain refinements, is also useful for obtaining (7.7) and (7.8).

Cheeger and Taylor [1] consider propagation of singularities and parametrices for the wave equation on $\mathbf{R}^n\backslash C$, where C is a cone, with smooth base, vertex at 0, or a body smooth except at a finite number of points p_j, near which it is a cone with vertex at p_j, and also the case where C is a polyhedron.

CHAPTER XI

L^p and Hölder Space Theory of Pseudodifferential Operators

In this chapter we study the continuity of pseudodifferential operators on L^p and C^α spaces. We obtain such results from classical theorems on Fourier multipliers, which results are collected in the first section. The next two sections, discuss results for operators in $OPS^m_{1,0}$ and $OPS^m_{1,\delta}$; most of these results have been obtained by various authors, scattered throughout the literature, though perhaps a few results are stated in sharper form here. The classes $OPS^m_{\rho,0}$, successful in examining L^2 regularity of the heat operator and other hypoelliptic operators, are not well adapted to L^p spaces. In Section 4 we examine two smaller classes, $OP\mathcal{M}^m_\rho$ and $OP\mathcal{N}^m_\rho$, which are well behaved on L^p. The results there contain as special cases well-known L^p results on the heat equation. In Section 5 we consider Besov spaces and boundary regularity of solutions to elliptic equations, including Schauder estimates.

§1. Fourier Multipliers on L^p and Hölder Spaces

In this section we describe some results, due to Marcinkiewicz, Mikhlin, Hörmander, Stein, and Taibleson, on the behavior of operators

$$P(D)u = \int e^{ix\cdot\xi}p(\xi)\hat{u}(\xi)\,d\xi$$

on $L^p(\mathbf{R}^n)$ and $C^\alpha(\mathbf{R}^n)$. Note that $P(D)$ simply multiplies the Fourier transform of u by $p(\xi)$; hence $P(D)$ is called a Fourier multiplier. It can also be written as a convolution operator

$$P(D)u = \hat{p} * u.$$

The basic results on continuity of such an operator on L^p and C^α are merely stated here, and the reader is referred to various places in the literature for proofs. We do take Taibleson's theorem and show it is equivalent to a condition which is somewhat parallel to Hörmander's version of the Marcinkiewicz multiplier theorem; our Theorem 1.7 may conceivably be new.

Marcinkiewicz [1] studied the L^p continuity of convolution operators on the torus $\mathbf{T}^n$. Mikhlin [1] translated some of these results to the $\mathbf{R}^n$ setting, and proved the following.

THEOREM 1.1. *$P(D): L^p \to L^p(\mathbf{R}^n)$, $1 < p < \infty$, provided*

$$|\xi|^{|\alpha|}|D_\xi^\alpha p(\xi)| \le C_\alpha, \qquad |\alpha| \le \left[\frac{n}{2}\right] + 1. \tag{1.1}$$

Actually, Mikhlin needed the estimate (1.1) for $|\alpha| \le n$, but Hörmander [4] reduced the number of derivatives one needs to estimate, and also replaced the pointwise estimate (1.1) by L^2 estimates. Hörmander's theorem is the following.

THEOREM 1.2. *$P(D): L^p(\mathbf{R}^n) \to L^p(\mathbf{R}^n)$, $1 < p < \infty$, provided*

$$R^{-n} \int_{R < |\xi| < 2R} | |\xi|^{|\alpha|} p^{(\alpha)}(\xi)|^2 \, d\xi < C, \qquad |\alpha| \le \left[\frac{n}{2}\right] + 1 \tag{1.2}$$

with C independent of R, $0 < R < \infty$.

We can restate this result as follows. Let Ω denote the annular region $\Omega = \{\xi \in \mathbf{R}^n : 1 \le |\xi| \le 4\}$, and let $p_r(\xi) = p(r\xi)$. The following is easily seen to be equivalent to Theorem 1.2.

COROLLARY 1.3. *$P(D): L^p(\mathbf{R}^n) \to L^p(\mathbf{R}^n)$, $1 < p < \infty$, provided*

$$\|p_r\|_{H^{[n/2]+1}(\Omega)} \le C, \qquad 0 < r < \infty \tag{1.3}$$

where $C < \infty$ is independent of r.

This result can be sharpened as follows:

THEOREM 1.4. *$P(D): L^p(\mathbf{R}^n) \to L^p(\mathbf{R}^n)$, $1 < p < \infty$, provided*

$$\|p_r\|_{H^{n/2+\epsilon}(\Omega)} \le C, \qquad 0 < r < \infty \tag{1.4}$$

for some $\epsilon > 0$, where C is independent of r.

The proof Hörmander gives in [4] of Theorem 1.2 can easily be modified to establish Theorem 1.4.

Note that the index $(n/2) + \epsilon$ could not be decreased, since $P(D): L^2 \to L^2$ implies $\sup_\xi |p(\xi)| < \infty$.

The conditions of Theorems 1.1 through 1.4 are invariant under rotations. Often one encounters symbols for which one or more directions are distinguished, such as parametrices for the heat equation. The following theorem, in some respects closer in spirit to the original theorem of Marcinkiewicz, may be found in Stein [1, p. 109]

THEOREM 1.5. *$P(D): L^p(\mathbf{R}^n) \to L^p(\mathbf{R}^n)$, $1 < p < \infty$, provided*

$$|\xi^\alpha p^{(\alpha)}(\xi)| \le C_\alpha, \qquad \xi \in \mathbf{R}^n \tag{1.5}$$

for all multi-indices $\alpha = (\alpha_1, \ldots, \alpha_n)$ with each α_j either 0 *or* 1.

In particular, it is more than sufficient that (1.5) hold for $|\alpha| \le n$. Note that (1.5) differs from (1.1) in that the homogeneous function $|\xi|^{|\alpha|}$ is replaced by the (smaller) *polynomial* ξ^α. Thus relatively nasty behavior of $p(\xi)$ is permitted on the hyperplanes $\xi_j = 0$. An even stronger theorem (Theorem 6′) is given in Stein [1].

We now consider the behavior of $P(D)$ on $C^\alpha(\mathbf{R}^n)$. For $0 < \alpha < 1$, we define $C^\alpha(\mathbf{R}^n)$ to consist of those functions u on $\mathbf{R}^n$ such that

$$\|u\|_{C^\alpha} = \|u\|_{L^\infty} + \sup_{\substack{x,h \in \mathbf{R}^n \\ |h| \le 1}} |h|^{-\alpha} |u(x+h) - u(x)| < \infty.$$

Taibleson [1] has found a necessary and sufficient condition that $P(D): C^\alpha(\mathbf{R}^n) \to C^\alpha(\mathbf{R}^n)$. Here we let, for $s \in \mathbf{R}$

$$L_{1,s} = (1 - \Delta)^{s/2} L^1(\mathbf{R}^n).$$

THEOREM 1.6. *$P(D): C^\alpha(\mathbf{R}^n) \to C^\alpha(\mathbf{R}^n)$, $0 < \alpha < 1$, if and only if $\hat{p} \in L_{1,s}$ for some $s > 0$, and the following holds*:

$$\sup_{0<t<1} \|t\, \Delta e^{t\Delta} \hat{p}\|_{L^1} + \|\hat{p}\|_{L_{1,s}} < \infty. \tag{1.6}$$

It turns out that the Hilbert transform doesn't even preserve $C^\alpha(\mathbf{R}^1)$, and similarly Calderon-Zygmund operators do not preserve $C^\alpha(\mathbf{R}^n)$, for more or less spurious reasons having to do with trouble at infinity; if such a convolution kernel $\hat{p}$ were truncated, so as to have compact support, it would act on $C^\alpha(\mathbf{R}^n)$. In fact, for

$$\hat{p} = \frac{k_0(x)}{|x|^n},$$

the important term in (1.6),

$$\sup_{0<t<1} \|t\, \Delta e^{t\Delta} \hat{p}\|_{L^1},$$

is finite; the "unimportant" term $\|\hat{p}\|_{1,s}$ is infinite in this case. Clearly, any $\hat{p} \in \mathscr{E}'(\mathbf{R}^n)$ belongs to $L_{1,s}$ for $s > 0$ sufficiently large. Thus, assuming $\hat{p} \in \mathscr{E}'(\mathbf{R}^n)$, we have $P(D): C^\alpha(\mathbf{R}^n) \to C^\alpha(\mathbf{R}^n)$ if, and only if,

$$\sup_{0<t<1} \|t\, \Delta e^{t\Delta} \hat{p}\|_{L^1} < \infty. \tag{1.7}$$

In order to recast Taibleson's theorem, let Ω be the annular region considered before, and pick $\psi \in C_0^\infty(\Omega)$ with $\psi(\xi) = 1$ for $3/2 \le \xi \le 7/2$. (we may as well suppose ψ only depends on $|\xi|$.) Let $p_r(\xi) = p(r\xi)$ as before.

THEOREM 1.7. *Suppose $\hat{p} \in \mathscr{E}'(\mathbf{R}^n)$. Then $P(D): C^\alpha(\mathbf{R}^n) \to C^\alpha(\mathbf{R}^n)$, $0 < \alpha < 1$, if, and only if, for some C independent of r,*

$$\|\psi p_r\|_{\mathfrak{F}L^1} \le C, \qquad 0 < r < \infty. \tag{1.8}$$

Here the $\mathfrak{F}L^1$ norm is defined by $\|u\|_{\mathfrak{F}L^1} = \|\hat{u}\|_{L^1}$. Note that $\mathfrak{F}L^1$ is a Banach algebra under multiplication, since L^1 is a convolution algebra. Before we prove Theorem 1.7, we remark on the pleasing parallel between condition (1.8) and condition (1.4). In fact, the Sobolev theorem that elements of $H^{n/2+\epsilon}(\mathbf{R}^n)$ are continuous is usually proved by showing that they have integrable Fourier transforms, i.e., $H^{n/2+\epsilon}(\mathbf{R}^n) \subset \mathfrak{F}L^1$. Thus condition (1.8) is just a little weaker than condition (1.4).

Proof. Granted Taibleson's theorem, we need only show that (1.8) and (1.7) are equivalent. So suppose (1.8) holds. Suppose $\int_{-\infty}^{\infty} \psi(e^{-y}\xi)\, dy = 1$. Let $\mu_y(\xi) = \psi(\xi)p(e^y\xi)$, so (1.8) is equivalent to $\|\mu_y\|_{\mathfrak{F}L^1} \le C < \infty$, $-\infty < y < \infty$. We have

$$p(\xi) = \int_{-\infty}^{\infty} \mu_y(e^{-y}\xi)\, dy.$$

We desire to estimate

$$\|t|\xi|^2 e^{-t|\xi|^2} p(\xi)\|_{\mathfrak{F}L^1},$$

$0 < t < 1$. This is dominated by

$$t\int_{-\infty}^{\infty} \|\mu_y(e^{-y}\xi)|\xi|^2 e^{-t|\xi|^2}\|_{\mathfrak{F}L^1}\, dy = t\int_{-\infty}^{\infty} e^{2y}\|\mu_y(\xi)|\xi|^2 e^{-te^{2y}|\xi|^2}\|_{\mathfrak{F}L^1}\, dy$$

since $\|g\|_{\mathfrak{F}L^1} = \|g_r\|_{\mathfrak{F}L^1}$, $0 < r < \infty$. We dominate the integrand by

$$Ce^{2y}\|\mu_y\|_{\mathfrak{F}L^1}\|\,|\xi|^2 e^{-te^{2y}|\xi|^2}\|_{H^s(\Omega)} \le C'e^{2y}(1 + t^s e^{2ys})e^{-te^{2y}}$$

where we have picked $s > n/2$. Thus

$$\|t|\xi|^2 e^{-t|\xi|^2} p(\xi)\|_{\mathfrak{F}L^1} \le Ct\int_{-\infty}^{\infty} e^{2y}e^{-te^{2y}}\, dy + Ct^{1+s}\int_{-\infty}^{\infty} e^{2y(1+s)}e^{-te^{2y}}\, dy$$
$$= C'' < \infty.$$

This shows that (1.8) implies (1.7).

For the converse, note that (1.7) is equivalent to

$$\|\,|\xi|^2 e^{-|\xi|^2}\, p_r\|_{\mathfrak{F}L^1} \le C < \infty \tag{1.9}$$

at least for $r > 1$; but since we're assuming $\hat{p} \in \mathscr{E}'$, $p(\xi)$ is smooth, so we have (1.9) for $0 < r < \infty$. But clearly

$$\|\psi p_r\|_{\mathfrak{F}L^1} \le \|\psi|\xi|^{-2}e^{|\xi|^2}\|_{\mathfrak{F}L^1}\|\,|\xi|^2 e^{-|\xi|^2}p_r\|_{\mathfrak{F}L^1} = C_0\|\,|\xi|^2 e^{-|\xi|^2}p_r\|_{\mathfrak{F}L^1},$$

so the converse is easy.

We need one more result about the action of $(1-\Delta)^{m/2}$ on Hölder spaces, whose proof may be found in Stein [1].

THEOREM 1.8. *If $k+\alpha$ and $k+\alpha-m$ are both positive and nonintegral, $m \in \mathbf{R}$,*

$$(1-\Delta)^{m/2}: C^{k+\alpha}_{\text{comp}}(\mathbf{R}^n) \to C^{k+\alpha-m}_{\text{loc}}(\mathbf{R}^n).$$

Here, for $k = 0, 1, 2, \ldots, \alpha \in (0, 1)$, we set $C^{k+\alpha}(\mathbf{R}^n) = \{u \in C^k(\mathbf{R}^n): D^\beta u \in C^\alpha(\mathbf{R}^n), |\beta| \leq k\}$.

This concludes our discussion of basic results on Fourier multipliers on $\mathbf{R}^n$. Note that if $p(\xi) \in S^0_{1,0}(\mathbf{R}^n)$, then (1.1), is certainly satisfied. All subsequent conditions are weaker than (1.1) (except that (1.5) involves a slightly different range of α's), so they are satisfied. This will lead us in the next section to L^p and Hölder continuity results for $p(x, D) \in OPS^0_{1,0}$. The Marcinkiewicz theorem 1.5 will be useful for certain operators not in the class $OPS^0_{1,0}$.

In Section 3 we shall need the following result, known as the Marcinkiewicz interpolation theorem, which also plays a crucial role in the proof of Theorems 1.1 through 1.4.

THEOREM 1.9. *Let $T: C_0^\infty(\mathbf{R}^n) \to L^\infty(\mathbf{R}^n)$ satisfy the conditions*

$$\text{meas}\,\{x: |Tu(x)| > \lambda\} \leq C_1 \lambda^{-p} \|u\|_{L^p}, \tag{1.10}$$

$$\text{meas}\,\{x: |Tu(x)| > \lambda\} \leq C_2 \lambda^{-q} \|u\|_{L^q}, \tag{1.11}$$

where $1 \leq p < q$. Then $T: L^r(\mathbf{R}^n) \to L^r(\mathbf{R}^n)$, $p < r < q$, with operator norm determined by C_1, C_2, r, n.

Note that $T: L^q(\mathbf{R}^n) \to L^q(\mathbf{R}^n)$ implies (1.11); one says T is of weak type (q, q) if (1.11) is satisfied. Further generalizations of Theorem 1.10 exist; see Bergh and Löfström [1]. We shall use the result with $p = 1$, $T: L^2 \to L^2$.

§2. L^p and C^α Behavior of Operators in $OPS^m_{1,0}$

To be brief, the main result of this section is the following.

THEOREM 2.1. *Let $p(x, D) \in OPS^0_{1,0}$. Then*

$$p(x, D): L^p_{\text{comp}} \to L^p_{\text{loc}}, \qquad 1 < p < \infty, \qquad \text{and}$$
$$p(x, D): C^\alpha_{\text{comp}} \to C^\alpha_{\text{loc}}, \qquad 0 < \alpha < 1.$$

Here L^p_{comp} is the space of L^p functions with compact support, L^p_{loc} is the space of locally L^p functions, etc.

Proof. Let u have support in B_R, the ball of radius R. Since $p(x, D)$ is pseudolocal, it suffices to show that, if $u \in L^p$ (resp. $u \in C^\alpha$), then $p(x, D)u \in L^p$

on B_{2R} (resp., $p(x, D)u \in C^\alpha$ on B_{2R}). We denote by $\| \ \|_{\mathscr{L}}$ the operator norm of $T: C_0^\alpha(B_R) \to C^\alpha(B_{2R})$ $0 < \alpha < 1$, or of $T: L^p(B_R) \to L^p(B_{2R})$, $1 < p < \infty$. Theorems 1.1 and 1.7 imply that, for a Fourier multiplier $p_\eta(D)$,

$$\|p_\eta(D)\|_{\mathscr{L}} \le C \sup_\xi \sum_{|\beta| \le [n/2]+1} |\xi|^{|\beta|} |p_\eta^{(\beta)}(\xi)| \tag{2.1}$$

where C depends on α or p, as the case may be.

Again by the pseudolocal property, we can suppose that $p(x, \xi)$ vanishes for $|x| \ge 2R$. Now write

$$p(x, \xi) = (2\pi)^{-n} \int_{\mathbf{R}^n} e^{-ix\cdot\eta} p_\eta(\xi)\, d\eta \tag{2.2}$$

where

$$p_\eta(\xi) = \int_{\mathbf{R}^n} p(x, \xi) e^{ix\cdot\eta}\, dx. \tag{2.3}$$

It follows that

$$p(x, D)u = (2\pi)^{-n} \int_{\mathbf{R}^n} e^{-ix\cdot\eta} p_\eta(D)u\, d\eta.$$

Now as a multiplication operator on L^p, $e^{-ix\cdot\eta}$ has norm 1, while as a multiplication operator on C^α, $e^{-ix\cdot\eta}$ has norm $< C(1+|\eta|)^\alpha$. Therefore,

$$\|p(x, D)\|_{\mathscr{L}} \le \int_{\mathbf{R}^n} C(\eta) \|p_\eta(D)\|_{\mathscr{L}}\, d\eta \tag{2.4}$$

where $C(\eta) = 1$ in the case of L^p and $C(\eta) = C(1+|\eta|)^\alpha$ in the case of C^α. It remains only to show that $\|p_\eta(D)\|_{\mathscr{L}}$ is rapidly decreasing, as $|\eta| \to \infty$. To establish this, we estimate the right-hand side of (2.1). In fact, by (2.3),

$$\begin{aligned} \eta^\gamma p_\eta^{(\beta)}(\xi) &= \int_{\mathbf{R}^n} D_\xi^\beta p(x, \xi) D_x^\gamma e^{ix\cdot\eta}\, dx \\ &= \int_{\mathbf{R}^n} D_x^\gamma D_\xi^\beta p(x, \xi) e^{ix\cdot\eta}\, dx. \end{aligned} \tag{2.5}$$

Therefore,

$$|\eta^\gamma|\, |p_\eta^{(\beta)}(\xi)| \le C_{\beta\gamma}(1+|\xi|)^{-|\beta|}. \tag{2.6}$$

Summing over $|\gamma| \le N$ and applying (2.1) yields

$$\|p_\eta(D)\|_{\mathscr{L}} \le C_N(1+|\eta|)^{-N}, \qquad C_N = C_N(p) \text{ or } C_N(\alpha)$$

as desired; this estimate together with (2.4) yields $\|p(x, D)\|_{\mathscr{L}} < \infty$.

Using Theorem 2.1 together with the pseudolocal properties of $OPS_{1,0}^0$ and invariance under coordinate transformations, one easily obtains the following.

THEOREM 2.2. *Let M be a compact manifold. Let $p(x, D) \in OPS^0_{1,0}$ on M. Then*

$$p(x, D): L^p(M) \to L^p(M), \qquad 1 < p < \infty \qquad \text{and}$$
$$p(x, D): C^\alpha(M) \to C^\alpha(M), \qquad 0 < \alpha < 1.$$

Furthermore, their operator norms can be bound by

$$\|p(x, D)\|_{\mathscr{L}} \leq C \max_{\substack{|\alpha| \leq [n/2]+1 \\ |\beta| \leq n+1}} \sup_{(x,\xi) \in T^*(M)} |D_x^\beta D_\xi^\alpha p(x, \xi)| \, |\xi|^{|\alpha|} \tag{2.7}$$

where C depends on p or α.

Here $D_x^\beta D_\xi^\alpha p(x, \xi)$ is defined on one of some finite number of selected coordinate systems covering M. The bound (2.7), which follows from (2.4) through (2.6), is very important. It has the following consequence.

COROLLARY 2.3. *Let $p_j(x, \xi)$ be a bounded set of symbols on M, in $S^0_{1,0}$. Then $p_j(x, D)$ form a bounded family of operators on $L^p(M)$, $1 < p < \infty$, and on $C^\alpha(M)$, $0 < \alpha < 1$.*

We also easily obtain results on $p(x, D) \in OPS^m_{1,0}$ acting on the Sobolev spaces $\mathscr{L}^s_p$ and the Hölder spaces $C^{k+\alpha}$. For $k = 1, 2, 3, \ldots, 0 < \alpha < 1$, $C^{k+\alpha} = \{u \in C^k : D^\beta u \in C^\alpha, |\beta| = k\}$. If M is a compact manifold and k is a positive integer, recall

$$\mathscr{L}^k_p(M) = \{u \in L^p(M) : Pu \in L^p(M), P \in OPD^k\}.$$

It is easy to see that for $1 < p < \infty$, $u \in \mathscr{L}^k_p(M)$ if, and only if, $p(x, D)u \in L^p(M)$ for all $p(x, D) \in OPS^k_{1,0}$. Indeed, if $p(x, D) \in OPS^k_{1,0}$, one can write $p(x, D) = \sum a_j(x, D)q_j(x, D)$ with $q_j(x, D) \in OPD^k$ and $a_j(x, D) \in OPS^0_{1,0}$, and the result follows since we know that $a_j(x, D): L^p(M) \to L^p(M)$, $1 < p < \infty$. Also, for $s \in \mathbf{R}$,

$$\mathscr{L}^s_p(M) = \{u \in \mathscr{D}'(M) : p(x, D)u \in L^p(M) \text{ for all } p(x, D) \in OPS^s_{1,0}\}.$$

Since both $L^p(M)$ and $OPS^s_{1,0}$ are invariant under coordinate transformations, so is $\mathscr{L}^s_p(M)$. We also have the following simple result.

PROPOSITION 2.4. *Let $u \in \mathscr{D}'(M)$, and let $q(x, D) \in OPS^s_{1,0}$ be elliptic. Then $u \in \mathscr{L}^s_p(M)$ if, and only if, $q(x, D)u \in L^p(M)$, if $1 < p < \infty$.*

Proof. Let $q(x, D)^{-1} \in OPS^{-s}_{1,0}$ denote any parametrix of $q(x, D)$. If $p(x, D) \in OPS^s_{1,0}$ is given, then $p(x, D)u = p(x, D)q(x, D)^{-1}q(x, D)u$ (mod C^∞) $= r(x, D)q(x, D)u \in L^p(M)$ since $r(x, D) \in OPS^0_{1,0}$ leaves L^p invariant.

In the next chapter we shall see that, for $s \in \mathbf{R}$, $(1-\Delta)^{s/2} \in OPS^s_{1,0}$. In consequence we have

$$\mathscr{L}^s_p(M) = (1-\Delta)^{-s/2} L^p(M).$$

We are now prepared for the following result.

THEOREM 2.5. *Let $p(x, D) \in OPS^m_{1,0}$ on M. Then*

$$p(x, D): \mathscr{L}^s_p \to \mathscr{L}^{s-m}_p, \qquad 1 < p < \infty \qquad \textit{and} \tag{2.8}$$

$$p(x, D): C^{k+\alpha} \to C^{k+\alpha-m}, \tag{2.9}$$

provided $k+\alpha$ and $k+\alpha-m$ are both positive and nonintegral.

Proof. Let $a(x, D) \in OPS^s_{1,0}$ and $b(x, D) \in OPS^{s-m}_{1,0}$ both be elliptic, with parametrices A^{-1} and B^{-1}, respectively. Then $b(x, D)p(x, D)u = b(x, D)p(x, D)A^{-1}a(x, D)u \bmod C^\infty$ and this belongs to L^p if $u \in \mathscr{L}^s_p$, since that yields $a(x, D)u \in L^p$, and $b(x, D)p(x, D)A^{-1} \in OPS^0_{1,0}$. This proves (2.8).

To prove (2.9), we use the invariance of $C^{k+\alpha}$ and of $OPS^m_{1,0}$ under coordinate transformation. Thus it suffices to show that if $u \in C^{k+\alpha}(\mathbf{R}^n)$ is supported in B_R and $p(x, \xi) = 0$ for $|x| \geq 2R$, then $p(x, D)u \in C^{k+\alpha-m}$, provided $k+\alpha$ and $k+\alpha-m$ are both positive and nonintegral. In fact, by Theorem 1.8 we have, for such u, with $\varphi \in C_0^\infty(B_{2R})$, $\varphi = 1$ on B_R,

$$\varphi(x)(1 - \Delta)^{m/2}u \in C^{k+\alpha-m}(\mathbf{R}^n).$$

But $p(x, D)u = p(x, D)(1-\Delta)^{-m/2}\varphi(x)(1-\Delta)^{m/2}u \bmod C^\infty$, and since $p(x, D)(1-\Delta)^{-m/2} \in OPS^0_{1,0}$, the assertion follows from Theorem 2.2.

It is interesting to note that we can improve the estimate (2.7), in the case of $p(x, D)$ acting on L^p. The following pretty argument is given by Strichartz [1]. Namely, one considers first the operator

$$Tu(x, y) = \int p(y, \xi)\hat{u}(\xi)e^{ix\cdot\xi}\, d\xi.$$

We think of Tu as a function of x taking values in some auxiliary Hilbert space, a space of functions of y. Now the Marcinkiewicz multiplier theorems, for example Theorem 1.4, are valid for convolution operators taking Hilbert space valued functions to Hilbert space valued functions; see Stein [1]. Here, our two spaces will be, respectively, $\mathbf{C}$ and $H^s(\mathbf{R}^n)$, where $s > n/2$. Therefore,

$$\|p(x, D)u\|_{L^p(\mathbf{R}^n)} \leq C\|Tu\|_{L^p(\mathbf{R}^n, H^s)} \leq C'\|u\|_{L^p}|||p|||_s, \qquad \text{where} \tag{2.10}$$

$$|||p|||_s = \sup_{0<r<\infty} \|p_r\|_{H^s(\Omega, H^s(\mathbf{R}^n))}. \tag{2.11}$$

$(p_r(x, \xi) = p(x, r\xi))$. In particular, we have the following strengthening of Theorem 2.2 for L^p.

THEOREM 2.6. *Let $u \in L^p$ be supported in B_R, and let $(n/2) + \epsilon = k + \sigma$, k an integer, $0 < \sigma < 1$. Then*

$$\|p(x, D)u\|_{L^p(B_{2R})} \leq C(p) \sup_{0<r<\infty} \|p_r\|_k^{1-\sigma}\|p_r\|_{k+1}^\sigma, \qquad 1 < p < \infty, \tag{2.12}$$

where

$$\|p_r\|_k = \max_{|\alpha|,|\beta|\le k}\ \sup_{(x,\xi)\in\mathbf{R}^n\times\Omega} |D_x^\beta D_\xi^\alpha p_r(x,\xi)|.$$

Proof. One need only show that

$$|||p|||_{n/2+\epsilon} \le C \sup_{0<r<\infty} \|p_r\|_k^{1-\sigma}\|p_r\|_{k+1}^\sigma,$$

which is an exercise in Sobolev space theory. We leave the details to the reader. (Assume $p(x,\xi)=0$ for large x.)

We conclude this section with a brief discussion of the behavior of $OPS_{\rho,\delta}^{-m}$ on L^p and Hölder spaces. If $\rho\ne 1$ one generally does not have continuity on L^p, or C^α for $m=0$; we need operators of negative order. The following result will be a simple consequence of Theorem 2.6.

THEOREM 2.7. *Suppose* $p(x,D)\in OPS_{\rho,\delta}^{-m}$. *Then*

$$p(x,D): L^p_{\text{comp}}\to L^p_{\text{loc}},\quad 1<p<\infty,\quad \text{and } p(x,D): C^\alpha_{\text{comp}}\to C^\alpha_{\text{loc}},\quad 0<\alpha<1,$$

provided $m>(n/2)(1-\rho+\delta)$.

Proof. Given that

$$|D_x^\beta D_\xi^\alpha p(x,\xi)| \le C_{\alpha\beta}(1+|\xi|)^{-m-\rho|\alpha|+\delta|\beta|},$$

we first want to show that the right-hand side of (2.12) is finite, granted $m>(n/2)(1-\rho+\delta)$. Indeed, we have $D_x^\beta D_\xi^\alpha p_r(x,\xi) = r^{|\alpha|}D_x^\beta D_\xi^\alpha p(x,r\xi)$, so

$$\begin{aligned}
&\|p_r\|_k^{1-\sigma}\|p_r\|_{k+1}^\sigma\\
&\le C\left[\sum_{\ell,j=0}^{k} r^j(1+r)^{-m-\rho j+\delta\ell}\right]^{1-\sigma}\left[\sum_{\lambda,\mu=0}^{k+1} r^\mu(1+r)^{-m-\rho\mu+\delta\lambda}\right]^\sigma\\
&\le C(1+r)^{-m}\left[\sum_{\ell,j=0}^{k}(1+r)^{\delta\ell+(1-\rho)j}\right]^{1-\sigma}\left[\sum_{\lambda,\mu=0}^{k+1}(1+r)^{\delta\lambda+(1-\rho)\mu}\right]^\sigma\\
&\le C(1+r)^{-m+(1-\rho+\delta)(k+\sigma)},
\end{aligned}$$

and this quantity is bounded on $0<r<\infty$ provided

$$m\ge(k+\sigma)(1-\rho+\delta) = \left(\frac{n}{2}+\epsilon\right)(1-\rho+\delta).$$

This handles the L^p case, by Theorem 2.6. The C^α case is obtained via the inclusions

$$C^\alpha\subset\bigcap_{p<\infty} W_p^{\alpha-\epsilon/2}\subset C^{\alpha-\epsilon}$$

the latter inclusion being one of Sobolev's imbedding theorems.

In the L^p case, we can improve this result, if $0\le\delta\le\rho\le 1$ $(\delta<1)$, as follows. In such a case, we have L^2 continuity for $OPS_{\rho,\delta}^0$. (For $\rho=\delta$

this uses the Calderon-Vaillancourt theorem, which we will prove in Chapter XIII.) Now an interpolation argument of Stein (see Stein and Weiss [1]) shows that one can interpolate between this result and the conclusion of Theorem 2.7. We shall not try to give the details of such an argument here, but the conclusion is the following.

THEOREM 2.8. *Suppose* $p(x, D) \in OPS_{\rho,\delta}^{-m}$ *with* $0 \leq \delta \leq \rho \leq 1 (\delta < 1)$. *Then*

$$p(x, D): L_{\text{comp}}^p \to L_{\text{loc}}^p, \qquad 1 < p < \infty, \qquad \text{if}$$

$$m = \frac{n}{2}(1 - \rho + \delta).$$

If $\delta = 0$, this result is sharp, even for symbols which are independent of x (see Wainger [1]). For $\delta > 0$, there exist sharper results, which should not be surprising since in the L^2 case the correct results for positive δ are completely inaccessible by such techniques as the proof of Theorem 2.1, or variants (see Chapter II, Section 6). Hörmander [12] shows that, for $0 \leq \delta < \rho < 1$, $OPS_{\rho,\delta}^{-(n/2)(1-\rho)}$ takes L_{comp}^p to L_{loc}^p, $1 < p < \infty$. Sharper results are given in a set of unpublished but widely circulated lecture notes by E. Stein, "Singular Integral and Pseudo-differential Operators," for a 1972 course at Princeton. Stein shows that if $p(x, D) \in OPS_{\delta,\rho}^m$ and either $0 \leq \delta < \rho = 1$ or $0 < \delta = \rho < 1$, then $p(x, D)$ is of weak type $(1, 1)$ if $m = -(1 - \rho)n/2$ and $p(x, D)$ is bounded on L^p $(1 < p < \infty)$ if $(1 - \rho)|(1/2) - (1/p)| \leq -(m/n)$. Also $p(x, D)$ is bounded on C^α if $0 < \rho$, $1 \geq \delta$, and $m = -(1 - \rho)(n/2)$. See also Fefferman [1] where the property $p(x, D): L^\infty \to BMO$ is examined. One case of Stein's results, also derived by Kagan [1] and others, is that, if $0 \leq \delta < 1$, $OPS_{1,\delta}^0$ is bounded on L^p, $1 < p < \infty$. We prove this in the next section.

In applications, the results of Theorems 2.7 and 2.8, or even their improvements, do not generally seem to be the right tools for the job. The general philosophy is that continuity on the Banach space of interest should hold for operators of order zero, and negative order should not be required. If such a result does not hold, it is probably because one has the wrong class of operators. In Section 4 we shall examine a subalgebra of $OPS_{\rho,0}^0$ consisting of continuous operators on L^p, $1 < p < \infty$, containing operators which arise in the study of the heat equation and in other applications.

Exercises

2.1. Let $p(x, D)$ be an elliptic operator of order m on an open region Ω, $u \in \mathscr{D}'(\Omega)$. Show that $p(x, D)u \in \mathscr{L}_{p,\text{loc}}^s(\Omega)$ $1 < p < \infty$, implies $u \in \mathscr{L}_{p,\text{loc}}^{s+m}(\Omega)$, and that $p(x, D)u \in C^{k+\alpha}_{\text{loc}}(\Omega)$ implies $u \in C^{k+m+\alpha}_{\text{loc}}(\Omega)$ if $k = 0, 1, 2, \ldots, 0 < \alpha < 1$. If $p(x, D)u$ is continuous on Ω, show that $u \in C^{m-1+\alpha}_{\text{loc}}(\Omega)$ for every $\alpha < 1$. If $p(x, D)u$ is a locally finite measure on Ω, show that $u \in \mathcal{L}^{m-\epsilon}_{1,\text{loc}}(\Omega)$, for every $\epsilon > 0$.

2.2. Let $u \in C(\bar{D})$ be harmonic in the unit disc $D = \{(x, y) \in \mathbf{R}^2 : x^2 + y^2 < 1\}$, and let $f = u|_{\partial D}$. Show that $u \in C^1(\bar{D})$ if, and only if, both f and its Hilbert transform are in $C^1(\partial D)$. Generalize to smooth bounded domains in $\mathbf{R}^n$.

2.3. Show that the definition of $\mathcal{L}^s_p(M)$ given in this chapter coincides with that given in Chapter I. Hint: Use the pseudolocal property of $OPS^s_{1,0}$ to reduce the problem to the case $M = \mathbf{T}^n$.

§3. L^p Behavior of $OPS^0_{1,\delta}$

The purpose of this section is to proven the following.

THEOREM 3.1. *If* $p(x, D) \in OPS^0_{1,\delta}$, $0 \le \delta < 1$, *then* $p(x, D): L^p_{\text{comp}} \to L^p_{\text{loc}}$, $1 < p < \infty$.

Note that Theorem 3.1 contains the part of Theorem 2.1 dealing with L^p behavior as a special case. The advantage of Theorem 2.1 is that its proof fell out directly as a consequence of the result, Theorem 1.1, for Fourier multipliers, whereas we shall have to work harder to prove Theorem 3.1, taking steps that would essentially prove (a weakened version of) Theorem 1.1. Our treatment follows Nagase [1]; see also Illner [1], Kagan [1], Kumano-Go and Nagase [1]. With the right conditions on $p(x, \xi)$ as $|x| \to \infty$, one can get global estimates on $L^p(\mathbf{R}^n)$, but for simplicity we shall omit these results.

To prove Theorem 3.1, it suffices to establish the following weak-type (1, 1) estimate.

PROPOSITION 3.2 *Suppose* $p(x, \xi) \in S^0_{1,\delta}$, $\delta < 1$, *has compact x-support. Then, for* $u \in L^1(\mathbf{R}^n)$, $\lambda \in \mathbf{R}^+$,

$$\text{meas}\,\{x : |p(x, D)u(x)| \ge \lambda\} \le \frac{C}{\lambda}\|u\|_{L^1}. \tag{3.1}$$

Indeed, for such $p(x, D)$, since $p(x, D): L^2 \to L^2$, the Marcinkiewicz interpolation theorem implies $p(x, D): L^p \to L^p$ for $1 < p \le 2$, proving Theorem 3.1 for $1 < p \le 2$. The result for $2 \le p < \infty$ follows by duality, since $p(x, D) \in OPS^0_{1,\delta}$ implies $p(x, D)^* \in OPS^0_{1,\delta}$. We now prepare to prove Proposition 3.2.

First, note there is no loss of generality in assuming $p(x, \xi) = 0$ for $|\xi| \le 1$. Now, let $\psi \in C_0^\infty$ be supported on $\{\xi : 1/2 \le |\xi| \le 2\}$ such that

$$\sum_{j=0}^{\infty} \psi(2^{-j}\xi) = 1$$

for $|\xi| \ge 1$. Set $q_j(x, \xi) = p(x, \xi)\psi(2^{-j}\xi)$,

$$p_N(x, \xi) = \sum_{j=0}^{N} q_j(x, \xi).$$

It suffices, of course, to prove (3.1) with $p(x, D)$ replaced by $p_N(x, D)$, provided the constant C is independent of N. Our first step is to derive some estimates on the kernels $k_j(x, x-y)$ of $q_j(x, D)$, given by

$$k_j(x, z) = \int e^{iz\cdot\xi} p(x, \xi)\psi(2^{-j}\xi)\, d\xi, \tag{3.2}$$

and on the kernel $K_N(x, x-y)$ of $p_N(x, D)$;

$$K_N(x, z) = \sum_{j=0}^{N} k_j(x, z).$$

LEMMA 3.3. *We have*

$$\int_{|x|\ge 2t} |K_N(x+x^0, x-y) - K_N(x+x^0, x)|\, dx \le C_0,$$

if $|y| \le t$.

Proof. From (3.2), with $\varkappa = [n/2]+1$, $|\alpha| \le \varkappa$, we have

$$(2^j x)^\alpha k_j(x+x^0, x)$$
$$= 2^{j|\alpha|} \sum_{\beta\le\alpha} C_{\alpha\beta} \int e^{ix\cdot\xi} D_\xi^\beta p(x+x^0, \xi)|\xi|^{|\beta|} \cdot D_\xi^{\alpha-\beta}\psi(2^{-j}\xi)|\xi|^{-|\beta|}\, d\xi.$$

Now $D_\xi^\beta p(x+x^0, \xi)|\xi|^{|\beta|}$ is a bounded subset of $S^0_{1,\delta}$, with x^0 as a parameter, so the L^2 estimates yield

$$\|(2^j x)^\alpha k_j(x+x^0, x)\|_{L^2}^2 \le C_\alpha 2^{2j|\alpha|} \sum_{\beta\le\alpha} \|D_\xi^{\alpha-\beta}\psi(2^{-j}\xi)\cdot|\xi|^{-|\beta|}\|_{L^2}^2 \tag{3.3}$$
$$\le C'_\alpha 2^{nj}, \qquad |\alpha| \le \left[\frac{n}{2}\right]+1.$$

Here C'_α is independent of j and x^0. Thus

$$\int |k_j(x+x^0, x)|\, dx \tag{3.4}$$
$$\le \left[\int (1+2^{2j}|x|^2)^\varkappa |k_j(x+x^0, x)|^2\, dx\right]^{1/2} \left[\int (1+2^{2j}|x|^2)^{-\varkappa}\, dx\right]^{1/2}$$
$$\le C_1,$$

independent of j, x^0. Also, by (3.3),

$$(3.5)\quad \int_{|x|\geq t} |k_j(x+x^0, x)|\, dx$$
$$\leq \left[\int_{|x|\geq t} (2^{2j}|x|^2)^{\varkappa}|k_j(x+x^0, x)|^2\, dx\right]^{1/2}\left[\int_{|x|\geq t} (2^{2j}|x|^2)^{-\varkappa}\, dx\right]^{1/2}$$
$$\leq C_2(2^jt)^{n/2-\varkappa}.$$

Hence, for $|y| \leq t$,

$$(3.6)\quad \int_{|x|\geq 2t} |k_j(x+x^0, x-y) - k_j(x+x^0, x)|\, dx$$
$$\leq \int_{|x|\geq 2t} |k_j(x+x^0+y, x)|\, dx + \int_{|x|\geq 2t} |k_j(x+x^0, x)|\, dx$$
$$\leq 2C_2(2^jt)^{n/2-\varkappa}.$$

When $2^jt \leq 1$ and $|y| \leq t$, then

$$|e^{-iy\cdot\xi} - 1| \leq |y|\,|\xi| \leq 2^jt \qquad \text{for} \qquad \xi \in \operatorname{supp} \psi(2^{-j}\xi),$$

and

$$|D_\xi^\beta(e^{-iy\cdot\xi} - 1)| \leq t^{|\beta|} \leq 2^jt\, 2^{-j|\beta|}, \qquad |\beta| \neq 0.$$

Thus, by the argument proving (3.4),

$$(3.7)\qquad \int |k_j(x+x^0, x-y) - k_j(x+x^0, x)|\, dx \leq C_3(2^jt)$$

for $|y| \leq t$ and $2^jt \leq 1$. Now, with

$$K_N(x, z) = \sum_{j=0}^{N} k_j(x, z),$$

it follows from (3.6), (3.7) that, for $|y| \leq t$,

$$\int_{|x|\geq 2t} |K_N(x+x^0, x-y) - K_N(x+x^0, x)|\, dx \leq C_4 \sum_{j=0}^{\infty} \min\{(2^jt)^{n/2-\varkappa}, 2^jt\}$$
$$\leq C_0,$$

proving the lemma.

We apply Lemma 3.3 in conjunction with the following covering lemma of Calderon and Zygmund, for whose proof we refer toStein [1].

LEMMA (Calderon and Zygmund). *Let $u \in L^1(\mathbf{R}^n)$ and $\lambda > 0$ be given. There exist v, w_k in $L^1(\mathbf{R}^n)$ and disjoint cubes I_k, $1 \leq k < \infty$, with centers $x^{(k)}$ such that*

$$\text{(i)}\ u = v + \sum_{k=1}^{\infty} w_k, \qquad \|v\|_{L^1} + \sum_{k=1}^{\infty} \|w_k\|_{L^1} \leq 3\|u\|_{L^1};$$

(ii) $|v(x)| \leq 2^n\lambda$;

(iii) $\int_{I_k} w_k(x)\,dx = 0$ *and* $\operatorname{supp} w_k \subset I_k$;

(iv) $\sum_{k=1}^{\infty} \operatorname{meas}(I_k) \leq \lambda^{-1}\|u\|_{L^1}$.

If u has compact support, the supports of v and w_k are contained in a fixed compact set.

It follows from this decomposition lemma that

$$\hat{u}(\xi) = \hat{v}(\xi) + \sum_k \hat{w}_k(\xi), \qquad |\hat{v}(\xi)| + \sum_k |\hat{w}_k(\xi)| \leq 3\|u\|_{L^1},$$

and

$$p_N(x, D)u = p_N(x, D)v + \sum_k p_N(x, D)w_k.$$

Set

$$I_k^* = \{x \in \mathbf{R}^n : x - x^{(k)} = 2\sqrt{n}(x' - x^{(k)}), \text{ some } x' \in I_k\},$$

so I_k^* is obtained from I_k by dilating it a factor of $2\sqrt{n}$, keeping the same center. Note that $\operatorname{meas}(I_k^*) = \gamma \operatorname{meas}(I_k)$, $\gamma = (2\sqrt{n})^n$. For some $t_k > 0$,

$$I_k \subset \{x : |x - x^{(k)}| \leq t_k\},$$
$$Y_k = \mathbf{R}^n \backslash I_k^* \subset \{x : |x - x^{(k)}| > 2t_k\}.$$

Now, using property (iii), we find that

$$\begin{aligned} p_N(x, D)w_k &= \int K_N(x, x-y)w_k(y)\,dy \\ &= \int_{I_k} \{K_N(x, x-y) - K_N(x, x-x^{(k)})\}w_k(y)\,dy. \end{aligned}$$

Thus, it follows from Lemma 3.3 that

$$\begin{aligned} &(3.8) \quad \int_{Y_k} |p_N(x, D)w_k(x)|\,dx \\ &\leq \int_{|y| \leq t_k} \int_{|x| \geq 2t_k} |K_N(x + x^{(k)}, x - y) - K_N(x + x^{(k)}, x)|\,|w_k(y + x^{(k)})|\,dx\,dy \\ &\leq C_0\|w_k\|_{L^1}. \end{aligned}$$

Now set

$$\mathcal{O}^* = \bigcup_{k=1}^{\infty} I_k^*, \qquad w = \sum_{k=1}^{\infty} w_k.$$

Note that

$$\operatorname{meas}(\mathcal{O}^*) \leq \frac{\gamma}{\lambda}\|u\|_{L^1},$$

by property (iv). Meanwhile (3.8) implies

$$\int_{\mathbf{R}^n \setminus \mathcal{O}} |p_N(x, D)w(x)|\,dx \le 3C_0\|u\|_{L^1}. \tag{3.9}$$

As for $p_N(x, D)v$, since $p_N(x, \xi)$ is bounded in $S^0_{1,\delta}$, we have

$$\|p_N(x, D)v\|^2_{L^2} \le C\|v\|^2_{L^2} \le C\lambda\|u\|_{L^1}, \tag{3.10}$$

where the last inequality uses property (ii). We are now in a position to estimate meas $\{x: |p_N(x, D)u(x)| > \lambda\}$. From (3.9) we have

$$\frac{\lambda}{2}\,\text{meas}\left\{x: |p_N(x, D)w(x)| > \frac{\lambda}{2}\right\} \le 3C_0\|^v\|_L{}^1 \tag{3.11}$$

while (3.10) implies

$$\left(\frac{\lambda}{2}\right)^2 \text{meas}\left\{x: |p_N(x, D)v(x)| > \frac{\lambda}{2}\right\} \le C\lambda\|u\|_{L^1}. \tag{3.12}$$

From (3.11) and (3.12), (3.1) is an immediate consequence, so the proof of Proposition 3.2, and hence the proof of Theorem 3.1, is complete.

Exercise

3.1. Recall from Chapter III, Section 3, the hypoelliptic operators

$$P_k = 1 - |x|^{2k}\Delta, \qquad (k \ge 2).$$

Show that if $u \in \mathscr{D}'$, $P_k u = f \in L^p$, $k \ge 2$, then $u \in L^p_{\text{loc}}$, $x^\alpha Du \in L^p_{\text{loc}}$, and $x^\beta D^2 u \in L^p_{\text{loc}}$, provided $|\alpha| \ge k$, $|\beta| \ge 2k$. What if $f \in \mathscr{L}^s_p$?

§4. The Algebras $OP\mathcal{M}_\rho^m$ and $OP\mathcal{N}_\rho^m$ on L^p

We define the following classes $\mathcal{M}_\rho^m$ of symbols which satisfy a variable coefficient version of the Marcinkiewicz condition (1.5) and which also yield a workable algebra of pseudodifferential operators. $\mathcal{M}_\rho^m$ will be a subalgebra of $S^m_{\rho,0}$.

DEFINITION. *The symbol $p(x, \xi) \in \mathcal{M}_\rho^m$ if, and only if, for all multiindices* α,

$$\xi^\alpha D_\xi^\alpha p(x, \xi) \in S^m_{\rho,0}. \tag{4.1}$$

Remark. Actually, one need only demand that (4.1) hold for $\alpha = (\alpha_1, \dots, \alpha_n)$ with $\alpha_j = 0$ or 1, but we won't insist on this greater generality.

If $p(x, \xi) \in \mathcal{M}_\rho^m$, we say $p(x, D) \in OP\mathcal{M}_\rho^m$.

Clearly, $\mathcal{M}_\rho^m \subset \mathcal{M}_{\rho'}^m$, if $\rho \ge \rho'$. Also $\mathcal{M}_1^m = S^m_{1,0}$. We also have the following simple results, assuming throughout that $0 < \rho \le 1$.

PROPOSITION 4.1. *Let* $p(x, \xi) \in \mathcal{M}_{\rho}^{m}$, $q(x, \xi) \in \mathcal{M}_{\rho}^{\mu}$. *Then*

(i) $pq \in \mathcal{M}_{\rho}^{m+\mu}$;

(ii) $D_x^{\beta} D_{\xi}^{\gamma} p \in \mathcal{M}_{\rho}^{m-\rho|\gamma|}$;

(iii) if $|p(x, \xi)| \geq C(1+|\xi|)^m$, then $p^{-1} \in \mathcal{M}_{\rho}^{-m}$.

Proof. Since

$$\xi^{\alpha} D_{\xi}^{\alpha}(pq) = \sum_{\alpha_1+\alpha_2=\alpha} C_{\alpha_1,\alpha_2}(\xi^{\alpha_1} D^{\alpha_1} p)(\xi^{\alpha_2} D^{\alpha_2} q),$$

(i) is an immediate consequence of the algebraic properties of $S_{\rho,0}^{m}$. To establish (ii), we may suppose $\beta = 0$ since clearly $D_x^{\beta} p \in \mathcal{M}_{\rho}^{m}$, and we may also suppose $|\gamma| = 1$, so $D_{\xi}^{\gamma} = D_{\xi_j}$ for some j. Now if $p \in \mathcal{M}_{\rho}^{m}$, $\xi^{\alpha} D_{\xi}^{\alpha} D_{\xi_j} p(x, \xi) = p_1(x, \xi) + (D_{\xi_j}\xi^{\alpha}) D_{\xi}^{\alpha} p(x, \xi)$ where $p_1 \in S_{\rho,0}^{m-\rho}$ and $(D_{\xi_j}\xi^{\alpha}) D_{\xi}^{\alpha} p = \alpha_j \xi^{\alpha'} D_{\xi}^{\alpha'} D_{\xi_j} p$ with $|\alpha'| = |\alpha| - 1$. An inductive argument shows that $\xi^{\alpha} D_{\xi}^{\alpha} D_{\xi_j} p(x, \xi)$ belongs to $S_{\rho,0}^{m-\rho}$, so (ii) is proved. As for (iii), we have

$$D_{\xi}^{\alpha} p(x, \xi)^{-1} = \sum_{\alpha_1+\cdots\alpha_\nu=\alpha} C_{\alpha_1\cdots\alpha_\nu} p^{(\alpha_1)}(x, \xi) \cdots p^{(\alpha_\nu)}(x, \xi) p(x, \xi)^{-1-\nu} \tag{4.2}$$

and hence

$$\xi^{\alpha} D_{\xi}^{\alpha} p(x, \xi)^{-1} = \sum C_{\alpha_1\cdots\alpha_\nu} \xi^{\alpha_1} p^{(\alpha_1)}(x, \xi) \cdots \xi^{\alpha_\nu} p^{(\alpha_\nu)}(x, \xi) p(x, \xi)^{-1-\nu}.$$

Since $\xi^{\alpha_j} p^{(\alpha_j)}(x, \xi) \in S_{\rho,0}^{m}$ while $p(x, \xi)^{-1-\nu} \in S_{\rho,0}^{-m(1+\nu)}$ by the ellipticity hypothesis, we see that $\xi^{\alpha} D_{\xi}^{\alpha} p(x, \xi)^{-1} \in S_{\rho,0}^{-m}$, which yields (iii).

Examples of elements of $\mathcal{M}_{\rho}^{m}$ which often arise occur as follows. Suppose $\mathbf{R}^n = \mathbf{R}^{\ell} \times \mathbf{R}^k$ has coordinates (ξ, η); $\xi \in \mathbf{R}^{\ell}$, $\eta \in \mathbf{R}^k$.

PROPOSITION 4.2. *Let* $r(\theta) \in S_{1,0}^{m}(\mathbf{R}^k)$, $m \geq 0$, *and let* $q(\xi, \eta) = r(|\xi|^{-\rho}\eta)$, $0 < \rho \leq 1$. *Then* $q \in \mathcal{M}_{\rho}^{(1-\rho)m}$, *in the conic set* $|\eta| < |\xi|$.

Here we microlocalize the symbol class $\mathcal{M}_{\rho}^{m}$ in the usual way; $q \in \mathcal{M}_{\rho}^{m}$ on a conic open set Γ if (4.1) holds for $(x, \xi) \in \Gamma$.

Proof. A simple induction argument shows that, if $|\alpha| \geq 1$,

$$D_{\xi}^{\alpha} r(|\xi|^{-\rho}\eta) = \sum_{1 \leq |\beta| \leq |\alpha|} C_{\beta\alpha}(\xi) r_{\beta}(|\xi|^{-\rho}\eta)|\xi|^{-|\alpha|} \tag{4.3}$$

where $r_{\beta}(\theta) = \theta^{\beta} r^{(\beta)}(\theta) \in S_{1,0}^{m}(\mathbf{R}^k)$ and $C_{\beta,\alpha}(\xi)$ is homogeneous of degree zero in ξ. It follows that

$$\begin{aligned} &\eta^{\gamma}\xi^{\alpha} D_{\eta}^{\gamma+\gamma'} D_{\xi}^{\alpha+\alpha'} r(|\xi|^{-\rho}\eta) \\ &= \sum_{1 \leq |\beta| \leq |\alpha+\alpha'|} C_{\beta,\alpha+\alpha'}(\xi)(\xi^{\alpha}|\xi|^{-|\alpha|})(|\xi|^{-\rho}\eta)^{\gamma} r_{\beta}^{(\gamma+\gamma')}(|\xi|^{-\rho}\eta)|\xi|^{-|\alpha'|-\rho|\gamma'|} \\ &= \sum_{1 \leq |\beta| \leq |\alpha+\alpha'|} C_{\beta,\alpha+\alpha'}(\xi)(\xi^{\alpha}|\xi|^{-|\alpha|}) r_{\beta+\gamma}^{(\gamma')}(|\xi|^{-\rho}\eta)|\xi|^{-|\alpha'|-\rho|\gamma'|}, \end{aligned}$$

and since $r_{\beta+\gamma}^{(\gamma')}(\theta) \in S_{1,0}^{m-|\gamma'|}(\mathbf{R}^k) \subset S_{1,0}^m(\mathbf{R}^k)$, it follows that

$$(4.4)\quad \begin{aligned} &|\eta^\gamma \xi^\alpha D_\eta^{\gamma+\gamma'} D_\xi^{\alpha+\alpha'} r(|\xi|^{-\rho}\eta)| \\ &\leq C_{\alpha\gamma\alpha'\gamma'}(1 + |\xi|^{-\rho}|\eta|)^{m-|\gamma'|}|\xi|^{-|\alpha'|-\rho|\gamma'|} \\ &\leq C(1 + |\xi|)^{(1-\rho)m-\rho(|\alpha'|+|\gamma'|)} \text{ on } |\eta| < |\xi|, \qquad |\xi| \geq 1. \end{aligned}$$

It is easy to see that a similar inequality holds for $D_\eta^{\gamma'} D_\xi^{\alpha'} \eta^\gamma \xi^\alpha D_\eta^\gamma D_\xi^\alpha r(|\xi|^{-\rho}\eta)$, which establishes the proposition.

As an example of a symbol to which Proposition 4.2 applies, consider the following, arising from a study of the heat operator

$$H = \frac{\partial}{\partial x} - \left(\frac{\partial^2}{\partial y_1^2} + \cdots + \frac{\partial^2}{\partial y_k^2}\right),$$

whose symbol is $i\xi + |\eta|^2$. Consider

$$q(\xi, \eta) = \frac{|\eta|^2 - i\xi}{|\eta|^2 + i\xi} = \frac{\xi^{-1}|\eta|^2 - i}{\xi^{-1}|\eta|^2 + i}.$$

This symbol microlocally belongs to $S_{1,0}^0$ except near the variety $\Sigma = \{\eta=0\}$, whose two connected components in $\mathbf{R}^{k+1} - \{0\}$ are $\Sigma_+ = \{\eta=0, \xi>0\}$ and $\Sigma_- = \{\eta=0, \xi<0\}$. On the cone $\Gamma_+ = \{|\eta|<\xi, \xi>0\}$, we have $q(\xi, \eta) = r(\xi^{-1/2}\eta)$ where

$$r(\theta) = \frac{|\theta|^2 - i}{|\theta|^2 + i} \in S_{1,0}^0(\mathbf{R}^k).$$

Thus q belongs to $\mathcal{M}_{1/2}^0$ on Γ_+. Similarly q belongs to $\mathcal{M}_{1/2}^0$ on a conic neighborhood of Σ_-.

Having considered the symbols in $\mathcal{M}_\rho^m$, we next show that $OP\mathcal{M}_\rho^m$ has decent algebraic properties.

PROPOSITION 4.3. *Assume* $0 < \rho \leq 1$. *If* $p \in \mathcal{M}_\rho^m$ *and* $q \in \mathcal{M}_\rho^\mu$, *we have* $p(x, D)q(x, D) \in OP\mathcal{M}_\rho^{m+\mu}$.

Proof. Since $\mathcal{M}_\rho^m \subset S_{\rho,0}^m$, it follows that $A = p(x, D)q(x, D) \in OPS_{\rho,0}^{m+\mu}$, and we have the asymptotic expansion

$$\sigma_A(x, \xi) \sim \sum_{\gamma \geq 0} \frac{1}{\gamma!} p^{(\gamma)}(x, \xi) q_{(\gamma)}(x, \xi).$$

Hence

$$(4.5)\quad \xi^\alpha D_\xi^\alpha \sigma_A(x, \xi) \sim \sum_{\gamma \geq 0} \sum_{\alpha_1+\alpha_2=\alpha} \frac{1}{\gamma!} C_{\alpha_1\alpha_2} \xi^{\alpha_1} p^{(\gamma+\alpha_1)}(x, \xi) \cdot \xi^{\alpha_2} q_{(\gamma)}^{(\alpha_2)}(x, \xi).$$

Since $\xi^{\alpha_1} p^{(\gamma+\alpha_1)} \in S_{\rho,0}^{m-\rho|\gamma|}$ while $\xi^{\alpha_2} q_{(\gamma)}^{(\alpha_2)}(x, \xi) \in S_{\rho,0}^\mu$, the proposition is established.

COROLLARY 4.4. *Assume $0 < \rho \leq 1$ and let $p \in \mathcal{M}_\rho^m$ satisfy, for large $|\xi|$,*

$$|p(x, \xi)| \geq C(1 + |\xi|)^m,$$

so p is elliptic. Let $q(x, D) \in OPS_{\rho,0}^{-m}$ be a parametrix for $p(x, D)$, so $p(x, D)q(x, D) = I + K_1$ and $q(x, D)p(x, D) = I + K_2$ with $K_j \in OPS^{-\infty}$. Then $q(x, D) \in OP\mathcal{M}_\rho^{-m}$.

Proof. This goes by the usual construction of parametrices for elliptic operators. Let $q_0(x, \xi) = \varphi(\xi)p(x, \xi)^{-1}$ where $\varphi = 1$ for large ξ and vanishes in a neighborhood of the zeros of p. Then $q_0(x, D) \in OP\mathcal{M}_\rho^{-m}$ by Proposition 4.1, and $q_0(x, D)p(x, D) = I + S$ with $S \in OPS_{\rho,0}^{-\rho}$ and

$$\sigma_S(x, \xi) \sim \sum_{\gamma > 0} \frac{1}{\gamma!} q_0^{(\gamma)}(x, \xi)p_{(\gamma)}(x, \xi)$$

for which a calculation similar to (4.5) shows that $S \in OP\mathcal{M}_\rho^{-\rho}$. Since we have $q(x, D) \sim (I - S + S^2 - \cdots)q_0(x, D)$, our assertion about $q(x, D)$ follows from Proposition 4.3.

The following is our basic L^p continuity result for operators in $OP\mathcal{M}_\rho^0$. The proof is similar to the proof of Theorem 2.1, but for completeness we give the details.

PROPOSITION 4.5. *If $p(x, D) \in OP\mathcal{M}_\rho^0$, $0 < \rho \leq 1$, then*

$$p(x, D): L^p_{\text{comp}} \to L^p_{\text{loc}}, \qquad 1 < p < \infty.$$

Proof. Since elements of $OP\mathcal{M}_\rho^0 \subset OPS_{\rho,0}^0$ are pseudolocal, we can assume without loss of generality that $p(x, \xi)$ has compact support in x. Now write

$$p(x, \xi) = (2\pi)^{-n} \int_{\mathbf{R}^n} e^{-ix\cdot\eta} p_\eta(\xi)\, d\eta$$

where

$$p_\eta(\xi) = \int_{\mathbf{R}^n} p(x, \xi)e^{ix\cdot\eta}\, dx. \tag{4.6}$$

It follows that

$$p(x, D)u = (2\pi)^{-n} \int_{\mathbf{R}^n} e^{-ix\cdot\eta} p_\eta(D)u\, d\eta.$$

Consequently,

$$\|p(x, D)u\|_{L^p} \leq (2\pi)^{-n} \int_{\mathbf{R}^n} \|p_\eta(D)u\|_{L^p}\, d\eta. \tag{4.7}$$

We estimate $\|p_\eta(D)u\|_{L^p}$ by the Marcinkiewicz multiplier theorem, Theorem 1.5; we have

$$\|p_\eta(D)u\|_{L^p} \leq C_p \sup_{\xi\in\mathbf{R}^n} \max_{|\alpha|\leq n} |\xi^\alpha p_\eta^{(\alpha)}(\xi)|\, \|u\|_{L^p}, \qquad 1 < p < \infty. \tag{4.8}$$

We estimate (4.8) using (4.6). Indeed,

$$\eta^\beta \xi^\alpha p_\eta^{(\alpha)}(\xi) = \int_{\mathbf{R}^n} [D_x^\beta \xi^\alpha D_\xi^\alpha p(x, \xi)] e^{ix \cdot \eta}\, dx,$$

and hence

$$|\eta^\beta|\, |\xi^\alpha p_\eta^{(\alpha)}(\xi)| \le C_{\alpha\beta}. \tag{4.9}$$

Summing (4.9) over $|\beta| \le n + 1$, we have $\|p_\eta(D)u\|_{L^p} \le C_p(1 + |\eta|)^{-n-1}\|u\|_{L^p}$, and hence, by (4.7), $\|p(x, D)u\|_{L^p} \le C_p^1\|u\|_{L^p}$, as desired.

Clearly, if $p(x, D) \in OP\mathcal{M}_\rho^0$ is properly supported we have $p(x, D)\colon L^p_{\text{loc}} \to L^p_{\text{loc}}$.

Finally, for $OP\mathcal{M}_\rho^m$ we have the following Sobolev-space results.

COROLLARY 4.6. *If* $p(x, D) \in OP\mathcal{M}_\rho^m$, $0 < \rho \le 1$, *then*

$$p(x, D)\colon \mathcal{L}^{s+m}_{p,\text{comp}} \to \mathcal{L}^s_{p,\text{loc}}, \qquad s \in \mathbf{R}, \qquad 1 < p < \infty.$$

Proof. We can assume $p(x, D)$ is properly supported. Taking $A \in OPS^s$ and $B \in OPS^{-s-m}$, both elliptic and properly supported, we see that it suffices to show that $C = Ap(x, D)B\colon L^p \to L^p$. But $C \in OP\mathcal{M}_\rho^0$, so this follows from Proposition 4.5.

One consequence of the extra structure of $\mathcal{M}_\rho^m$ is that one can say more about

$$q(x, \lambda) = e^{-i\lambda\psi(x)} p(x, D)(a(x, \lambda)e^{i\lambda\psi(x)}) \tag{4.10}$$

given $p \in OP\mathcal{M}_\rho^m$ than we could with $p \in OPS^m_{\rho,0}$ with $0 < \rho < 1/2$. We suppose $a(x, \lambda) \in S^\mu_{1,0}$, compactly supported, ψ is real valued, and $\nabla\psi \ne 0$ on supp a.

LEMMA 4.7. *We have* $|q(x, \lambda)| \le C_\varepsilon|\lambda|^{m+\mu+\varepsilon}$, *for any* $\varepsilon > 0$.

Proof. Choosing an elliptic $A \in OPS^{m+\varepsilon}$ we have $e^{-i\lambda\psi}A(ae^{i\lambda\psi}) \in S^{m+\mu+\varepsilon}_{1,0}$, which shows that $(1 + |\lambda|)^{-m-\mu-\varepsilon}A(a(x, \lambda)e^{i\lambda\psi})$ is bounded in $L^\infty \subset \mathcal{L}^0_p$ for all $p < \infty$. Thus $(1 + |\lambda|)^{-m-\mu-\varepsilon}a(x, \lambda)e^{i\lambda\psi}$ is bounded in $\mathcal{L}^{m+\varepsilon}_p$, for all $p < \infty$. It follows from Corollary 4.6 that

$$(1 + |\lambda|)^{-m-\mu-\varepsilon}p(x, D)(ae^{i\lambda\psi})$$

is bounded in $\mathcal{L}^\varepsilon_p$, for all $p < \infty$. But for p large enough, $\mathcal{L}^\varepsilon_p \subset L^\infty$, which proves the proposition.

Remark. If one keeps track of the growth of the operator norm of an element of $OP\mathcal{M}_\rho^0$ on L^p, as $p \to \infty$, one can obtain the improved result

$$|q(x, \lambda)| \le c|\lambda|^{m+\mu}(\log \lambda)^k (\lambda \text{ large}).$$

Also, as Professor Hörmander has pointed out to the author, one can obtain the above inequality directly, with $k = n - 1$, using some of the structure of $\mathcal{M}_\rho^m$ that enters into the proof of the Marcinkiewicz multiplier theorem, rather than the theorem itself.

Lemma 4.7 immediately leads to the following more precise description.

PROPOSITION 4.8. *We have* $q(x, \lambda) \in S_{\rho,1-\rho}^{m+\mu+\varepsilon}$.

Proof. We want to estimate $D_x^\beta D_\lambda^j q(x, \lambda)$. First consider a couple of special cases.

$$D_\lambda q(x, \lambda) = -e^{-i\lambda\psi}\psi P(ae^{i\lambda\psi}) + ie^{-i\lambda\psi}P(\psi ae^{i\lambda\psi}) \\ = ie^{-i\lambda\psi}[P, \psi](ae^{-i\lambda\psi}). \tag{4.11}$$

Note that $[P, \psi] \in OP\mathcal{M}_\rho^{m-\rho}$, since

$$\sigma_{[P,\psi]} \sim \sum_{|\alpha|\geq 1} \frac{1}{\alpha!}(p^{(\alpha)}\psi_{(\alpha)} - \psi^{(\alpha)}p_{(\alpha)}).$$

Now Lemma 4.7 applied to (4.11) yields $|D_\lambda q(x, \lambda)| \leq c\lambda^{m+\mu-\rho+\varepsilon}$. Similarly we have

$$D_{x_j}q(x, \lambda) = -i\lambda\psi_{x_j}e^{-i\lambda\psi}P(ae^{i\lambda\psi}) + e^{-i\lambda\psi}P(i\lambda\psi_{x_j}ae^{i\lambda\psi} + a_{x_j}e^{i\lambda\psi}) \\ + e^{-i\lambda\psi}[P, D_{x_j}](ae^{-i\lambda\psi}) \\ = i\lambda e^{-i\lambda\psi}[P, \psi_{x_j}](ae^{i\lambda\psi}) + e^{-i\lambda\psi}P(a_{x_j}e^{i\lambda\psi}) \\ + e^{-i\lambda\psi}[P, D_{x_j}](ae^{i\lambda\psi}). \tag{4.12}$$

Now $[P, \psi_{x_j}] \in OP\mathcal{M}_\rho^{m-\rho}$ while $[P, D_{x_j}] \in OP\mathcal{M}_\rho^m$, with symbol

$$\sigma_{[P,Dx_j]} \sim \frac{\partial p}{\partial x_j}.$$

Consequently, Lemma 4.7 applied to (4.12) yields $|D_{x_j}q(x, \lambda)| \leq c\lambda^{m+\mu+\varepsilon+1-\rho}$.

Iterating (4.11), we see that

$$D_\lambda^j q(x, \lambda) = ie^{-i\lambda\psi}Ad(\psi)^j \cdot P(ae^{i\lambda\psi})$$

where $Ad(A) \cdot P = [P, A]$. To see how to iterate (4.12), write (4.12) as

$$D_{x_j}q(x, \lambda) = e^{-i\lambda\psi}(\lambda T_{\psi,j}(P) + S_j(P))(ae^{i\lambda\psi}) + e^{-i\lambda\psi}P(a_{x_j}e^{-i\lambda\psi})$$

where $T_{\psi,j}(P) = [P, \psi_{x_j}]$ and $S_j(P) = [P, D_{x_j}]$. Note that $T_{\psi,j}: OP\mathcal{M}_\rho^s \to OP\mathcal{M}_\rho^{s-\rho}$ and $S_j: OP\mathcal{M}_\rho^s \to OP\mathcal{M}_\rho^s$. Now write

$$D_x^\beta D_\lambda^j q(x, \lambda) = e^{-i\lambda\psi}(\lambda T_{\psi,1} + S_1)^{\beta_1} \cdots (\lambda T_{\psi,n} + S_n)^{\beta_n} \cdot Ad(\psi)^j \cdot P(ae^{i\lambda\psi}) + \cdots \\ = e^{-i\lambda\psi} \sum_{k=0}^{|\beta|} \lambda^k U_{k,j} \cdot P(ae^{i\lambda\psi}) + \cdots$$

where $U_{k,j}: OP\mathcal{M}_\rho^s \to OP\mathcal{M}_\rho^{m-k\rho-j\rho}$, so

$$|\lambda^k U_{k,j} \cdot P(ae^{i\lambda\psi})| \leq c\lambda^{m+\mu+\varepsilon-j\rho+k(1-\rho)}.$$

Consequently, $|D_x^\beta D_\lambda^j q(x, \lambda)| \le c_\varepsilon \lambda^{m+\mu+\varepsilon-j\rho+|\beta|(1-\rho)}$, and the proposition is proved.

Using Proposition 4.8, we analyze $p(x, D)K$ as a Fourier integral operator, given K, the solution operator to a hyperbolic equation $(\partial/\partial t)u = i\lambda(t, x, D_x)u$, i.e. $Ku(0) = u(t)$, say

$$Ku = \int a(x, \xi)e^{i\varphi(x,\xi)}\hat{u}(\xi)\, d\xi. \tag{4.13}$$

In fact, we have, for $p(x, D) \in OP\mathcal{M}_\rho^m$, $a(x, \xi) \in S_{1,0}^0$,

$$p(x, D)Ku = \int q(x, \xi)e^{i\varphi(x,\xi)}\hat{u}(\xi)\, d\xi, \tag{4.14}$$

$$q(x, \xi) = e^{-i\varphi}p(x, D)(ae^{i\varphi}) \in S_{\rho,1-\rho}^{m+\varepsilon}.$$

An attempt to apply K^{-1} to both sides of (4.14) does not meet with immediate success, if $\rho < 1/2$, since the amplitude $q(x, \xi)$ does not possess much structure. Nevertheless, $K^{-1}p(x, D)K$ is a pseudodifferential operator, as we now show.

Clearly $T = K^{-1}p(x, D)K$ preserves wave front sets, i.e. $WF(Tu) \subset WF(u)$. Consequently we can localize the operator and assume without loss of generality that T operates on distributions on the torus $\mathbf{T}^n$.

THEOREM 4.9. *If $p(x, D) \in OP\mathcal{M}_\rho^m$ and the Fourier integral operator K is given, as a solution operator $Ku(0) = u(t)$ to a hyperbolic equation $(\partial/\partial t)u = i\lambda(y, x, D_x)u$, then*

$$K^{-1}p(x, D)K \in OPS_{\rho,1-\rho}^{m+\varepsilon}.$$

Proof. We first consider the case when $p(x, \xi)$ is independent of x, i.e. $p(x, \xi) = p(\xi)$, so $p(x, D) = p(D)$.

Now replacing $i\lambda$ by its skew adjoint part has the effect of multiplying K by an elliptic operator in OPS^0, so without loss of generality, we can suppose K is unitary. Now write

$$K^{-1}p(D)K = p(A_1, \ldots, A_n) \tag{4.15}$$

where $A_j = K^{-1}D_jK$. The A_j are commuting self-adjoint operators in OPS^1, and furthermore $\sum A_j^2 = -K^{-1}\,\Delta K$ is elliptic. In order to analyze the right-hand side of (4.15), we use a method which will be further developed in Chapter XII. Namely, write

$$p(A_1, \ldots, A_2)u = \int \hat{p}(s)e^{i(s_1A_1+\cdots+s_nA_n)}u\, ds, \tag{4.16}$$

and note that the ellipticity of $\sum A_j^2$ allows one to suppose $\hat{p}(s)$ is supported near $s = 0$, since $p(\xi) \in S_{\rho,0}^m$ implies $\hat{p}(s)$ is smooth and all its derivatives are rapidly decreasing, away from $s = 0$. Now replace $e^{is\cdot A}$ by its geometrical optics approximation,

$$e^{is\cdot A}u = e^{is_1A_1}\cdots e^{is_nA}\,u = \int b(s,x,\xi)e^{i\varphi(s,x,\xi)}\hat{u}(\xi)\,d\xi, \tag{4.17}$$

by, for example, first so expressing $e^{is_nA_n}$ and then iterating this procedure. Thus φ solves the eikonal equations $(\partial/\partial s_j)\varphi = a_j(x, \nabla_x\varphi)$, $a_j(x,\xi)$ being the principal symbol of A_j, and $\varphi(0,x,\xi) = x\cdot\xi$. Similarly, $b(s,x,\xi)$ satisfies certain transport equations, and $b(0,x,\xi)=1$. Substitution of (4.17) into (4.16) yields

$$p(A_1,\ldots,A_n)u = \int p(D_s)[b(s,x,\xi)e^{i\varphi(s,x,\xi)}]|_{s=0}\hat{u}(\xi)\,d\xi.$$

Now Proposition 4.8 yields $p(D_s)b(s,x,\xi)e^{i\varphi(s,x,\xi)} = c(s,x,\xi)e^{i\varphi(s,x,\xi)}$ with $c\in S^{m+\varepsilon}_{\rho,1-\rho}$. Taking $q(x,\xi)=c(0,x,\xi)\in S^{m+\varepsilon}_{\rho,1-\rho}$, we get

$$p(A_1,\ldots,A_n)u = \int q(x,\xi)e^{ix\cdot\xi}\hat{u}(\xi)\,d\xi,$$

which proves the theorem in the case $p(x,D)=p(D)$.

In general, write

$$p(x,\xi) = \sum_{\nu\in\mathbf{Z}^n} e^{i\nu\cdot x}p_\nu(\xi)$$

with $p_\nu(\xi)\in\mathcal{M}^m_\rho$ a rapidly decreasing sequence. Thus

$$\begin{aligned} K^{-1}p(x,D)K &= \sum_\nu T_\nu K^{-1}p_\nu(D)K \\ &= \sum_\nu T_\nu q_\nu(x,D) \end{aligned} \tag{4.18}$$

where $q_\nu(x,D)\in OPS^{m+\varepsilon}_{\rho,1-\rho}$ is a rapidly decreasing sequence, and $T_\nu f = K^{-1}(e^{i\nu\cdot x}Kf)$, so $T_\nu\in OPS^0$. One easily sees that $|D_\xi^\alpha D_x^\beta\sigma_{T_\nu}(x,\xi)| \le c(1+|\nu|)^{|\alpha|+|\beta|}\langle\xi\rangle^{-|\alpha|}$, and hence

$$|D_\xi^\alpha D_x^\beta \sigma_{T_\nu q_\nu}(x,\xi)| \le c_{k\alpha\beta}(1+|\nu|)^{-k}\langle\xi\rangle^{m+\varepsilon-\rho|\alpha|+(1-\rho)|\beta|}$$

which implies that the last sum in (4.18) converges to an element of $OPS^{m+\varepsilon}_{\rho,1-\rho}$. The proof of the theorem is now complete.

We shall now define a subalgebra of $OP\mathcal{M}^m_\rho$, with a little more structure. Like the examples arising in Proposition 4.2, such as pseudodifferential operators giving a parametrix for the heat equation, symbols in $\mathcal{N}^m_\rho$ will belong to $S^m_{1,0}$ except in a conic neighborhood of a variety $\{\eta=0\}$, a linear variety, independent of x. We shall be able to cut symbols off fairly close to this variety, and obtain information from the resulting decomposition. The need for such information has already arisen in Chapter X. We parametrize $T^*_x(X)$ by $(\xi,\eta)\in\mathbf{R}^\ell\times\mathbf{R}^n$, letting

$$\Sigma = \{(x,\xi,\eta)\in T^*(X)\backslash 0 : \eta = 0\}.$$

DEFINITION. *The symbol* $p(x, \xi, \eta) \in \mathcal{N}_\rho^m(\Sigma)$ *if, and only if, for* $|\eta| < |\xi|$,

$$|D_x^\beta D_\eta^\gamma D_\xi^\alpha p(x, \xi, \eta)| \le c_{\alpha\beta\gamma}|\xi|^{m-|\alpha|}(|\xi|^\rho + |\eta|)^{-|\gamma|} \tag{4.19}$$

and $p \in S_{1,0}^m$ on any conic set not intersecting Σ.

It is fairly easy to see that $\mathcal{N}_\rho^m(\Sigma)$ depends only on the linear variety Σ and not on the particular (linear) splitting of variables. We usually take Σ to be understood, and write $\mathcal{N}_\rho^m$ for short. Note that 4.19 implies

$$\begin{aligned} |\eta^\gamma D_x^\beta D_\eta^{\gamma+\gamma'} D_\xi^\alpha p(x, \xi, \eta)| &\le C|\xi|^{m-|\alpha|}|\eta|^{|\gamma|}(|\xi|^\rho + |\eta|)^{-|\gamma|-|\gamma'|} \\ &= C|\xi|^{m-|\alpha|}(|\xi|^\rho + |\eta|)^{-|\gamma'|}\frac{|\eta|^{|\gamma|}}{(|\xi|^\rho + |\eta|)^{|\gamma|}}, \end{aligned}$$

so as an immediate consequence we have

PROPOSITION 4.10. $\mathcal{N}_\rho^m \subset \mathcal{M}_\rho^m$.

Our defining inequality (4.19) is motivated by the inequality (4.4). In fact, (4.4) leads immediately to the following.

PROPOSITION 4.11. *Let* $r(\theta) \in S_{1,0}^m(\mathbf{R}^k)$, $m \ge 0$, *and let* $q(\xi, \eta) = r(|\xi|^{-\rho}\eta)$, $0 < \rho \le 1$. *Then* $q \in \mathcal{N}_\rho^{(1-\rho)m}$ *on the conic set* $|\eta| < |\xi|$.

By analogy with Proposition 4.1, elements of $\mathcal{N}_\rho^m$ have the following properties.

PROPOSITION 4.12. *Let* $p \in \mathcal{N}_\rho^m$, $q \in \mathcal{N}_\rho^\mu$, $0 < \rho \le 1$. *Then*

(i) $pq \in \mathcal{N}_\rho^{m+\mu}$;

(ii) $D_x^\beta D_\eta^\gamma D_\xi^\alpha p \in \mathcal{N}_\rho^{m-|\alpha|-\rho|\gamma|}$;

(iii) *if* $|p(x, \xi, \eta)| \ge C(1 + |\xi| + |\eta|)^m$, *then* $p(x, \xi, \eta)^{-1} \in \mathcal{N}_\rho^{-m}$.

Proof. To prove (i), write

$$D_\eta^\gamma D_\xi^\alpha(pq) = \sum_{\substack{\gamma_1+\gamma_2=\gamma \\ \alpha_1+\alpha_2=\alpha}} C_{\alpha_1\alpha_2\gamma_1\gamma_2}(D_\eta^{\gamma_1} D_\xi^{\alpha_1} p)(D_\eta^{\gamma_2} D_\xi^{\alpha_2} q)$$

so, for $|\eta| < |\xi|$,

$$\begin{aligned} |D_\eta^\gamma D_\xi^\alpha(pq)| &\le C \sum |\xi|^{m-|\alpha_1|}(|\xi|^\rho + |\eta|)^{-|\gamma_1|}|\xi|^{\mu-|\alpha_2|}(|\xi|^\rho + |\eta|)^{-|\gamma_2|} \\ &= C'|\xi|^{m+\mu-|\alpha|}(|\xi|^\rho + |\eta|)^{-|\gamma|}, \end{aligned}$$

and (i) follows easily. Assertions (ii) and (iii) are similarly simple variants of the other assertions of Proposition 4.1.

We denote the set of operators $p(x, D)$ associated with $p(x, \xi, \eta) \in \mathcal{N}_\rho^m$ by $OP\mathcal{N}_\rho^m$. Thus $OP\mathcal{N}_\rho^m \subset OP\mathcal{M}_\rho^m$, and so the L^p continuity results of Proposition 4.5 and Corollary 4.6 apply. In exact analogy with the proofs of Proposition 4.3 and Corollary 4.4 one has the following.

PROPOSITION 4.13. *If* $p \in \mathcal{N}_\rho^m$, $q \in \mathcal{N}_\rho^\mu$, $0 < \rho \le 1$, *then* $p(x, D)q(x, D) \in OP\mathcal{N}_\rho^{m+\mu}$. *Furthermore if* $|p(x, \xi, \eta)| \ge C(1 + |\xi| + |\eta|)^m$ *for* $|\xi| + |\eta|$ large (*on an open conic set* Γ) *and E is a (microlocal) parametrix for* $p(x, D)$, *then* $E \in OP\mathcal{N}_\rho^{-m}$.

So far we have seen that the theory of $OP\mathcal{N}_\rho^m$ can be developed along lines parallel to the development of $OP\mathcal{M}_\rho^m$ given in the previous section. We now examine what happens to elements of $\mathcal{N}_\rho^m$ when certain cut-offs are applied. Let $\psi(\theta) \in S_{1,0}^0(\mathbf{R}^k)$ and suppose that $\psi(\theta) = 0$ for $|\theta| \le 1$. By Proposition 4.11, if $0 < a \le 1$, $\psi(|\xi|^{-a}\eta) \in \mathcal{N}_a^0$.

PROPOSITION 4.14. *If* $p(x, \xi, \eta) \in \mathcal{N}_\rho^m$ *and* $a > \rho$, *then*

$$\psi(|\xi|^{-a}\eta)p \in \mathcal{N}_a^m.$$

Proof. It suffices to note that $|D_x^\beta D_\eta^\gamma D_\xi^\alpha p| \le C|\xi|^{m-|\alpha|}(|\xi|^a + |\eta|)^{-|\gamma|}$ for $|\xi|^a \le |\eta| \le |\xi|$, which is an immediate consequence of (4.19).

The way Proposition 4.14 can be useful is the following: often the remainder term $p - \psi(|\xi|^{-a}\eta)p$ is of lower order, and the nicer properties of $\mathcal{N}_a^m$ (compared to $\mathcal{N}_\rho^m$) can be exploited in handling the "principal term" $\psi(|\xi|^{-a}\eta)p$. This phenomenon arose in Chapter X, Section 5, where we had to analyze the behavior of

$$Q(a(x)e^{i\lambda\psi(x)}), \qquad \lambda \to \infty \tag{4.20}$$

where $Q \in OP\mathcal{N}_{1/3}^0$ has symbol

$$q(\xi, \eta) = |\xi|^{-1/3}r(|\xi|^{-1/3}\eta), \tag{4.21}$$

$r(\theta) \in S_{1,0}^{1/2}(\mathbf{R}), (k = 1)$; in the example there, $r(\theta) = A'/A(\theta)$, $A(\theta) = Ai(e^{2/3\pi i}\theta)$, where $Ai(z)$ is the Airy function. Together, Propositions 4.14 and 4.15 prove the formulas (5.24) and (5.25) of Chapter X. The next result, of course, generalizes, in fashions we leave to the reader.

PROPOSITION 4.15. *With* $\varphi \in C_0^\infty(\mathbf{R})$, $Q \in OP\mathcal{N}_{1/3}^0$ *with symbol* $q(\xi, \eta)$ *given by* (4.21), *we have*

$$q\varphi(|\xi|^{-a}\eta) \in \mathcal{N}_{1/3}^{-(1-a)/2} \qquad \text{if} \qquad a > \tfrac{1}{3}.$$

Proof. It suffices to show that $q_1\varphi_1 = r(|\xi|^{-1/3}\eta)\varphi(|\xi|^{-a}\eta) \in \mathcal{N}_{1/3}^{(a-1/3)/2}$ on $|\eta| < |\xi|$. But

$$D_\eta^k D_\xi^\alpha(q_1\varphi_1) = \sum_{\substack{k_1+k_2=k\\ \alpha_1+\alpha_2=\alpha}} C_{\alpha_1\alpha_2k_1k_2}(D_\eta^{k_1}D_\xi^{\alpha_1}q_1)(D_\eta^{k_2}D_\xi^{\alpha_2}\varphi_1),$$

so picking a $\tilde{\varphi} \in C_0^\infty(\mathbf{R})$ positive, with $\tilde{\varphi} = 1$ on supp φ, we have, using inequality (4.4),

$$\begin{aligned}|D_\eta^k D_\xi^\alpha(q_1\varphi_1)| &\le C\sum(1+|\xi|^{-1/3}|\eta|)^{1/2-k_1}|\xi|^{-|\alpha_1|-k_1/3}\\ &\quad\cdot|\xi|^{-|\alpha_2|}(|\xi|^a+|\eta|)^{-k_2}\tilde{\varphi}(|\xi|^{-a}\eta)\\ &\le C|\xi|^{-|\alpha|}\tilde{\varphi}(|\xi|^{-a}\eta)(1+|\xi|^{-1/3}|\eta|)^{1/2}\\ &\quad\cdot\sum_{k_1+k_2=k}(|\xi|^{1/3}+|\eta|)^{-k_2}(|\xi|^a+|\eta|)^{-k_2}\\ &\le C|\xi|^{-|\alpha|}|\xi|^{(a-1/3)/2}(|\xi|^{1/3}+|\eta|)^{-k}\text{ if }|\xi|\ge 1,\qquad a>\tfrac{1}{3},\end{aligned}$$

and the proof is complete.

Thus, writing $Q=Q_1+Q_2$ with $\sigma_{Q_1}=\sigma_Q\psi(|\xi|^{-a}\eta)$ and $\sigma_{Q_2}=\sigma_Q\varphi(|\xi|^{-a}\eta)$, we have $Q_1\in OP\mathcal{N}_a^0\subset OPS_{a,0}^0$ and $Q_2\in OP\mathcal{N}_{1/3}^{-(1-a)/2}\subset OP\mathcal{M}_{1/3}^{-(1-a)/2}$.

Exercises

4.1. Consider the heat operator

$$H=\frac{\partial}{\partial x}-\left(\frac{\partial^2}{\partial y_1^2}+\cdots+\frac{\partial^2}{\partial y_k^2}\right)$$

on a region $\Omega\in\mathbf{R}^{k+1}$. Suppose $u\in\mathscr{D}'(\Omega)$ and $Hu\in\mathscr{L}_{p,\text{loc}}^s(\Omega)$, for some $s\in\mathbf{R}$, $1<p<\infty$. Show that $D_y^\alpha u\in\mathscr{L}_{p,\text{loc}}^s(\Omega)$, $|\alpha|\le 2$ and $D_x u\in\mathscr{L}_{p,\text{loc}}^s(\Omega)$. Generalize this result to second order parabolic equations with variable coefficients $(\partial/\partial x)-a(x,y,D_y)$.

4.2. Let us say $p(x,\xi)\in\mathcal{M}_{\rho,\delta}^m$ if $\xi^\alpha D_\xi^\alpha p(x,\xi)\in S_{\rho,\delta}^m$ for all multi-indices α. If $0\le\delta<\rho\le 1$, consider assertions about $\mathcal{M}_{\rho,\delta}^m$ and $OP\mathcal{M}_{\rho,\delta}^m$ similar to Proposition 4.1 through Corollary 4.4. Examine the behavior of $OP\mathcal{M}_{\rho,\delta}^m$ on $\mathscr{L}_p^s$, $1<p<\infty$.

4.3. Show that $\sigma_{Q_1}^{(\alpha)}(\xi,\eta)\in\mathcal{N}_a^{-a|\alpha|-(1-a)/2}$ if $|\alpha|\ge 1$, an improvement over the usual implication from $\sigma_{Q_1}(\xi,\eta)\in\mathcal{N}_a^0$.

4.4. Show that $e^{\xi_1-|\xi|}\in\mathcal{N}_{1/2}^0$.

4.5. Let $q(\xi,\eta)=r(|\xi|^{-1/3}\eta)\in\mathcal{N}_{1/3}^{1/3}$ (on $|\eta|<|\xi|$) where $r(s)\in S_{1,0}^{1/2}(\mathbf{R})$. Show that, if $|\gamma|\ge 1$, then

$$D_\eta^\gamma D_\xi^\alpha q(\xi,\eta)\in\mathcal{N}_{1/3}^{-|\alpha|-|\gamma|/3},$$

rather than merely belonging to $\mathcal{N}_{1/3}^{1/3-|\alpha|-|\gamma|/3}$.

4.6. Let $q(\xi,\eta)=(i\xi-|\eta|^2)^{-1}\in\mathcal{N}_{1/2}^{-1}$ on $|\eta|<|\xi|$. Show that $\eta_j(i\xi-|\eta|^2)^{-1}\in\mathcal{N}_{1/2}^{-1/2}$.

4.7. Show that $(\partial/\partial x-\Delta_y)u\in\mathscr{L}_p^s$ implies

(i) $u\in\mathscr{L}_p^{s+1}$;

(ii) $\Delta_y u\in\mathscr{L}_p^s$;

(iii) $\dfrac{\partial}{\partial y_j} u \in \mathscr{L}_p^{s+1/2}$.

Hint: Use Exercise 4.6 to get (iii).

§5. Besov Spaces and Boundary Regularity

If $u \in \mathscr{L}_p^s(\mathbf{R}_+^{n+1})$, $s > 1/p$, the trace theorem asserts that

$$u\big|_{x_1=0} \in B_p^{s-1/p}(\mathbf{R}^n),$$

where the Besov spaces $B_p^s(\mathbf{R}^n)$ are defined as follows. If $0 < s \le 1$, we say $f \in B_p^s(\mathbf{R}^n)$ if, and only if, $f \in L^p(\mathbf{R}^n)$ and

$$(5.1) \qquad \int_{\mathbf{R}^n}\int_{\mathbf{R}^n} |f(x+y) - 2f(x) + f(x-y)|^p |y|^{-n-sp}\, dy\, dx < \infty.$$

If $s = k+\sigma$, $0 < \sigma \le 1$, we say $f \in B_p^s(\mathbf{R}^n)$ if, and only if, $D_x^\alpha f \in B_p^\sigma(\mathbf{R}^n)$ for all α with $|\alpha| \le k$. The restriction map $\rho: \mathscr{L}_p^s(\mathbf{R}_+^{n+1}) \to B_p^{s-1/p}(\mathbf{R}^n)$ is onto, for $s > 1/p$. In fact, the following is true.

Proposition 5.1. *If $f \in B_p^\sigma(\mathbf{R}^n)$, $\sigma > 0$, then $u(y, x) = e^{-y\sqrt{-\Delta}}f = P^y f$, the Poisson integral of f, belongs to $\mathscr{L}_p^{\sigma+1/p}([0, T] \times \mathbf{R}^n)$, for any $T < \infty$.*

For this result and other results about Besov spaces, we refer the reader to Stein [1] (where B_p^s is denoted $\Lambda_s^{p,p}$) and also to Bergh and Löfström [1]. The reader can take Proposition 5.1, as a characterization of B_p^s. From the characterization of $B_p^s(\mathbf{R}^n)$ as the space of traces of elements of $\mathscr{L}_p^{s+1/p}$ it easily follows that if $\varphi \in C_0^\infty(\mathbf{R}^n)$, $f \in B_p^s(\mathbf{R}^n)$, then $\varphi f \in B_p^s(\mathbf{R}^n)$. Also, since $\mathscr{L}_p^s$ is invariant under coordinate transformations, we are led to conclude that B_p^s is invariant under coordinate transformations. Consequently $B_p^s(M)$ is well defined on any compact manifold M, for $s > 0$, $1 < p < \infty$. Furthermore, if Ω is any compact manifold with boundary, the restriction map $\rho: \mathscr{L}_p^s(\Omega) \to B_p^{s-1/p}(\partial\Omega)$ is well defined, for $s > 1/p$, and onto.

Proposition 5.2. *If $p(x, D) \in OPS_{1,0}^0$, M compact, then*

$$p(x, D): B_p^s(M) \to B_p^s(M), \qquad s > 0, \qquad 1 < p < \infty.$$

Proof. First, if $q(\xi) \in S_{1,0}^0(\mathbf{R}^n)$, the characterization (5.1) and the L^p estimates for $q(D)$ show that $q(D): B_p^s(\mathbf{R}^n) \to B_p^s(\mathbf{R}^n)$, assuming $0 < s \le 1$, and since $q(D)$ commutes with D^α, we have this for all $s > 0$. The proof that $q(x, D): B_p^s(\mathbf{R}^n) \to B_p^s(\mathbf{R}^n)$ provided $q(x, \xi) \in S_{1,0}^0$, compactly supported in x, is an exact duplication of the proof of Theorem 2.1, and our result follows as usual via partitions of unity subordinate to coordinate patches.

Remark. Exercise 5.1 indicates a proof of Proposition 5.2 where Proposition 5.1 is taken to characterize B_p^s.

The following important result generalizes Proposition 5.1.

THEOREM 5.3. *Let u solve the elliptic evolution equation*

$$(5.2)\qquad \begin{aligned} \frac{\partial}{\partial y} u &= K(y, x, D_x)u, \\ u(0) &= f \in \mathscr{D}'(M) \end{aligned}$$

where $K(y, x, \xi) \in S^1$ is a $k \times k$ matrix, whose principal symbol $K_1(y, x, \xi)$ has only eigenvalues with real part $\le -c_0|\xi|$, $c_0 > 0$. If $f \in B_p^s(M)$, $s > 0$, $1 < p < \infty$, it follows that $u \in \mathscr{L}_p^{s+1/p}([0, T] \times M)$.

Proof. We use the parametrix for (5.2) given in Chapter IX, Section 2. There it was shown that, if $0 < c_1 < c_0$,

$$u(y) = A(y, x, D_x)P^{c_1 y}f, \qquad \operatorname{mod} C^\infty([0, T] \times M)$$

where $D_y^j A(y, x, D_x)$ is bounded in $OPS_{1,0}^j$, $0 \le y \le T$. Therefore,

$$u(y) = A(y, x, D_x)v(y)$$

where by Proposition 5.1 we know that $v(y) = P^{c_1 y}f \in \mathscr{L}_p^{s+1/p}([0, T) \times M)$. The proof will be complete if we show that, for $A(y, x, D_x)$ as above, $1 < p < \infty$, $\sigma \ge 0$,

$$(5.3)\qquad A(y, x, D_x)\colon \mathscr{L}_p^\sigma([0, T] \times M) \to \mathscr{L}_p^\sigma([0, T] \times M).$$

Indeed, suppose $\sigma = k$ is an integer. If $w \in \mathscr{L}_p^k$, $j + |\beta| \le k$, we have

$$D_y^j D_x^\beta A(y, x, D)w = \sum_{\ell + m = j} \binom{j}{\ell} A_{\ell,\beta}(y, x, D) \frac{\partial^m w}{\partial y^m}$$

where

$$A_{\ell,\beta}(y, x, D) = D_x^\beta \frac{\partial^\ell A}{\partial y^\ell}(y, x, D) \in OPS_{1,0}^{|\beta| + \ell},$$

bounded for $0 \le y \le T$. Consequently, by Theorem 2.5, we have

$$\begin{aligned} \left\| A_{\ell,\beta}(y, x, D_x) \frac{\partial^m w}{\partial y^m} \right\|_{L^p([0,T] \times M)} &\le c \left\| \frac{\partial^m w}{\partial y^m} \right\|_{\mathscr{L}_p^{\ell + |\beta|}} \\ &\le c\|w\|_{\mathscr{L}_p^k}. \end{aligned}$$

This proves (5.1) for $\sigma = k = 0, 1, 2, \ldots$, and the general result follows by interpolation. The proof of Theorem 5.2 is complete.

We are now ready for the first main result of this section.

THEOREM 5.4. *Let*

$$A = \sum_{|\alpha| \le m} a_\alpha(x) D^\alpha$$

be an elliptic operator on a compact manifold Ω, *with smooth boundary* $\partial\Omega$. *Suppose* $u \in L^p(\Omega)$ *satisfies*

$$Au = f \in \mathscr{L}_p^s(\Omega), \tag{5.4}$$

$$B_j u\big|_{\partial\Omega} = g_j \in B_p^{m+s-\mu_j-1/p}(\partial\Omega) \tag{5.5}$$

where B_j *are differential operators of order* μ_j. *Suppose* (5.5) *is a regular elliptic boundary value problem (in the sense of Chapter IX, Section* 3, *or of Chapter V). Assume* $s > 1/p$, $1 < p < \infty$. *Then we have*

$$u \in \mathscr{L}_p^{s+m}(\Omega). \tag{5.6}$$

Proof. Extending f to an element $\tilde{f}$ of $W_p^s(M)$ for some neighborhood M of Ω and solving $Av = \tilde{f} \bmod C^\infty$, we can assume, without loss of generality, that in (5.4), $f \in C^\infty(\bar{\Omega})$. Also, writing

$$g_j = \sum_\ell \varphi_\ell(x, D) g_j$$

where each $\varphi_\ell(x, \xi) \in S^0$ has symbol of order $-\infty$ outside some tiny conic subset of $T^*(\partial\Omega)$, and using Proposition 5.2, we can assume without loss of generality that the g_j have wave front set so small that the complete decoupling procedure described in Chapter IX, Section 1, for the first order system obtained from A, works.

If $f \in C^\infty(\bar{\Omega})$, we only have to worry about the smoothness of u in a neighborhood of $\partial\Omega$, which by the collar neighborhood theorem we write as $[0, T] \times \partial\Omega$. (5.4) gives rise to a first order system, in completely decoupled form,

$$\frac{\partial}{\partial y} w = \begin{pmatrix} E & \\ & F \end{pmatrix} w + R(y)w, \qquad \bmod C^\infty([0, T] \times \partial\Omega),$$

$$\beta w(0) = h$$

where $(\partial/\partial y) - E$ is an elliptic evolution equation, $(\partial/\partial y) - F$ is a backwards elliptic evolution equation, $R(y) \in OPS^{-\infty}$, bounded, for $0 \le y \le T$, and $\beta \in OPS^0$ on $\partial\Omega$. The hypothesis (5.5) guarantees $h \in B_p^{s-1/p+1}(\partial\Omega)$, and the conclusion (5.6) is equivalent to saying

$$w = \begin{pmatrix} w^I \\ w^{II} \end{pmatrix} \in \mathscr{L}_p^{s+1}([0, T] \times \partial\Omega).$$

Automatically $w^{II} \in C^\infty([0, T] \times \partial\Omega)$. On the other hand, by Theorem 5.3, we have $w^I \in \mathscr{L}_p^{s+1}([0, T] \times \partial\Omega)$ provided $w^I(0) \in B_p^{s-1/p+1}(\partial\Omega)$.

Now recall that the boundary value problem (5.4), (5.5) is regular precisely when the map

$$f \mapsto h = \beta\begin{pmatrix} f \\ 0 \end{pmatrix}$$

is an elliptic operator in OPS^0. Consequently $h \in B_p^{s-1/p+1}(\partial\Omega)$ implies $w^I(0) \in B_p^{s-1/p+1}(\partial\Omega)$, and the proof of Theorem 5.4 is complete.

Finally, we derive results on boundary regularity in Hölder classes. To do this, we shall use the following classical result on the Poisson integral of a Hölder continuous function, whose proof the reader can find in Stein [1].

LEMMA 5.5. *If $f \in C^\alpha(\mathbf{R}^n)$, then $v(y, x) = P^y f(x)$ satisfies the condition*

$$|\nabla_{y,x} v| \le c y^{\alpha-1}, \qquad 0 < y < T,$$

assuming $0 < \alpha < 1$.

COROLLARY 5.6. *If $u \in C^\alpha(\mathbf{R}^n)$, $A \in OPS^1_{1,0}$, both compactly supported in x, then $|P^y Au| \le C y^{\alpha-1}$, $0 < y < T$.*

Proof. $P^y Au = (1 - D_y)P^y((1+\sqrt{-\Delta})^{-1}A)$ with $(1+\sqrt{-\Delta})^{-1}Au \in C^\alpha(\mathbf{R}^n)$.

We can now prove the following result, analogous to Theorem 5.3.

THEOREM 5.7. *Assume u is as given in Theorem 5.3. If $f \in C^\alpha(M)$, $0 < \alpha < 1$, then $u \in C^\alpha([0, T] \times M)$. More generally, if $f \in C^{k+\alpha}(M)$, $0 < \alpha < 1$, $k = 0, 1, 2, \ldots$, then $u \in C^{k+\alpha}([0, T] \times M)$.*

Proof. It will be convenient to write the solution u in the form

$$u(y) = P^{c_1 y} B(y, x, D_x) f, \qquad \text{mod } C^\infty([0, T] \times M) \tag{5.7}$$

with $D_y^j B(y, x, \xi)$ bounded in $S^j_{1,0}$. The representation (5.7) is obtained as follows. Let the solution to

$$\frac{\partial}{\partial y} v = K(y)^* v$$

$$v(0) = g$$

be given (mod C^∞) by $v(y) = C(y, x, D_x)P^{c_1 y} g$. Then let $B(y, x, D_x) = C(y, x, D_x)^*$.

Now suppose $f \in C^\alpha(M)$, $0 < \alpha < 1$. Then, for $0 < y < T$, $B(y, x, D_x)f$ is bounded in $C^\alpha(M)$. Since

$$D_{x_j} u = P^{c_1 y} D_{x_j} B(y, x, D_x) f$$

and

$$D_y u = -c_1 P^{c_1 y}\sqrt{-\Delta} B(y, x, D_x) f + P^{c_1 y}\sqrt{-\Delta}((-\Delta)^{-1/2} B'(y)) f,$$

we see from Corollary 5.6 that

$$|\nabla_{y,x} u| \le c y^{\alpha-1}, \qquad 0 < y < T. \tag{5.8}$$

The estimate (5.8) implies $u \in C^\alpha([0, T] \times M)$. In fact, suppose $|y_1 - y_2| \le |x_1 - x_2|$ and write, with $h = |x_1 - x_2|$,

$$\begin{aligned}|u(y_1, x_1) - u(y_2, x_2)| &\le |u(y_1, x_1) - u(y_1+h, x_1)| \\ &\quad + |u(y_1+h, x_1) - u(y_1+h, x_2)| \\ &\quad + |u(y_1+h, x_2) - u(y_2, x_2)| \\ &\le c\int_0^h t^{\alpha-1}\,dt + ch^{1-\alpha}|x_1 - x_2| + c\int_0^{2h} t^{\alpha-1}\,dt \\ &\le c'|x_1 - x_2|^\alpha.\end{aligned}$$

A similar argument treats the case $|x_1 - x_2| \le |y_1 - y_2|$. The result for $f \in C^{k+\alpha}(M)$ now follows standard arguments, which we leave to the reader.

From Theorem 5.7 we obtain in the same fashion as Theorem 5.4, the result on boundary regularity in the C^α category:

THEOREM 5.8. *In the context of Theorem 5.4, suppose $u \in L^2(\Omega)$ solves the regular elliptic boundary value problem*

$$Au = f \in C^{k+\alpha}(\bar{\Omega}),$$
$$B_j u|_{\partial\Omega} = g_j \in C^{m+k-\mu_j+\alpha}(\partial\Omega)$$

where $0 < \alpha < 1$, $k = 0, 1, 2, \ldots$, the boundary values existing in some weak sense. Then

$$u \in C^{m+k+\alpha}(\bar{\Omega}).$$

Exercises

5.1. Using the characterization of $B_p^s(\mathbf{R}^n)$, $s > 0$, $1 < p < \infty$, given in Proposition 5.1, show that, if $a(x, \xi) \in S_{1,0}^0(\mathbf{R}^n)$ has compact support in x, then $a(x, D): B_p^s \to B_p^s$. Hint: Write $P^y a(x, D)f = p(y, x, D)P^y f + R(y)f$ where $R(y)$ has large negative order and $p(y, x, D_x)$ has the properties exploited in the proof of Theorem 5.2.

5.2. Prove Lemma 5.5. Hint: Note that $D_y P^y f$ or $D_{x_j} P^y f$ can be written in the form $t^{-n-1}\int q(t^{-1}z) f(x-z)\,dz$ with $|q(z)| \le c(1+|z|)^{-n-1}$, and $\int q(z)\,dz = 0$. This is thus equal to

$$t^{-n-1}\int q(t^{-1}z)(f(x-z) - f(x))\,dz,$$

so

$$\begin{aligned}|\nabla_{y,x} P^y f| &\le ct^{-n-1}\int_{\mathbf{R}^n} \langle t^{-1}z\rangle^{-n-1}|z|^\alpha\,dz \\ &= ct^{\alpha-1}\int_{\mathbf{R}^n} \langle x\rangle^{-n-1}|x|^\alpha\,dx \\ &= c't^{\alpha-1}.\end{aligned}$$

5.3. Let $f \in C^\alpha(S^1)$, $0 < \alpha < 1$, and let $u \in C(\bar{\Omega})$ be the Poisson integral of f, so $\Delta u = 0$ and $u|_{S^1} = f$. $\Omega = \{(x, y): x^2 + y^2 < 1\}$.

(i) Following the argument of Exercise 5.2, show that $|\nabla u| \le c(1-r)^{\alpha-1}$, and deduce that $u \in C^\alpha(\bar{\Omega})$ (following an argument used in the proof of Theorem 5.7).

(ii) If v denotes the harmonic conjugate of u, use the Cauchy-Riemann equations to establish $|\nabla v| \le c(1-r)^{\alpha-1}$, and deduce that $v \in C^\alpha(\bar{\Omega})$.

(iii) Deduce from parts (a) and (b) that the Hilbert transform maps $C^\alpha(S^1)$ to itself, $0 < \alpha < 1$.

5.4. Pick $\phi(\xi) \in C_0^\infty(\mathbf{R}^n)$, supported on $1 \le |\xi| \le 4$, positive, such that

$$\sum_{\nu=-\infty}^{\infty} \phi(2^{-\nu}\xi) \equiv 1.$$

Let $\phi_t(\xi) = \phi(t\xi)$. If $u \in \mathscr{E}'(\mathbf{R}^n)$, show that $u \in C^\alpha(\mathbf{R}^n)$ if, and only if,

$$|\phi_t(D)u| \le ct^\alpha, \qquad 0 \le t \le 1.$$

Hint: Derive the estimates

$$\|\phi_t(D)u\|_{L^\infty} \le Ct \sum_j \|D_{x_j} e^{-t\sqrt{-\Delta}} u\|_{L^\infty},$$

$$\|D_{x_j} e^{-t\sqrt{-\Delta}} u\|_{L^\infty} \le C \int_t^\infty \tau^{-1} e^{-t/2} \|\phi_\tau(D)u\|_{L^\infty}\, d\tau + C\|u\|_{H^{-\sigma}}.$$

5.5. Using Exercise 5.4, prove Theorem 1.7 directly. Hint: Show that (1.8) implies

$$\|\phi_t(D)P(D)u\|_{L^\infty} \le C\|\phi_t(D)u\|_{L^\infty}.$$

§6. References to Further Work

The L^p and Hölder space theory of linear elliptic equations is important in the study of nonlinear elliptic equations. See Morrey [1], Gilbarg and Trudinger [1]. Our treatment of boundary regularity in Section 5 is a little different from that of Agmon, Douglis, and Nirenberg [1]. For a nice treatment of the initial-boundary value problem for parabolic equations, see Hamilton [1]. The role of L^p estimates for $OPS_{1,0}^0$ in harmonic analysis will be discussed in the next chapter.

Regarding further generalizations of the basic results for $OPS_{1,0}^m$, other than the classes $OPS_{1,\delta}^m$ and $OP\mathscr{M}_p^m$ considered here, there has been some very important work by Stein and his colleagues (see Folland and Stein [1], Nagel and Stein [1], and Rothschild and Stein [1]) on classes of operators whose kernels arise from fundamental solutions to certain hypoelliptic operators on nilpotent Lie groups. These results are not at all amenable to analysis via Euclidean results, though the kernels that arise fall within the framework of operators studied by Coifman and Weiss [1]. These operators provide parametrices for the $\bar{\partial}$-Neumann problem on strictly pseudo-convex domains, among other things. Such parametrices have also been constructed by Boutet de Monvel [2] by another method; they belong to $OPS^m_{1/2,1/2}$. More precisely, Boutet de Monvel introduces a class of symbols having a lot in common with our class $\mathcal{N}^m_{1/2}$, but it seems unlikely that his classes of operators are bounded on L^p in general.

Beals [4] has L^p estimates for subclasses of Hörmander's [23] general classes of pseudodifferential operators, along lines similar to Nagel and Stein [1].

There has also been work on estimates for operators with minimal smoothness hypotheses on the symbols. Nagase [1] shows that $p(x, D)$ is continuous on $L^p(\mathbf{R}^n)$, $1 < p < \infty$, provided

$$|D_\xi^\alpha p(x, \xi)| \leq C_\alpha(1+|\xi|)^{-|\alpha|} \tag{6.1}$$

and

$$|D_\xi^\alpha p(x, \xi) - D_\xi^\alpha p(y, \xi)| \leq C_\alpha |x-y|^\sigma (1+|\xi|)^{-|\alpha|+\sigma\delta} \tag{6.2}$$

for $|\alpha| \leq n+2$, for some $\sigma \in (0, 1]$, $\delta \in [0, 1)$. This generalizes the L^p boundedness of $OPS^0_{1,\delta}$, $0 \leq \delta < 1$. To generalize the L^p boundedness of $OPS^0_{1,0}$, one needs even less regularity, and Coifman and Meyer [1] show that $p(x, D)$ is continuous on $L^p(\mathbf{R}^n)$, $1 < p < \infty$, provided (6.1) holds for all α and

$$|D_\xi^\alpha p(x, \xi) - D_\xi^\alpha p(y, \xi)| \leq C_\alpha \omega(|x-y|)(1+|\xi|)^{-|\alpha|} \tag{6.3}$$

where the modulus of continuity $\omega(t)$ satisfies the sub-Dini condition

$$\sum_{j=0}^{\infty} \omega(2^{-j})^2 < \infty. \tag{6.4}$$

Such a condition on $\omega(t)$ cannot be weakened.

Since a powerful method in linear *PDE* is to obtain parametrices as sums of operators of the form $T^{-1}AT$, A a pseudodifferential operator and T a Fourier integral operator, it would be good to have classes of operators known to be continuous on L^p, invariant under conjugation by (perhaps subclasses of) Fourier integral operators. A related problem would be to obtain L^p estimates for operators in a class like $\mathcal{N}^m_\rho(\Sigma)$, but allowing the variety Σ to twist as x varies (and perhaps to be non-flat for each x). A great deal of work remains to be done on the L^p theory of pseudodifferential operators.

CHAPTER XII

Spectral Theory and Harmonic Analysis of Elliptic Self-Adjoint Operators

If $P \in OPS^m$ is an elliptic positive self-adjoint operator on a compact manifold X, of order $m > 0$, then $(I + P)^{-1} \in OPS^{-m}$ is compact, so P has a complete orthonormal set of eigenfunctions u_j; $Pu_j = \lambda_j u_j$, with $\lambda_j \to \infty$. Several topics naturally occur.

(i) *The asymptotic behavior of λ_j as $j \to \infty$.* We examine this in Section 2.

(ii) *The convergence of eigenfunction expansions.* We have

$$f = \sum_{j=1}^{\infty} (f, u_j)u_j = \sum_{j=1}^{\infty} P_j f.$$

Clearly this expansion converges in L^2 if $f \in L^2$. From this it follows easily that you get convergence in $H^s(X)$ for $f \in H^s(X)$. For $f \in L^p(X)$, $p \neq 2$, convergence in $L^p(X)$ fails even on the torus $\mathbf{T}^2$, as a celebrated result of Fefferman shows, but if one uses various summability methods, one has results for $L^p(X)$, $1 < p < \infty$, and also C^α and pointwise almost everywhere results. We look at this in Section 4.

(iii) *Functional calculus.* If $p(\lambda)$ is a bounded continuous function on $\mathbf{R}$, then $p(P) = \sum p(\lambda_j)P_j$ defines a bounded continuous operator on $L^2(X)$. It is useful to obtain a more precise description of $p(P)$ for $p(\lambda)$ with additional structure. In particular, if $p(\lambda) \in S^0_{\rho,0}(\mathbf{R})$, we show that $p(P) \in OPS^0_{\rho,1-\rho}$, provided $\rho > 1/2$, and its principal symbol is $p(\sigma_P)$.

Indeed, Topic (iii) is the first one we tackle, in Section 1, and we use this functional calculus as the basis for our subsequent analysis in Sections 2 and 3. We deal only with first order operators. It turns out that $A = P^{1/m} \in OPS^1$. This is proved in Seeley [3]; we include a proof for $m = 2$ at the end of Section 1, which would extend without difficulty to any integer m. We have not gone into further details since our principal operator of interest is $A = \sqrt{-\Delta}$.

We give two derivations of the asymptotic behavior of λ_j, the first via the analysis of $tr\ p_t(A)$ for a class of functions $p(\lambda)$ equal to 1 for $\lambda \in [0, 1]$ and vanishing for $\lambda > 1 + \epsilon$ (with $p_t(\lambda) = p(t\lambda)$). This is similar in spirit to the analysis by Hörmander in [13], but we do not obtain the sharp estimate on the remainder term. Since Hörmander's first work on Fourier integral operators was in this paper on the sharp estimate for the remainder, it might seem pretty perverse to omit this discussion in this chapter, but I decided to refer the reader to Hörmander [13] and Duistermaat and Guillemin [1] for elegant expositions of this topic and concentrate in Section 2 on showing that the classical estimate on λ_j is a very quick consequence of the analysis of e^{itA}. Our second derivation uses the analysis of e^{-tA} obtained in Chapter VIII, Section 4, via complex geometrical optics. This approach has a lot in common with the heat equation approach which has been popular in Riemannian geometry (see Berger, Gauduchon, and Mazet [1], Gilkey [1], and Mckean and Singer [1]). This approach produces an asymptotic analysis of

$$\sum_j e^{-t\lambda_j}$$

as $t \downarrow 0$; from there one needs to apply a Tauberian theorem. In the first approach no Tauberian theorem is needed, due to the fact that our set of functions $p_t(\lambda)$ is much richer than $e^{-t\lambda}$.

In Section 3 we analyze the kernels $K_t(x, y)$ of operators $p(tA)$ for $p(\lambda) \in S^{-s}_{1,0}(\mathbf{R})$; $p(tA)u = \int K_t(x, y)u(y)\,dy$. It is shown that, for $s > 0$, these kernels have a lot in common with Poisson kernels. The method used here is a further development of the techniques used in Section 2 to analyze the spectrum of A. At the end of Section 3 we deduce some basic properties of the spectral function of A. Our analysis of the Poisson-like kernels also provides the tool for analyzing pointwise convergence of eigenfunction expansions in Section 4.

In Section 5 we consider some special properties of eigenfunction expansions of measures, and in Section 6 we consider some applications of the results of Sections 1 through 5 to the study of representation and harmonic analysis on compact Lie groups. This section will be of interest only to people familiar with basic theory of compact Lie groups.

Even though the machinery we develop here allows one to dispense with Tauberian theorems, we present some alternative approaches to several propositions, requiring the use of Tauberian theorems, partly in order to elucidate the original role of Tauberian theorems in spectral theory. We include a section on Tauberian theorems, proving the Karamata theorem used in Section 2 and the Tauberian theorem arising in Section 5, which follows easily from Karamata's theorem. We give a self-contained treatment of Weiner's Tauberian theorem, using the beautiful derivation due to Korevaar [1], via distribution theory. Also our Theorem

7.11 is more general than Wiener's theorem XI′ [1] (specialized to the positive case), and one can obtain the classical Tauberian theorems of Hardy-Littlewood and Karamata more directly from Theorem 7.11 than from Wiener's theorem.

§1. Functions of Elliptic Self-adjoint Operators

Let $A \in OPS^1$ be an elliptic positive self-adjoint operator on a compact manifold M; assume $\sigma_A(x, \xi)$ has an asymptotic expansion in functions homogeneous of degree ≤ 1 in ξ, in principal part being a eal scalar valued symbol. If $p(\lambda)$ is a Borel function on $\mathbf{R}$, then $p(A)$ is defined by the spectral theorem. The purpose of this section is to prove that, for a large class of functions $p(\lambda)$, $p(A)$ is a pseudodifferential operator. A special case of this is a result of Seeley [3] that A^γ is a pseudodifferential operator of order $Re\ \gamma$, for every $\gamma \in \mathbf{C}$. Our analysis starts with the formula

$$p(A)f = \frac{1}{\sqrt{2\pi}} \int_{-\infty}^{\infty} \hat{p}(t) e^{itA} f\, dt \tag{1.1}$$

which is a simple consequence of the Fourier inversion formula and the spectral theorem. Note that (1.1) is valid for every $f \in \mathscr{D}'(M)$. Note that $u(t, x) = e^{itA}f(x)$ solves the hyperbolic equation $(\partial/\partial t)u = iAu$, so for small $|t|$, we can replace $e^{itA}f$ by its geometrical optics approximation. Thus it would be desirable to split the integral in (1.1) into two pieces, $\int q_1(t)e^{itA}f\, dt + \int q_2(t)e^{itA}f\, dt$, with q_1 supported on a small interval $(-\epsilon, \epsilon)$, and $\int q_2(t)e^{itA}f\, dt$ giving no trouble. To accomplish this, we need two simple lemmas.

LEMMA 1.1. *If $q_2(t), q_2'(t), \ldots, q_2^{(k)}(t)$ all belong to $L^1(\mathbf{R})$, then*

$$\int_{-\infty}^{\infty} q_2(t) e^{itA}\, dt$$

maps $H^s(M)$ to $H^{s+k}(M)$, for each s.

Proof. We have

$$\int_{-\infty}^{\infty} q_2(t)e^{itA}\, dt = (-iA)^{-k} \int_{-\infty}^{\infty} q_2^{(k)}(t) e^{itA}\, dt;$$

the latter integral is bounded on L^2 since $q_2^{(k)} \in L^1$ and $\|e^{itA}\| \equiv 1$, the operator norm being on any H^s, while $A^{-k}: H^s(M) \to H^{s+k}(M)$.

LEMMA 1.2. *Let $p(\lambda) \in S^m_{\rho,0}(\mathbf{R})$, $0 < \rho \leq 1$; i.e., suppose*

$$|p^{(j)}(\lambda)| \leq c_j(1 + |\lambda|)^{-j\rho+m}, \qquad j = 0, 1, 2, \ldots$$

Then $\hat{p}(t)$ and all its t derivatives are integrable on $\mathbf{R}^1\backslash(-\epsilon,\epsilon)$ for any $\epsilon>0$. In fact, they are all rapidly decreasing on $\mathbf{R}\backslash(-\epsilon,\epsilon)$.

Proof.

$$t^j\hat{p}^{(\ell)}(t) = c\int_{-\infty}^{\infty} e^{i\lambda t}D_\lambda^j(\lambda^\ell p(\lambda))\,d\lambda$$
$$= c\int_{-\infty}^{\infty} e^{i\lambda t}q_{j,\ell}(\lambda)\,d\lambda$$

where $q_{j,\ell}\in S_{\rho,0}^{m+\ell-j\rho}$. Note that $q_{j,\ell}(\lambda)$ is integrable if $m+\ell-j\rho<-1$, so

$$|t^j\hat{p}^{(\ell)}(t)|\le c_{j,\ell}<\infty, \qquad j\rho>m+\ell+1.$$

We remark that, using Plancherel's theorem, one obtains $t^j\hat{p}^{(\ell)}(t)\in L^2(\mathbf{R})$ provided $j\rho>m+\ell+(1/2)$, which leads to $\hat{p}^{(j)}(t)\in L^1(\mathbf{R}-(-\epsilon,\epsilon))$.

Now for the main theorem of this section.

THEOREM 1.3. *If A is a self-adjoint elliptic operator as described above, and if $p(\lambda)\in S_{\rho,0}^m(\mathbf{R})$ with $1/2<\rho\le 1$, then $p(A)\in OPS_{\rho,1-\rho}^m$.*

Proof. We have

$$p(A)f=\frac{1}{\sqrt{2\pi}}\int_{-\infty}^{\infty}\hat{p}(t)e^{itA}f\,dt.$$

Without loss of generality, we can assume $\hat{p}$ is supported in a small neighborhood of $t=0$, for by Lemmas 1.1 and 1.2, altering $p(\lambda)$ to have this property would only change $p(A)$ by a smoothing operator. Next, for small $|t|$ we can replace $e^{itA}f$ by its geometric optics approximation

$$e^{itA}f=\int b(t,x,\xi)e^{i\varphi(t,x,\xi)}\hat{f}(\xi)\,d\xi \qquad (\text{mod } C^\infty)$$

where φ solves the eikonal equation $\varphi_t=a_1(x,\nabla_x\varphi)$, $\varphi(0,x,\xi)=x\cdot\xi$, and $b(t,x,\xi)$ solves a certain transport equation with $b(0,x,\xi)=1$. Thus, up to a smooth error, we can write

$$p(A)f=\frac{1}{\sqrt{2\pi}}\iint\hat{p}(t)b(t,x,\xi)e^{i\varphi(t,x,\xi)}\hat{f}(\xi)\,d\xi\,dt$$
$$=\int p(D_t)[b(t,x,\xi)e^{i\varphi(t,x,\xi)}]|_{t=0}\hat{f}(\xi)\,d\xi$$
$$=\int q(x,\xi)e^{ix\cdot\xi}\hat{f}(\xi)\,d\xi$$

where

$$q(x,\xi)=e^{-ix\cdot\xi}p(D_t)[b(t,x,\xi)e^{i\varphi(t,x,\xi)}]|_{t=0}$$

can be asymptotically evaluated by the stationary phase method. By the fundamental asymptotic expansion lemma (see Chapter VIII, Section 7), it follows that $q(x, \xi) \in S^m_{\rho,1-\rho}$ if $\rho > 1/2$, and in fact

$$q(x, \xi) \sim b(0, x, \xi)p(\varphi_t(0, x, \xi)) + \cdots$$
$$= p(a_1(x, \xi)) + \cdots$$

In the case $\rho = 1$, this result has been obtained by Strichartz [1]. His method is somewhat different from ours, but it probably extends to $\rho > 1/2$ also. We should note that even for $0 < \rho < 1/2$, if $p(\lambda) \in S^0_{\rho,0}(\mathbf{R})$ then $p(A)$ is microlocal, i.e., $WF(p(A)u) \subset WF(u)$. Indeed, with $\varphi_1 \in C_0^\infty(-\epsilon, \epsilon)$, $\varphi_2 = 0$ near 0, $\varphi_1 + \varphi_2 = 1$, writing

$$p(A) = \frac{1}{\sqrt{2\pi}} \int \hat{p}(t)e^{itA}\varphi_1(t)\,dt + \frac{1}{\sqrt{2\pi}} \int \hat{p}(t)e^{itA}\varphi_2(t)\,dt$$
$$= T_1 + T_2,$$

we see that $T_2 \in OPS^{-\infty}$ and $WF(T_1u) \cap S^*(X)$ lies within a $C\epsilon$ neighborhood of $WF(u) \cap S^*(X)$. Thus $WF(p(A)u) \subset WF(u)$. Consequently $p(A)$ ought to be some sort of pseudodifferential operator, even though the analysis in Theorem 1.3 completely breaks down, and one cannot expect $p(A)$ to belong to $OPS^m_{\rho,1-\rho}$. This is similar to the phenomenon one sometimes encounters of considering $Ip(x, D)I^{-1}$ where I is an elliptic Fourier intergral operator and $p(x, D) \in OPS^m_{\rho,0}$ with $\rho < 1/2$. In either case, we may wish to regard the operators so obtained as pseudodifferential operators in search of a symbol class. We might remark that, as a consequence of the work of Melin and Sjöstrand [1] and others on Fourier integral operators with amplitude in $S^m_{1/2,1/2}$, it follows that for $p(\lambda) \in S^m_{1/2,0}(\mathbf{R})$ one has $p(A) \in OPS^m_{1/2,1/2}$, due to the fact that the symbol $q(x, \xi)$ arising in the proof of Theorem 1.3 belongs to $S^m_{1/2,1/2}$ in this case. However, the formula for the principal symbol breaks down.

It will be useful to note that Theorem 1.3 continues to hold for functions $p(A_1, \ldots, A_n)$ of several self-adjoint operators, provided

(1.2) $A_1, \ldots A_n \in OPS^1$ are *commuting* self-adjoint operators on M,

(1.3) $B = A_1^2 + \cdots + A_n^2$ is *elliptic*, and

(1.4) $p(\lambda) \in S^m_{\rho,0}(\mathbf{R}^n)$.

In fact, with $y \cdot A = y_1A_1 + \cdots + y_nA_n$, we have

$$p(A_1, \ldots, A_n)f = (2\pi)^{-n/2} \int \hat{p}(y)e^{iy \cdot A}f\,dy.$$

Lemma 1.2 generalizes to show that all derivatives of $\hat{p}(y)$ are rapidly decreasing, and consequently if $\varphi_1 + \varphi_2 = 0$, we have, for all k,

$$\int \hat{p}(y)\varphi_2(y)e^{iy \cdot A}f\,dy = (1+B)^{-k} \int e^{iy \cdot A}f(1 - \Delta_y)^k(\varphi_2(y)\hat{p}(y))\,dy$$

which shows that $p(A_1, \ldots, A_n)$ is pseudolocal on M, and the rest of the proof of Theorem 1.3 goes through:

$$p(A_1, \ldots, A_n)f = \int p(D_y)[b(y, x, \xi)e^{i\varphi(y,x,\xi)}]|_{y=0}\hat{f}(\xi)\,d\xi$$
$$= \int q(x, \xi)e^{ix\cdot\xi}\hat{f}(\xi)\,d\xi$$

with $q(x, \xi) \sim p(a_{(1)1}(x, \xi), \ldots, a_{(n)1}(x, \xi)) + \cdots$

The case of Theorem 1.3 which we shall primarily exploit is the case $p \in S^m_{1,0}$, which leads to $p(A) \in OPS^m_{1,0}$. It follows that, if $\{p_t: 0<t<1\}$ is a *bounded* family of symbols in $S^0_{1,0}$, say, then $\{p_t(A): 0<t<1\}$ is a bounded family of operators on M in $OPS^0_{1,0}$. A particular example arises from fixing $p \in S^0_{1,0}$ and letting $p_t(\lambda) = p(t\lambda)$; for instance, one could have $p_t(\lambda) = e^{-t\lambda}$ for $\lambda \geq 0$. Since we obtained the boundedness on $L^p(M)$ of elements of $OPS^0_{1,0}$ in Chapter IV, unform operator boundedness of the $p_t(A)$ on $L^p(M)$ follows. Of course, uniform boundedness of $p_t(A)$ on $L^2(M)$ is a trivial consequence of the spectral theorem, but even in L^2 we get some interesting results which do not drop out of the spectral theorem, such as the following simple consequence of Theorem 1.3.

COROLLARY 1.4. *If $\{p_t(\lambda): t \in I\}$ is a bounded set of symbols in $S^0_{1,0}(\mathbf{R})$, then for any first order operator $T \in OPS^1_{1,0}$, the family of commutators*

$$[p_t(A), T]$$

is uniformly bounded on $L^2(M)$.

In particular, we get L^2 operator boundedness of $[e^{-tA}, T]$, $0 < t < 1$, so e^{-tA} could play the role of a Friedrichs mollifier. Of course, we also get L^p boundedness and $C^{k+\alpha}$ boundedness.

A particular example of an elliptic self-adjoint operator of order one is $\sqrt{-\Delta}$, where Δ is the Laplacian on M, endowed with a Riemannian metric. The fact that $\sqrt{-\Delta}$ is a pseudodifferential operator is contained in the results of Seeley, but since this case is fairly straightforward, we indicate a proof. First we construct an elliptic pseudodifferential operator A of order 1 which is positive self adjoint and such that $A^2 + \Delta \in OPS^{-\infty}$. We then deduce that $A - \sqrt{-\Delta} \in OPS^{-\infty}$. Let $A_1 \in OPS^1$ be any positive self-adjoint operator whose principal symbol is $\sqrt{\sigma_{-\Delta}(x, \xi)}$. We have $A_1^2 = -\Delta + R_1$ with $R_1 \in OPS^1$. We now look for a self adjoint $A_0 \in OPS^0$ such that $(A_1 + A_0)^2 + \Delta = R_0 \in OPS^0$. In fact, this condition is equivalent to $A_0 = -(1/2)A_1^{-1}R_1 \bmod OPS^0$. Since we want A_0 to be self adjoint, take

$$A_0 = -\tfrac{1}{4}A_1^{-1}R_1 - \tfrac{1}{4}R_1A_1^{-1}.$$

We continue in this fashion, obtaining

$$(A_1 + A_0 + \cdots + A_{-j})^2 + \Delta = R_{-j} \in OPS^{-j}$$

with $A_{-j} = -(1/4)A_1^{-1}R_{1-j} - (1/4)R_{1-j}A_1^{-1}$. Now there exists an operator $A \in OPS^1$ such that

$$A \sim A_1 + A_0 + \cdots + A_{-j} + \cdots$$

in the sense that $A - (A_1 + \cdots + A_{-j}) \in OPS^{-j-1}$ for all j. Pick one such A. Clearly it differs from its adjoint by an element of OPS^{-j} for all j, so we may suppose A is self adjoint. It is elliptic, and by Gårding's inequality it can have at most finitely many nonpositive eigenvalues counting multiplicities. Hence we can assume A is positive. Clearly $A^2 + \Delta = R \in OPS^{-\infty}$. It remains only to show that $A - \sqrt{-\Delta}$ is a smoothing operator. We use the formulas

$$(-\Delta)^{-1/2} = \frac{1}{2\pi i}\int_\gamma z^{-1/2}(z+\Delta)^{-1}\,dz,$$

$$A^{-1} = \frac{1}{2\pi i}\int_\gamma z^{-1/2}(z - A^2)^{-1}\,dz$$

$$= \frac{1}{2\pi i}\int_\gamma z^{-1/2}(z+\Delta-R)^{-1}\,dz,$$

where γ is the obvious sort of curve circling the spectrum of $-\Delta$. Strictly speaking one should modify $-\Delta$ on its zero eigenspace, to make it strictly positive. This yields

$$A^{-1} - (-\Delta)^{-1/2} = \frac{1}{2\pi i}\int_\gamma z^{-1/2}\{(z+\Delta-R)^{-1} - (z+\Delta)^{-1}\}\,dz$$

$$= \frac{1}{2\pi i}\int_\gamma z^{-1/2}(z+\Delta-R)^{-1}R(z+\Delta)^{-1}\,dz.$$

Now for any $u \in \mathscr{D}'(M)$, $(z+\Delta)^{-1}u$ is a bounded subset of $\mathscr{D}'(M)$, as z varies over γ; hence $R(z+\Delta)^{-1}u$ is a bounded subset of $C^\infty(M)$, since $R \in OPS^{-\infty}$; hence $z^{-1/2}(z+\Delta-R)^{-1}R(z+\Delta)^{-1}u$ is integrable on γ in any $H^k(M)$. Thus $A^{-1} - (-\Delta)^{-1/2} = R^1 \in OPS^{-\infty}$, and hence $A - \sqrt{-\Delta} = -AR^1\sqrt{-\Delta}: \mathscr{D}'(M) \to C^\infty(M)$.

Exercises

1.1. Let $A: \mathscr{D}'(R^n) \to \mathscr{D}'(R^n)$ be properly supported and suppose that, denoting $[T, S]$ by $Ad\ T(S)$, we know that

$$(Ad\ x)^\alpha(Ad\ D)^\beta A: H^s(R^n) \to H^{s+|\alpha|}(R^n). \tag{1.5}$$

Show that $A \in OPS^0_{1,0}$. Hint: Define $a(x, \xi) = e^{-ix\cdot\xi}A(e^{ix\cdot\xi})$, so for $u \in \mathscr{E}'(R^n)$, $Au = \int a(x, \xi)e^{ix\cdot\xi}\hat{u}(\xi)\,d\xi$. Then $D_\xi^\alpha D_x^\beta a(x, \xi) = e^{-ix\cdot\xi}B_{\alpha\beta}(e^{ix\cdot\xi})$ where $B_{\alpha\beta} = (Ad\ x)^\alpha(Ad\ D)^\beta A$. Thus (1.5) leads to estimates on $D_x^\beta D_\xi^\alpha a(x, \xi)$. This result was communicated to me by H. O. Cordes.

1.2. Let $A:\mathcal{D}'(M) \to \mathcal{D}'(M)$, M compact, and suppose that, given any $P_j \in OPS^1_{1,0}$, you have

$$Ad(P_1)\cdots Ad(P_k)A:L^2(M) \to L^2(M).$$

Show that $A \in OPS^0_{1,0}$. Hint: Use Exercise 1.1.

1.3. Let $U \in OPS^0_{1,0}$ be a unitary operator on a compact manifold M. If $f \in C^\infty(S^1)$, define $f(U)$ by

$$f(U) = \frac{1}{2\pi}\sum_{n=-\infty}^{\infty} \hat{f}(n)U^n. \tag{1.6}$$

Show that $\hat{f}(n)U^n$ is rapidly decreasing in $OPS^0_{1,0}$, and hence that (1.6) gives a convergent series for $f(U) \in OPS^0_{1,0}$. Hint: It suffices to show that each x, ξ derivative of the symbol of U^n is polynomially bounded in n. Deduce this from Exercise 1.2 and a polynomial bound, in n, on the operator norm on $L^2(M)$ of

$$Ad(P_1)\cdots Ad(P_k)U^n$$

given any $P_j \in OPS^1_{1,0}$.

1.4. Let $f(U)$ be as in Exercise 1.3, assuming $U \in OPS^0$. Show that

$$\sigma_{f(U)}(x, \xi) = f(\sigma_U(x, \xi)) \bmod S^{-1}_{1,0}.$$

Hint: It suffices to show that $\sigma_{U^n(x,\xi)} - V_n(x, \xi)$ is polynomially bounded in n, in $S^{-1}_{1,0}$, where $V_n(x, \xi) = U_0(x, \xi)^n\phi(\xi)$, $\phi(\xi) = 0$ for $|\xi| \leq 1$, $\phi(\xi) = 1$ for $|\xi| \geq 2$. Establish this by writing $V_j = V_j(x, D)$ and

$$\begin{aligned} U^n &= V_n - (V_n - UV_{n-1}) - (UV_{n-1} - U^2V_{n-2}) - \cdots - (U^{n-1}V_1 - U^nV_0) \\ &\quad + U^n - U^nV_0 \\ &= V_n - \sum_{j=1}^{n} U^{j-1}(V_{n-j+1} - UV_{n-j}) + U^n(I - V_0). \end{aligned}$$

Now estimate the operator norms, from $L^2(M)$ to $H^1(M)$, of $Ad(P_1)\cdots Ad(P_k)(U^n - V_n)$, making use of the already established "polynomial boundedness" of U^n.

1.5. Let $A \in OPS^0$ be self adjoint. Given $g \in C^\infty(R)$, define $g(A)$ by the spectral theorem. Show that $g(A) \in OPS^0$ and that

$$\sigma_{g(A)}(x, \xi) - g(\sigma_A(x, \xi)) \in S^{-1}.$$

Hint: Use Exercises 1.3 and 1.4.

1.6. Let $A \in OPS^\sigma_{1,0}$ be self adjoint, on a compact manifold M, $0 \leq \sigma < 1/2$.

(i) If A has scalar principal symbol and is elliptic, show that $e^{iA} \in OPS^0_{1-\sigma,\sigma}$. Hint: Write $e^{iA} = f(B)$ where $B = A^{1/\sigma} \in OPS^1_{1,0}$ and $f(s) = e^{is^\sigma} \in S^0_{1-\sigma,0}(\mathbf{R})$.

(ii) Show that $e^{iA} \in OPS^0_{1-\sigma,\sigma}$ generally, for $\sigma < 1/2$, i.e. with no ellipticity hypothesis. Hint: Construct a parametrix P^t for the equation

$$\frac{\partial}{\partial t} u = iAu; \qquad u(0) = g$$

in the form $P^t g = \int p(t, x, \xi) e^{ix \cdot \xi} \hat{g}(\xi)\, d\xi$, in the spirit of the proof of Proposition 2.1 in Chapter IX; note that such a construction works precisely for $\sigma < 1/2$.

(iii) Show that the conclusion of (ii) holds for $0 \leq \sigma < 1$, by constructing a parametrix P^t for $(\partial/\partial t)u = iAu$ in the form

$$P^t u(x) = \int b(t, x, \xi) e^{i\psi(t,x,\xi) + ix \cdot \xi} \hat{v}(\xi)\, d\xi$$

where, for $|\xi|$ large, ψ solves the "eikonal equation" (with $A = a(x, D)$)

$$\frac{\partial}{\partial t} \psi = a(x, \nabla_x \psi + \xi), \qquad \psi(0, x, \xi) = 0$$

which gives $\psi \in S^\sigma_{1,0}$, real valued, so $e^{i\psi} \in S^0_{1-\sigma,\sigma}$.

1.7. Let

$$f(t) = \sum_{j=-\infty}^{\infty} a_j e^{i\lambda_j t}$$

with $\lambda_j \in R$ and $\lambda_{j+1} \geq (1+c)\lambda_j$ for $j > 0$, $\lambda_{j-1} \leq -(1+c)|\lambda_j|$ for $j < 0$, $c > 0$. Suppose $\sum |a_j| < \infty$.

(i) If I is a nonempty interval on R and $f \in C^\infty(I)$, show that $f \in C^\infty(R)$.

(ii) If f is real analytic on I, show that f is analytic on R.

(iii) If $f = 0$ on I, show that $f \equiv 0$.

Hint for (i): If $T_a u(t) = u(t + a)$, note that $T_a f(t) = \sum a_j e^{i\lambda_j a} e^{i\lambda_j t}$. Show that $T_a f = P_a(D) f$ for some $P_a(\xi) \in S^0_{1,0}(R)$, i.e. there is a $P_a(\xi) \in S^0_{1,0}(R)$ such that $P_a(a\lambda_j) = e^{i\lambda_j a}$. Hint for (ii): Show that, indeed, you can choose $P_a(\xi) \in S^0_{1,0}(R)$ such that $P_a(a\lambda_j) = e^{i\lambda_j a}$, and such that

$$(1.7) \qquad |D^k P_a(\xi)| \leq C_1 (C_1 k)^k |\xi|^{-k} \qquad \text{for } |\xi| \geq C_2$$

with $C_j = C_j(a)$. Show that (1.7) leads to $\hat{P}_a(t)$ being real analytic for $t \neq 0$, and in fact

$$(1.8) \qquad |D_t^k \hat{P}_a(t)| \leq C_{\mu,a} (C_{\mu,a} k)^k t^{-\mu}, \qquad \mu \geq 2.$$

Now apply (1.8) to $f(t+a) = \hat{P}_a * f(t)$. In order to find such a $P_a(\xi)$, construct $P_a(\zeta)$, bounded and holomorphic in the wedge $-(\pi/2) < \arg \zeta < \pi/2$, such that $P_a(a\lambda_j) = e^{i\lambda_j a}$. (That such an interpolation problem can be solved is demonstrated in Carleson [1].) Make a similar construction for $\xi < 0$.

Let $u \in \mathscr{D}'(M)$ have eigenfunction expansion

$$u = \sum_{j=0}^{\infty} a_j u_j$$

where the u_j are the eigenfunctions of $\sqrt{-\Delta}$, with eigenvalues $\lambda_j \uparrow$. We say u has a gap series expansion if $a_j = 0$ except for j of the form n_k with $n_k \geq (1+c)n_{k-1}$, $c > 0$.

1.8. Let $u \in \mathscr{D}'(M)$ have a gap series expansion, M a compact, connected, Riemannian manifold. Show that if $u = 0$ on Ω, a nonempty open subset of M, then $u \equiv 0$. Assume $u \in H^s(M)$, $s > (3/2)n$. Hint: Define $w(t, x)$ on $R \times M$ by

$$\left(\frac{\partial^2}{\partial t^2} - \Delta\right) w = 0,$$

$$w(0, x) = u; \qquad w_t(0, x) = 0.$$

By finite propagation speed, $w = 0$ on $(-\varepsilon, \varepsilon) \times \mathcal{O}$ for some nonempty open $\mathcal{O} \subset \Omega$. Since

$$w = \frac{1}{2} \sum_{j=0}^{\infty} a_j u_j (e^{i\lambda_j t} + e^{-i\lambda_j t}),$$

show that 1.7 applies, to yield $w(t, x) = 0$ on $R \times \mathcal{O}$. Use a unique continuation principle to deduce from this that $u \equiv 0$. (See Exercise 2.6 of Chapter XIV.)

1.9. Let $u \in \mathscr{D}'(M)$ have a gap series expansion, as before. Show that $WF(u)$ is invariant under the geodesic flow on $T^*(M)$. Assume $s > (3/2)n$. Hint: Use a variant of the method for Exercise 1.8, showing that if $WF(u) \cap \operatorname{supp} \psi = \emptyset$, so $\psi(x, D)u \in C^\infty(M)$, then $\psi(x, D)u \in C^\infty(R \times M)$, and replacing the unique continuation argument by propagation of singularities for solutions to the wave equation on $R \times M$.

1.10. In Exercises 1.8 and 1.9, remove the restriction $u \in H^s(M)$ for $s > (3/2)n$. Hint: In 1.8, replace u by $u_j = \int_{-\infty}^{\infty} j\phi(jt)w(t)\,dt$ where $\phi \in C_0^\infty(R)$, $\int \phi = 1$. In 1.9, replace u by $(1-\Delta)^{-m}u$, $m > (3/4)n - (s/2)$.

1.11. Generalize Exercises 1.7 through 1.10 to the case where the gaps between the λ_j satisfy the weaker condition

$$|\lambda_{j+1} - \lambda_j| \geq c|\lambda_j|^\alpha$$

where $0 < \alpha \leq 1$.

§2. The Asymptotic Behavior of the Spectrum

If $A \in OPS^1$ is a positive self-adjoint elliptic operator on a compact manifold M, it has a complete set of eigenvalues $0 < \lambda_1 \leq \lambda_2 \leq \cdots$. We shall obtain information on the asymptotic nature of λ_j as $j \to \infty$ by studying the behavior of

$$tr\, p(tA) = \sum_j p(t\lambda_j)$$

as $t \downarrow 0$, for various $p(\lambda) \in \mathcal{S}(\mathbf{R})$. Note that for $p \in \mathcal{S}$, $p(tA)$ is a smoothing operator on M, $t > 0$, and hence is of trace class. (See Exercise 2.4 at the end of this section).

As shown in Section 1, we have

$$p(tA)u = \int_{-\infty}^{\infty} \int_{\mathbf{R}^n} \hat{p}_t(s) b(s, x, \xi) e^{i\varphi(s,x,\xi)} \hat{u}(\xi)\, d\xi\, ds + R(t)u$$

where $b(s, x, \xi)$ is supported in a small neighborhood of $s = 0$ and $R(t)$ is a smoothing operator,

$$\begin{aligned} R(t)u &= \int_{-\infty}^{\infty} \hat{p}_t(s) \varphi_2(s) e^{isA} u\, ds \\ &= (iA)^{-k} \int_{-\infty}^{\infty} t^{-1} D_s^k \left(\hat{p}\left(\frac{s}{t}\right) \varphi_2(s) \right) e^{isA} u\, ds \qquad (2.1) \\ &= (iA)^{-k} \sum_{\ell=0}^{k} \binom{k}{\ell} t^{-\ell-1} \int_{-\infty}^{\infty} \hat{p}^{(\ell)}\left(\frac{s}{t}\right) \varphi_2^{(k-\ell)}(s) e^{isA} u\, ds. \end{aligned}$$

Now the estimate on $\hat{p}^{(\ell)}$ derived in Lemma 1.2 of Section 1 shows that

$$t^{-\ell-1} \int_{-\infty}^{\infty} \left| \hat{p}^{(\ell)}\left(\frac{s}{t}\right) \varphi_2^{(k-\ell)}(s) \right| ds \le C_K t^K.$$

Thus (2.1) yields

$$\|R(t)u\|_{H^{s+k}} \le C_{s,k,K} t^K \|u\|_{H^s}.$$

It follows that $R(t)$ has a kernel $R(t, x, y)$ which is $0(t^\infty)$ as $t \downarrow 0$. Consequently, $p(tA)$ has the kernel $K_t(x, y)$, satisfying

$$p(tA)u(x) = \int K_t(x, y) u(y)\, dy,$$

and, modulo $0(t^\infty)$, we have

$$\begin{aligned} K_t(x, y) &= \int_{\mathbf{R}^n} \int_{-\infty}^{\infty} \hat{p}_t(s) b(s, x, \xi) e^{i\varphi(s,x,\xi) - iy\cdot\xi}\, ds\, d\xi \qquad (2.2) \\ &= \int_{\mathbf{R}^n} p_t(D_s)[b(s, x, \xi) e^{i\varphi(s,x,\xi) - iy\cdot\xi}]\big|_{s=0}\, d\xi \\ &= \int_{\mathbf{R}^n} q_t(x, \xi) e^{i(x-y)\cdot\xi}\, d\xi \end{aligned}$$

where

$$q_t(x, \xi) = e^{-ix\cdot\xi} p_t(D_s)(b(s, x, \xi) e^{i\varphi(s,x,\xi)})\big|_{s=0}.$$

The fundamental asymptotic expression lemma yields

$$(2.3)\quad q_t(x,\xi) = a(\xi)p(a_1(x,t\xi)) + a(\xi)\sum_{j=1}^{2k} t^j p^{(j)}(a_1(x,t\xi))D_\sigma^j(b(\sigma,x,\xi)e^{i\rho(\sigma,x,\xi)})\big|_{\sigma=0} + r_k(t,x,\xi)$$

where $r_k(t, x, \xi)$ is a bounded subset of $S_{1,0}^{-k-1}$ for $0 < t < 1$. Here we have picked $a(\xi) = a_0(|\xi|)$ equal to 1 for $|\xi| \geq 2$, equal to 0 for $|\xi| \leq 1$, and $\rho(\sigma, x, \xi) = x \cdot \xi - \varphi(\sigma, x, \xi) + \sigma\varphi_\sigma(\sigma, x, \xi)$. Note that

$$c_j(x,\xi) = D_\sigma^j(b(\sigma,x,\xi)e^{i\rho(\sigma,x,\xi)})\big|_{\sigma=0}$$
$$= \sum_{\ell+m=j}\ \sum_{\substack{m_1+\cdots+m_\nu=m \\ m_\mu\geq 2}} c_{\ell m_1\cdots m_\nu}(D_\sigma^\ell b)(D_\sigma^{m_1}\varphi)\cdots(D_\sigma^{m_\nu}\varphi)\big|_{\sigma=0}.$$

Thus $c_j(x, \xi)$ is asymptotic to a sum of terms homogeneous in ξ of degree $\leq [j/2]$. Say

$$c_j(x,\xi) \sim \sum_{k=-\infty}^{[j/2]} c_{jk}(x,\xi); \qquad (c_0(x,\xi)=1).$$

Setting $y = x$ in (2.2), we get

$$K_t(x,x) = \int_{\mathbf{R}^n} q_t(x,\xi)\,d\xi$$
$$= \sum_{j=0}^{2\ell} \int_{\mathbf{R}^n} a(\xi)c_j(x,\xi)t^j p^{(j)}(a_1(x,t\xi))\,d\xi + \int_{\mathbf{R}^n} r_\ell(t,x,\xi)\,d\xi.$$

If we pick $\ell \geq n$, it follows that

$$\int_{\mathbf{R}^n} r_\ell(t,x,\xi)\,d\xi$$

is bounded independently of t, as $t \to 0$. We now determine the asymptotic behavior of the sum in (2.4), as $t \to 0$.

$$(2.5)\quad \int_{\mathbf{R}^n} a(\xi)c_{jk}(x,\xi)t^j p^{(j)}(a_1(x,t\,\xi))\,d\xi = \int_{S^{n-1}}\int_0^\infty a_0(r)r^k c_{jk}(x,\omega)t^j p^{(j)}(rta_1(x,\omega))r^{n-1}\,dr\,d\omega.$$

A routine calculation yields, with $\alpha_\ell(n+k, j)$ and $\beta_\ell(n+k, j)$ depending on the behavior of $p(s)$, that as $t \downarrow 0$,

$$(2.6)\quad \int_0^\infty r^{n+k-1}a_0(r)p^{(j)}(ta_1 r)\,dr \sim \sum_{\ell=-(n+k)}^{\infty} \alpha_\ell(n+k,j)(a_1 t)^\ell + \log(2a_1 t)\sum_{\ell=0}^{\infty}\beta_\ell(n+k,j)(a_1 t)^{\ell-(n+k)}$$

where the term containing $\log(2a_1 t)$ only appears if $n+k-1<0$ (recall that k may be negative).

If we integrate (2.6) over $\omega \in S^{n-1}$, we see from (2.4) and (2.5) that

$$K_t(x, x) \sim c_{-n}t^{-n} + \cdots + c_{-1}t^{-1} + \sum_{\nu=1}^{\infty} (c_\nu t^\nu + d_\nu t^\nu \log t) + r_\ell(t, x) \tag{2.7}$$

where

$$c_{-n} = \left(\int_0^\infty p(s)\, ds\right) \int_{S^{n-1}} a_1(x, \omega)^{-n}\, d\omega$$

and where

$$r_\ell(t, x) = \int_{\mathbf{R}^n} r_\ell(t, x, \xi)\, d\xi = 0(1), \qquad t \downarrow 0.$$

This remainder term is too crude for the sum

$$\sum_{\nu=1}^{\infty} (c_\nu t^\nu + d_\nu t^\nu \log t)$$

to be meaningful, but one can similarly analyze $(\partial/\partial t)K_t(x, y)$, the kernel of $Ap(tA)$, obtaining

$$\frac{\partial}{\partial t} K_t(x, x) \sim c'_{-n-1}t^{-n-1} + \cdots + c'_{-1}t^{-2} + d'_0 \log t + 0(1),$$

and integrating yields

$$K_t(x, x) \sim c_{-n}t^{-n} + \cdots + c_{-1}t^{-1} + c_0 + d_0 t \log t + 0(t).$$

Repetition of such an argument justifies the complete asymptotic expansion (2.7), with c_0 included, with a remainder that is $0(t^N)$ for any N.

We shall derive the asymptotic behavior of λ_j from

$$\begin{aligned} tr\, p(tA) &= \int_M K_t(x, x) d \operatorname{vol}(x) \\ &= \left(\int_{S^*(x)} a_1(x, \omega)^{-n}\right)\left(\int_0^\infty p(s)\, ds\right) t^{-n} + 0(t^{-n+1}) \\ &= ct^{-n} + 0(t^{-n+1}). \end{aligned} \tag{2.8}$$

THEOREM 2.1 *The eigenvalues λ_j of A have the asymptotic behavior*

$$\#\{j : \lambda_j \le \lambda\} = c\lambda^n + o(\lambda^n), \qquad \lambda \to \infty, \tag{2.9}$$

$$\lambda_j = c' j^{1/n} + o(j^{1/n}), \qquad j \to \infty. \tag{2.10}$$

Here

$$c = \int_{S^*M} a_1(x, \omega)^{-n}$$

and $c' = c^{-1/n}$.

Proof. Pick $p(\lambda)$ such that $p(\lambda) = 1$ on $[0, 1], 0 \leq p(\lambda) \leq 1$, and $p(\lambda) = 0$ for $\lambda > 1 + \epsilon$, where $\epsilon > 0$ is prescribed. Note that

$$\#\{j:\lambda_j \leq t^{-1}\} \leq tr\ p(tA) \leq \#\{j:\lambda_j \leq (1 + \epsilon)t^{-1}\}.$$

From (2.8) we see that, given any $\epsilon > 0$, there is a μ_0 such that, for $\lambda > \mu_0$,

$$c(1 - \epsilon)\lambda^n \leq \#\{j:\lambda_j \leq \lambda\} \leq c(1 + \epsilon)\lambda^n.$$

From this, (2.9) is an immediate consequence, and (2.10) follows from (2.9).

Just for the sake of comparison, let us analyze the trace of e^{-tA}, $t \to 0$. This time, we assume A^2 is a differential operator, e.g., $A^2 = -\Delta$, and use the parametrix for e^{-tA} constructed in Chapter VIII, Section 4, via complex phase functions. This will yield an approach to (2.9) similar in spirit to the heat equation method, used for example in Berger, Gauduchon, and Mazet [1]. As we saw in Chapter VIII, we can write $e^{-tA}u$ in the form

$$e^{-tA}u(x) = (2\pi)^{-n/2} \int_{\mathbf{R}^n} b(t, x, \xi)e^{i\varphi(t,x,\xi)}\hat{u}(\xi)\, d\xi$$

where $\varphi(t, x, \xi) = x \cdot \xi + it\gamma(t, x, \xi)$ with $\gamma(t, x, \xi) = a_1(x, \xi) + 0(t|\xi|)$. Also $b(0, x, \xi) = 1$ for $|\xi| \geq 1$. Consequently, we see that the kernel of e^{tA} is

$$E_t(x, y) = (2\pi)^{-n} \int_{\mathbf{R}^n} b(t, x, \xi)e^{-t\gamma(t,x,\xi)}e^{i(x-y)\cdot\xi}\, d\xi. \tag{2.11}$$

Setting $y = x$, we have

$$\begin{aligned} E_t(x, x) &= (2\pi)^{-n} \int_{\mathbf{R}^n} b(t, x, \xi)e^{-t\gamma(t,x,\xi)}\, d\xi \\ &= (2\pi)^{-n} \int_{S^{n-1}} \int_0^\infty b(t, x, r\omega)e^{-rt\gamma(t,x,\omega)}r^{n-1}\, dr\, d\omega. \end{aligned}$$

We can suppose

$$b(t, x, \xi) \sim a_0(|\xi|) \sum_{j=0}^{\infty} b_j(t, x, \xi),$$

with b_j homogeneous in ξ of degree $-j$. As in (2.6), we have

$$\int_0^\infty r^{n-1-j}a_0(r)e^{-t\gamma r}\, dr$$

$$\sim \sum_{\ell=-(n-j)}^{\infty} \alpha_\ell(n - j)(t\gamma)^\ell + \log(\gamma t) \sum_{\ell=0}^{\infty} \beta_\ell(n - j)(t\gamma)^{\ell-(n-j)}$$

where the term containing $\log(\gamma t)$ only appears if $n-j \leq 0$. Consequently, if we integrate (2.12) over $\omega \in S^{n-1}$ we get

$$E_t(x, x) \sim \sum_{\ell=-n}^{\infty} c_\ell t^\ell + \sum_{\ell=1}^{\infty} d_\ell t^\ell \log t \tag{2.13}$$

where

$$c_{-n} = (2\pi)^{-n} \int_{S^{n-1}} \gamma(0, x, \omega)^{-n}\, d\omega = (2\pi)^{-n} \int_{S^{n-1}} a_1(x, \omega)^{-n}\, d\omega.$$

Thus we have rederived a special case of (2.7). This yields

$$tr\, e^{-tA} = \int_M E_t(x, x) d \operatorname{vol}(x) = (2\pi)^{-n}\left(\int_{S^*M} a_1(x, \omega)^{-n}\right) t^{-n} + 0(t^{-n+1})$$

or, what is the same thing,

$$\sum_{j=1}^{\infty} e^{-t\lambda_j} = (2\pi)^{-n}\left(\int_{S^*M} a_1(x, \omega)^{-n}\right) t^{-n} + 0(t^{-n+1}). \tag{2.14}$$

Now the expansion (2.9) can be derived from (2.14), but not as quickly as we derived (2.9) from (2.8). What is involved is a Tauberian theorem, due to Karamata, which says precisely that (2.14) implies (2.9). We shall give the proof of this Tauberian theorem in section 7.

Exercises

2.1. Let $f \in \mathscr{D}'(M)$ and write $\hat{f}(j) = (f, u_j)$, where u_j is the normalized eigenfunction associated with λ_j; $Au_j = \lambda_j u_j$. Show that the following are equivalent, for any $s \in \mathbf{R}$.

(i) $f \in H^s(M)$;

(ii) $\sum_{j=1}^{\infty} |\hat{f}(j)|^2 j^{2s/n} < \infty$;

(iii) $\sum_{j=1}^{N} \hat{f}(j) u_j$ converges to f in the H^s topology, as $N \to \infty$.

2.2. Show that

$$\sum_{j=1}^{\infty} a_j u_j$$

is the eigenfunction expansion of some $f \in \mathscr{D}'(M)$ if, and only if, a_j has at most polynomial growth as $j \to \infty$. Deduce that $\mathscr{D}'(M)$ is topologically isomorphic to the sequence space s' of "slowly increasing" sequences, with its obvious LF topology.

2.3. Show that

$$\sum_{j=1}^{\infty} a_j u_j = f \in C^\infty(M)$$

if, and only if, a_j is rapidly decreasing as $j \to \infty$. Deduce that $C^\infty(M)$ is naturally isomorphic with the space s of rapidly decreasing sequences, with its obvious Frechet space topology.

2.4. Show that any $p(x, D) \in OPS^{-\infty}$ is of trace class. More generally, show that, P is of trace class for all $P \in OPS_{1,0}^{-m}$ if, and only if, $m > n$, and P is Hilbert-Schmidt for all $P \in OPS_{1,0}^{-m}$ if, and only if, $m > n/2$.

§3. Poisson-like Kernels

We derived information on the spectrum of A, the first order elliptic self-adjoint operator on the compact manifold M considered in the last section, by examining the kernels $K_t(x, y)$ of $p(tA)$, $0 < t < 1$. In this section we shall examine these kernels more closely, with a view toward further applications in Sections 4 and 5. The special case $p(\lambda) = e^{-\lambda}$ leads to the Poisson kernel for e^{-tA}; it turns out that $p(tA)$ shares many of its nice properties, given $p(\lambda) \in S_{1,0}^{-s}(\mathbf{R})$, $s > 0$.

As we have seen, modulo $0(t^\infty)$,

$$K_t(x, y) = \int_{\mathbf{R}^n} q_t(x, \xi) e^{i(x-y)\cdot\xi}\, d\xi \qquad \text{where}$$

$$q_t(x, \xi) = a(\xi) p(ta_1(x, \xi)) + a(\xi) \sum_{j=1}^{2k} t^j p^{(j)}(ta_1(x, \xi)) c_j(x, \xi) + r_k(t, x, \xi) \tag{3.1}$$

where $r_k(t, x, \xi)$ is a bounded subset of $S_{1,0}^{-k-1}$, $0 < t < 1$, $a(\xi) = a_0(|\xi|) = 1$ for $|\xi| \geq 2$, 0 for $|\xi| \leq 1$, and $c_j(x, \xi)$ is asymptotic to a sum of terms homogeneous of degree $\leq [j/2]$. Cutting $c_j(x, \xi)$ off near $\xi = 0$, we may suppose $c_j(x, \xi) \in S^{[j/2]}$. We begin our analysis with the principal term in (3.1), and a definition is appropriate.

DEFINITION. *Let $p(\lambda) \in S_{1,0}^{-s}(\mathbf{R})$. An elementary Poisson-like kernel associated with p is defined to be*

$$K_p(t, x, x-y) = \int_{\mathbf{R}^n} a(\xi) p(ta_1(x, \xi)) e^{i(x-y)\cdot\xi}\, d\xi. \tag{3.2}$$

Here $a(\xi) = 1$ for large ξ and 0 for small ξ.

Note that K_p also depends on the principal symbol of A, but we suppress this dependence. For $s \leq n$, the integral (3.2) is not absolutely convergent, but the usual integration by parts procedure makes sense out of it.

To analyze (3.2), write $b(\xi) = 1 - a(\xi) \in C_0^\infty(\mathbf{R}^n)$, so

$$K_p(t, x, z) = \int_{\mathbf{R}^n} a(\xi)p(ta_1(x, \xi))e^{iz\cdot\xi}\,d\xi \tag{3.3}$$

$$= \int_{\mathbf{R}^n} p(ta_1(x, \xi))e^{iz\cdot\xi}\,d\xi$$

$$+ \int_{\mathbf{R}^n} b(\xi)p(ta_1(x, \xi))e^{iz\cdot\xi}\,d\xi$$

$$= t^{-n}\int_{\mathbf{R}^n} p(a_1(x, \xi))e^{i(z/t)\xi}\,d\xi + \Psi(t, x, z)$$

$$= t^{-n}\Phi(x, t^{-1}z) + \Psi(t, x, z).$$

Clearly $\Psi(t, x, z)$ is smooth, up to $t = 0$. It remains to analyze

$$\Phi(x, z) = \int_{\mathbf{R}^n} p(a_1(x, \xi))e^{iz\cdot\xi}\,d\xi \tag{3.4}$$

$$= \int_{\mathbf{R}^n} a(\xi)p(a_1(x, \xi))e^{iz\cdot\xi}\,d\xi$$

$$+ \int_{\mathbf{R}^n} b(\xi)p(a_1(x, \xi))e^{iz\cdot\xi}\,d\xi$$

$$= \Phi_0(x, z) + \Phi_1(x, z).$$

We first analyze $\Phi_0(x, z)$.

LEMMA 3.1. *Let $q(x, \xi) \in S_{1,0}^{-s}$ and let*

$$\tilde{\Phi}(x, z) = \int_{\mathbf{R}^n} q(x, \xi)e^{iz\cdot\xi}\,d\xi. \tag{3.5}$$

Then $\tilde{\Phi}$ is smooth for $z \neq 0$ and rapidly decreasing, together with all its derivatives, as $|z| \to \infty$. Furthermore, for $|z| \le C$,

$$|\tilde{\Phi}(x, z)| \le \begin{cases} C|z|^{-n+s}, & s < n \\ C|\log|z||, & s = n. \end{cases} \tag{3.6}$$

Proof. Note that $\tilde{\Phi}(x, z) = q(x, D_z)\delta(z)$; we can regard x as a parameter and hence omit it from the estimates. Thus let $q(\xi) \in S_{1,0}^{-s}$ and consider

$$\tilde{q}(z) = \int_{\mathbf{R}^n} q(\xi)e^{iz\cdot\xi}\,d\xi = q(D_z)\delta.$$

The smoothness and rapid decrease as $|z| \to \infty$ follow from

$$|z|^{2|\alpha| + 2k}D_z^\alpha\tilde{q}(z) = \int_{\mathbf{R}^n} \xi^\alpha\Delta^{|\alpha|+k}q(\xi)e^{iz\cdot\xi}\,d\xi.$$

It remains to prove (3.6). Now $q(\xi) \in S_{1,0}^{-s}$ implies that $q_r(\xi) = r^s q(r\xi)$ is bounded on C^∞ $(1 \le |\xi| \le 2)$ for $1 \le r < \infty$. Hence we can write

$$q(\xi) = q_0(\xi) + \int_0^\infty p_\tau(e^{-\tau}\xi)\,d\tau$$

with $q_0 \in C_0^\infty$ and $e^{s\tau}p_\tau(\xi)$ bounded in the Schwartz space $\mathcal{S}(\mathbf{R}^n)$, $0 \leq \tau < \infty$. Thus

$$q(D)\delta = \hat{q}_0(z) + \int_0^\infty e^{n\tau}\hat{p}_\tau(e^\tau z)\,d\tau,$$

and $e^{s\tau}\hat{p}_\tau(z)$ is bounded in $\mathcal{S}(\mathbf{R}^n)$. In particular,

$$e^{s\tau}|\hat{p}_\tau(z)| \leq C_N(1+|z|)^{-N}, \qquad C_N \text{ independent of } \tau.$$

Consequently,

$$\begin{aligned}|q(D)\delta| &\leq |\hat{q}_0(z)| + C_N \int_0^\infty e^{(n-s)\tau}(1+|e^\tau z|)^{-N}\,d\tau \\ &\leq C + C_N|z|^{s-n}\int_0^\infty |e^\tau z|^{n-s}(1+|e^\tau z|)^{-N}\,d\tau \\ &= C + C_N|z|^{-n+s}\int_{\log|z|}^\infty e^{(n-s)\tau}(1+e^\tau)^{-N}\,d\tau\end{aligned}$$

which immediately yields (3.6).

Since $\Phi_0(x, z)$ is given by (3.5) with $q(x, \xi) = a(\xi)p(a_1(x, \xi))$, the conclusion of Lemma 3.1 yields our analysis of $\Phi_0(x, z)$. We turn now to the analysis of $\Phi_1(x, z)$.

LEMMA 3.2. *The function $\Phi_1(x, z)$ is smooth in x and z and we have*

$$|\Phi_1(x, z)| \leq c(1+|z|)^{-n-1}. \tag{3.7}$$

Proof. Since

$$\Phi_1(x, z) = \int_{\mathbf{R}^n} b(\xi)p(a_1(x, \xi))e^{iz\cdot\xi}\,d\xi$$

and $b \in C_0^\infty(\mathbf{R}^n)$, the smoothness is clear. The estimate (3.7) arises from the nature of the singularity of $p(a_1(x, \xi))$ at $\xi = 0$. In fact, as $|z| \to \infty$, $\Phi_1(x, z)$ is asymptotic to a sum of terms homogeneous in z of degree $-n-1, -n-2, \ldots$. See Titchmarsh [1].

Remark. Suppose $p(\lambda) \equiv 1$ for λ near 0, or suppose $p(\lambda)$ is an even function of λ (e.g. $p(\lambda) = e^{-\lambda^2}$) and $a_1(x, \xi)^2$ is a *polynomial* in ξ, e.g. A^2 is a differential operator. Then $p(a_1(x, \xi))$ is smooth, and it follows that $\Phi_1(x, z)$ is rapidly decreasing on $z \to \infty$.

We now show how (3.1) yields an analysis of $p(tA)$ in terms of elementary Poisson-like kernels.

THEOREM 3.3. *Suppose $p(\lambda) \in S_{1,0}^{-s}(\mathbf{R})$. Then*

$$\begin{aligned}p(tA)u &= \int K_p(t, x, x-y)u(y)\,dy \\ &\quad + t\sum_{j=1}^{\ell}\sum_{\nu=1}^{n(j)}\int K_{q_{j\nu}}(t, x, x-y)(B_{j\nu}u(y))\,dy \\ &\quad + r_\ell(t, x, D)u\end{aligned} \tag{3.8}$$

where $q_{j\nu}(\lambda) \in S_{1,0}^{-s-1}(\mathbf{R})$, *and* $B_{j\nu} \in OPS^{-j+1}$, *and* $r_\ell(t, x, D)$ *is bounded in* $OPS_{1,0}^{-\ell}$, *for* $0 < t < 1$.

In order to prove this theorem it suffices, in view of (3.1), to show that, for $j \geq 1$,

$$a(t, x, D)u = \int a(\xi)t^j p^{(j)}(ta_1(x, \xi))c_j(x, \xi)e^{ix\cdot\xi}\hat{u}(\xi)\, d\xi$$

can be written in the form of the last two terms in (3.8), given $c_j(x, \xi) \in S^{-[j/2]}$. Let $\nu = [j/2]$; so $j - [j/2] = j - \nu = \mu \geq 1$. Thus

$$a(t, x, D)u = t\int a(\xi)q_j(ta_1(x, \xi))b_j(x, \xi)e^{ix\cdot\xi}\hat{u}(\xi)\, d\xi \tag{3.10}$$

where $q_j(\lambda) = \lambda^{j-1}p^{(j)}(\lambda) \in S_{1,0}^{-s-1}(\mathbf{R})$ and $b_j(x, \xi) = c_j(x, \xi)a_1(x, \xi)^{-j+1} \in S^{-\mu+1}$. From (3.10) we derive

$$\begin{aligned} a(t, x, D)u = {} & t\int K_{q_j}(t, x, x-y)(b_j(y, D)u(y))\, dy \\ & - t\sum_{|\alpha|=1}^{\ell} \int \frac{1}{\alpha!} D_\xi^\alpha(a(\xi)q_j(ta_1(x, \xi)))D_x^\alpha b_j(x, \xi)\hat{u}(\xi) \\ & \cdot e^{ix\cdot\xi}\, d\xi + tR_\ell(t, x, D)u \end{aligned} \tag{3.11}$$

where $R_\ell(t, x, D)$ is bounded in $OPS_{1,0}^{-\ell-\mu+1}$, for $0 < t < 1$. Now for $|\alpha| \geq 1$,

$$t\int D_\xi^\alpha(a(\xi)q_j(ta_1(x, \xi)))D_x b_j(x, \xi)\hat{u}(\xi)e^{ix\cdot\xi}\, d\xi$$

is seen to be a finite sum of terms of the form

$$t\int \tilde{b}_j(x, \xi)q_{j+\nu}(ta_1(x, \xi))\hat{u}(\xi)e^{ix\cdot\xi}\, d\xi \tag{3.12}$$

with $q_{j+\nu}(\lambda) = \lambda^{j+\nu-1}p^{(j+\nu)}(\lambda)$, $1 \leq \nu \leq |\alpha|$, and with $\tilde{b}_j(x, \xi) \in S^{-\mu+1-|\alpha|}$, vanishing for $|\xi| \leq 1$. Thus (3.12) is of the same form as (3.10), except $\tilde{b}_j(x, \xi)$ has *lower order* than $b_j(x, \xi)$. An inductive argument shows that $a(t, x, D)$ has the desired form, and this completes the proof of the theorem.

It is convenient to alter the terms containing operators $B_{j\nu}$ of order zero. We easily obtain the following.

COROLLARY 3.4. *If* $p(\lambda) \in S_{1,0}^{-s}(\mathbf{R})$, *and* $0 \leq \sigma \leq 1$, *then the sum over* $j = 1$ *in* (3.8) *can be replaced by*

$$t^{1-\sigma}\sum_{\nu=1}^{n(1)} \int K_{\tilde{q}_\nu}(t, x, x-y)(\tilde{B}_\nu u(y))\, dy \tag{3.13}$$

where $\tilde{B}_\nu \in OPS^{-\sigma}$, *and* $\tilde{q}_\nu(\lambda) \in S_{1,0}^{-s-1+\sigma}(\mathbf{R})$, *provided we assume* $p(\lambda) \equiv 1$ *for* λ *near* 0.

Proof. As one can see from the proof of Theorem 3.3, it suffices to handle operators of the form

$$a(t, x, D) = t \int a(\xi)q(ta_1(x, \xi))b(x, \xi)e^{ix\cdot\xi}\hat{u}(\xi)\, d\xi$$

with $b(x, \xi) \in S^0$ vanishing for $|\xi| \leq 1$, with $q(\lambda) = p'(\lambda)$ or $q(\lambda) = \lambda p''(\lambda)$. We can rewrite this as

$$a(t, x, D) = t^{1-\sigma} \int a(\xi)\tilde{q}(ta_1(x, \xi))\tilde{b}(x, \xi)e^{ix\cdot\xi}\hat{u}(\xi)\, d\xi \tag{3.14}$$

with $q(\lambda) = \lambda^{\sigma}q(\lambda) \in S_{1,0}^{-s-1+\sigma}(\mathbf{R})$ and with $\tilde{b}(x, \xi) = a_1(x, \xi)^{-\sigma}b(x, \xi) \in S^{-\sigma}$. We can analyze (3.14) in the same manner we analyzed (3.10), which proves the corollary.

Remark. If we don't assume $p(\lambda) \equiv 1$ near $\lambda = 0$, $\lambda^{\sigma}q(\lambda)$ won't be smooth at $\lambda = 0$. However, this doesn't cause grave difficulties, since $a(\xi)\tilde{q}(ta_1(x, \xi))$ is still bounded in $S_{1,0}^{-s-1+\sigma}$, for $0 < t < 1$. We could have incorporated such mild singularities into symbols $\tilde{q}(\lambda)$ from the beginning, with only minor changes, and then this extra condition on $p(\lambda)$ would not be needed in the corollary.

So far we have analyzed the kernel $K_t(x, y)$ of $p(tA)$, up to a remainder term which is smooth in t, x, y for $t \geq 0$. Such an analysis omits one question: does $K_t(x, y) \to 0$ as $t \downarrow 0$, for $x \neq y$? In fact, it is easy to see that is is the case.

PROPOSITION 3.5. *If $p(\lambda) \in S_{1,0}^{-s}(\mathbf{R})$, $p(0) = 1$, as above, and $K_t(x, y)$ is the kernel of $p(tA)$, then $K_t(x, y) \to 0$ as $t \downarrow 0$, uniformly on compact subsets of $M \times M\backslash\Delta$, where $\Delta = \{(x, x): x \in M\}$ is the diagonal. In fact, $K_t(x, y) \to 0$ in $C^\infty(M \times M\backslash\Delta)$ as $t \to 0$.*

Proof. $K_t(x, y) = p(tA)\delta_y$. Now if $\varphi, \psi \in C^\infty(M)$, and $\operatorname{supp}\varphi \cap \operatorname{supp}\psi = \phi$, we have $\varphi(x)K_t(x, y)\psi(y) = L_t(x, y)$ is the kernel of the operator $\varphi(x)p(tA)(\psi u) = L_t u$. But since $\{p(tA): 0 < t < 1\}$ is bounded in $OPS_{1,0}^0$, it follows that $\{L_t: 0 < t < 1\}$ is bounded in $OPS^{-\infty}$. $u \in \mathscr{D}'(M) \Rightarrow p(tA)(\psi u) \to \psi u$ in $\mathscr{D}'(M)$ as $t \downarrow 0 \Rightarrow L_t u \to 0$ in $\mathscr{D}'(M)$. But we know that $\{L_t u: 0 < t < 1\}$ is bounded in $C^\infty(M)$, so in fact $L_t u \to 0$ in $C^\infty(M)$. This happens uniformly on compact sets of u, e.g. on $\{\delta_y: y \in M\}$. This shows that $L_t(x, y) \to 0$ uniformly as $t \downarrow 0$, which proves the proposition.

In general, one cannot expect $K_t(x, y)$ to vanish to higher order on $M \times M\backslash\Delta$, as $t \downarrow 0$. In fact, suppose $p(tA) = e^{-t\sqrt{-\Delta}}$, so $K_t(x, y) = e^{-t\sqrt{-\Delta}}\delta_y$. For y fixed, we have seen that $u(t, x) = e^{-t\sqrt{-\Delta}}\delta_y \to 0$ as $t \downarrow 0$ on $U = M\backslash\{y\}$. Suppose also that $(\partial/\partial t)u(t, x) \to 0$ as $t \downarrow 0$ on some open subset $U_1 \subset U$. But $u(t, x)$ solves the elliptic equation

$$\left(\frac{\partial^2}{\partial t^2} + \Delta\right)u = 0,$$

and if $u(0, x)$ and $(\partial/\partial t)u(0, x)$ both vanished in U_1, this would imply that u is identically zero, by uniqueness in the Cauchy problem (see Chapter XIV), which is impossible.

However, we have the following.

PROPOSITION 3.6. *Suppose* $p(\lambda) \in S_{1,0}^{-s}(\mathbf{R})$ *is as before,* $p(0) = 1$, $K_t(x, y)$ *the kernel of* $p(tA)$.

(i) *If* $p(\lambda) - 1$ *vanishes to order* j *at* $\lambda = 0$, *then* $K_t(x, y)$ *vanishes to order* j *as* $t \downarrow 0$, *in* $C^\infty(M \times M \backslash \Delta)$.

(ii) *If* $p(\lambda)$ *is an even function of* λ *and* $B = A^2$ *is a differential operator, then* $K_t(x, y)$ *vanishes to infinite order as* $t \downarrow 0$, *in* $C^\infty(M \times M \backslash \Delta)$.

Proof. The function $(\partial^j/\partial t^j)K_t(x, y)$ is the kernel of $(\partial^j/\partial t^j)p(tA) = A^j p^{(j)}(tA)$. If $p^{(j)}(0) = 0$, then $p^{(j)}(tA)u \to 0$ in $H^s(M)$ for any $u \in H^s(M)$. Picking $\varphi, \psi \in C^\infty(M)$ with $\text{supp}\,\varphi \cap \text{supp}\,\psi = \phi$, then

$$\tilde{L}_t(x, y) = \varphi(x)(\partial^j/\partial t^j)K_t(x, y)\psi(y)$$

is the kernel of the operator $\tilde{L}_t u = \varphi(x)A^j p^{(j)}(tA)(\psi u)$, $\{\tilde{L}_t : 0 < t < 1\}$ is bounded in $OPS^{-\infty}$; $u \in \mathscr{D}'(M)$ implies $\tilde{L}_t u \to 0$ in $\mathscr{D}'(M)$ as $t \to 0$; thus $\tilde{L}_t u \to 0$ in $C^\infty(M)$ as $t \downarrow 0$. This proves (i) and also shows that $(\partial^j/\partial t^j)K_t(x, y) \to 0$ in $C^\infty(X \times X \backslash \Delta)$ for all odd j, in case (ii). On the other hand, if $j = 2k$, $(\partial^j/\partial t^j)K_t(x, y)$ is the kernel of $B^k p^{(2j)}(tA)$, and we know that the kernel of $p^{(2j)}(tA)$ vanishes as $t \downarrow 0$ on $C^\infty(M \times M \backslash \Delta)$, by Proposition 3.5. Since B^k is a local operator, this shows that $(\partial^{2k}/\partial t^{2k})K_t(x, y) \to 0$ as $t \downarrow 0$, in $C^\infty(X \times X \backslash \Delta)$, establishing (ii).

Finally, we will combine our information on the kernel of $p(tA)$ for $p(\lambda) \in S_{1,0}^{-\infty}$ with the spectral analysis of Section 2 to obtain estimates on the spectral function $e(x, y, \lambda)$, defined to be the kernel of $E_\lambda = e_\lambda(A)$, where $e_\lambda(\mu) = 1$ for $\mu \le \lambda$, 0 for $\mu > \lambda$. Note that

$$e(x, y, \lambda) = \sum_{\lambda_j \le \lambda} u_j(x)\overline{u_j(y)}$$

if $\{u_j\}$ is the orthonormal basis of eigenfunctions of A.

THEOREM 3.7. *We have for the spectral function*

$$\lambda^{-n} e(x, x, \lambda) \to (2\pi)^{-n} k(x), \qquad \lambda \to \infty, \tag{3.15}$$

$$\lambda^{-n} e(x, y, \lambda) \to 0, \qquad \lambda \to \infty, x \neq y, \tag{3.16}$$

where $k(x) = \text{vol}\,\{\xi \in T_x^*(M) : a_1(x, \xi) \le 1\}$.

Proof. The proof of (3.15) is contained in the proof of Theorem 2.1, if one neglects to perform the integration in (2.8).

Now, for any given $\epsilon > 0$, pick $p(\lambda) \in C_0^\infty(\mathbf{R})$ such that $0 \le p(\lambda) \le 1$, $p(\lambda) = 1$ on $[0, 1]$, and $p(\lambda) = 0$ for $\lambda > 1 + \epsilon$. Let $\tilde{e}(x, y, \lambda)$ denote the kernel of $p(tA)$, with $\lambda = t^{-1}$. Using Theorem 3.3 and Corollary 3.4, together with the structure of elementary Poisson-like kernels as given by (3.3), (3.4) and Lemmas 3.1 and 3.2, one sees that

$$\tilde{e}(x, y, \lambda) - \lambda^n \Phi(x, \lambda(x-y)) = 0(\lambda^{n-1+\sigma}) \tag{3.17}$$

where

$$\Phi(x, z) = \int_{\mathbf{R}^n} p(a_1(x, \xi)) e^{iz \cdot \xi} \, d\xi.$$

In particular,

$$\Phi(x, 0) = \int_{\mathbf{R}^n} p(a_1(x, \xi)) \, d\xi,$$

so we have $k(x) \leq \Phi(x, 0) \leq (1+\epsilon)^n k(x)$. We wish to estimate the difference $\tilde{e}(x, y, \lambda) - e(x, y, \lambda) = d(x, y, \lambda)$. Note that $d(x, y, \lambda)$ is the kernel of a positive operator, so we have

$$|d(x, y, \lambda)| \leq d(x, x, \lambda)^{1/2} d(y, y, \lambda)^{1/2}.$$

On the other hand, $e(x, y, \lambda) - \tilde{e}(x, y, (1-\epsilon)\lambda) = \tilde{d}(x, y, \lambda)$ is also the kernel of a positive operator. It follows that

$$\begin{aligned} d(x, x, \lambda) &\leq \tilde{e}(x, x, \lambda) - \tilde{e}(x, x, (1-\epsilon)\lambda) \\ &= (\lambda^n - (1-\epsilon)^n \lambda^n) \Phi(x, 0) + 0(\lambda^{n-1+\sigma}) \\ &\leq n\epsilon\lambda^n \Phi(x, 0) + 0(\lambda^{n-1+\sigma}). \end{aligned}$$

It follows that $\lambda^{-n} d(x, y, \lambda) \to 0$ as $\lambda \to \infty$. Thus (3.17) yields

$$\lambda^{-n} e(x, y, \lambda) - \Phi(x, \lambda(x-y)) \to 0, \qquad \lambda \to \infty$$

which reproves (3.15) and also establishes (3.16).

Remark. Since $|e(x, y, \lambda)| \leq e(x, x, \lambda)^{1/2} e(y, y, \lambda)^{1/2}$, it follows that $\{\lambda^{-n} e(x, y, \lambda) : x, y \in M, \lambda \geq 1\}$ is bounded.

We mention two generalizations of our treatment of $p(tA)$ which will prove useful in Section 6. First, we relax the assumption that $p(\lambda) \in S_{1,0}^{-s}(\mathbf{R})$ to the hypothesis that, for $j \geq 1$

$$|\lambda^{j-1} p^{(j)}(\lambda)| \leq c_j (1 + |\lambda|)^{-s-1}. \tag{3.18}$$

Thus a mild singularity of p at the origin is permitted. We call this slightly enlarged class of symbols $\tilde{S}_{1,0}^{-s}$. It is easy to see that, if $p \in \tilde{S}_{1,0}^{-s}$, then $P_t(\lambda) = p(t\lambda)$ is bounded in $\tilde{S}_{1,0}^{0}$, for $0 < t < 1$. Consequently, if we fix a $q(\lambda)$, equal to 1 for $|\lambda| > 2$ and equal to 0 for $|\lambda| < 1$, it follows that $\tilde{p}_t(\lambda) = q(\lambda) p(t\lambda)$ is bounded in $S_{1,0}^{0}(\mathbf{R})$, $0 < t < 1$. Note that

$$p(tA) = \tilde{p}_t(A) + R(t)$$

where $R(t)$ is bounded in $OPS^{-\infty}$, $0 < t < 1$. Consequently the analysis leading to Theorem 3.3 goes through with very little change. The conclusion of Lemma 3.2 must be slightly weakened in this case, due to the slightly nastier behavior of $p(a_1(x, \xi))$ at $\xi = 0$. We have

$$|\Phi_1(x, z)| \le c_\epsilon(1 + |z|)^{-n-1+\epsilon},$$

which is not an important change. In fact, $K_p(t, x, x-y)$ could be analyzed, making the weaker assumption that

$$|\lambda^j p^{(j)}(\lambda)| \le c_j(1 + |\lambda|)^{-s}. \tag{3.19}$$

However, the analysis of further terms brings in $q_j(\lambda) = \lambda^{j-1}p^{(j)}(\lambda)$, which satisfy an analogue of (3.19) only if p satisfies (3.18).

The second generalization is to consider functions of several commuting operators, $A_1, \ldots, A_K$, with $A_1^2 + \cdots + A_K^2$ elliptic. In such a case $p(tA_1, \ldots, tA_K)$ can be analyzed precisely as function of one operator, with the elementary Poisson-like kernels of the form

$$K_p(t, x, x-y) = \int_{\mathbf{R}^n} a(\xi)p(ta_1(x, \xi), \ldots, ta_K(x, \xi))e^{i(x-y)\cdot\xi}\, d\xi.$$

Exercises

3.1. Suppose $p(\lambda) \in C_0^\infty(\mathbf{R})$ and $p(\lambda) = 1$ for $|\lambda| \le 1$. Show that

$$\begin{aligned} p(tA)u = &\int K_p(t, x, x-y)u(y)\, dy \\ &+ t^{1/2}\sum_{\nu=1}^{n(1)} \int K_{q_{1\nu}}(t, x, x-y)B_{1\nu}u(y)\, dy \\ &+ t\sum_{j=2}^{\ell}\sum_{\nu=1}^{n(j)} \int K_{q_{j\nu}}(t, x, x-y)B_{j\nu}u(y)\, dy + \int r_\ell(t, x, y)u(y)\, dy \end{aligned}$$

where

(i) $B_{j\nu} \in OPS_{1,0}^{-a}, \quad a > 0$;

(ii) $q_{j\nu}(\lambda) = 0$ for $\lambda \le 1$; $q_{j\nu}(\lambda) \in C_0^\infty(\mathbf{R})$;

(iii) $r_\ell(t, x, y) \in C^k(I \times M \times M)$ if ℓ is large enough; $\quad I = [0, 1]$.

3.2. If $q(\lambda) \in C_0^\infty(\mathbf{R})$, $q(\lambda) = 0$ for $|\lambda| \le 1$, show that, for $v \in C^k(M)$,

$$\left|\int K_q(t, x, x-y)v(y)\, dy\right| \le ct^k\|v\|_{C^k(M)}.$$

Hint: Write $K_q(t, x, z) = t^{-n}F(x, t^{-1}z)$ where

$$F(x, z) = \int_{\mathbf{R}^n} q(a_1(x, \xi))e^{iz\cdot\xi}\, d\xi,$$

and note that $\int z^\beta F(x, z)\, dz = D_\xi^\beta q(a_1(x, \xi))|_{\xi=0} = 0$, for all β.

3.3. With $p(\lambda)$ as in Exercise 3.1, write, as in (3.3)

$$K_p(t, x, x-y) = t^{-n}\Phi(x, t^{-1}(x-y)) + \Psi(t, x, x-y).$$

Show that

(i) $t^{-n}\int \Phi(x, t^{-1}(x-y))u(y)\, dy \to u(x)$ as $t \to 0, \qquad u \in C(M)$;

(ii) $\Psi(t, x, z) \in C^\infty(I \times M \times M)$;

(iii) $\Phi(x, z)$ smooth, and rapidly decreasing as $|z| \to \infty$.

3.4. Suppose $w \in C(M)$ vanishes on an open subset Ω of M. Show that $t^{-n}\int\Phi(x, t^{-1}(x-y))w(y)\, dy = 0(t^\infty)$ and $p(tA)w(x) = 0(t^\infty)$, $x \in \Omega$. From Exercise 3.2, deduce that, if $u \in C^k(M)$ vanishes on Ω,

$$\int R_\ell(t, x, y)u(y)\, dy = 0(t^k), \qquad x \in \Omega$$

where $R_\ell(t, x, y) = r_\ell(t, x, y) + \Psi(t, x, x-y)$. From the smoothness of R_ℓ, deduce that, if ℓ is sufficiently large,

$$|R_\ell(t, x, y)| \le ct^k. \tag{3.20}$$

3.5. If $p(\lambda) \in C_0^\infty(\mathbf{R})$ and $p(\lambda) = 1$ for $|\lambda| \le 1$, deduce that

$$\begin{aligned} p(tA)u(x) = t^{-n}\int \Phi(x, t^{-1}(x-y))u(y)\, dy \\ + \Bigg(t^{1/2}\sum_{\nu=1}^{n(1)} \int K_{q_{1\nu}}(t, x, x-y)B_{1\nu}u(y)\, dy \\ + t\sum_{j=2}^{\ell}\sum_{\nu=1}^{n(j)} \int K_{q_{j\nu}}(t, x, x-y)B_{j\nu}u(y)\, dy\Bigg) + \int R_\ell(t, x, y)u(y)\, dy \end{aligned}$$

where $R_\ell(t, x, y)$ satisfies (3.20).

3.6. Prove the following generalization of a theorem of Jackson. Let V_η denote the linear span of the eigenspaces of A with eigenvalues $\le \eta$. Show that, for $u \in C^k(M)$,

$$\inf_{g \in V_\eta} \|u-g\|_{L^\infty(M)} \le c\eta^{-k}\|u\|_{C^k(M)}.$$

Hint: Pick $p(\lambda)$ as in Exercises 3.1 through 3.5 and show that

$$\|u-p(tA)u\|_{L^\infty(M)} \le ct^k\|u\|_{C^k(M)}. \tag{3.21}$$

To prove (3.21), use the results of Exercises 3.5 and 3.2.

3.7. Prove the following generalization of an inequality of Bernstein.

$$\|f\|_{C^k(M)} \le C_k\eta^k\|f\|_{L^\infty(M)}, \qquad f \in V_\eta. \tag{3.22}$$

Hint: Pick $p(\lambda)$ as above, with $p(\lambda) = 0$ for $\lambda \geq 2$. Thus $p(tA)$ coincides with the identity on V_η, for $\eta \leq 1/2t$. To prove (3.22), estimate

$$\int_M |D_x^\alpha K_t(x, y)|\, dy$$

for $|\alpha| \leq k$.

3.8. Suppose that $u \in C(M)$ and that

$$\inf_{g \in V_\eta} \|u - g\|_{L^\infty} \leq C\eta^{-k-\alpha},$$

$k \in \mathbf{Z}^+$, $0 < \alpha < 1$. Show that $u \in C^{k+\alpha}(M)$. Hint: Look up the proof of Bernstein's theorem for trigonometric polynomials in Butzer and Berens [1] and deduce this result from (3.22).

3.9. Generalize Exercise 3.6 to obtain a converse to the result of Exercise 3.8.

3.10. Show that, if $n = \dim M$, $0 \leq s < n/2$,

$$\|u\|_{L^\infty(M)} \leq C\eta^{n/2-s}\|u\|_{H^s(M)}, \qquad u \in V_\eta.$$

What kind of estimate can you get when $s = n/2$?

3.11. If $f(\lambda)$ is even, note that

$$f(A) = \frac{1}{\sqrt{2\pi}} \int_{-\infty}^{\infty} \hat{f}(t) \cos tA\, dt.$$

If $A^2 = B$ is a differential operator, note that $\cos tA\delta_y$ is supported in a ball of radius $|t|$ about y (finite propagation speed for hyperbolic *PDE*). Use this to give a direct proof of Proposition 3.6, part (ii).

§4. Convergence of Eigenfunction Expansions

In this section we examine the convergence of weighted means of eigenfunction expansions of $f \in \mathscr{D}'(M)$:

$$E_m f = \sum_{j=1}^{\infty} p_m(\lambda_j) P_j f.$$

Here $P_j f$ is the projection of f onto the λ_j eigenspace of A defined by $P_j f = (f, u_j)u_j$. As usual, A is an elliptic self-adjoint element of OPS^1, on a compact manifold M, with scalar principal symbol. We obtain conditions on the $p_m(\lambda)$ which will yield norm or almost everywhere convergence of $E_m f$ to f, given $f \in L^p(M)$, $C^\alpha(M)$, etc.

The way we obtain results is to realize that $E_m = p_m(A)$, an operator which was discussed in Section 1. From the analysis there, plus the result of Chapter XI, Section 2, it follows that if $f \in L^p(M)$ $(1 < p < \infty)$ (resp. if $f \in C^\infty(M)$, $0 < \alpha < 1$), then $\|p_m(A)f\|_{L^p(M)}$ (resp. $\|p_m(A)\|_{C^\alpha(M)}$) is bounded independent of m, provided $\{p_m\}$ is bounded in $S^0_{1,0}(\mathbf{R})$. A particularly important case of this is $p_m(\lambda) = p(\lambda/m)$, given $p \in S^0_{1,0}(\mathbf{R})$. As a consequence of these observations, we have the following.

THEOREM 4.1. *Suppose $\{p_m\}$ is a bounded subset of $S^0_{1,0}(\mathbf{R})$ and that, for each λ, $p_m(\lambda) \to 1$ as $m \to \infty$. Then*

(i) *$p_m(A)f \to f$ in $L^p(M)$ as $n \to \infty$, given $f \in L^p(M)$, $1 < p < \infty$;*

(ii) *$p_m(A)f \to f$ in $C^{\alpha-\epsilon}(M)$ as $n \to \infty$, given $f \in C^\alpha(M)$, $0 < \alpha < 1$.*

Proof. Since

$$\sup_\lambda |p_m(\lambda)| \le C < \infty$$

and $p_m(\lambda) \to 1$ as $m \to \infty$, it is clear that, given $f \in C^\infty(M)$, $p_m(A)f \to f$ in $C^\infty(M)$. Since $C^\infty(M)$ is dense in $L^p(M)$, (i) follows from the uniform boundedness of the operators $p_m(A)$ on $L^p(M)$, $1 < p < \infty$. Now $C^\infty(M)$ is not dense in $C^\alpha(M)$, but the closure of $C^\infty(M)$ in $C^{\alpha-\epsilon}(M)$ contains $C^\alpha(M)$, from which (ii) follows, by the uniform boundedness of the operators $p_m(A)$ on $C^{\alpha-\epsilon}(M)$.

Remark. The "correct" C^α result is given in Exercise 4.4.

The above theorem generalizes without trouble as follows.

COROLLARY 4.2. *Suppose $\{p_m\}$ is a bounded subset of $S^0_{1,0}(\mathbf{R})$ and that, for each λ, $p_m(\lambda) \to 1$ as $m \to \infty$. Then*

(i) *$p_m(A)f \to f$ in $\mathscr{L}^s_p(M)$ as $m \to \infty$, given $f \in \mathscr{L}^s_p(M)$, $1 < p < \infty$.*

(ii) *$p_m(A)f \to f$ in $C^{k+\alpha-\epsilon}(M)$ as $m \to \infty$, given $f \in C^{k+\alpha}(M)$, $0 < \alpha < 1$, $k = 0, 1, 2, \ldots$.*

Proof. Since $A \in OPS^1$ is elliptic, we know that $A^s \in OPS^s_{1,0}$, $s \in \mathbf{R}$. Thus if $f \in \mathscr{L}^s_p$, we have $A^s p_m(A)f = p_m(A)(A^s f) \to A^s f$ in $L^p(1 < p < \infty)$, which shows that $p_m(A)f \to$ in $\mathscr{L}^s_p$. Similarly, the equation $p_m(A)A^k = A^k p_m(A)$ plus the fact that $A^k: C^{k+\alpha} \to C^\alpha$ and $A^{-k}: C^{\alpha-\epsilon} \to C^{k+\alpha-\epsilon}$ yield assertion (ii).

The following will be our result on pointwise convergence.

THEOREM 4.3. *Let $p(\lambda) \in S^{-s}_{1,0}(\mathbf{R})$ for some $s > 0$ and let $p_m(\lambda) = p(\alpha_m \lambda)$, where α_m is any sequency tending to zero. Suppose $p(0) = 1$. Then $p_m(A)f \to f$ almost everywhere on M, given $f \in L^1(M)$.*

We shall prove this theorem as a consequence of an estimate on the maximal function associated with $p_m(A)f$,

$$f^*(x) = \sup_m |p_m(A)f(x)|. \tag{4.1}$$

We have the following estimate.

PROPOSITION 4.4. *Granted the hypotheses of Theorem 4.3, we have*

(i) $\|f^*\|_{L^p(M)} \le C_p\|f\|_{L^p(M)}, \qquad 1 < p < \infty.$

(ii) $\text{meas}\,\{x \in M : f^*(x) \ge \lambda\} \le \dfrac{C}{\lambda}\|f\|_{L^1(M)}.$

Theorem 4.3 follows easily from the estimate (ii) as follows. We already know that $p_m(A)g \to g$ in $C^\infty(M)$ (and a fortiori at every point), given $g \in C^\infty(M)$, which is dense in $L^1(M)$. Consequently, if $\delta > 0$ is given and if S_δ is the subset of $x \in M$ where $|p_m(A)f(x) - f(x)| \ge \delta$ for infinitely many m, then taking $g \in C^\infty(M)$ with $\|g - f\|_{L^1} < \epsilon$ we have $S_\delta = \{x \in M: |p_m(A)(f-g)(x) - (f-g)(x)| \ge \delta$ for infinitely many $m\}$, so

$$\begin{aligned}\text{meas}\,S_\delta &\le \text{meas}\left\{x \in M:(f-g)^*(x) \ge \frac{\delta}{2}\right\} + \text{meas}\left\{x \in M:(f-g)(x) \ge \frac{\delta}{2}\right\}\\ &\le \frac{2C}{\delta}\|f-g\|_{L^1} + \frac{2}{\delta}\|f-g\|_{L^1}\\ &\le \frac{2(C+1)}{\delta}\epsilon.\end{aligned}$$

and taking ϵ arbitrarily small we conclude that meas $S_\delta = 0$. Thus the set of points $x \in M$ where $p_m(A)f$ does *not* converge to f, which is equal to

$$\bigcup_{j=1}^{\infty} S_{1/j},$$

has measure zero. Thus all the work will be involved in proving Proposition 4.4. Assertion (i) of this proposition is not needed for the proof of Theorem 4.3, but for $f \in L^p$ with $1 < p < \infty$ it is the correct result, and the proof of (i) arises simultaneously with the proof of (ii).

With $t = \alpha_m$ we have, by Theorem 3.3,

$$\begin{aligned}p_m(A)f = p(tA)f &= \int K_p(t, x, x-y)f(y)\,dy\\ &\quad + t\sum_{j=1}^{\ell}\sum_{\nu=1}^{n(j)}\int K_{q_{j\nu}}(t, x, x-y)(B_{j\nu}f(y))\,dy\\ &\quad + r_\ell(t, x, D)f.\end{aligned} \tag{4.2}$$

This fact, together with the estimates on $K_p(t, x, x-y)$ contained in Lemmas 3.1 and 3.2, will enable us to estimate $f^*(x)$ in terms of the Hardy-Littlewood maximal function $M(f)$ defined by

$$M(f)(x) = \sup_{r>0}\frac{1}{\text{vol}\,B_r}\int_{|y-x|\le r}|f(y)|\,dy.$$

The importance of the Hardy-Littlewood maximal function arises from the following classical result, for whose proof the reader is referred to Stein [1] or Garsia [1].

LEMMA 4.5. *The Hardy-Littlewood maximal function satisfies the estimates*

(i) $\|M(f)\|_{L^p} \le C_p \|f\|_{L^p}, \qquad 1 < p \le \infty;$

(ii) $\text{meas}\,\{x \in \mathbf{R}^n : M(f)(x) \ge \lambda\} \le \dfrac{C}{\lambda} \|f\|_{L^1(\mathbf{R}^n)}.$

As our first step in the analysis of (4.2), we have the following.

LEMMA 4.6. *If $K_p(t, x, x-y)$ is an elementary Poisson-like kernel for $p(\lambda) \in S_{1,0}^{-s}(\mathbf{R})$, $s > 0$, we have the estimate*

$$\sup_{0<t<1} \left| \int K_p(t, x, x-y) f(y)\, dy \right| \le CM(f)(x). \tag{4.3}$$

Proof. We are working on a coordinate patch, but supposing f has compact support, and (4.3) is to hold for $x \in K$, a compact subset of $\mathbf{R}^n$. By the analysis of K_p given in Section 3, i.e., by (3.3) and Lemmas 3.1 and 3.2, we know $K_p(t, x, z) = t^{-n}\Phi(x, t^{-1}z) + \Psi(t, x, z)$ where Ψ is smooth and, for $s < n$,

$$|\Phi(x, z)| \le \begin{cases} C|z|^{-n+s}, & |z| \le 1 \\ C|z|^{-n-1}, & |z| \ge 1. \end{cases}$$

Thus it suffices to show that, for $0 < t \le 1$,

$$|\tilde{\Phi}_t * f|(x) \le CM(f)(x) \tag{4.4}$$

where $\tilde{\Phi}_t(z) = t^{-n}\tilde{\Phi}(t^{-1}z)$ and

$$\tilde{\Phi}(z) = \begin{cases} |z|^{-n+s}, & |z| \le 1 \\ |z|^{-n-1}, & |z| \ge 1. \end{cases}$$

Since M commutes with dilation, it suffices to prove (4.4) for $t = 1$. Now $|\tilde{\Phi} * f(x)| \le \int |\tilde{\Phi}(x-y)|\, |f(y)|\, dy \le \int \psi(x-y) |f(y)|\, dy$ where $\psi \in L^1$ is any function such that $\psi \ge |\tilde{\Phi}|$. Thus to prove (4.4) (for $t = 1$) it suffices to have

$$|\tilde{\Phi}(x)| \le c \int_{|x|}^{\infty} r^{-n}\beta(r)\, dr = c \int_0^{\infty} r^{-n}\chi_{B_r(0)}(x)\beta(r)\, dr$$

for some $\beta \in L^1(\mathbf{R}^+)$. Taking $\beta(r) = r^{-2}$ on $[1, \infty)$ and $\beta(r) = r^{-1+s}$ on $[0, 1]$ yields the desired result.

We can now prove Proposition 4.4. Indeed, applying Lemma 4.6 to (4.2) we see that

$$f^*(x) \leq CM(f)(x) + c \sum_{j=1}^{\ell} \sum_{\nu=1}^{n(j)} M(B_{j\nu}f)(x) + c \sup_{0<t<1} |r_\ell(t, x, D)f|.$$

Since $B_{j\nu} \in OPS^{-j+1}$, $B_{j\nu}: L^p \to L^p$, $1 < p < \infty$, so part (i) of Proposition 4.4 follows immediately from part (i) of Lemma 4.5. For $f \in L^1$, we want to get rid of the operators $B_{j\nu}$ of order zero, so use Corollary 3.4, and the remark following it, to write

$$f^*(x) \leq CM(f)(x) + C \sum_{\nu=1}^{n(1)} M(\tilde{B}_\nu f)(x) + C \sum_{j=2}^{\ell} \sum_{\nu=1}^{n(j)} M(B_{j\nu}f) + c \sup_{0<t<1} |r_\ell(t, x, D)f|$$

where now all operators $\tilde{B}_\nu$, $B_{j\nu}$ have negative order; hence map $L^1(M)$ to $L^1(M)$. Now part (ii) of Proposition 4.4 follows from part (ii) of Lemma 4.5. This completes the proof of our pointwise convergence result, Theorem 4.3.

Remark. In our statement of Theorem 4.3, we took a countable sequence $\alpha_m \downarrow 0$ in order to avoid measurability difficulties. In general if $p(\lambda) \in S_{1,0}^{-s}(\mathbf{R})$ and s is close to zero, $f \in L^1(M)$, for each fixed t_0, $p(t_0 A)f$ will only be defined almost everywhere. On the other hand $p(tA)f$ will be continuous on M provided $s > n$, $f \in L^1(M)$, or more generally, if $s > n/p$, $f \in L^p(M)$. In such a case, you have almost everywhere convergence of $p(tA)f$ as $t \to 0$.

As a scan of the proofs of our theorem shows, one does not need estimates on *all* the derivatives of $p(\lambda)$ in order to get the desired results. Only some finite number of derivatives (depending on the dimension n of M) are needed. Thus the conclusions of these theorems apply for $p(\lambda) \in C_0^k(\mathbf{R})$ if k is large enough, and in particular to the Riesz means

$$R_m^\alpha u = \sum_{j=1}^{\infty} \left(1 - \left(\frac{\lambda_j}{m}\right)^2\right)_+^\alpha P_j u$$

(obtained by setting $p(\lambda) = (1-\lambda^2)^\alpha$ for $|\lambda| \leq 1$, $p(\lambda) = 0$ for $|\lambda| > 1$), provided α is large enough. If one tightens up the arguments in the proofs of Theorem 4.1, using the estimate on L^p norms given in (2.10) and (2.11) of Chapter IV, plus further interpolation arguments to sharpen up results obtained by differentiating $p_m(\lambda)$ an integral number of times, one can obtain the L^p norm convergence of $R_m^\alpha u$, $1 < p < \infty$, provided $\alpha \geq n - (1/2)$.

This argument is lengthy, though not without points of interest. At any rate, Hörmander has obtained L^p norm convergence for $\alpha \geq n - 1$, for all $p \in (1, \infty)$, and also pointwise almost everywhere convergence. Any straightforward tightening of the argument in Theorem 4.3 would probably require α in the neighborhood of $(3/2)n$. In the special case when M is a compact Lie group and Δ a bi-invariant Laplacian, $A = \sqrt{-\Delta}$, people have gotten results for $\alpha \geq (n-1)/2$, the "critical index" which occurs in the case of the torus $\mathbf{T}^n$. See Stanton [1] and Weiss [1]. I do not know whether such sharp results are valid for general M.

Exercises

4.1. If $p(\lambda) \in S_{1,0}^{-s}(\mathbf{R})$, $s > 0$, and if $u \in C(M)$, show that $p(tA)u \to u$ uniformly on M, as $t \downarrow 0$.

4.2. If $p(\lambda) \in S_{1,0}^{0}(\mathbf{R})$ and if $u \in C^\alpha(M)$, $\alpha > 0$, show that $p(tA)u \to u$ uniformly.

4.3. If $p(\lambda) \in S_{1,0}^{-s}(\mathbf{R})$, $s > 0$, show that $p(tA)u \to u$ in $C^\alpha(M)$ if, and only if, u belongs to the closure of $C^\infty(M)$ in $C^\alpha(M)$. $(0 < \alpha < 1)$

4.4. If $p(\lambda) \in S_{1,0}^{-s}(\mathbf{R})$, $s > 0$, and if $u \in C^\alpha(M)$, $0 < \alpha < 1$, define

$$v(t, x) = p(tA)u(x).$$

Show that $v \in C^\alpha(\overline{\mathbf{R}}_+ \times M)$.

4.5. Prove that the eigenfunction expansion $f = \sum a_j u_j$ is absolutely and uniformly convergent for all $f \in H^s(M)$, if and only if $s > n/2$, where $n = \dim M$.

4.6. Suppose $f \in L^1(M)$ has eigenfunction expansion $f = \sum a_j u_j$ and suppose that, for almost all x, $a_j u_j(x) = 0(1/j)$. Show that

$$\sum_{j=1}^{\infty} a_j u_j(x)$$

converges almost everywhere to $f(x)$. Hint: Get convergence by a summability method and use a Tauberian theorem.

4.7. If $A, B \in OPS^1$ are both elliptic positive self-adjoint operators on M, show that $(\log A)(\log B)^{-1} \in OPS_{1,0}^{0}$. (Assume 1 is not an eigenvalue of B.)

4.8. Prove that the eigenfunction expansion of $f = \sum a_j u_j$ is absolutely and uniformly convergent provided $A^{n/2}(\log A)^{1/2+\epsilon} f \in L^2(M)$. (Hint: Show that you can reduce the problem to $A = \sqrt{-\Delta}$ on the torus.) Note that this condition is

$$\sum_{j=2}^{\infty} j(\log j)^{1+2\epsilon} |a_j|^2 < \infty.$$

4.9. Show that the eigenfunction expansion of $f = \sum a_j u_j$ is absolutely and uniformly convergent provided $|a_j| \le Cj^{-1}(\log j)^{-1-\epsilon}$. Compare this result with Exercise 4.8. Hint: Start with

$$\sum_{j=2}^{\infty} j^{-1}(\log j)^{-\sigma}|u_j(x)| \le \sum_{\nu=1}^{\infty}\left(\sum_{j=2^\nu}^{2^{\nu+1}} j^{-2}(\log j)^{-2\sigma}\right)^{1/2}\left(\sum_{j=2^\nu}^{2^{\nu+1}} |u_j(x)|^2\right)^{1/2}$$

and use the fact that

$$\sum_{j=1}^{N} |u_j(x)|^2 \le CN,$$

which follows from Theorem 3.7.

4.10. Generalize the results of Exercises 4.8 and 4.9. Which method is the more powerful? In particular, show that, if $\varphi \in C_0^\infty((0, 1))$,

$$\sum_{j=2}^{\infty} |a_j|\,|u_j(x)| \le c\int_0^1 \|\varphi(sA)A^n u\|_{L^2}\,ds$$

for $u = \sum a_j u_j(x)$.

4.11. If $u \in C^\infty(M, E)$ where E is a Banach space and if

$$a_j = \int_M u(x)\overline{u_j(x)}\,dx \in E,$$

show $\sum a_j u_j$ is absolutely convergent to $u(x)$, uniformly in x.

§5. Eigenfunction Expansions of Measures

In this section we shall discuss generalizations and variations of a classical result of Wiener regarding finite measures on the line. In fact, if μ is a finite measure on $\mathbf{R}$ with Fourier transform $\mu(\xi) = \int e^{ix\xi}\,d\mu(x)$, Wiener's theorem says that

$$\lim_{r\to\infty} \frac{1}{2r}\int_{-r}^{r} |\hat{\mu}(\xi)|^2\,d\xi = \sum |a_j|^2 \tag{5.1}$$

where all the point masses of μ are of the form $a_j\delta_{x_j}$.

More generally, let μ be a finite measure on $\mathbf{R}^n$, all of whose point masses are $a_j\delta_{x_j}$. Denote by γ the sum $\sum |a_j|^2$. We claim that if you average $|\hat{\mu}(\xi)|^2$ over the dilates of any set S of positive Lebesgue measure, you get γ in the limit. In fact, given $f \in L^1(\mathbf{R}^n)$ with $\int f(\xi)\,d\xi = 1$, and setting $f_r(\xi) = f(\xi/r)$, you get

$$\lim_{r\to\infty} r^{-n}\int f_r(\xi)|\hat{\mu}(\xi)|^2\,d\xi = \gamma. \tag{5.2}$$

Here's an easy proof of (5.2).

$$r^{-n}\int f_r(\xi)|\hat{\mu}(\xi)|^2\,d\xi = r^{-n}\int f_r(\xi)\int e^{ix\cdot\xi}\,d\mu(x)\int e^{-iy\cdot\xi}\,d\overline{\mu(y)}\,d\xi$$

$$= \iint\left\{\int f(\xi)e^{ir(x-y)\cdot\xi}\,d\xi\right\}d\mu(x)\,d\overline{\mu(y)}.$$

Since the expression in brackets is bounded by $\|f\|_{L^1}$, we can pass to the limit under the integral sign. Now $\int f(\xi)e^{ir(x-y)\cdot\xi}\,d\xi$ tends to 0 as $r\to\infty$ if $x\neq y$, by the Riemann-Lebesgue lemma, while the expression is 1 if $x=y$. Thus

$$\lim_{r\to\infty} r^{-n}\int f_r(\xi)|\hat{\mu}(\xi)|^2\,d\xi = \iint_{x=y} d\mu(x)\,d\overline{\mu(y)} = \gamma.$$

More generally still, $f(\xi)$ could be replaced by a finite measure ν such that $\hat{\nu}(\xi)\to 0$ as $|\xi|\to\infty$. For example, ν could be a piecewise smooth density on a hypersurface of nonzero curvature. Thus one could average over spheres, pieces of spheres, etc., for $n>1$.

Before we get down to serious business, let us consider the following amusing variation. Let G be a locally compact abelian group and let $\hat{G}$ be its dual group, the group of homomorphisms $\lambda: G\to S^1$, where S^1 is the circle group. Then

$$\hat{\mu}(\lambda) = \int_G \lambda(x)\,d\mu(x)$$

is a bounded continuous function on $\hat{G}$, if μ is a finite measure on G. Now let M_λ be any invariant mean on $\hat{G}$; for the existence of such means, see Hewitt and Ross [1]. Since $M_\lambda(\lambda(x)) = M_\lambda(\lambda'\lambda(x)) = \lambda'(x)M_\lambda(\lambda(x))$, it follows that $M_\lambda(\lambda(x)) = 0$ if $x\neq e$. Of course, $M_\lambda(\lambda(e)) = M_\lambda(1) = 1$. We can write

$$M_\lambda(|\hat{\mu}(\lambda)|^2) = \int_{\beta\hat{G}} |\hat{\mu}(\lambda)|^2\,dm(\lambda)$$

where m is a probability measure on the Stone-Cech compactification $\beta\hat{G}$ of $\hat{G}$. Our analogue of (5.1) is

$$M_\lambda(|\hat{\mu}(\lambda)|^2) = \gamma. \tag{5.3}$$

Proof. We have

$$M_\lambda(|\hat{\mu}(\lambda)|^2) = \int_{\beta\hat{G}}\int_G \lambda(x)\,d\mu(x)\int_G \overline{\lambda(y)}\,d\overline{\mu(y)}\,dm(\lambda)$$

$$= \int_G\int_G\left\{\int_{\beta\hat{G}} \lambda(xy^{-1})\,dm(x)\right\}d\mu(x)\,d\overline{\mu(y)}$$

$$= \iint_{x=y} d\mu(x)\,d\overline{\mu(y)} = \gamma.$$

From now on, we consider the behavior of the eigenfunction expansion $\mu = \sum \hat{\mu}(j)u_j$ of a finite measure μ on a compact manifold M, with respect to an elliptic self-adjoint operator $A \in OPS^1$. We obtain the following, where all the point masses of μ are of the form $a_j\delta_{x_j}$.

THEOREM 5.1. *We have*

$$\lim_{N\to\infty} \frac{1}{N} \sum_{j=1}^{N} |\hat{\mu}(j)|^2 = \left(\int_M k(x)\,dx\right)^{-1} \sum_j k(x_j)|a_j|^2$$

where $k(x) = \operatorname{vol}\{\xi \in T_x^*(M) : a_1(x, \xi) \le 1\}$.

Proof. If $e(x, y, \lambda)$ is the spectral function considered in Section 3, the kernel of the projection operator E_λ, we have

$$\sum_{\lambda_j \le \lambda} |\hat{\mu}(j)|^2 = (E_\lambda \mu, \mu) = \iint e(x, y, \lambda)\,d\mu(x)\,d\overline{\mu(y)}.$$

But by Theorem 3.7 and the remark following it, we see that

$$\begin{aligned}\lim_{\lambda\to\infty} \lambda^{-n} \iint e(x, y, \lambda)\,d\mu(x)\,d\overline{\mu(y)} &= (2\pi)^{-n} \iint_{x=y} k(x)\,d\mu(x)\,d\overline{\mu(y)} \\ &= (2\pi)^{-n} \sum k(x_j)|a_j|^2.\end{aligned}$$

On the other hand, since

$$\lambda_N^n \sim (2\pi)^n N \left(\int_M k(x)\,dx\right)^{-1},$$

by Theorem 2.1, our result follows.

To give some feeling for this result, we consider the following example. Let G be a compact Lie group with a bi-invariant Riemannian metric, and consider the Laplacian Δ on G; let $A = \sqrt{-\Delta}$, whose eigenvalues are $0 = \lambda_1 < \lambda_2 \le \lambda_3 \le \ldots$. Now, it is known that, if $\{\pi^k\}$ is a complete set of irreducible unitary representations of G, onto vector spaces V_k of dimension d_k, then the entries of each matrix valued function π^k span an eigenspace of A, and $\{\sqrt{d_k}\pi_{ij}^k\}$ is a complete orthogonal set of eigenfunctions for A on G; see, for example, Wallach [1]. If μ_e is the point mass concentrated at the identity element of G, then $\hat{\mu}_e(i, j, k) = \sqrt{d_k}\delta_{ij}$, so if we fix G, we easily see that the mean of $|\hat{\mu}_e(i, j, k)|^2$ is 1, a result consistent with Theorem 5.1. On the other hand, $\hat{\mu}_e$ fails to enjoy some of the nice properties of the Fourier transforms of measures on $\mathbf{R}^n$. For example, it is not bounded. As another instance, for a measure μ on $\mathbf{R}^n$ we have

$$\lim_{r\to\infty} (\operatorname{vol} B_r)^{-1} \int_{|\xi| \le r} \hat{\mu}(\xi)\,d\xi = \mu(\{0\}) \tag{5.4}$$

a result that can be proved in a manner similar to (5.2). This result will not in general hold on compact manifolds. In fact, if we fix k, the mean value of $\hat{\mu}_e(i, j, k)$ is $d_k^{-1/2}$. If $G = SO(3)$, for example, $d_k \to \infty$ as $k \to \infty$, so

$$\left(\sum_{k=1}^{\ell} d_k^2\right)^{-1} \sum_{k=1}^{\ell} \sum_{i,j} \hat{\mu}_e(i, j, k) \to 0$$

as $\ell \to \infty$, instead of tending to $\mu_e(\{0\}) = 1$. For a correct analogue of (5.4), see Exercise 5.2. We shall see some applications of Theorem 5.1 and also the following results, to the theory of representations of compact Lie groups, in the next section.

We next consider the behavior of $(p(tA)\mu, \mu)$ as $t \downarrow 0$, where the measure μ is a smooth density $\mu = f(x)\, d\, \mathrm{vol}_X$ on a compact k-dimensional submanifold X of M (whose dimension is n), and $p(\lambda) \in S_{1,0}^{-s}(\mathbf{R})$ with $s > n-k$. We begin by analyzing what one would guess to be the principal term,

$$A(t) = \iint K_p(t, x, x-y) f(y) \overline{f(x)}\, d\, \mathrm{vol}_X(y)\, d\, \mathrm{vol}_X(x). \tag{5.5}$$

Recall that $K_p(t, x, z) = t^{-n}\Phi(x, t^{-1}z) + \Psi(t, x, z)$ where Ψ is smooth and

$$\Phi(x, z) = \int_{\mathbf{R}^n} p(a_1(x, \xi)) e^{iz\cdot\xi}\, d\xi.$$

Suppose that, in the local coordinate system, X is defined by $x_{k+1} = \cdots = x_n = 0$, and $d\, \mathrm{vol}_X(x) = \Omega(x)\, dx_1 \cdots dx_k$. Therefore,

$$A(t) = t^{-n} \int_{\mathbf{R}^k} \int_{\mathbf{R}^k} \Phi(x, t^{-1}(x-y)) \times f(y)\Omega(y)\overline{f(x)}\Omega(x)\, dy'\, dx' + 0(1). \tag{5.6}$$

We are writing $x = (x', x'')$ with $x' = (x_1, \ldots, x_k)$, $x'' = (x_{k+1}, \ldots, x_n)$. The x and y that appear in (5.6) satisfy $x = (x', 0)$, $y = (y', 0)$. Clearly

$$t^{-k} \int_{\mathbf{R}^k} \Phi(x, t^{-1}(x-y)) f(y)\Omega(y)\, dy' \to c_0(x) f(x)\Omega(x) \qquad \text{as } t \downarrow 0 \tag{5.7}$$

if $x \in X$, where

$$c_0(x) = \int_{\mathbf{R}^k} \Phi(x, z)\, dz' = \int_{\mathbf{R}^{n-k}} p(a_1(x, 0, \xi''))\, d\xi''.$$

Therefore,

$$A(t) \sim t^{k-n} \int_{\mathbf{R}^k} c_0(x)|f(x)|^2 \Omega(x)^2\, dx' = t^{k-n} \int_X |f(x)|^2 \Omega(x) c_0(x)\, d\, \mathrm{vol}_X(x).$$

Note that

$$\Omega(x) c_0(x) = \Omega(x) \int_{\mathbf{R}^{n-k}} p(a_1(x, 0, \xi''))\, d\xi''$$

is merely the integral of the principal symbol $p(a_1)$ over the fiber in the normal bundle to X, with volume element induced from the Riemannian metric on M, and hence is invariantly defined, independently of the coordinate system.

Now $(p(tA)\mu, \mu)$ is, modulo $0(1)$, equal to $A(t)$ plus a finite sum of terms of the form

$$B(t) = t^a \iint K_q(t, x, x - y) B_j \mu(y)\, dy\, d\bar{\mu}(x) \tag{5.8}$$

where $0 < a \leq 1$ and $B_j \in OPS_{1,0}^{-\sigma}$, $\sigma > 0$. We show that such an expression is $o(t^{k-n})$ as $t \downarrow 0$. As a tool in analyzing (5.8) we need the following lemma, whose proof we leave as an exercise.

LEMMA 5.2. *If $B_j \in OPS_{1,0}^{-\sigma}(\mathbf{R}^n)$, μ is a smooth distribution on $\mathbf{R}^k$, then there exists a smooth family $c_j(x'', x', D_{x'}) \in OPS_{1,0}^{-\sigma}(\mathbf{R}^k)$ such that $B_j\mu = c_j(x'', x', D_{x'})\mu \bmod C^\infty$.*

COROLLARY. *If B_j, μ are as above, $\sigma > 0$, then $B_j\mu(y) = m(y', y'')$ is C^∞ in y'', with values in $L^1(\mathbf{R}^k)$, as a function of y'.*

Consequently, we have

$$\begin{aligned} B(t) &= t^a \int_X \int_M K_q(t, x, x - y) m(y', y'')\, dy\, f(x)\, d\,\mathrm{vol}_X(x) \\ &= t^{a-n} \int_{\mathbf{R}^k} \int_{\mathbf{R}^k} \int_{\mathbf{R}^{n-k}} \tilde{\Phi}(x, t^{-1}(x - y)) m(y', y'') f(x) \Omega(x)\, dy''\, dy'\, dx' + 0(1), \end{aligned}$$

and the same method used to estimate (5.6) shows that

$$|B(t)| \leq c t^{a+k-n} = o(t^{k-n})$$

as asserted. We summarize as follows.

THEOREM 5.3. *Let X be a smooth k-dimensional submanifold of a manifold M of dimension n, and suppose $\mu \in \mathcal{D}'(M)$ is a smooth multiple $f(x)$ of the volume element of X. If $p(\lambda) \in S_{1,0}^{-s}(\mathbf{R})$, $s > n - k$, we have*

$$(p(tA)\mu, \mu) = t^{-(n-k)} \int_X |f(x)|^2 \pi(x)\, d\,\mathrm{vol}_X(x) + o(t^{-(n-k)}) \tag{5.9}$$

where

$$\pi(x) = \Omega(x) \int_{\mathbf{R}^{n-k}} p(a_1(x, 0, \xi''))\, d\xi''.$$

From this we easily obtain the following analogue of Theorem 5.1, regarding the Fourier coefficients

$$\hat{\mu}(j) = (\mu, u_j) = \int_X u_j(x) f(x)\, d\,\mathrm{vol}_X(x).$$

Here $N^*(X)$ is the conormal bundle to X.

COROLLARY 5.4. *Under the hypotheses of Theorem 5.3,*

$$\sum_{j=1}^{N} |\hat{\mu}(j)|^2 \sim c N^{1-k/n} \qquad \text{as } N \to \infty \tag{5.10}$$

where

$$c = (c')^{n-k} \int_X |f(x)|^2 \pi_0(x) \, d\,\mathrm{vol}_X(x)$$

with c' *the constant in* (2.10) *and* $\pi_0(x) = \mathrm{vol}\,\{\xi \in N_x^*(X) : a_1(x, \xi) \le 1\}$.

Proof. Pick $p(\lambda) \in C_0^\infty(\mathbf{R})$ equal to 1 for $0 \le \lambda \le 1$, equal to 0 for $\lambda \ge 1 + \epsilon$, $0 \le p(\lambda) \le 1$. Apply Theorem 5.3 to $(p(tA)\mu, \mu)$ and pass to the limit as $\epsilon \to 0$, recalling the argument in the proof of Theorem 2.1. This produces a formula for the asymptotic behavior of

$$\sum_{\lambda_j \le t^{-1}} |\hat{\mu}(j)|^2.$$

Using $\lambda_j \sim c' j^{1/n}$ completes the proof.

Let's linger a minute over a possible alternate proof of Corollary 5.4, in the special case when $A = \sqrt{-\Delta}$. Namely, suppose we analyze the behaviour of $(e^{t\Delta}\mu, \mu)$ where $e^{t\Delta}$ is the heat kernel. Now

$$e^{t\Delta} u(x) = \int_M H(t, x, y) u(y) \, dy$$

where

$$H(t, x, y) \sim (4\pi t)^{-n/2} e^{-\gamma^2/4t} (u_0(x, y) + t u_1(x, y) + \cdots) \tag{5.11}$$

where $\gamma = \mathrm{dist}\,(x, y)$, and $u_0(x, x) = 1$. Since $e^{t\Delta} = p(sA)$ with $p(\lambda) = e^{-\lambda^2}$, $s = \sqrt{t}$, (5.11) can be deduced from (3.8), but indeed there are several other ways of establishing (5.11), discussed in [Berger, Gauduchon, and Mazet], Gilkey [1], and Mckean and Singer [1]. From (5.11) we immediately obtain

$$(e^{t\Delta}\mu, \mu) \sim (4\pi t)^{-n/2} \int_X \int_X e^{-\gamma^2/4t} u_0(x, y) f(x) \overline{f(y)} \, d\,\mathrm{vol}_X(x) \, d\,\mathrm{vol}_X(y),$$

and since

$$\int_X g(x, y) e^{-\gamma^2/4t} \, d\,\mathrm{vol}_X(y) \sim (4\pi t)^{k/2} g(x, x)$$

for $g \in C^\infty(X \times X)$,

$$(e^{t\Delta}\mu, \mu) \sim (4\pi t)^{-(n-k)/2} \int_X |f(x)|^2 \, d\,\mathrm{vol}_X(x), \qquad t \to 0.$$

This implies

$$\sum_{j=1}^{\infty} |\hat{\mu}(j)|^2 e^{-t\lambda_j^2} \sim (4\pi t)^{-(n-k)/2} \int_X |f(x)|^2 \, d\,\mathrm{vol}_X(x), \qquad t \to 0. \tag{5.12}$$

That (5.12) implies (5.10) is a Tauberian theorem, which we prove in Section 7.

Thus we have obtained a second proof of Corollary 5.4, using lighter analytical machinery than the first. However, the approach via the functional calculus admits greater flexibility than the heat equation approach, as we shall see in the next section.

Exercises

5.1. If μ and ν are finite measures on $\mathbf{R}^n$, show that, with $B_r = \{\xi : |\xi| \le r\}$,

$$\lim_{r \to \infty} (\text{vol } B_r)^{-1} \int_{B_r} \hat{\mu}(\xi)\overline{\hat{\nu}(\xi)}\, d\xi = \sum_{p \in \mathbf{R}^n} \mu(\{p\})\overline{\nu(\{p\})},$$

where, of course, the terms in the sum are nonzero for only denumerably many p. From this, deduce (5.4).

5.2. If μ and ν are measures on M, show that

$$\lim_{N \to \infty} \frac{1}{N} \sum_{j=1}^{N} \hat{\mu}(j)\overline{\hat{\nu}(j)} = \left(\int k(x)\, dx\right)^{-1} \sum_{p \in M} k(p)\mu(\{p\})\overline{\nu(\{p\})}.$$

5.3. For a distribution $\omega \in \mathscr{D}'(M)$, consider the possibility that

$$\frac{1}{N} \sum_{j=1}^{N} |\hat{\omega}(j)|^2 \le c < \infty \qquad \text{for all } N. \tag{5.13}$$

(i) Show that (5.13) implies $\omega \in H^{-n/2-\epsilon}(M)$.
(ii) Show that $\omega \in H^{-n/2}(M)$ implies (5.13).
(iii) Find an $\omega \in \mathscr{D}'(M)$ satisfying (5.13) which is not a measure.

5.4. Prove lemma 5.2.

§6. Harmonic Analysis on Compact Lie Groups

In this section we apply some of the results of the previous sections to study a couple of topics in harmonic analysis on Lie groups. We have no space to introduce this subject, and the reader unfamiliar with the theory of representations of compact Lie groups can skip this section without loss of continuity. A couple of good treatments of this subject are given in Wallach [2] and Zelobenko [1]. We fix notations and recall a few basic facts in the next few paragraphs.

The irreducible unitary representations π_λ of a compact Lie group G are naturally indexed by $\lambda \in \mathbf{Z}^k \cap C$, where C is a convex cone in $\mathbf{R}^k$ called a Weyl chamber. Here k is the dimension of a maximal torus of G. The entries $\pi_\lambda^{ij}(x)$ of the matrix π_λ are functions on G. If P is any bi-invariant differential operator on G (we say $P \in \mathfrak{U}$), then $\{\pi_\lambda^{ij}\}$ belong to an eigenspace for P, for any fixed λ. An example of this is $P = \Delta$, the Laplacian on G, endowed with a bi-invariant Riemannian metric (which induces a metric on $\mathbf{R}^k$, the tangent space to a maximal torus $\mathbf{T}^k$). In this case we have

$$-\Delta\pi_\lambda^{ij} = (\|\lambda + \delta\|^2 - \|\delta\|^2)\pi_\lambda^{ij}.$$

Here $\delta \in \mathbf{R}^k$ is half the sum of the positive roots. More generally, let $q_m(\lambda)$ be any homogeneous polynomial on $\mathbf{R}^k$, of degree m, which is invariant under the action of the Weyl group. The following is true.

THEOREM 6.1. *There exists a bi-invariant differential operator Q_m, of order m, such that, for some polynomial $\tilde{q}_m(\lambda)$ of degree m, invariant under the Weyl group, whose m^{th} order principal part is $q_m(\lambda)$, we have*

$$Q_m\pi_\lambda^{ij} = \tilde{q}_m(\lambda+\delta)\pi_\lambda^{ij}.$$

For a proof of this, we refer to Zelebenko [1 p. 369].

For our purposes, it is convenient to reexpress this result as follows.

COROLLARY 6.2. *There exists a bi-invariant differential operator Q, of order m, such that*

$$Q\pi_\lambda^{ij} = q_m(\lambda+\delta)\pi_\lambda^{ij}. \tag{6.1}$$

Proof. Pick Q_m and $\tilde{q}_m$ according to Theorem 6.1. Write $\tilde{q}_m(\lambda) = q_m(\lambda) + q_{m-1}(\lambda) + r(\lambda)$ where $r(\lambda)$ has degree $m-2$ and $q_{m-1}(\lambda)$ is homogeneous of degree $m-1$. Now apply Theorem 6.1 to the polynomial $q_{m-1}(\lambda)$, which is also invariant under the Weyl group. An operator Q_{m-1} of order $m-1$ is produced. Continue this process, and let $Q = Q_m - Q_{m-1} - Q_{m-2} \cdots - Q_0$.

The particular case $q_2(\lambda) = \|\lambda\|^2$ yields the operator $Q = -\Delta + \|\delta\|^2$, as mentioned before. Let $A = (-\Delta + \|\delta\|^2)^{1/2} \in OPS^1$. Then we have

$$A^s\pi_\lambda^{ij} = \|\lambda+\delta\|^s\pi_\lambda^{ij}. \tag{6.2}$$

It follows from the treatment of Zelobenko [1] that the relation between $q_m(\lambda)$ and $q_m(e, \xi)$, the principal symbol of Q, is the following. We think of $\lambda \in T_e^*(\mathbf{T}^k)$ included in $T_e^*(G)$, and then $q_m(\lambda) = q_m(e, \lambda)$. Since $q_m(e, \xi)$ is invariant under the adjoint action of G on $T_e^*(G)$, this uniquely specifies $q_m(e, \xi)$.

A. Multipliers on G

An operator $T: C^\infty(G) \to \mathcal{D}'(G)$ with the property that

$$T\pi_\lambda^{ij} = \tau(\lambda)\pi_\lambda^{ij}, \qquad \tau(\lambda) \in \mathbf{C} \tag{6.3}$$

is called a *multiplier*. Clearly $|\tau(\lambda)| \le C(1 + \|\lambda\|)^K$ for some K; thus $T: C^\infty(G) \to C^\infty(G)$ and $T: \mathcal{D}'(G) \to \mathcal{D}'(G)$. Furthermore, if $\omega \in \mathcal{D}'(G)$ is defined by $\langle u, \omega\rangle = Tu(e)$, $u \in C^\infty(G)$, we see that $Tu = \check{\omega} * u = u * \check{\omega}$, where $\check{\omega}(y) = \omega(y^{-1})$. Thus T is also a convolution operator: convolution by $\check{\omega}$, which belongs to the center of the convolution algebra $\mathcal{D}'(G)$. In this subsection we wish to show that large classes of multipliers are pseudo-differential operators.

THEOREM 6.3. *Let $p(\lambda) \in S_{1,0}^m(\mathbf{R}^k)$ be invariant under the Weyl group. Let $P: \mathcal{D}'(G) \to \mathcal{D}'(G)$ be defined by*

$$P\pi_\lambda^{ij} = p(\lambda+\delta)\pi_\lambda^{ij}. \tag{6.4}$$

Then $P \in OPS_{1,0}^m$ on G.

Proof. Let P_w be the set of polynomials on $\mathbf{R}^k$ invariant under the Weyl group. It is known that the algebra P_w is generated by some finite number of real valued homogeneous polynomials $p_1(\lambda), \ldots, p_K(\lambda)$. Denote by $C_w^\infty(\mathbf{R}^k)$ the C^∞ functions on $\mathbf{R}^k$ invariant under the Weyl group. We have a map

$$T: C^\infty(\mathbf{R}^K) \to C_w^\infty(\mathbf{R}^k)$$

given by $(Tf)(\lambda_1, \ldots, \lambda_k) = f(p_1(\lambda), \ldots, p_k(\lambda))$. We shall use the following result, due to J. Mather [1].

(6.5) $\quad T$ is split surjective.

Thus, there is a continuous linear map $E: C_w^\infty(\mathbf{R}^k) \to C^\infty(\mathbf{R}^K)$ such that $TEg = g$ for all $g \in C_w^\infty(\mathbf{R}^k)$.

If $p_j(\lambda)$ is homogeneous of degree m_j, let $q_j(\lambda) = \|\lambda\|^{1-m_j} p_j(\lambda)$, homogeneous of degree 1, and let $Q_j = A^{1-j}\tilde{Q}_j$ where $\tilde{Q}_j$ is the bi-invariant differential operator of order m_j such that $\tilde{Q}_j \pi_\lambda^{ik} = p_j(\lambda+\delta)\pi_\lambda^{ik}$, whose existence is guaranteed by Corollary 6.2. Thus $Q_j \in OPS^1, j = 1, 2, \ldots, K$, and we see that

$$Q_j \pi_\lambda^{ik} = q_j(\lambda+\delta)\pi_\lambda^{ik}.$$

In particular, each Q_j is self adjoint. We may as well suppose that $Q_1 = A$. Now our plan of attack is to show that P can be written as a function of the operators $Q_1, \ldots, Q_K$, and apply the functional calculus from Section 1. To do this, we need (6.5).

We modify the map T as follows. Let

$$\tilde{T}: C^\infty(\mathbf{R}^K \backslash 0) \to C_w^\infty(\mathbf{R}^K \backslash 0).$$

be defined by $(\tilde{T}f)(\lambda) = f(q_1(\lambda), \ldots, q_k(\lambda))$. Then the results of Mather [1] also imply that $\tilde{T}$ is split surjective. Let $\tilde{E}: C_w^\infty(\mathbf{R}^k \backslash 0) \to C^\infty(\mathbf{R}^K \backslash 0)$ be a right inverse: $\tilde{T}\tilde{E} = I$. Note that

$$C^{-1}\|\lambda\|^2 \leq \sum_{j=1}^{K} |q_j(\lambda)|^2 \leq C\|\lambda\|^2,$$

so f vanishes near 0 (resp., near ∞) implies $\tilde{T}f$ vanishes near 0 (resp., near ∞). The converse need not hold, but note that, if $g(\lambda)$ is supported in $1 \leq |\lambda| \leq 4$, and if $\tilde{\psi} \in C_0^\infty(\mathbf{R}^K)$ is equal to 1 for $C^{-1} \leq |\xi| \leq 4C$, 0 for $|\xi| < (1/2)C^{-1}$, then $\tilde{T}\tilde{\psi}\tilde{E}g = g$.

We construct a new right inverse $\tilde{E}_0$ for $\tilde{T}$, guaranteed to be well behaved near ∞, as follows. Given $g \in C^\infty(\mathbf{R}^k \backslash 0)$, write

$$g = \sum_{\ell=-\infty}^{\infty} g_\ell$$

where $g_\ell(\lambda) = \psi_\ell(\lambda)g(\lambda)$. Here $\{\psi_\ell\}$ is a partition of unity such that $\psi_\ell(\lambda) = \psi_0(2^{-\ell}\lambda)$ and $\psi_0 \in C_0^\infty(1 \leq |\lambda| \leq 4)$. Thus g_ℓ is supported in $2^\ell \leq |\lambda| \leq$

$2^{\ell+2}$. If $\rho_r g(\lambda) = g(r\lambda)$, $R_r f(\xi) = f(r\xi)$, let

$$\tilde{E}_0 g = \sum_{\ell=-\infty}^{\infty} R_{2^{-\ell}} \tilde{\psi} \tilde{E} \rho_{2^\ell} g_\ell . \tag{6.6}$$

Clearly (6.6) is a finite sum over any compact set in $\mathbf{R}^K \backslash 0$. Furthermore, since $\tilde{T} R_r = \rho_r \tilde{T}$, we see that $\tilde{T}\tilde{E}_0 g = g$.

Now $g \in S^m_{1,0}(\mathbf{R}^k \backslash 0)$ is equivalent to $2^{-m\ell}\rho_{2^\ell} g_\ell$ being *bounded* in $C_0^\infty(1 \leq |\lambda| \leq 4)$, which yields $\tilde{E}_0 g \in S^m_{1,0}(\mathbf{R}^K \backslash 0)$.

We can now easily complete the proof of Theorem 6.3. Taking $G(\xi) = \tilde{E}_0 g(\xi)$ as above (and altering near $\xi = 0$ to make it smooth), we see that $G(Q_1, \ldots, Q_K)\pi^{ij}_\lambda = \tilde{p}(\lambda + \delta)\pi^{ij}_\lambda$ where $\tilde{p}(\lambda) = p(\lambda)$ for $|\lambda| \geq C$. Thus $G(Q_1, \ldots, Q_K)$ differs from P by a smoothing operator. Since $G(Q_1, \ldots, Q_K) \in OPS^m_{1,0}$ by Section 1, the proof is complete.

In general, if $g \in S^m_{\rho,0}(\mathbf{R}^k \backslash 0)$, we do not expect $\tilde{E}g \in S^m_{\rho,0}(\mathbf{R}^K \backslash 0)$, as the following simple example shows. Let $p(s) = q(s^2)$, $q(s) \in C_0^\infty(\mathbf{R})$, and consider $a(\xi, \eta) = p(|\xi|^{-\sigma}\eta) \in S^0_{\sigma,0}$. This symbol is invariant under the transformation $(\xi, \eta) \to (\xi, -\eta)$, so we want to write $a(\xi, \eta) = b(\xi, \zeta)$, where $\zeta = |\xi|^{-1}\eta^2$. Clearly we have $b(\xi, \zeta) = q(|\xi|^{-(2\sigma-1)}\zeta)$, which only belongs to $S^0_{2\sigma-1,0}$. If $b(\xi, \zeta) \in S^m_{\rho,0}$, this requires $\sigma = (1/2)(1+\rho)$. An elaboration of the argument proving Theorem 6.3 easily proves the following.

THEOREM 6.4. *Let $p(\lambda) \in S^m_{\rho,0}(\mathbf{R}^k)$ be invariant under the Weyl group, $1/2 < \rho$. Assume furthermore that*

(i) *$p(\lambda) \in S^m_{(1+\rho)/2,0}$ in a conic neighborhood of the walls of a Weyl chamber;*

(ii) *$p(\lambda) \in S^m_{1,0}$ in a conic neighborhood of the corners of a Weyl chamber.*

If $P: \mathcal{D}'(G) \to \mathcal{D}'(G)$ is given by $P\pi^{ij}_\lambda = p(\lambda + \delta)\pi^{ij}_\lambda$, then $P \in OPS^m_{\rho,1-\rho}$.

Remark. If one keeps track of the loss of finite order differentiability of $\tilde{E}g$, hypothesis (ii) can probably be weakened.

If one combines Theorem 6.3 with the L^p and Hölder results of Chapter XI, continuity of multipliers (6.4) on L^p and C^α is obtained. The L^p results are contained in the work of Weiss [1] and various subsequent authors, though the C^α results may not have been considered before.

B. Asymptotic Behavior of Multiplicities

Let G be a compact Lie group as we have been considering, with irreducible representations π_λ. Let K be a compact subgroup of G, and fix an irreducible unitary representation ρ of K. Each π_λ, restricted to K, splits up into a direct sum of irreducible representations of K, and we denote by $\nu(\rho, \lambda)$ the multiplicity with which ρ is contained in $\pi_\lambda|_K$. We wish to consider the asymptotic behavior of $\nu(\rho, \lambda)$, at least in some averaged sense, as $\lambda \to \infty$.

Let d_λ denote the dimension of the vector space V_λ on which π_λ operates and δ_ρ the dimension of the vector space on which ρ operates. The Weyl orthogonality relations lead to the fact that

$$\delta_\rho \int_K \pi_\lambda(x) tr\, \rho(x)\, dx = P_{\lambda,\rho} \tag{6.7}$$

is the projection onto the subspace of V_λ where K acts like copies of ρ. If $\pi_\lambda = (\pi_\lambda^{ij})$, then $\{\sqrt{d_\lambda}\pi_\lambda^{ij}\}$ form a complete orthonormal basis of $L^2(G)$, and consequently, since $\delta_\rho \nu(\rho, \lambda)$ is easily seen to be the square of the Hilbert-Schmidt norm of $P_{\lambda,\rho}$, we have

$$\begin{aligned} \nu(\rho, \lambda) &= \delta_\rho \sum_{i,j} \left| \int_K tr\, \rho(x) \pi_\lambda^{ij}(x)\, dx \right|^2 \\ &= \frac{\delta_\rho}{d_\lambda} \sum_{i,j} |((tr\, \rho)\mu_K, \sqrt{d_\lambda}\pi_\lambda^{ij})|^2 \end{aligned}$$

where μ_K is Haar measure on K, considered as a measure on G. Consequently, if p_1 is a sufficiently rapidly decreasing function on $\mathbf{Z}^k \cap C$, we have

$$\begin{aligned} \sum_{\lambda \in \mathbf{Z}^k \cap C} d_\lambda \nu(\rho, \lambda) p_1(\lambda) &= \delta_\rho \sum_\lambda \sum_{i,j} p_1(\lambda) |((tr\, \rho)\mu_K, \sqrt{d_\lambda}\pi_\lambda^{ij})|^2 \\ &= \delta_\rho (P_1 \mu, \mu) \end{aligned} \tag{6.8}$$

where $\mu = (tr\, \rho)\mu_K$ and the operator P_1 is defined by

$$P_1 \pi_\lambda^{ij} = p_1(\lambda) \pi_\lambda^{ij}. \tag{6.9}$$

We can similarly apply this argument to $p(t(\lambda + \delta)) = p_t(\lambda)$, given $p(\lambda) = p_1(\lambda - \delta) \in S_{1,0}^{-s}(\mathbf{R}^k)$, invariant under the Weyl group, $s > n - k$. By part A we know that the operators P_t defined by

$$P_t \pi_\lambda^{ij} = p(t(\lambda + \delta)) \pi_\lambda^{ij}$$

can be analyzed as pseudodifferential operators. Indeed, by part A there exists $G(\xi) \in S_{1,0}^{-s}(\mathbf{R}^K \backslash 0)$ such that $p(\lambda) = G(q_1(\lambda), \dots, q_K(\lambda))$, which yields $P_t = G(tQ_1, \dots, tQ_K)$. Note that, more precisely, $G(\xi) \in \tilde{S}_{1,0}^{-s}(\mathbf{R}^K)$, so the comments at the end of Section 3 apply. Consequently, we can analyze $(P_t \mu, \mu)$ after the fashion of Section 5, obtaining

$$(P_t \mu, \mu) = c_0 t^{-(n-k)} + o(t^{-(n-k)}), \qquad t \downarrow 0 \tag{6.10}$$

where, with $\tilde{q}_j(x, \xi)$ denoting the principal symbol of Q_j

$$\begin{aligned} c_0 &= (\text{vol } K)^{-2} \int_{N^*(K)} |tr\, \rho(x)|^2 G(\tilde{q}_1(x, \xi), \dots, \tilde{q}_K(x, \xi))\, d\, \text{vol}_{N^*(K)}(x, \xi) \\ &= (\text{vol } K)^{-1} \int_{N_e^*(K)} G(\tilde{q}_1(e, \xi), \dots, \tilde{q}_K(e, \xi))\, d\, \text{vol}_{N_e^*(K)}\, \xi, \end{aligned}$$

since the bi-invariant operators Q_j have bi-invariant symbols, and

$$\int_K |tr\, \rho(x)|^2\, dx = 1$$

by the irreducibility of ρ.

The connection between $q_j(\lambda)$ and $q_j(e, \xi)$ is the following. We think of $\lambda \in T_e^*(\mathbf{T}^k)$ included in $T_e^*(G)$, and then $q_j(\lambda) = \tilde{q}_j(e, \xi)$. Since $\tilde{q}_j(e, \xi)$ is invariant under the action of the adjoint representation of G on $T_e^*(G)$, this uniquely specifies $\tilde{q}_j(e, \xi)$. If we define $F: \mathbf{R}^n \to \mathbf{R}^k/W$ by $F(\xi) =$ any $\lambda \in \mathbf{R}^k$ such that $\xi = \mathscr{A}d(x)\lambda$ for some $x \in G$, we have $\tilde{q}_j(e, \xi) = q_j(F(\xi))$, which yields $p(F(\xi)) = G(\tilde{q}_1(e, \xi), \ldots, \tilde{q}_K(e, \xi))$. Thus we obtain

$$c_0 = (\text{vol } K)^{-1} \int_{N_e^*(K)} p(F(\xi))\, d\, \text{vol}_{N_e^*(K)}(\xi). \tag{6.11}$$

Note that, if $p(\lambda)$ is the characteristic function of some domain $U \subset \mathbf{R}^k$, then $p(F(\xi)$ is the characteristic function of the orbit $\mathscr{A}d\, G(U)$ of U under $\mathscr{A}d\, G$, in $\mathbf{R}^n = T_e^*(G)$.

We must distinguish between $d\, \text{vol}_K$, induced by the Riemannian metric on G (which makes vol $G = 1$), and Haar measure on K, which assigns to K a total mass of 1; $d\, \text{vol}_K = (\text{vol } K)\mu_K$. If K is discrete, for example, vol $K = 0(K)$.

We now obtain our main result.

THEOREM 6.5. *With G, K, π_λ, ρ as above, let U be any bounded open domain in $C \cap \mathbf{R}^k$ with sufficiently nice boundary, and denote by U_R the dilated domain $U_R = \{R\lambda : \lambda \in U\}$. We have the following asymptotic formula for the multiplicity $\nu(\rho, \lambda)$.*

$$\sum_{\lambda+\delta \in U_R} d_\lambda \nu(\rho, \lambda) \sim c\delta_\rho R^{n-k}, \qquad R \to \infty \tag{6.12}$$

where $c = \text{vol}\,(N_e^*(K) \cap \mathscr{A}d\, G(U))(\text{vol } K)^{-1}$.

Proof. Let $V \Supset U$, and pick $p(\lambda) \in C_0^\infty(\mathbf{R}^k)$ such that $p(\lambda) = 1$ on U and $p(\lambda) = 0$ on $C \backslash V$. Assume also $p(\lambda)$ is invariant under the Weyl group, and $0 \le p(\lambda) \le 1$ for all λ. Now (6.8) and (6.10) imply that

$$\sum_\lambda d_\lambda \nu(\rho, \lambda) p(R^{-1}\lambda) \sim c_0 \delta_\rho R^{n-k}, \qquad R \to \infty$$

where

$$c_0 = (\text{vol } K)^{-1} \int_{N_e^*(K)} p(F(\xi))\, d\, \text{vol}_{N_e^*(K)}(\xi).$$

Passing to the limit as $V \to U$ proves the theorem provided ∂U is nice.

We note that the heat equation method applied to this problem yields the weaker result that, with $\tau_\lambda = \|\lambda + \delta\|^2 - \|\delta\|^2$,

$$\sum d_\lambda \nu(\rho, \lambda) e^{-t\tau_\lambda} \sim t^{-(n-k)/2} \delta_\rho (\text{vol } K)^{-1} \Gamma\left(\frac{n}{2} + 1\right)^{1-k/n},$$

or, upon applying the Tauberian theorem mentioned at the end of Section 5, if we order the representations $\pi_1 = \pi_{\lambda_1}, \pi_2 = \pi_{\lambda_2}, \ldots$ such that $\|\lambda_j + \delta\|$ is increasing,

$$\sum_{j=1}^{N} d_j \nu(\rho, \lambda_j) \sim c\delta_\rho \left(\sum_{j=1}^{N} d_j^2 \right)^{1-k/n} \tag{6.13}$$

with

$$c = \Gamma(\tfrac{1}{2}(n-k)+1)^{-1}\Gamma\left(\frac{n}{2}+1\right)^{1-k/n} (\text{vol } K)^{-1}.$$

The conclusion of Theorem 6.5 is sharper than this in so far as a more local analysis within the Weyl chamber of $d_\lambda \nu(\rho, \lambda)$ is obtained. The heat equation method of obtaining (6.13) is fairly similar to the method of obtaining a formula of Gelfand and Gangolli used in Wallach [2].

Further significance of (6.12) and (6.13) is discussed in Cahn and Taylor [1]. See also Cahn [1] for an application of these techniques to the behavior of K-types of a principal series representation of a noncompact semisimple Lie group.

Exercises

6.1. (i) If $f(x) \in C^\infty(\mathbf{R})$ is even, show that for some smooth g, $f(x) = g(x^2)$. In fact, show that $f(\sqrt{x})$ is smooth.

(ii) If $f \in C^\infty(\mathbf{R}^2)$, $z = x + iy$, write $f(z)$ for $f(x, y)$. Suppose $f(z) = f(e^{2/3\pi i}z)$. Show that, for some $g \in C^\infty(\mathbf{R}^2)$, you have $f(z) = g(u, v)$, where $u = Re\, z^3$, $v = Im\, z^3$.

(iii) Suppose $f(x) \in C^\infty(\mathbf{R}^n)$ is symmetric in $(x_1, \ldots, x_n)$. Try to show that, for some $g \in C^\infty(\mathbf{R}^n)$, $f(x) = g(\sigma_1, \ldots, \sigma_n)$ where $\sigma_1, \ldots, \sigma_n$ are the elementary symmetric polynomials on $\mathbf{R}^n$. (At least do $n = 2$.)

(iv) Now read Mather [1].

6.2. Check out Theorem 6.5 when $G = SO(3)$ and $K = SO(2)$.

§7. Some Tauberian Theorems

We shall prove Karamata's Tauberian theorem and also the Tauberian theorem used in Section 4 here. As noticed by Wiener [1], such Tauberian theorems and numerous others follow from a single principle, a "closure of translates" theorem. To be precise, if f is a function on $\mathbf{R}$, let

$$T_f = \left\{ \sum_{j=1}^{n} c_j f(x - x_j) : c_j \in C, x_j \in \mathbf{R} \right\}.$$

Theorems 7.1 and 7.6 tell when T_f is dense in L^1 and in W, another space defined shortly which is useful for establishing Tauberian theorems. The approach we use to the proof of Theorem 7.1 is due to Korevaar [1].

THEOREM 7.1. (Wiener): *Let $f \in L^1$. Then the following are equivalent:*

(i) *T_f is dense in $L^1(\mathbf{R})$.*

(ii) *$g \in L^\infty, f * g = 0 \Rightarrow g = 0$.*

(iii) *$\hat{f}(\xi) \neq 0, \xi \in \mathbf{R}$.*

It is easy to see that (i) and (ii) are equivalent and that (i) $\Rightarrow$ (iii). To conclude that (iii) $\Rightarrow$ (ii), we use the following proposition.

PROPOSITION 7.2. *Let μ be a finite measure on $\mathbf{R}$. Suppose $\hat{\mu}(\xi_0) \neq 0$. Then there is a finite measure ν such that, if $\alpha = \nu * \mu$,*

(i) $\hat{\alpha} \in C_0^\infty(\mathbf{R})$;

(ii) $\hat{\alpha}(\xi_0) \neq 0$.

Granted this proposition, it is very easy to prove (iii) $\Rightarrow$ (ii). Suppose $\hat{f}$ never vanishes. For any given $\xi_0 \in \mathbf{R}$, choose ν such that $\alpha = \nu * f \Rightarrow \alpha \in C_0^\infty$ and $\hat{\alpha}(\xi_0) \neq 0$. Now if $g \in L^\infty$, $f * g = 0$, we have

$$\alpha * g = 0.$$

The deal is that we can take the Fourier transform of this ($\hat{\alpha} \in \mathcal{S}, \hat{g} \in \mathcal{S}'$, the Schwartz space of tempered distributions) to get

$$\hat{\alpha}\hat{g} = 0.$$

Hence $\hat{g}$ vanishes on a neighborhood of ξ_0. Since ξ_0 is arbitrary, $\hat{g} \equiv 0$, so $g = 0$.

To prove the proposition, we need a couple of lemmas.

LEMMA 7.3. *Let $\phi \in \mathcal{S}$, $\phi_n(x) = (1/n)\phi(x/n)$; let μ be a finite measure on $\mathbf{R}$. Then $\|\mu * \phi_n - \hat{\mu}(0)\phi_n\|_{L^1} \to 0$ as $n \to \infty$.*

Proof. Note that

$$\int |\mu * \phi_n(x) - \hat{\mu}(0)\phi_n(x)|\, dx \leq \iint |\phi_n(x-y) - \phi_n(x)|\, d|\mu|(y)\, dx$$
$$= \iint \left|\phi\left(t - \frac{y}{n}\right) - \phi(t)\right| d|\mu|(y)\, dt \to 0$$

as $n \to \infty$, by Lebesgue's dominated convergence theorem.

LEMMA 7.4. *Let μ, ν be finite measures on $\mathbf{R}$, $\|\nu\| < 1$. Then*

$$\frac{\hat{\mu}}{1+\hat{\nu}}$$

is the Fourier transform of a finite measure ω.

Proof. Let $\omega = \mu - \nu * \nu + \mu * \nu * \nu - \cdots$.

Proof of Proposition 7.2. Pick $\epsilon > 0$ such that $\hat{\mu}(\xi) \neq 0$ if $|\xi - \xi_0| < \epsilon$. Let $\beta = e^{i\xi_0 x}\mu$, so $\hat{\beta}(\xi) = \hat{\mu}(\xi - \xi_0)$. Hence $\hat{\beta}(\xi) \neq 0$ if $|\xi| < \epsilon$. Choose

$w \in \mathcal{S}$, the Schwartz space of rapidly decreasing functions, such that

(i) $\hat{w} = 1$ on $[-\frac{1}{4}\epsilon, \frac{1}{4}\epsilon]$;

(ii) $\operatorname{supp} w \subset (-\frac{1}{2}\epsilon, \frac{1}{2}\epsilon)$.

If $w_n(x) = (1/n)w(x/n)$, then $\hat{w}_n(\xi) = \hat{w}(n\xi)$. Note also that

$$\frac{\hat{w}_{2n}}{\hat{\beta}} = \frac{\hat{w}_{2n}}{\hat{\beta}(0) + \hat{\beta}\hat{w}_n - \hat{\beta}(0)\hat{w}_n} = \frac{\hat{u}}{1+\hat{v}}, \qquad \text{where}$$

$$u = \frac{1}{\hat{\beta}(0)} w_{2n} \qquad \text{and} \qquad v = \frac{1}{\hat{\beta}(0)}(\beta * w_n - \hat{\beta}(0)w_n).$$

By Lemma 7.3 $\|\beta * w_n - \hat{\beta}(0)w_n\|_{L^1} \to 0$ as $n \to \infty$, so if n is large, $\|v\|_{L^1} < 1$. Hence by Lemma 7.4, $\hat{u}/(1+\hat{v})$ is the Fourier transform of a finite measure v_0; $\hat{v}_0\hat{\beta} = \hat{w}_{2n}$. Thus if $\alpha_0 = v_0 * \beta$, we see that $\hat{\alpha}_0 \in C_0^\infty$ and $\hat{\alpha}_0(0) = \hat{w}_{2n}(0) = 1$.

Finally, if $v = e^{-i\xi_0 x}v_0$, we see that $\alpha = v * \mu$ satisfies the conclusions of the proposition.

Proposition 7.2 enables us to establish other "closure of translates" theorems. We state a few of these, omitting the details of the proof.

Proposition 7.5. *Let f belong to the appropriate space.*

(i) *T_f is dense in $C_0(\mathbf{R})$, continuous functions vanishing at infinity, if, and only if, $\hat{f}$ vanishes on no open set. ($\hat{f}$ considered as a tempered distribution.)*

(ii) *T_f is dense in $L^\infty(\mathbf{R})$, with the weak *topology, if, and only if, $\hat{f}$ vanishes on no open set.*

(iii) *If $f \in L^\infty$, let Λ_f be the set of common zeros of $\hat{q}$ such that $q \in L^1$ and $q * f = 0$. Then $\Lambda_f = \operatorname{supp} \hat{f}$.*

(iv) *T_f is dense in L^2 if, and only if, $\hat{f}$ vanishes on no set of positive measure.*

Of these, (iv) is particularly easy and doesn't require Proposition 7.2. Only in (i) do we need Proposition 7.2 for measures. All other results here only use Proposition 7.2 for L^1 functions.

To derive Tauberian theorems, Wiener found it convenient to investigate certain other spaces of functions.

Definition. *We set*

$$M(\mathbf{R}) = \left\{ f \in L^\infty(\mathbf{R}) \colon \sum_{k=-\infty}^{\infty} \operatorname*{ess\,sup}_{k \le x \le k+1} |f(x)| = \|f\|_w < \infty \right\}.$$

$$W(\mathbf{R}) = \{ f \in M; f \text{ continuous} \}.$$

With the norm $\|f\|_w$, $W(\mathbf{R})$ is a Banach space, and its dual is easily seen to be

$$E = \{\mu \text{ measure}: |\mu|([x, x+1]) \le C, \text{ independent of } x\}.$$

Note that $E \subset \mathscr{S}'$. Unfortunately, the dual of $M(\mathbf{R})$ is not a space of distributions.

THEOREM 7.6 (Wiener). *If $f \in W$, the following are equivalent.*
(i) *T_f is dense in W,*
(ii) $\mu \in E, \quad f * \mu = 0 \Rightarrow \mu = 0;$
(iii) $\hat{f}(\xi) \neq 0, \quad \xi \in \mathbf{R}.$

Proof. As before, the only difficult implication is (iii) $\Rightarrow$ (ii) (or (i)). As before, fix $\xi_0 \in \mathbf{R}$, and let ν be a measure such that if $\alpha = \nu * f$, then $\hat{\alpha} \in C_0^\infty$ and $\hat{\alpha}(\xi_0) \neq 0$. Then $f * \mu = 0 \Rightarrow \alpha * \mu = 0 \Rightarrow \hat{\alpha}\hat{\mu} = 0$, so $\hat{\mu} = 0$ in a neighborhood of ξ_0. Hence $\hat{\mu} \equiv 0$, so $\mu = 0$.

The purpose of such theorems as we have proved is to replace one type of averaging procedure by another. The following, for example, is a simple consequence of Theorem 7.6. We leave the proof to the reader.

THEOREM 7.7. *Let $f \in W$, and suppose $\hat{f}(\xi) \neq 0$, $\xi \in \mathbf{R}$. Let $\mu \in E$, and suppose*

(i) $\displaystyle\lim_{x \to \infty} \int_{-\infty}^{\infty} f(x-y)\, d\mu(y) = A \int_{-\infty}^{\infty} f(x)\, dx.$

Then for any $g \in W$ we have

(ii) $\displaystyle\lim_{x \to \infty} \int_{-\infty}^{\infty} g(x-y)\, d\mu(y) = A \int_{-\infty}^{\infty} g(x)\, dx.$

Such a theorem is applied to study the behavior of a certain measure μ, but usually we are not given that $\mu \in E$. Fortunately there is a very simple criterion for this if μ is positive which we now state. The proof is left to the reader.

PROPOSITION 7.8. *Let μ be a positive measure. Suppose there is a single positive measurable function f, bounded away from zero on some open set, such that $\int_{-\infty}^{\infty} f(x-y)\, d\mu(y) \leq C$, independent of x. Then $\mu \in E$.*

Thus if $f \in W$ is positive, $\hat{f}(\xi) \neq 0$, then if μ is positive and supported on $[-N, \infty]$ we can apply Theorem 7.7 whenever (i) holds.

In applications, it is often desirable to apply conclusion (ii) to a *discontinuous* g. If μ is positive, this is accomplished by a simple limiting argument, which we now discuss.

Suppose $g \in M(\mathbf{R})$ has the property that, for any $\epsilon > 0$, there exist $u \in M$, $h \in W$ such that
(i) $g + u \in W$;
(ii) $|u| \leq h$;
(iii) $\int_{-\infty}^{\infty} h(x)\, dx < \epsilon$;
We say $g \in \tilde{W}$.

PROPOSITION 7.9. *A function $g \in M(\mathbf{R})$ belongs to $\tilde{W}$ if, and only if, g is locally Riemann integrable.*

Proof. We leave this as an exercise.

Suppose f satisfies the condition of Theorem 7.7, and suppose μ is positive. If $g \in \tilde{W}$, let $\epsilon > 0$, and pick u, h as above. Let

$$\begin{aligned}\gamma(y) &= \int_{-\infty}^{\infty} g(y-x)\, d\mu(x) \\ &= \int_{-\infty}^{\infty} (g+u)(y-x)\, d\mu(x) - \int_{-\infty}^{\infty} u(y-x)\, d\mu(x).\end{aligned}$$

Now $\int_{-\infty}^{\infty} (g+u)(y-x)\, d\mu(x) \to A\int g + A\int u$ by (ii) of Theorem 7.7 since $g + u \in W$. Note that $|A\int u| < A\epsilon$. Also, if μ is positive, $|\int_{-\infty}^{\infty} u(y-x)\, d\mu(x)| \leq \int_{-\infty}^{\infty} h(y-x)\, d\mu(x) \to A\int h$ as $y \to \infty$, and $|A\int h| < A\epsilon$. In conclusion,

$$\limsup_{y\to\infty} \gamma(y) - 2A\epsilon < A \int g(x)\, dx < \liminf_{y\to\infty} \gamma(y) + 2A\epsilon.$$

Since $\epsilon > 0$ is arbitrary, we conclude that $\gamma(y) \to A\int g(x)\, dx$. To summarize:

THEOREM 7.10. *Let $f \in W$ be ≥ 0 and suppose $\hat{f}(\xi) \neq 0$, $\xi \in \mathbf{R}$. Let μ be a positive measure, and suppose* $\operatorname{supp} \mu \subset [-N, \infty)$ *and*

(i) $\displaystyle\lim_{x\to\infty} \int_{-\infty}^{\infty} f(x-y)\, d\mu(y) = A \int_{-\infty}^{\infty} f(x)\, dx.$

Then for any $g \in \tilde{W}$ we have

(ii) $\displaystyle\lim_{x\to\infty} \int_{-\infty}^{\infty} g(x-y)\, d\mu(y) = A \int_{-\infty}^{\infty} g(x)\, dx.$

Finally, we translate Theorem 7.10 to the situation where we are convoluting functions and measures on the multiplicative group $\mathbf{R}^+$. Thus we make the substitutions $y = \log s$, $\mu([-\infty, y]) = (1/s)\phi([0, s])$, $f(y) = sN_1(s)$, $g(y) = sN_2(s)$.

DEFINITION. *We set*

$$M(\mathbf{R}^+) = \left\{ f \in L^{\infty}_{\mathrm{loc}}(\mathbf{R}^+) \colon \sum_{k=-\infty}^{\infty} \operatorname*{ess\,sup}_{2^k \leq \lambda \leq 2^{k+1}} \lambda |f(\lambda)| < \infty \right\}.$$

We let $W(\mathbf{R}^+)$ and $\tilde{W}(\mathbf{R}^+)$ consist of those elements of $M(\mathbf{R}^+)$ which are, respectively, continuous, or locally Riemann integrable. Theorem 7.10 translates as follows.

THEOREM 7.11. *Let $N_1 \in W(\mathbf{R}^+)$ be ≥ 0 and suppose $\int_0^\infty N_1(\lambda)\lambda^{i\xi}\, d\lambda \neq 0$, $\xi \in \mathbf{R}$. Let ϕ be a positive measure on $\mathbf{R}^+$, and suppose*

(i) $\displaystyle\lim_{\lambda\to\infty} \frac{1}{\lambda} \int_0^{\infty} N_1\left(\frac{s}{\lambda}\right) d\phi(s) = A \int_0^{\infty} N_1(s)\, ds.$

Then for any $N_2 \in \tilde{W}(\mathbf{R}^+)$, *we have*

$$\text{(ii)}\quad \lim_{\lambda\to\infty} \frac{1}{\lambda}\int_0^\infty N_2\left(\frac{s}{\lambda}\right) d\phi(s) = A\int_0^\infty N_2(s)\, ds.$$

Let us now give a specific example of a Tauberian theorem. Let $w(\sigma)$ be an increasing function on $\mathbf{R}^+$, and suppose

$$\int_0^\infty \frac{dw(\sigma)}{\lambda+\sigma} = P\lambda^{a-1} + o(\lambda^{a-1}) \text{ as } \lambda\to\infty, \qquad 0<a<1.$$

We rewrite this so it looks like the hypothesis of the previous theorem:

$$\frac{1}{\lambda}\int_0^\infty \left(\frac{\sigma}{\lambda}\right)^{a-1} \frac{d\tilde{w}(\sigma)}{1+\sigma/\lambda} \to P \text{ as } \lambda\to\infty, \qquad d\tilde{w}(\sigma) = \sigma^{1-a}\, dw(\sigma).$$

Thus we take

$$N_1(\lambda) = \frac{\lambda^{a-1}}{1+\lambda}.$$

It follows that

$$\int_0^\infty N_1(\lambda)\lambda^{iu}\, d\lambda = \frac{\pi}{\sin \pi(a+iu)} \neq 0, \qquad u\in\mathbf{R}, \qquad 0<a<1.$$

Therefore,

$$\lim_{\lambda\to\infty}\frac{1}{\lambda}\int_0^\infty N_2\left(\frac{\sigma}{\lambda}\right) d\tilde{w}(\sigma) = \frac{P}{\int_0^\infty N_1(\lambda)\, d\lambda}\int_0^\infty N_2(\lambda)\, d\lambda.$$

We pick N_2 so that $w(\lambda)$ is recovered on the left-hand side. Thus let

$$N_2(\lambda) = \begin{cases} \lambda^{a-1} & 0\le\lambda\le 1 \\ 0 & \lambda>1 \end{cases}.$$

The conclusion follows simply:

$$\lambda^{-a}w(\lambda) \to P\frac{\sin \pi a}{\pi a} \qquad \text{as } \lambda\to\infty.$$

This is a Tauberian theorem of Hardy and Littlewood.

This argument is easily abstracted, as follows.

Suppose

$$\frac{1}{s^\alpha}\int_0^\infty N_1\left(\frac{x}{s}\right) dw(x) \to A, \qquad s\to\infty, \qquad (0<\alpha)$$

where w is increasing. We can rewrite this as

$$\frac{1}{s}\int_0^\infty N_1\left(\frac{x}{s}\right)\left(\frac{x}{s}\right)^{\alpha-1} d\tilde{w}(x) \to A, \qquad \text{with } d\tilde{w}(x) = x^{1-\alpha}dw(x).$$

Thus if $N_0(\lambda) = N_1(\lambda)\lambda^{\alpha-1} \in W(\mathbf{R}^+)$ and $\int_0^\infty N_1(\lambda)\lambda^{\alpha-1+iu}\,d\lambda \neq 0$ for $u \in \mathbf{R}$, we take $N_2(\lambda)$ as above and conclude that

$$s^{-\alpha}w(s) \to \frac{A}{\alpha \int_0^\infty N_1(\lambda)\lambda^{\alpha-1}d\lambda} \qquad \text{as } x \to \infty.$$

Such a result produces other Tauberian theorems, such as the following theorem of Karamata. Suppose w is increasing and

$$\int_0^\infty e^{-sx}\,dw(x) = As^{-\alpha} + o(s^{-\alpha}) \text{ as } s \to 0, \qquad (0 < \alpha).$$

This means

$$\frac{1}{s^\alpha}\int_0^\infty e^{-x/s}\,dw(x) \to A$$

as $s \to \infty$, so we take $N_1(\lambda) = e^{-\lambda}$. The relevant identity is

$$\int_0^\infty e^{-\lambda}\lambda^{\alpha-1+iu}\,d\lambda = \Gamma(\alpha + iu) \neq 0, \qquad u \in \mathbf{R}.$$

Hence the conclusion is

$$s^{-\alpha}w(s) \to \frac{A}{\alpha\Gamma(\alpha)} \qquad \text{as } s \to \infty.$$

We now establish the Tauberian theorem needed in Section 5.

PROPOSITION 7.12. *If $\alpha_j \geq 0$, the following are equivalent.*

(i) $\sum_{j=1}^{\infty} e^{-\epsilon\lambda_j}\alpha_j \sim C_0\epsilon^{-(n-\ell)}$ as $\epsilon \downarrow 0$;

(ii) $\sum_{j=1}^{N} \alpha_j \sim C_0\Gamma(n-\ell+1)^{-1}N^{1-\ell/n}$,

assuming $\lambda_j = j^{1/n} + o(j^{1/n})$.

Proof. That (ii) implies (i) is an Abelian theorem which we leave to the reader. To prove that (i) implies (ii), let $\mu = \sum \alpha_j\delta_{\lambda_j}$. Thus (i) means that $\int_0^\infty e^{-\epsilon s}\,d\mu(s) \sim C_0\epsilon^{-(n-\ell)}$. Karamata's theorem applies, so $s^{-(n-\ell)}\mu(0, s) \to C_0\Gamma(n-\ell+1)^{-1}$, $s \to \infty$, or

$$s^{-(n-\ell)} \sum_{\lambda_j \leq s} \alpha_j \to C_0\Gamma(n-\ell+1)^{-1}.$$

Therefore,

$$\sum_{j=1}^{N} \alpha_j = \sum_{\lambda_j \leq \lambda_N} \alpha_j \sim C_0\Gamma(n-\ell+1)^{-1}\lambda_N^{n-\ell} = C_0\Gamma(n-\ell+1)^{-1}N^{1-\ell/n},$$

as asserted.

Remark. In the above, n need not be an integer, nor need ℓ.

One should not think that a magical path is provided to all Tauberian theorems. Suppose

$$f(s) = \int_0^\infty e^{-sx}\, dw(x), \qquad Re\, s > 1$$

with w increasing. If $F(s) - A/(s-1)$ is bounded as $s \to 1$, so $(s-1)$ $\int_0^\infty e^{-sx}\, dw(x) \to A$ as $s \to 1$, let $\sigma = 1/(s-1)$ and write

$$\frac{1}{\sigma}\int_0^\infty e^{-(1/\sigma+1)x}\, dw(x) \to A \qquad \text{as } \sigma \to \infty.$$

We set $d\tilde{w}(x) = e^{-x} dw(x)$ and apply Theorem 7.11 to get

$$\frac{1}{\sigma}\int_0^\infty N_2\left(\frac{s}{\sigma}\right) e^{-s}\, dw(s) \to A \int_0^\infty N_2(\lambda)\, d\lambda, \qquad \sigma \to \infty$$

for all $N_2 \in \tilde{W}(\mathbf{R}^+)$. The hitch here is that, unlike in the previous cases, we cannot pick N_2 so as to recover $w(\sigma)$ on the left-hand side. We can derive such conclusions as $(1/\sigma)\int_0^\sigma e^{-s}\, dw(s) \to A$ as $\sigma \to \infty$, but we would like to make a stronger conclusion. In fact, Ikehara's theorem says that if $f(s) - A/(s-1)$ is continuous on the closed half plane $Re\, s \geq 1$, then $e^{-s}w(s) \to A$ as $s \to \infty$. This theorem, used in proving the prime number theorem, seems to require an additional argument. We refer the reader to Donoghue's *Theory of Distributions*, Donoghue [1] for a nice proof, and also to N. Wiener's "Tauberian Theorems", Wiener [1], which has been the source of much of the material of this section.

CHAPTER XIII

The Calderon-Vaillancourt Theorem and Hörmander-Melin Inequalities

In 1972, Calderon and Vaillancourt [2] published a proof of the continuity of operators $p(x, D)$ on $L^2(\mathbf{R}^n)$ assuming the symbol $p(x, \xi)$ satisfies, for some $\rho \in [0, 1)$,

$$|D_x^\beta D_\xi^\alpha p(x, \xi)| \leq c_{\alpha\beta}(1+|\xi|)^{-\rho(|\alpha|-|\beta|)},$$

i.e. assuming $p(x, \xi) \in S_{\rho,\rho}^0$, $0 \leq \rho < 1$. This very quickly obtained a spectacular application, in the case $\rho = 1/2$, to the problem of local solvability, in Beals and Fefferman [1]. Since then, the L^2 boundedness of $OPS_{1/2,1/2}^0$ has found a number of applications to the study of linear *PDE*. Calderon and Vaillancourt used, as a tool in their argument, a lemma due to Cotlar and Stein (Lemma 0.1 below) on sums of "almost orthogonal" operators. This lemma has been useful in proving L^2 estimates for other classes of integral operators, and further implications of it are discussed in Coifman and Weiss [1] and in Folland and Stein [1].

More recently, Cordes [2] noticed that L^2 continuity of $OPS_{0,0}^0$ could be deduced by a synthesis of $p(x, D)$ from trace class operators. Kato [1] extended this argument to the general case $OPS_{\rho,\rho}^0$, $\rho < 1$, and abstracted the functional analysis involved in Cordes' argument (Lemma 0.2 below).

We shall use Lemma 0.2 to deduce the L^2 continuity of $OPS_{0,0}^0$. From this result we shall obtain the L^2 continuity of $OPS_{\rho,\rho}^0$, using Lemma 0.1. Our treatment of this follows an argument of Beals [2]. This approach allows us to exploit both of the elegant operator theory lemmas, which we now state and prove.

LEMMA 0.1. (Cotlar-Stein). *Let Y be a σ-finite measure space, $A(y)$ a weakly measurable family of bounded operators on a Hilbert space H, such that, for all $x \in Y$,*

$$\int \|A(x)A(y)^*\|^{1/2}\, dy \leq C; \qquad \int \|A(x)^*A(y)\|^{1/2}\, dy \leq C.$$

Then the integral $A = \int A(x)\, dx$ is weakly convergent, and $\|A\| \leq C$.

Proof. It suffices to obtain the estimate $\|A\| \le c$ assuming $\|A(x)\| \le M$ and meas $Y = m < \infty$. In such a case we have $\|A\|^{2n} = \|(A^*A)^n\|$, and writing

$$(A^*A)^n = \int_{Y^{2n}} A(x_1)^*A(x_2)A(x_3)^* \cdots A(x_{2n})\,dx_1 \cdots dx_{2n}$$

and using the inequality

$$\|A_1 \cdots A_{2n}\| \le \|A_1\|^{1/2}\|A_1A_2\|^{1/2} \cdots \|A_{2n-1}A_{2n}\|^{1/2}\|A_{2n}\|^{1/2},$$

we see that

$$\|A\|^{2n} \le \int_{Y^{2n}} \|A(x_1)^*\|^{1/2}\|A(x_1)^*A(x_2)\|^{1/2} \cdots \|A(x_{2n-1})^*A(x_{2n})\|^{1/2}\,dx_1 \cdots dx_{2n}$$
$$\le mMC^{2n-1}.$$

Thus $\|A\| \le C^{1-1/2n}(mM)^{1/2n}$, and letting $n \to \infty$, we obtain $\|A\| \le C$, as desired.

LEMMA 0.2. (Cordes-Kato). *Let Y be a σ-finite measure space and $U(y)$ a weakly measurable family of bounded operators on a Hilbert space H such that*

$$\int_Y |(U(y)f, g)|^2\,dy \le c\|f\|^2\|g\|^2, \qquad \forall f, g \in H. \tag{0.1}$$

If $b(y) \in L^\infty$, then, for any trace class G on H,

$$B = b\{G\} = \int_Y b(y)U(y)^*GU(y)\,dy$$

satisfies $\|B\| \le c\|b\|_{L^\infty}\|G\|_{tr}$, the integral converging weakly.

Proof. Writing

$$Y = \bigcup_j Y_j$$

with $Y_1 \subset Y_2 \subset \cdots$ and meas $Y_j < \infty$, one sees that it suffices to derive the desired estimate for the norm of

$$\int_{Y_{jk}} b(y)U(y)^*GU(y)\,dy$$

where $Y_{jk} = \{y \in Y_j : \|U(y)\| \le k\}$.

We do this in several steps.

Step 1. Suppose $G \geq 0$ and $b(y) = 1$. Write $Gu = \sum \lambda_j(u, u_j)u_j$ with u_j an orthonormal basis of H, so $\lambda_j \geq 0$, $\sum \lambda_j = \|G\|_{tr}$. Then

$$\begin{aligned}(Bf, f) &= \int (GU(y)f, U(y)f)\, dy \\ &= \sum \lambda_j \int |(U(y)f, u_j)|^2\, dy \\ &\leq c\|f\|^2 \sum \lambda_j\end{aligned}$$

which is the desired inequality in this case.

Step 2. We have $b(y) \in L^\infty$, $G \geq 0$. In this case,

$$\begin{aligned}(b\{G\}f, f) &\leq \|b\|_{L^\infty}(1\{G\}f, f) \\ &\leq c\|b\|_{L^\infty}\|G\|_{tr}\|f\|^2.\end{aligned}$$

Step 3. We have $b(y) \in L^\infty$, G of trace class. In this case, write $G = G_1 - G_2 + iG_3 - iG_4$ with $G_j \geq 0$ of trace class, and appeal to Step 2. This concludes the proof.

The family $U(y)$ used in Section 1 is described as follows. We take $Y = \mathbf{R}^n \times \mathbf{R}^n$ with Lebesgue measure, $y = (x, \xi)$. We let $H = L^2(\mathbf{R}^n)$, $X_j u(x) = x_j u(x)$, $D_j u(x) = (1/i)(\partial/\partial x_j)u(x)$. Then we take

$$U(y) = U(x, \xi) = e^{ix \cdot D} e^{-i\xi \cdot X} \tag{0.2}$$

where $\xi \cdot X = \sum \xi_j X_j$ and $x \cdot D = \sum x_j D_j$.

LEMMA 0.3. *If $U(y)$ is given by* (0.2), *then* (0.1) *is satisfied.*

Proof. We have

$$\begin{aligned}\int_{\mathbf{R}^{2n}} |(e^{ix \cdot D}e^{-i\xi \cdot X}f, g)|^2\, d\xi\, dx &= \int_{\mathbf{R}^{2n}} |(e^{i\xi \cdot X}f, e^{-ix \cdot D}g)|^2\, d\xi\, dx \\ &= \int_{\mathbf{R}^{2n}} \left| \int e^{-i\xi \cdot y} f(y)\overline{g(y-x)}\, dy \right|^2 d\xi\, dx \\ &= (2\pi)^{-n} \int_{\mathbf{R}^{2n}} |f(y)\overline{g(y-x)}|^2\, dy\, dx \\ &= (2\pi)^{-n}\|f\|_{L^2}^2\|g\|_{L^2}^2.\end{aligned}$$

The next to the last identity uses Plancherel's theorem. This completes the proof.

Finally, we obtain one simple lemma which will be useful.

LEMMA 0.4. *Let Y be a measure space and suppose $k(x, y)$ satisfies*

$$\int_Y |k(x, y)|\, dy \leq c_0, \qquad \int_Y |k(x, y)|\, dx \leq c_1.$$

Then $Tu(x) = \int k(x, y)u(y)\,dy$ *defines a continuous operator on* $L^2(Y)$ *and* $\|T\| \le (c_0 c_1)^{1/2}$.

Proof. We have

$$\begin{aligned}
|(Tu, v)| &\le \int |k(x, y)||u(x)|\,|v(y)|\,dx\,dy \\
&= \int |k(x, y)|^{1/2}|u(x)|\,|k(x, y)|^{1/2}|v(y)|\,dx\,dy \\
&\le \left(\int |k(x, y)|\,|u(x)|^2\,dy\,dx\right)^{1/2}\left(\int |k(x, y)||v(y)|^2\,dx\,dy\right)^{1/2} \\
&\le c_0^{1/2}\|u\|_{L^2} \cdot c_1^{1/2}\|v\|_{L^2}.
\end{aligned}$$

This proves the lemma.

Exercises

0.1. Let A_j be compact operators on a Hilbert space such that $\|A_j A_k^*\| \le a(j, k)2^{-\epsilon|j-k|}$ and $\|A_k^* A_j\| \le a(j, k)2^{-\epsilon|j-k|}$. If $a(j, k) \to 0$ as $\sup(j, k) \to \infty$, show that

$$A = \sum_{j=0}^{\infty} A_j$$

is compact.

0.2. In the context of Lemma 0.2, suppose

$$\sup_{Y-Y_j} |b(y)| \le \beta_j \to 0$$

as $j \to \infty$. Prove that $b\{G\}$ is compact, provided

$$\sup_y \|U(y)\| \le M < \infty.$$

0.3. Apply Lemma 0.2 to the case when Y is a semi-simple Lie group and $U(y)$ is a discrete series representation.

0.4. If $Tu(x) = \int k(x, y)u(y)\,dy$ satisfies the hypothesis of Lemma 0.4, show that $T: L^p \to L^p$, $1 \le p \le \infty$.

§1. L^2 Continuity of $OPS^0_{0,0}(\mathbf{R}^n)$

In this section we show that $p(x, D): L^2(\mathbf{R}^n) \to L^2(\mathbf{R}^n)$ provided

$$(1.1) \qquad |D_x^\beta D_\xi^\alpha p(x, \xi)| \le c_{\alpha\beta}, \qquad (x, \xi) \in \mathbf{R}^{2n}, \qquad |\alpha|, |\beta| \le 3n + 4.$$

In order to prove this result, we use the following decomposition.

LEMMA 1.1 *Suppose $p = b * g$, convolution on $\mathbf{R}^{2n}$, with $b \in L^\infty(\mathbf{R}^{2n})$, $g \in L^1(\mathbf{R}^{2n})$. Let $G = g(x, D)$. Then, for $u \in \mathscr{S}(\mathbf{R}^n)$,*

$$(1.2) \qquad p(x, D)u = \iint b(x, \xi) e^{i\xi \cdot X} e^{-ix \cdot D} G e^{ix \cdot D} e^{-i\xi \cdot X} u \, dx \, d\xi.$$

Proof. The proof is straightforward.

We shall apply (1.2) to the case

$$b = (1-\Delta_x)^{s/2}(1-\Delta_\xi)^{s/2} p,$$
$$g = \psi(x)\psi(\xi)$$

where $\psi(x) = (1-\Delta)^{-s/2}\delta$, i.e. $\hat{\psi}(y) = (1+|y|^2)^{-s/2}$. We pick $s > 3n + 2$ to be an even integer, $s \leq 3n + 4$. In such a case, hypothesis (1.1) guarantees that $b(x, \xi) \in L^\infty(\mathbf{R}^{2n})$. It remains to examine $G = g(x, D) = \psi(x)\psi(D)$.

LEMMA 1.2. *Given $s > 3n + 2$, G is a trace class operator on $L^2(\mathbf{R}^n)$.*

Proof. Applying the inverse Fourier transform to $\hat{\psi}(y) = (1+|y|^2)^{-s/2}$, one easily sees that $\psi(x) \in C^\infty(\mathbf{R}^n \backslash 0)$ and is exponentially decreasing as $|x| \to \infty$, together with all its derivatives. Also $\psi(x) \in C^\ell(\mathbf{R}^n)$ provided $s > n + \ell$. Consequently, for $u \in L^2(\mathbf{R}^n)$,

$$(1.3) \qquad D_x^\beta |x|^{2j} \psi(x)\psi(D)u \in L^2(\mathbf{R}^n), \qquad |\beta| \leq \ell, j = 0, 1, 2, \ldots$$

for $s > n + \ell$. From (1.3) we see that

$$(1.4) \qquad (\Delta - |x|^2)^k \psi(x)\psi(D) : L^2(\mathbf{R}^n) \to L^2(\mathbf{R}^n)$$

provided $s > n + 2k$. Now $-(\partial^2/\partial x^2) + x^2$ is the Hamiltonian for the harmonic oscillator and, as is well known, its eigenvalues are $2m+1$, $m = 0, 1, 2, \ldots$, all simple. Since $-\Delta + |x|^2$ is a sum of such operators, we see that $(-\Delta+|x|^2)^{-\sigma}$ is of trace class if and only if $\sigma > n$. From (1.4) it follows that $\psi(x)\psi(D)$ is of trace class, provided $s > 3n + 2$.

We can now easily prove the main result of this section.

THEOREM 1.3. *Let $p(x, \xi) \in S^0_{0,0}(\mathbf{R}^n)$ and denote by $|||p|||_0$ the quantity*

$$\sup_{|\alpha|, |\beta| \leq 3n+4} C_{\alpha\beta},$$

where $C_{\alpha\beta}$ are given in (1.1). *Then*

$$\|p(x, D)u\|_{L^2(\mathbf{R}^n)} \leq C |||p|||_0 \|u\|_{L^2(\mathbf{R}^n)}$$

where C depends only on n.

Proof. In view of the decomposition (1.2) and Lemma 1.2, this result is an immediate consequence of Lemma 0.2 and Lemma 0.3.

Remark. In order to shorten the proof of Theorem 1.3, we have assumed that (1.1) holds over a much wider range of α and β than is necessary. Cordes [2] has obtained L^2 boundedness assuming only that (1.1) holds for $|\alpha|, |\beta| \le [n/2] + 1$.

Exercise

1.1. If $|p(x, \xi)| \to 0$ as $|x| + |\xi| \to \infty$ and $p(x, \xi)$ satisfies (1.1), show that $p(x, D)$ is compact on $L^2(\mathbf{R}^n)$. Hint: Use Exercise 0.2.

§2. L^2 Boundedness of $OPS^0_{\rho,\rho}(\mathbf{R}^n)$, $0 < \rho < 1$

In this section we prove that $p(x, D): L^2(\mathbf{R}^n) \to L^2(\mathbf{R}^n)$ provided

$$|D_x^\beta D_\xi^\alpha p(x, \xi)| \le C_{\alpha\beta}(1+|\xi|)^{\rho(|\beta|-|\alpha|)}, \qquad (x, \xi) \in \mathbf{R}^{2n} \tag{2.1}$$

where $\rho \in (0, 1)$. In order to accomplish this, we write $p(x, D)$ as a sum of almost orthogonal operators, as follows.

Choose a partition of unity f_j on $[0, \infty)$, $j = -1, 0, 1, 2, \ldots$, such that f_{-1} is supported on $[0, 1)$, f_j is supported on $(2^{j-1}, 2^{j+1})$, $j \ge 0$, and so that

$$f_j(t) = 1 \text{ if } |t-2^j| \le \tfrac{1}{4}2^j, \qquad j \ge 0,$$
$$|f_j^{(k)}(t)| \le C_k 2^{-jk}, \qquad j \ge 0.$$

Such a partition can be found. Let $\langle\xi\rangle = (1+|\xi|^2)^{1/2}$ and set

$$p_j(x, \xi) = f_j(c\langle\xi\rangle^\rho)p(x, \xi),$$

with some constant $c > 0$.

We want to apply Lemma 0.1 with $Y = \{-1, 0, 1, \ldots\}$, $A(y) = A_j = p_j(x, D)$.

First, we estimate the operator norm of $p_j(x, D)$. On the support of $p_j(x, \xi)$ we have $2^{j-1} \le c\langle\xi\rangle^\rho \le 2^j$. Consequently, (2.1) yields

$$|D_x^\beta D_\xi^\alpha p_j(x, \xi)| \le C^1_{\alpha\beta} 2^{j(|\beta|-|\alpha|)} \tag{2.2}$$

with $C^1_{\alpha\beta}$ independent of j. Now let V_j be the *unitary* operator on $L^2(\mathbf{R}^n)$ defined by

$$V_j u(x) = 2^{nj/2}u(2^j x).$$

It follows that $B_j = V_j^* A_j V_j$ is a pseudodifferential operator with symbol $b_j(x, \xi) = p_j(2^{-j}x, 2^j\xi)$, and (2.2) implies

$$|D_x^\beta D_\xi^\alpha b_j(x, \xi)| \le C^1_{\alpha\beta}. \tag{2.3}$$

From Theorem 1.3 we conclude that

$$\|A_j\| \le CM$$

where

$$M = \sup_{|\alpha|, |\beta| \le 3n+4} C^1_{\alpha\beta}.$$

We now estimate the operator norms of $A_k^*A_j$ and $A_jA_k^*$, assuming $|k-j|\geq 4$. In such a case, the symbols of A_j and A_k have disjoint supports, so $A_k^*A_j$ and $A_jA_k^*$ have smooth kernels. Consequently, one might expect to obtain decent bounds on their operator norms by elementary means.

For $|j-k|\geq 4$, if $\overline{p_k(x,\eta)}p_j(y,\xi)\neq 0$, this implies $\langle\eta\rangle^\rho\sim 2^k$ and $\langle\xi\rangle^\rho\sim 2^j$ simultaneously, which yields

$$|\xi-\eta|\geq c(2^j+2^k)^{1+\epsilon}\langle\xi-\eta\rangle^\epsilon \quad\text{with}\quad \epsilon=\frac{1-\rho}{1+\rho}\frac{1}{\rho}. \tag{2.4}$$

Now

$$A_k^*A_ju(x)=\int k(x,y)u(y)\,dy$$

where $k(x,y)=\int\overline{p_k(x,\xi)}p_j(z,\eta)e^{i(x\cdot\xi-z\cdot\xi+z\cdot\eta-y\cdot\eta)}\,dz\,d\xi\,d\eta$. Integrating by parts yields

$$k(x,y)=\int b_N(x,y,z,\xi,\eta)e^{i(x\cdot\xi-z\cdot\xi+z\cdot\eta-y\cdot\eta)}\,dz\,d\xi\,d\eta, \tag{2.5}$$

with

$$\begin{aligned} b_N=\langle x-z\rangle^{-2N}&\langle z-y\rangle^{-2N}(1-\Delta_\xi)^N(1-\Delta_\eta)^N\\ &\cdot\big[|\xi-\eta|^{-2N}(-\Delta_z)^N\overline{p_k(z,\xi)}p_j(z,\eta)\big].\end{aligned}$$

Therefore,

$$|b_N|\leq c[\langle x-z\rangle^{-1}\langle z-y\rangle^{-1}|\xi-\eta|^{-1}(2^j+2^k)]^{2N}. \tag{2.6}$$

Also, on supp b_N, if $|j-k|\geq 4$, then (2.4) holds. Substituting into (2.6), we get

$$|b_N|\leq c[\langle x-z\rangle^{-1}\langle y-z\rangle^{-1}\langle\xi-\eta\rangle^{-\epsilon}(2^j+2^k)^{-\epsilon}]^{2N}. \tag{2.7}$$

If

$$N>\max\left(\frac{n}{2},\frac{3}{2\epsilon},\frac{n}{2\epsilon}\right),$$

we may integrate (2.7), deducing from (2.5) that

$$|k(x,y)|\leq c\langle x-y\rangle^{-2N}(2^j+2^k)^{-\epsilon}.$$

In view of Lemma 0.4, this yields

$$\|A_k^*A_j\|\leq c(2^j+2^k)^{-\epsilon}$$

provided $|j-k|\geq 4$. But for $|j-k|\leq 4$ we have $\|A_k^*A_j\|\leq\|A_k\|\,\|A_j\|\leq M^2$, so in any event we obtain

$$\|A_k^*A_j\|\leq c2^{-\epsilon|j-k|}. \tag{2.8}$$

The estimate for $A_jA_k^*$ is a little easier. In fact,

$$(A_jA_k^*u)\hat{}(\xi) = \int \varkappa(\xi, \eta)\hat{u}(\eta)\, d\eta$$

where $\varkappa(\xi, \eta) = \int a_j(x, \zeta)\overline{a_k(y, \zeta)}e^{i(-x\cdot\xi + x\cdot\zeta - y\cdot\zeta + y\cdot\eta)}\, dx\, d\zeta\, dy$. But clearly $|j-k| \geq 4$ implies $\varkappa(\xi, \eta) \equiv 0$, so $A_jA_k^* = 0$ for $|j-k| \geq 4$. For $|j-k| \leq 4$ use $\|A_jA_k^*\| \leq \|A_j\|\,\|A_k\| \leq M^2$, to obtain

$$\|A_jA_k^*\| \leq C2^{-\epsilon|j-k|}. \tag{2.9}$$

From (2.8) and (2.9), the Cotlar-Stein lemma immediately yields the L^2 boundedness of $p(x, D) = \sum A_j$. We summarize this result. Let

$$N_0 = \max\left(3n+4, \frac{3}{2}\frac{1+\rho}{1-\rho}, \frac{n}{2}\frac{1+\rho}{1-\rho}\right).$$

THEOREM 2.1. *Suppose* $p(x, \xi) \in S^0_{\rho,\rho}(\mathbf{R}^n)$, $0 \leq \rho < 1$, *and*

$$|D_x^\beta D_\xi^\alpha p(x, \xi)| \leq C_{\alpha\beta}(1+|\xi|)^{\rho(|\beta|-|\alpha|)}, \qquad (x, \xi) \leq \mathbf{R}^{2n}, \quad |\alpha|, |\beta| \leq N_0. \tag{2.10}$$

Then $p(x, D): L^2(\mathbf{R}^n) \to L^2(\mathbf{R}^n)$ *and, if*

$$|||p|||_\rho = \max_{|\alpha|,|\beta| \leq N_0} C_{\alpha\beta}$$

with $C_{\alpha\beta}$ *satisfying* (2.10), *we have*

$$\|p(x, D)u\|_{L^2} \leq C|||p|||_\rho\|u\|_{L^2}$$

where C *depends only on* n *and* ρ.

Remark. As in Section 1, we require (2.10) for too large a set of α and β. Kato [1] has shown one needs (2.10) only for $|\alpha| \leq [n/2] + 1$, $|\beta| \leq [n/2] + 2$.

Exercises

2.1. Consider three Hilbert spaces, H, H_1, H_2. Let $\mathscr{L}$ denote the set of bounded operators $A: H_1 \to H_2$; $\mathscr{L} = \mathscr{L}(H_1, H_2)$. Let $U(y)$ satisfy the conditions of Lemma 0.2, and let $b(y) \in L^\infty(Y, \mathscr{L})$. With G a trace class operator on H, let

$$B = \int_Y b(y) \otimes U(y)^*GU(y)\, dy.$$

Prove B is a continuous operator from $H \otimes H_1$ to $H \otimes H_2$ (Hilbert space tensor product), with operator norm estimated by

$$\|B\| \leq c\left(\sup_y \|b(y)\|_{\mathscr{L}}\right)\|G\|_{tr}.$$

Similarly generalize Lemmas 0.1 and 0.4 to operators on Hilbert space-valued objects.

2.2. Generalize Theorem 2.1 to the case where $p(x, \xi)$ takes values in the set of bounded linear operators from a Hilbert space H_1 to a Hilbert space H_2.

2.3. Suppose $p(x, \xi) \in S^0_{0,\rho}(\mathbf{R}^n)$, $0 < \rho < 1$. Show that (2.6) holds, and hence, for $|j-k| \geq 4$, (2.8) and (2.9) hold.

2.4. Suppose $p(x, \xi) \in C^\infty(\mathbf{R}^{2n})$ satisfies estimates of the form

$$|D_x^\beta D_\xi^\alpha p(x, \xi)| \leq C_{\alpha\beta}\Psi_{\alpha\beta}(\langle\xi\rangle).$$

And suppose it is known that $q(x, D): L^2(\mathbf{R}^n) \to L^2(\mathbf{R}^n)$ provided $q(x, \xi)$ satisfies estimates of the form

$$|D_x^\beta D_\xi^\alpha q(x, \xi)| \leq C^1_{\alpha\beta}\langle\xi\rangle^{(\sigma-1)(|\alpha|-|\beta|)}\Psi_{\alpha\beta}(\langle\xi\rangle^\sigma), \qquad \sigma > 1.$$

Assume the $\Psi_{\alpha\beta}(t)$ satisfy the estimates

$$\Psi_{\alpha'\beta}(t) \leq C''_{\alpha\beta}t^{\rho|\alpha''|}\Psi_{\alpha\beta}(t), \qquad \alpha' + \alpha'' = \alpha, \qquad t \geq 1.$$

Deduce that $p(x, D): L^2(\mathbf{R}^n) \to L^2(\mathbf{R}^n)$ provided also $p(x, \xi) \in S^0_{0,\rho}(\mathbf{R}^n)$, with $\rho = (\sigma - 1)/\sigma$.

2.5. If $|p(x, \xi)| \to 0$ as $|x| + |\xi| \to \infty$, and $p(x, \xi)$ satisfies (2.10), show that $p(x, D)$ is compact on $L^2(\mathbf{R}^n)$. Hint: Use Exercises 1.1 and 0.1.

2.6. Say $p(x, \xi) \in S^m_{1/2*}$ if $p(x, \xi) \in S^m_{1/2,1/2}$ and furthermore, for $|\alpha| + |\beta| > 0$,

$$(1 + |\xi|)^{(|\alpha|-|\beta|)/2}|D_x^\beta D_\xi^\alpha p(x, \xi)| \to 0 \qquad \text{as } |x| + |\xi| \to \infty.$$

Show that elements of $OPS^0_{1/2*}$ have compact commutators on L^2.

§3. L^2 Continuity of Other Sets of Operators

In this section we prove the following result.

THEOREM 3.1. *Let $p(x, \xi) \in S^{\rho-\delta}_{\rho,\delta}(\mathbf{R}^n)$, $0 \leq \delta < \rho \leq 1$, and suppose*

$$|p(x, \xi)| \leq C, \qquad (x, \xi) \in \mathbf{R}^{2n}. \tag{3.1}$$

Then $p(x, D): L^2_{\text{comp}} \to L^2_{\text{loc}}$.

Proof. This result is a simple consequence of the sharp Gårding inequality, proved in Chapter VII. Indeed, (3.1) implies

$$Re(c - e^{i\theta}p(x, \xi)) \geq 0,$$

and the sharp Gårding inequality yields, for $u \in C_0^\infty(B_R)$,

$$Re((c - e^{i\theta}p(x, D))u, u) \geq -C'\|u\|^2.$$

Therefore,

$$Re\, e^{i\theta}(p(x, D)u, u) \leq C''\|u\|^2, \qquad \text{or}$$

$$|(p(x, D)u, u)| \leq C''\|u\|^2$$

which yields the desired continuity.

The case of greatest interest is $p(x, \xi) \in S^1_{1,0}(\mathbf{R}^n)$, $|p(x, \xi)| \leq C$. This case arises from time to time in proving pseudodifferential inequalities, usually in conjunction with the Calderon-Vaillancourt theorem (see, for example, Beals and Fefferman [1]). It is appropriate to regard the Calderon-Vaillancourt theorem as a limiting case of Theorem 3.1, as $\rho \downarrow \delta$. One may be tempted to speculate that such a result continues "analytically" to $\delta > \rho$, but in fact a stronger condition is necessary for L^2 continuity in this case, as was shown by Hörmander [17]; Calderon and Vaillancourt [2] establish the sufficiency of such a condition.

§4. Hörmander-Melin Inequalities

The sharp Gårding inequality of Chapter VII established the boundedness from below of first order pseudodifferential operators, $P \in OPS^1$, with nonnegative principal symbol. Here we investigate the boundedness from below of second order operators, $A \in OPS^2$, on a compact manifold X, with nonnegative principal symbol:

$$a_2(x, \xi) \geq 0. \tag{4.1}$$

It will be necessary to make some assumptions about the term $a_1(x, \xi)$ in the asymptotic expansion of the complete symbol of A, $a(x, \xi) \sim a_2(x, \xi) + \cdots$, and about the Hessian of $a_2(x, \xi)$ at points in the characteristic variety

$$\sum = \{(x, \xi): a_2(x, \xi) = 0\}. \tag{4.2}$$

First, we assume $a_2(x, \xi)$ vanishes to exactly second order on $\sum$, assumed to be a manifold; so $a_2(x, \xi) \geq c[\text{dist}((x, \xi), \sum)]^2$. By the Morse lemma this implies that, locally, if $\sum$ has codimension μ,

$$a_2(x, \xi) = \sum_{j=1}^{\mu} b_j(x, \xi)^2, \qquad \nabla_{x,\xi} b_j \neq 0 \text{ on } \sum \tag{4.3}$$

where $b_j(x, \xi)$ are homogeneous of degree 1 in ξ and vanish to precisely first order on $\sum$.

Let $Q = Q_{(x,\xi)}$ denote the Hessian of $a_2(x, \xi)$ at a point $(x, \xi) \in \sum$. Q is a quadratic form on $T(T^*(X))$, $Q(u)$, and we denote the associated bilinear form by $Q(u, v)$. If σ denotes the symplectic form on $T^*(X)$, we define the associated *Hamilton map* F by

$$\sigma(u, Fv) = Q(u, v). \tag{4.4}$$

Since σ is skew-symmetric, it is not hard to see that all eigenvalues of F are purely imaginary, for semi-definite Q; say the nonzero eigenvalues of F are $\pm i\mu_\nu$ ($\mu_\nu > 0$). We denote by tr^+F the sum

$$tr^+F = \sum_\nu \mu_\nu. \tag{4.5}$$

A. Melin has shown that A satisfies the estimates

$$Re(Au, u) \geq -\varepsilon\|u\|^2_{H^{1/2}} - C(\varepsilon)\|u\|^2_{L^2} \tag{4.6}$$

provided $A \in OPS^2$, (4.1) holds, and

$$Re\, a_1(x, \xi) + tr^+F \geq 0 \text{ on } \Sigma. \tag{4.7}$$

We refer to Melin [1] for a proof of this. Our goal is to establish the following semiboundedness results, the first of which is due to Hörmander [20].

THEOREM 4.1. *Let $A \in OPS^2$ satisfy* (4.1) *through* (4.3), *and* (4.7), *and also assume that the symplectic form, restricted to Σ, has constant rank. Then, for some constant C,*

$$Re(Au, u) \geq -C\|u\|^2_{L^2}, \qquad u \in C^\infty(X). \tag{4.8}$$

We also establish the following simpler result.

THEOREM 4.2. *Let $A \in OPS^2$ satisfy* (4.1) *through* (4.3) *as above and assume*

$$Re\, a_1(x, \xi) + tr^+F > 0 \text{ on } \Sigma. \tag{4.9}$$

Then A satisfies the semiboundedness estimate (4.8).

As a first step in proving these theorems, we put the quadratic form Q in a standard form.

LEMMA 4.3. *Let Q be a positive semidefinite quadratic form on a symplectic vector space $V \approx \mathbf{R}^n + \mathbf{R}^n$. Then there exists a symplectic basis $\{e_j, f_j\}$ satisfying $\sigma(e_j, e_k) = \sigma(f_j, f_k) = 0$ and $\sigma(e_j, f_k) = \delta_{jk}$, such that, with $(x, \xi) = \sum x_j e_j + \xi_j f_j$,*

$$Q(x, \xi) = \sum_{j=1}^{k} \mu_j(x_j^2 + \xi_j^2) + \sum_{j=k+1}^{k+\ell} x_j^2. \tag{4.10}$$

The eigenvalues of the Hamilton map F are $\pm i\mu_j$.

Proof. Denote by V_λ the space of generalized λ-eigenvectors of F on $V_{\mathbf{C}} = V \otimes \mathbf{C}$. We claim $V_0^r = V_0 \cap V$ is a symplectic space. Indeed, since $F+\mu: V_\lambda \to V_\lambda$ isomorphically if $\lambda+\mu \neq 0$ and $\sigma((F+\mu)^N V_\lambda, V_\mu) = \sigma(V_\lambda, (-F+\mu)^N V_\mu) = 0$ for N sufficiently large, we have

$$\sigma(V_\lambda, V_\mu) = 0 \qquad \text{if} \qquad \lambda+\mu \neq 0$$

which implies that the orthogonal complement of V_0^r with respect to σ has zero intersection with V_0^r. Thus we can choose symplectic coordinates so that V_0^r is defined by $x_j = \xi_j = 0$, $j \leq k$. Thus $Q(x, \xi) = Q_1(x', \xi') + Q_2(x'', \xi'')$, $x' = (x_1, \ldots, x_k)$, $x'' = (x_{k+1}, \ldots, x_{k+\ell})$, the Hamilton map F_1 of Q_1 is invertible, and that of Q_2 is nilpotent. We deal now with such Q_1 and Q_2.

Consider Q_1, a quadratic form on $V_1 \approx \mathbf{R}^k + \mathbf{R}^k$, which is positive definite. Then $F_1^{-1} = K_1$ satisfies $Q_1(u, K_1 v) = \sigma(u, v)$, so K_1 is skew symmetric with respect to the inner product Q_1. Thus there exists a basis $\{E_1, E_2, \ldots, F_k\}$ of V, orthonormal with respect to Q_1, such that

$$K_1 E_j = -\lambda_j F_j, \qquad K_1 F_j = \lambda_j E_j, \qquad (\lambda_j > 0).$$

Set

$$e_j = \frac{1}{\sqrt{\lambda_j}} E_j, \qquad f_j = \frac{1}{\sqrt{\lambda_j}} F_j.$$

Then $\{e_j, f_j\}$ is a symplectic basis of V_1, since

$$\sigma(\lambda_j^{-1} E_j, F_k) = Q_1(\lambda_j^{-1} E_j, \lambda_k E_k) = \delta_{jk},$$

etc. Furthermore, with $\mu_j = \lambda_j^{-1}$,

$$Q_1(x_1 e_1 + \cdots + \xi_k f_k) = \sum \mu_j (x_j^2 + \xi_j^2).$$

Finally, consider Q_2; F_2 is nilpotent. We claim this implies $F_2^2 = 0$. Indeed, suppose $F_2^3 u = 0$. Note that, generally,

$$\sigma(v, Fu) = Q(v, u) = Q(u, v) = \sigma(u, Fu) = -\sigma(Fv, u),$$

and hence

$$Q(v, Fu) = \sigma(v, F^2 u) = -\sigma(Fv, Fu) = -Q(Fv, u).$$

Hence, for Q_2, F_2, if $F_2^3 u = 0$, we have $0 = \sigma(u, F_2^3 u) = Q_2(u, F_2^2 u) = -Q_2(F_2 u, F_2 u)$. Since Q_2 is by hypothesis positive semidefinite, this implies $F_2 u \in \ker Q_2$, and hence $F_2^2 u = 0$. Thus $F_2^2 = 0$, which implies $\sigma(F_2 u, F_2 v) = 0$, and also $Q(F_2 u, F_2 v) = 0$. Choosing a symplectic basis on $V_2 \approx \mathbf{R}^\ell + \mathbf{R}^\ell$ with respect to which $Q_2(u, v) = \sigma(u, F_2 v)$ has the desired form is now elementary.

Lemma 4.3 plays an important role in Melin's theorem (see Melin [1]) and also in subelliptic estimates (see Hörmander, [19], Menikoff [3]). Here, its importance is to help establish Lemma 4.4. Let Q be a positive semidefinite quadratic form on $V = \mathbf{R}^n + \mathbf{R}^n = T^*(\mathbf{R}^n)$ as above, with Hamilton map F, and let V_λ be the generalized λ-eigenspace of F on $V_{\mathbf{C}} = V \otimes \mathbf{C}$. Let $V_0^r = V_0 \cap V$, and let

$$V^+ = \bigoplus_{\mu > 0} V_{i\mu}.$$

LEMMA 4.4. *Choose an orthonormal basis $v_1, \ldots, v_k$ for V^+ with the quadratic form $(1/2)Q(v, \bar{v})$ and an orthonormal basis $v_{k+1}, \ldots, v_{k+\ell} \in V_0^r$ for V_0^r/N, $N = \ker F$, with the quadratic form induced by Q there. Set*

$$L_j(x, \xi) = Q((x, \xi), v_j). \tag{4.11}$$

Then

$$Q(x, \xi) = \sum |L_j(x, \xi)|^2 \tag{4.12}$$

and

$$\sum \{Re\, L_j, Im\, L_j\} = -tr^+ F. \tag{4.13}$$

Proof. If Q has the form (4.10), then (4.12) and (4.13) are valid if one takes the particular choice of v_j such that $L_j(x, \xi) = \sqrt{\mu_j}(\xi_j - ix_j)$, $1 = j \leq k$, and $L_j(x, \xi) = x_j$, $k+1 \leq j \leq k+\ell$. However, (4.12), (4.13) is invariant under a unitary transformation of $v_1, \ldots, v_k$ and $v_{k+1}, \ldots, v_{k+\ell}$, so (4.12), (4.13) are always valid when Q has the form (4.10). But the lemma is invariant under symplectic changes of coordinates in V, so the proof is complete.

We can now prepare to prove Theorem 4.1. For each $(x, \xi) \in \sum$, the hessian $Q_{(x,\xi)}$ is defined. Then $V^+_{(x,\xi)}$ form a vector bundle over $\sum$, with hermitian structure given by $(1/2)Q_{(x,\xi)}(v, \bar{v})$, provided σ has constant rank on $\sum$. In fact, the null space of $Q_{(x,\xi)}$ is the tangent space of $\sum$ so it has constant dimension and the symplectic form has constant rank there by hypothesis, which implies that $V^+_{(x,\xi)}$ has constant dimension. Furthermore, V_0^r/N, $N = T(\sum)$, is a real vector bundle with Euclidean metric induced by Q. Near any given $(x_0, \xi_0) \in \sum$, we can choose orthonormal bases $v_1, \ldots, v_k$ and $v_{k+1}, \ldots, v_\ell$ for these bundles as in Lemma 4.4. Then $L_j(v) = Q_{(x,\xi)}(v, v_j)$ is a complex cotangent vector of $T^*(X)$ at (x, ξ) which is normal to $\sum$. Thus

$$Q_{(x,\xi)}(v) = \sum_{j=1}^{k+\ell} |L_j(v)|^2.$$

Now, locally, we have the representation (4.3) for $a_2(x, \xi)$. $Q_{(x,\xi)} = \sum db_j(x, \xi)^2$, so, on $\sum$ near (x_0, ξ_0), there is a $(2k+\ell) \times (2k+\ell) C^\infty$ orthogonal matrix $0_{jk}(x, \xi)$ such that

$$\Lambda_{x,\xi,j} = \sum_k 0_{jk}(x, \xi)\, db_k(x, \xi) \tag{4.14}$$

where $\Lambda_{x,\xi,j} = (Re\ L_1, Im\ L_1, \dots, Re\ L_k, Im\ L_k, L_{k+1}, \dots, L_{k+\ell})$. Let

$$c_j(x, \xi) = \sum_k 0_{jk}(x, \xi) b_k(x, \xi). \tag{4.15}$$

Then

$$a_2(x, \xi) = \sum c_j(x, \xi)^2, \tag{4.16}$$

and $dc_j(x, \xi) = \Lambda_{x,\xi,j}$ on $\sum$. Microlocally near (x_0, ξ_0), let

$$\begin{aligned} X_j(x, \xi) &= c_{2j-1}(x, \xi) + ic_{2j}(x, \xi), \qquad j = 1, \dots, k, \\ X_j(x, \xi) &= c_{k+j}(x, \xi), \qquad j = k+1, \dots, k+\ell. \end{aligned} \tag{4.17}$$

Then, in view of (4.16), we have

$$a_2(x, \xi) = \sum_{j=1}^{k+\ell} |X_j(x, \xi)|^2 \tag{4.18}$$

and hence

$$A - \sum_{j=1}^{k+\ell} X_j(x, D)^* X_j(x, D) = B(x, D) \in OPS^1. \tag{4.19}$$

$B(x, D)$ is self adjoint, and hence has real principal part $b_1(x, \xi)$. On $\sum$, the principal part is

$$a_1(x, \xi) - \frac{1}{i} \sum_j \frac{\overline{\partial X_j}}{\partial \xi_j} \frac{\partial X_j}{\partial x_j}.$$

Since we know $b_1(x, \xi)$ is real, we can take the real part and get

$$Re\ a_1(x, \xi) + \sum_{j=1}^{k} \{Im\ X_j, Re\ X_j\} = Re\ a_1(x, \xi) + tr^+ F. \tag{4.20}$$

Hypothesis (4.7) thus is equivalent to

$$b_1(x, \xi) \geq 0 \text{ on } \sum. \tag{4.26}$$

Now we can extend (4.20) from $\sum$ to a positive smooth function $q \in S^1$ on a conic neighborhood of (x_0, ξ_0) and write there

$$b_1(x, \xi) = q(x, \xi) + \sum \overline{(r_j(x, \xi)} X_j(x, \xi) + r_j(x, \xi) \overline{X_j(x, \xi)}) \tag{4.22}$$

with $q \geq 0$ and r_j homogeneous of degree 0. Thus, on a conic neighborhood of (x_0, ξ_0),

$$\begin{aligned} A &= q(x, D) + \sum X_j^* X_j + \sum R_j^* X_j + \sum X_j^* R_j + R_0 \\ &= Q + \sum (X_j + R_j)^*(X_j + R_j) \end{aligned} \tag{4.23}$$

where R_j are of order 0 and Q of order 1 with principal symbol $q \geq 0$.

We paste together the local representations (4.23) with a pseudodifferential partition of unity. Pick a cover of T^*X by conic sets U_j on each of which a decomposition (4.23) holds, and let $\psi_j(x, \xi)^2$ be a partition of unity supported on these open conic sets, ψ_j smooth, homogeneous of of degree 0,

$$\sum \psi_j(x, \xi)^2 = 1 \text{ on } T^*X.$$

Pick $\psi_j(x, D) \in OPS^0$ self-adjoint, principal symbol $\psi_j(x, \xi)$, essential support in such conic sets U_j, with $\sum \psi_j(x, D)^2 - 1 \in OPS^{-\infty}$. Now

$$(4.24)\quad \sum (A\psi_j(x, D)u, \psi_j(x, D)u) - (Au, u)$$
$$= Re \sum ([A, \psi_j]u, \psi_j u) + \left(\left(\sum_j \psi_j^* \psi_j - 1\right)Au, u\right).$$

Here $[A, \psi_j] \in OPS^1$ has pure imaginary principal symbol. Thus the right side of (4.24) is bounded by $C\|u\|_{L^2}^2$. Now, by (4.23) and the sharp Gårding inequality of Chapter VII applied to Q, we have

$$\sum_j (A\psi_j(x, D)u, \psi_j(x, D)u) \geq -C\|u\|_{L^2}^2,$$

and together with (4.24) this establishes the lower semiboundedness (4.8) and hence proves Theorem 4.1.

The proof of Theorem 4.2 is similar, and a little simpler. In fact, suppose (4.1) through (4.3) and (4.9) hold. Fix $(x_0, \xi_0) \in \sum$. Then one can find $c_j(x, \xi)$ such that, in a conic neighborhood U_j of (x_0, ξ_0), (4.15), (4.16) holds, and, at (x_0, ξ_0), (4.14) holds, which implies $dc_j(x_0, \xi_0) = \Lambda_{x_0,\xi_0,j}$. Define $X_j(x, \xi)$ by (4.17). Then (4.18) holds on U_j and, at (x_0, ξ_0), (4.20) holds. By hypothesis (4.9), this means

$$(4.25)\qquad b_1(x, \xi) > 0 \quad \text{on} \quad \sum \cap U_j,$$

if the conic neighborhood U_j is suitably shrunk. Consequently, we can write, as in (4.22),

$$b_1(x, \xi) = q(x, \xi) + \sum \overline{(r_j(x, \xi)}X_j(x, \xi) + r_j(x, \xi)X_j(x, \xi))$$

with $q(x, \xi) > 0$ on U_j, and the analogue of (4.23) obtains, microlocally on U_j. Using the pseudodifferential partition of unity $\psi_j(x, D)^2$ as before and applying the Gårding inequality to Q again yields the estimate (4.8) and completes the proof of Theorem 4.2.

CHAPTER XIV

Uniqueness in the Cauchy Problem

In this chapter we shall address the following question. Suppose u solves a partial differential equation $p(x, D)u = 0$ on a domain $\Omega \subset \mathbf{R}^n$, and suppose $u = 0$ for $\varphi(x) > 0$, where $\varphi:\Omega \to \mathbf{R}$ is a smooth function with $\nabla\varphi \neq 0$ on $\varphi = 0$, which thus defines a smooth hypersurface in Ω. When can we conclude that $u = 0$ on a neighborhood of $\varphi = 0$?

When $p(x, D)$ has analytic coefficients, a strong positive result is given by Holmgren's uniqueness theorem, which asserts that, provided the surface $S = \{\varphi = 0\}$ is noncharacteristic, i.e. $p_m(x, \nabla_x\varphi) \neq 0$ on S, then $u = 0$ on $\varphi > 0$, $p(x, D)u = 0$, implies $u = 0$ on a neighborhood of S. In particular, if $p(x, D)$ has analytic coefficients and is elliptic, then $p(x, D)u = 0$ on Ω, and $u = 0$ on any nonempty open set $U \subset \Omega$ implies $u \equiv 0$ on Ω, if Ω is connected. For operators with merely C^∞ coefficients, the results are not so clean, and there even exist fourth order elliptic operators annihilating some nontrivial functions with compact support (see Pliś [1]). We shall show here that there is such a unique continuation of u across a noncharacteristic S, in the case when $p_m(x, \xi)$ is real, provided that the complex characteristics of $p_m(x, \zeta)$ are *simple*, and assuming a further technical condition (given in Section 2) which implies that the real null bicharacteristics hitting S either cross transversally or, if they hit tangentially, have exactly second order contact and stay in $\{\varphi \geq 0\}$, so $\{\varphi \leq 0\}$ is convex with respect to such tangential bicharacteristics. In fact, for second order differential operators with real principal symbol, the technical condition is equivalent to this convexity requirement, as will be shown in Section 2. Note that, under such a convexity hypothesis, null bicharacteristics over points close to S must pass into $\{\varphi > 0\}$, so propagation of singularities results imply $u \in C^\infty$ near S. Such a uniqueness result was first obtained by Hörmander (see Hörmander [16]). We shall follow the approach of Trèves [5], as worked out by Menikoff [21].

Before Hörmander's result, Calderon [1] had proved a unique continuation theorem, assuming there were only transversal null bicharacteristics to S, giving the first successful application of pseudodifferential operators (then called singular integral operators) to nonelliptic *PDE*.

Further work along these lines is contained in Nirenberg [1] and Kumanogo [1].

Besides the obvious intrinsic interest, unique continuation properties play an important role in some solvability theorems for *PDE*'s, and in certain other qualitative studies. We discuss this briefly in Section 3, referring to the literature for details.

Uniqueness in the Cauchy problem still is not terribly well understood, and it is not clear that the method of Carleman estimates, used here as in other places, is really the correct tool for the natural results. At any rate, most of the above cited references have certain results not obtained here, and Trèves [5] has further conjectures. Even when $p(x,D)$ has analytic coefficients, Holmgren's uniqueness theorem does not tell the whole story, and the results on propagation of the analytic wave front set of a solution to $p(x, D)u = 0$ of Sato, Kashiwara, and Kawai [21] and Hörmander[18] yield circumstances under which there is unique continuation across some hypersurfaces S which fail to be noncharacteristic, as shown by Hörmander [18] and Kawai. There are also global situations where one looks for unique continuation, such as forward and backward unique continuation for solutions to parabolic equations.

§1. Carleman Estimates

We shall deduce uniqueness in the Cauchy problem from estimates of the form

$$\sum_{|\alpha|+k<m} h(\tau)^{2(m-|\alpha|-k)}\tau^{2k} \int e^{2\tau\psi(x)}|D^\alpha v|^2\,dx \leq c\int e^{2\tau\psi(x)}|p(x, D)v|^2\,dx \tag{1.1}$$

given $v \in C_0^\infty(\Omega)$. Here $h(\tau) = (1+\tau)^\delta$, $0 < \delta \leq 1$, $\tau \geq \tau_0 > 0$. Estimates of this form were first used by Carleman [1], in work on *UCP* in two variables. The function ψ will be slightly different from φ, and the set-up is the following. Suppose $u \in C^\infty$ and $p(x, D)u = 0$ on Ω and $u = 0$ for $\varphi(x) > 0$, and we want to prove that $u = 0$ in a neighborhood of a point x_0 where $\varphi(x_0) = 0$. Suppose (1.1) holds for a ψ with $\psi(x_0) = 0$, $\psi(x) \leq \varphi(x) - c|x-x_0|^2$ near x_0, where c is a small constant; assume $\nabla\psi \neq 0$ on $\psi = 0$. Thus the level sets of φ and ψ are as pictured in Figure XIV.1. Under such a condition, we can show that $u = 0$ in a neighborhood of x_0.

Indeed, let $v = \rho(\psi(x))u$, where $\rho(y) = 1$ for $y \geq -y_0$ (y_0 being a small positive number), $\rho(y) = 0$ for $y \leq -2y_0$. Consequently (1.1) applies, to yield

$$\tau^{2\delta m}\int_{\psi(x)\geq -y_0} e^{2\tau\psi}|u|^2\,dx \leq Ce^{-2\tau y_0}, \qquad \tau \geq \tau_0. \tag{1.2}$$

Considering how fast the right-hand side of (1.2) is decreasing as $\tau \uparrow \infty$,

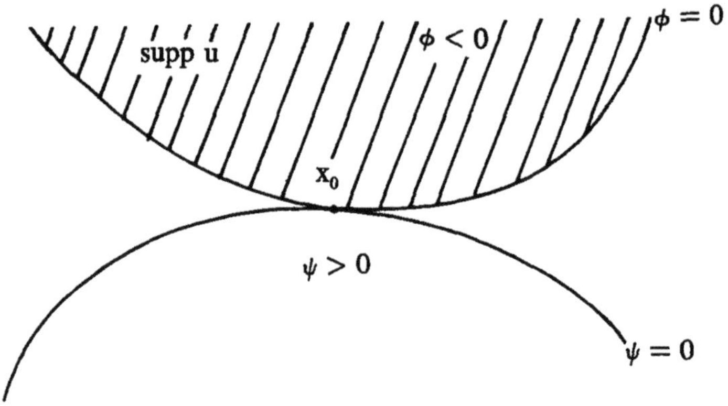

Figure XIV.1

we see immediately that $u = 0$ for $\psi(x) > -y_0$, which implies $u = 0$ in a neighborhood of x_0.

Clearly (1.2) would follow from a much weaker estimate than (1.1), but the strong estimate (1.1) has the advantage of stability. In particular, (1.1) remains valid if $p(x, D)$ is perturbed by a lower order operator, so we may assume $p(x, \xi)$ is homogeneous of order m.

§2. Reduction to Subelliptic Estimates, and Proof of UCP

We assume $p(x, \xi)$ is homogeneous in ξ of degree m. The purpose of this section is to analyze (1.1) in terms of a subelliptic estimate for the operator

$$P_\psi = p_\psi(x, D_x, D_t) = p(x, D_x + i|D_t|\, \nabla\psi) \tag{2.1}$$

on $\mathbf{R}_t \times \Omega$. The operator P_ψ is defined on $v \in \mathscr{E}'(\mathbf{R} \times \Omega)$ by

$$P_\psi v(t, x) = \int e^{i(x \cdot \xi + t\sigma)} p(x, \xi + i|\sigma|\, \nabla\psi)\hat{v}(\sigma, \xi)\, d\sigma\, d\xi. \tag{2.2}$$

Note that P_ψ is not quite a pseudodifferential operator, since $p_\psi(x, \xi, \sigma) = p(x, \xi + i|\sigma|\, \nabla\psi)$ is singular at $\sigma = 0$ (not at $\xi = 0$ since $p(x, \xi)$ is a *polynomial* in ξ). This will not cause grave difficulties.

PROPOSITION 2.1. *Let U be an open subset of Ω. Suppose there exist constants C and T such that*

$$\sum_{|\alpha|+k<m} \|h(D_t)^{m-k-|\alpha|} D_x^\alpha D_t^k u\|_{L^2(\mathbf{R}\times U)} \leq C\|P_\psi u\|_{L^2(\mathbf{R}\times U)} \tag{2.3}$$

for all $u \in C_0^\infty([-T, T] \times U)$. Then (1.1) holds for all $u \in C_0^\infty(U)$.

Proof. Let $v \in C_0^\infty(\mathbf{R} \times \Omega)$, $\tau \in \mathbf{R}$. Then

$$e^{-it\tau} P_\psi(e^{it\tau} v) = \int e^{i(\sigma-\tau)t} p(x, D_x + i|\sigma|\, \nabla\psi)\tilde{v}(\sigma - \tau, x)\, d\sigma \tag{2.4}$$

where $\tilde{v}(\sigma, x) = (1/2\pi)\int v(t, x)e^{-i\sigma t}\, dt$.

Leibnitz' formula yields

$$(2.5)\quad \begin{aligned} p(x, D_x+i|\sigma|\,\nabla\psi)w &= e^{|\sigma|\psi}p(x, D_x)(e^{-|\sigma|\psi}w) \\ &= e^{(|\sigma|-|\tau|)\psi}e^{|\tau|\psi}p(x, D_x)(e^{-(|\sigma|-|\tau|)\psi}e^{-|\tau|\psi}w) \\ &= \sum_{|\alpha|\le m} e^{(|\sigma|-|\tau|)\psi}D_x^\alpha e^{-(|\sigma|-|\tau|)\psi}p^{(\alpha)}(x, D_x+i|\tau|\,\nabla\psi)w \\ &= p(x, D_x+i|\tau|\,\nabla\psi)w + \sum_{\substack{j+k+|\alpha|<m \\ j>0}} C_{k,j,\alpha}(x)(|\tau|-|\sigma|)^j\tau^k D_x^\alpha w. \end{aligned}$$

$C_{k,j,\alpha}(x)$ are smooth functions of x depending only on $p(x, \xi)$. Apply (2.5) to $w = f(x)\hat{g}(\sigma-\tau)$ and use (2.4), to get

$$(2.6)\qquad \begin{aligned} e^{-\tau t}P_\psi(e^{i\tau t}f(x)g(t)) &= g(t)p(x, D_x+i|\tau|\,\nabla\psi)f \\ &\quad + \sum_{\substack{j+k+|\alpha|<m \\ j>0}} C_{j,k,\alpha}G_j(\tau)g(t)\tau^k D_x^\alpha f \end{aligned}$$

where

$$G_j(\tau)g(t) = \int e^{i\sigma t}(|\sigma+\tau| - |\sigma|)^j\hat{g}(\sigma)\,d\sigma.$$

Applying Plancherel's theorem yields

$$(2.7)\qquad \|G_j(\tau)g\|^2 \le c\int |\sigma|^{2j}|\hat{g}(\sigma)|^2\,d\sigma \le c\|g\|^2_{H^j(\mathbf{R})}.$$

Take a fixed $g \in C_0^\infty([-T, T])$ which does not vanish identically. By (2.6) and (2.7),

$$(2.8)\quad \begin{aligned} \|P_\psi(e^{i\tau t}(f(x)g(t))\| &\le c\|g\|\,\|p(x, D_x+i|\tau|\,\nabla\psi)f\| \\ &\qquad + c\sum_{\substack{k+|\alpha|<m \\ j<m}} \tau^k\|g\|_{H^j}\|D_x^\alpha f\|. \end{aligned}$$

Consider

$$\begin{aligned} e^{-i\tau t}h(D_t)^j D_t^k(e^{i\tau t}g(t)) &= \int e^{i(\sigma-\tau)t}h(\sigma)^j\hat{g}(\sigma-\tau)\,d\sigma \\ &= \sum_{j'=0}^{j}\sum_{k'=0}^{k}\binom{j}{j'}\binom{k}{k'}h(\tau)^{j-j'}\tau^{k-k'}H_{j',k'}(\tau)g(t) \end{aligned}$$

where

$$H_{j',k'}(\tau)g(t) = \int e^{i\sigma t}[h(\sigma+\tau) - h(\tau)]^{j'}\sigma^{k'}\hat{g}(\sigma)\,d\sigma.$$

Therefore,

$$\|H_{j',k'}(\tau)g\| \le C_{j'}\|g\|_{H^{aj'+k'}(\mathbf{R})}.$$

Thus we have

$$\|e^{i\tau t}h(D_t)^j D_t^k(e^{i\tau t}g) - h(\tau)^j\tau^k g\| \le B_{jk}h(\tau)^j(h(\tau)^{-1}+\tau^{-1}), \tag{2.9}$$

$$\|h(D_t)(e^{i\tau t}g)\| \le B_{jk}h(\tau)^j\tau^k. \tag{2.10}$$

Combining (2.3), (2.8), (2.9), and (2.10) gives

$$(1-Bh(\tau)^{-1})\sum_{|\alpha|+k<m} h(\tau)^{m-|\alpha|-k}\tau^k\|D_x^\alpha f\| \le C\|p(x, D_x+i|\tau|\,\nabla\psi)f\| \tag{2.11}$$
$$+ Ch(\tau)^{-1}\sum_{|\alpha|+k<m} h(\tau)^{m-|\alpha|-k}\tau^k\|D_x^\alpha f\|.$$

If τ is chosen so large that

$$h(\tau)^{-1} < \frac{1}{2}\frac{1}{B+C},$$

we can absorb the last term on the left side. Applying (2.11) to $f = e^{\tau\psi}u$ and keeping in mind that

$$\sum_{k+|\alpha|<\ell} \tau^k\|e^{\tau\psi}D_x^\alpha(e^{-\tau\psi}f)\| \le C\sum_{k+|\alpha|<\ell}\tau^k\|D^\alpha f\|,$$

we obtain (1.1).

We now establish the estimate (2.3), under certain hypotheses on $p(x,\zeta)$. In particular, we will suppose $p(x,\xi)$ is real valued and a homogeneous polynomial of degree m in ξ. First, we show that it suffices to work microlocally. Indeed, suppose there is a finite open covering of the sphere $|\xi|^2+\tau^2 = 1$ and a subordinate smooth partition of unity $\chi_\nu(\xi,\tau)$, homogeneous of degree 0, such that $\sum \chi_\nu^2 \equiv 1$ and

$$\|\chi_\nu(D_x, D_t)h(D_t)u\|_{H^{m-1}}^2 \le C\|P_\psi\chi_\nu u\|_{L^2}^2 + C\|u\|_{H^{m-1}}^2 \tag{2.12}$$

for all $u \in C_0^\infty([-T, T]\times U)$ where U is a sufficiently small neighborhood of x_0, T small. Thus

$$\begin{aligned}\|h(D_t)u\|_{H^{m-1}}^2 &= \sum_\nu \|\chi_\nu h(D_t)u\|_{H^{m-1}}^2 \\ &\le C\sum \|P_\psi\chi_\nu u\|_{L^2}^2 + C\|u\|_{H^{m-1}}^2 \\ &\le C\sum\|\chi_\nu P_\psi u\|_{L^2}^2 + C\sum\|[\chi_\nu, P_\psi]u\|_{L^2}^2 + C\|u\|_{H^{m-1}}^2 \\ &\le C\|P_\psi u\|_{L^2}^2 + C\|u\|_{H^{m-1}}^2\end{aligned} \tag{2.13}$$

where we used the fact

$$[\chi_\nu, P_\psi]: H^s \to H^{s+m-1} \tag{2.14}$$

which is valid even though $P_\psi \notin OPS_{1,0}^m$, as is easy to verify. Now if T is sufficiently small, then

$$\|u\|_{H^{m-1}} \leq \epsilon\|h(D_t)u\|_{H^{m-1}}; \tag{2.15}$$

this is a special case of the following lemma.

LEMMA 2.2. *Let $f(y)$ be a Hilbert space valued distribution, $y \in \mathbf{R}^n$; let $\epsilon > 0$, $\delta > 0$, and $s \geq -(n/2)$ be given.*

If f is supported in a sufficiently small neighborhood of y_0, then

$$\|f\|_{H^s} \leq \epsilon\|f\|_{H^{s+\delta}}.$$

Proof. If $f = \sum f_j e_j$ where e_j is an orthonormal basis of the Hilbert space, $\|f\|^2_{H^s}$ is comparable to

$$\sum_j \|f_j\|^2_{H^s}.$$

Thus it suffices to consider the scalar case. If the conclusion is false, there exist f_ν supported on $|y-y_0| \leq 1/\nu$, $C_0 > 0$, with $\|f_\nu\|_{H^s} = 1$, $\|f_\nu\|_{H^{s+\delta}} \leq C_0$. Picking a subsequence, we can assume $f_\nu \to f_0$ weakly in $H^{s+\delta}$. Rellich's theorem implies $\|f_\nu - f_0\|_{H^s} \to 0$. Now f_0 must be supported on $\{y_0\}$; since $f_0 \in H^s$, $s \geq -(n/2)$, this implies $f_0 = 0$, which contradicts $\|f_\nu\|_{H^s} = 1$. This proves the lemma.

Applying (2.14) and (2.15) to (2.13), we get

$$\|h(D_t)u\|_{H^{m-1}} \leq C\|P_\psi u\|_{L^2} \tag{2.16}$$

provided U and T are sufficiently small. This implies (2.3), under the hypothesis (2.12).

We shall prove (2.12), under the following hypotheses on the differential operator $p(x, D)$:

(2.17) $\quad p_m(x, \xi)$ is real, for $\xi \in \mathbf{R}^n$;

(2.18) $\quad \nabla_\zeta p_m(x, \zeta) \neq 0$ for $x \in \Omega$, $\quad \zeta = \xi + i|\sigma|\,\nabla\psi$,
$\xi \in \mathbf{R}^n$, $\quad |\xi|^2 + \sigma^2 \neq 0$;

(2.19) $\quad \nabla_\xi\, Re\, p_\psi(x, \xi, \sigma) \neq 0$ on $\Omega \times (\mathbf{R}^{n+1}\backslash 0)$, where
$p_\psi(x, \xi, \sigma) = p_m(x, \xi + i|\sigma|\,\nabla\psi)$;

(2.20) $\quad$ if $p_\psi(x, \xi, \sigma) = 0$, $\quad \sigma \neq 0$, then
$\{Re\, p_\psi, Im\, p_\psi\} \geq C|\sigma|(|\xi| + |\sigma|)^{2(m-1)}$.

Assumption (2.18) says that the real characteristics and certain of the complex characteristics of $p_m(x, \xi)$ are simple. Assumption (2.20) will be interpreted later in this section as a convexity hypothesis.

We first show that (2.20) implies (2.12) in the case when supp χ_ν is bounded away from the hyperplane $\sigma = 0$. Indeed, in this case we have the subelliptic estimate

$$\|\chi_\nu u\|^2_{m-1/2} \leq C\|P_\psi \chi_\nu u\|^2_{L^2} + C\|u\|^2_{H^{m-1}}. \tag{2.21}$$

Such an estimate was discussed in Chapter V, and we recall that argument.

PROPOSITION 2.3. *Let $A = a(x, D) \in OPS^m$. Suppose that, for $(x, \xi) \in U$, a conic open subset of $T^*(\Omega)$, $a_m(x, \xi) = 0$ implies $\{Re\ a_m, Im\ a_m\} > 0$. If $\chi_\nu \in S^0$ is supported in U, then*

$$\|\chi_\nu u\|^2_{H^{m-1/2}} \leq C\|A\chi_\nu u\|^2_{L^2} + C\|u\|^2_{H^{m-1}}. \tag{2.22}$$

Proof. We estimate $\|A\chi_\nu u\|^2_{L^2} = (A\chi_\nu u, A\chi_\nu u)$ from below by

$$\begin{aligned}(A\chi_\nu u, A\chi_\nu u) &= (A^*A\chi_\nu u, \chi_\nu u)\\ &= (AA^*\chi_\nu u, \chi_\nu u) + ([A^*, A]\chi_\nu u, \chi_\nu u).\end{aligned}$$

Since $[A^*, A] \in OPS^{2m-1}$ has principal symbol $\{\bar{a}_m, a_m\} = 2\{Re\ a_m, Im\ a_m\}$, and the hypothesis implies that $[A^*, A] + \eta\Lambda^{-1}AA^*$ is strongly elliptic of order $2m-1$ on supp χ_ν for large η, we use Gårding's inequality to conclude that

$$\begin{aligned}\|A\chi_\nu u\|^2_{L^2} &= ((I-\eta\Lambda^{-1})AA^*\chi_\nu u, \chi_\nu u) + (([A^*, A]+\eta\Lambda^{-1}AA^*)\chi_\nu u, \chi_\nu u)\\ &\geq (([A^*, A]+\eta\Lambda^{-1}AA^*)\chi_\nu u, \chi_\nu u) - C\|u\|^2_{H^{m-1}}\\ &\geq C_1\|\chi_\nu u\|^2_{H^{m-1/2}} - C_2\|u\|^2_{H^{m-1}}.\end{aligned}$$

This proves the proposition.

It remains to show that (2.17) through (2.20) implies (2.12) in the case when χ_ν is supported in a small conic neighborhood of $\sigma = 0$. First, if $p_m(x_0, \xi_0) \neq 0$, so P_ψ is elliptic at $(x_0, \xi_0, 0)$, we have

$$\|\chi_\nu u\|_{H^m} \leq C\|P_\psi \chi_\nu u\|_{L^2} + C\|u\|_{H^{m-1}}, \tag{2.23}$$

which is stronger than (2.12). (The fact that the symbol of p_ψ is not smooth in σ is counterbalanced by the fact that it is independent of t, so the usual proof of the elliptic estimate gives through.)

It remains to consider points (x_0, ξ_0) where $p_m(x_0, \xi_0) = 0$. By assumption (2.18), we can write

$$p_m(x, \zeta) = q(x, \zeta)(\zeta_j - \lambda(x, \zeta'))$$

for (x, ζ) in a neighborhood $V \times \Gamma$ of (x_0, ξ_0), where $\zeta' = (\zeta_1, \ldots, \zeta_{j-1}, \zeta_{j+1}, \ldots, \zeta_n)$, V is a small neighborhood of x_0, and Γ is a conic neighborhood of x_0, and Γ is a conic neighborhood of ξ_0 in $\mathbf{C}^n$. We can make q and λ homogeneous in ζ of degree $m-1$ and 1, respectively. Relabel the coordinates so that $y = x_j$, $\eta = \xi_j$, and denote $(x_1, \ldots, \hat{x}_j, \ldots, x_n)$ by x, $(\xi_1, \ldots, \hat{\xi}_j, \ldots, \xi_n)$ by ξ. Let $\mathscr{L}$ be the operator with symbol

$\zeta_j - \lambda(x, \zeta')$, i.e.

$$\mathscr{L} = D_y + i\psi_y|D_t| - \lambda(x, y, D_x + i|D_t| \nabla_x\psi). \tag{2.24}$$

Estimate (2.12) will hold if

$$\|h(D_t)\chi_v u\|_{L^2} \leq C\|\chi_v \mathscr{L} u\|_{L^2} + C\|u\|_{L^2}, \tag{2.25}$$

for $u \in C_0^\infty([-T, T] \times U')$, $U' \subset\subset U$. Indeed, let $A \in C_0^\infty(U)$, $A = 1$ on U', and apply (2.25) to $A(x)q(x, t, D_x, D_t)u$ to get

$$\begin{aligned}\|\chi_v hAqu\|_{L^2} &\leq C\|\chi_v \mathscr{L} Aqu\|_{L^2} + C\|qu\|_{L^2} \\ &\leq C\|\chi_v \mathscr{L} qAu\|_{L^2} + C\|u\|_{H^{m-1}} \\ &\leq C\|\chi_v P_\psi u\| + C\|u\|_{H^{m-1}}.\end{aligned}$$

But since q is elliptic on supp χ_v,

$$\|\chi_v hu\|_{H^{m-1}} \leq C\|\chi_v hAqu\|_{L^2} + C\|u\|_{H^{m-1}},$$

so (2.25) yields (2.12). We may ignore χ_v altogether by extending $\mathscr{L}$ outside Γ, without violating conditions (2.17) through (2.20).

We shall now use a Fourier integral operator to simplify the form (2.24) of $\mathscr{L}$, to $L = D_y + ib(y, x, D_x, |D_t|)$ for a real valued function b. Let $a = a(y, x, \xi, \sigma)$ and $b = b(y, x, \xi, \sigma)$ be the real and imaginary parts of $i|\sigma|\psi_y - \lambda(x, y, \xi + i|\sigma| \nabla_x\psi)$. Indeed, let $W(y)$ be the one parameter family of operators such that

$$D_y W + a(y, x, D_x, |D_t|)W = 0, \qquad W(y_0) = I. \tag{2.26}$$

Let $Wf(y, x) = W(y)f(y, x)$, so $(D_y + a)W = WD_y$. Let $L = W^* \mathscr{L} W$. (The amplitude of W is singular at $\sigma = 0$, but as in previous cases the translation invariance with respect to t saves the day.) Note that

$$L = D_y + iW^*bW, \tag{2.27}$$

and Egorov's theorem extends to show that the symbol of W^*bW, which we will again call b, is the old symbol of b composed with a canonical transformation. Since $\|\mathscr{L}u\|_{L^2} \leq C\|LWu\|_{L^2} + C\|Wu\|_{L^2}$, we see that (2.25) will follow from

$$\|h(D_t)u\|_{L^2} \leq C\|Lu\|_{L^2} + C\|u\|_{L^2}. \tag{2.28}$$

For $L = D_y + ib(y, x, \xi, |\sigma|)$, condition (2.20) becomes the following.

(2.29) If $b(y, x, \xi, |\sigma|) = 0, \qquad |\sigma| \neq 0, \qquad$ then $\dfrac{\partial}{\partial y} b(y, x, \xi, |\sigma|) \geq C|\sigma|$.

Since $p_m(x, \xi)$ is real and $\nabla_\xi p_m \neq 0$ when $\sigma = 0$, we have $Im\ \lambda = 0$ when $\sigma = 0$, so $b(y, x, \xi, 0) = 0$. Thus we can write $b(y, x, \xi, \tau) = \tau\tilde{b}(y, x, \xi, \tau)$

where $\tilde{b}$ is C^∞. We claim that

$$\text{If } \tilde{b}(y, x, \xi, 0) = 0, \qquad \frac{\partial}{\partial y}\tilde{b}(y, x, \xi, 0) \geq C. \tag{2.30}$$

In order to deduce (2.30) from (2.29), we need the following lemma.

LEMMA 2.4. *Let $P(\xi)$ be a polynomial. Suppose $P(\xi_0) = 0$, $\nabla_\xi P(\xi_0) \neq 0$, and $N \cdot \nabla_\xi P(\xi_0) = 0$ for some $N \in \mathbf{C}^n$. Then there exist smooth functions $\xi(r) \in \mathbf{R}^n$, $\tau(r) \in \mathbf{R}^1$, defined for r near 0 in $\mathbf{R}$, such that $\xi(0) = \xi_0$, $\tau(0) = 0$, $\tau(r) > 0$ for $r > 0$, and $P(\xi(r) + i\tau(r)N) = 0$.*

The proof is a simple consequence of the implicit function theorem.

Now, since $p_m(x, \xi)$ is real, we have

$$p_m(x, \xi + i\tau\nabla\psi) = p_m(x, \xi) + 0(\tau^2) + i\tau\left(\sum \frac{\partial}{\partial \xi_j} p_m(x, \xi)\frac{\partial}{\partial x_j}\psi + 0(\tau^2)\right).$$

Thus if $\tilde{b}(y, x, \xi, \sigma) = 0$ at $\sigma = 0$,

$$\sum \frac{\partial p_m}{\partial \xi_j}\frac{\partial \psi}{\partial x_j} = 0,$$

and for x fixed we can apply the lemma. After the canonical transformation, this implies there is a sequence $(y_v, x_v, \xi_v, \sigma_v)$ at which $\tilde{b} = 0$ and $\sigma_v \downarrow 0$, which yields (2.30). We now deduce (2.28) from (2.30).

First assume $\tilde{b}(y_0, x_0, \xi_0, 0) \neq 0$. Thus $|b| \geq C|\sigma|$ in a conic neighborhood of (x_0, y_0, ξ_0). Let $\hat{u}(x, y, \sigma)$ be the Fourier transform of $u(x, y, t)$ with respect to t. If, for example, $b > 0$, we have

$$(L\hat{u}, \hat{u}) = i\sigma(b(y, x, D_x, |\sigma|)\hat{u}, \hat{u}),$$

so Gårding's inequality yields

$$\|L\hat{u}\|_{L^2}\|\hat{u}\|_{L^2} \geq |\sigma|(C_1\|\hat{u}\|_{L^2} - C_1\|u\|^2_{H^{-1/2}}).$$

We get rid of the $\|u\|_{H^{-1/2}}$ term by making the diameter of U small (using lemma 2.2). Integrating with respect to σ yields $\|D_t u\| \leq C\|Lu\|$, which is stronger than (2.28).

Finally, if $\tilde{b}(y_0, x_0, \xi_0, 0) = 0$, we have $(\partial/\partial y)\tilde{b}(y_0, x_0, \xi_0) > 0$, so $(\partial/\partial y)b > C|\sigma|$ in a neighborhood of (x_0, y_0, ξ_0). Thus

$$\|L\hat{u}\|^2 = \|L^*u\|^2 + ([L^*, L]\hat{u}, \hat{u}) \geq ([L^*, L]\hat{u}, \hat{u}).$$

The symbol of $[L^*, L]$ is $2(\partial/\partial y)b \geq C|\sigma|$. Again by Gårding's inequality we have $\|Lu\| \geq C\| |D_t|^{1/2}u\|$, which yields (2.28).

To summarize, we have shown that hypotheses (2.17) through (2.20) imply the subelliptic estimate (2.3), which leads to the Carleman estimate

(1.1), and hence, to uniqueness in the Cauchy problem. We summarize the result in the following theorem:

THEOREM 2.5. *Let P be a differential operator of order m whose principal symbol $p = p_m(x, \xi)$ is real valued. Suppose $u \in C^\infty$ near $x_0 \in S = \{\varphi = 0\}$ and $u = 0$ on $\{\varphi > 0\}$. We assume S is noncharacteristic. Suppose $Pu = 0$. Then, if there is a smooth function ψ, with $\{\psi = 0\}$ a regular hypersurface tangent to S at x_0, $\psi(x) \geq \varphi(x) - C|x - x_0|$, such that hypotheses* (2.17) *through* (2.20) *are satisfied, it follows that $u = 0$ near x_0.*

We now relate condition (2.20) to geometric conditions on S. With $A = Re\, p(x, \xi + i|\sigma|\, \nabla\psi)$, $B = Im\, p(x, \xi + i|\sigma|\, \nabla\psi)$, observe that

$$\lim_{|\sigma| \to 0} \frac{1}{|\sigma|} B = \nabla_\xi p \cdot \nabla_x \varphi.$$

On the other hand $H_p\varphi = \nabla_\xi p \cdot \nabla_x \varphi$, so

$$\lim_{|\sigma| \to 0} \frac{1}{|\sigma|} B = H_p \varphi. \tag{2.31}$$

Now (2.20) says

$$\left\{A, \frac{1}{|\sigma|} B\right\} \geq C(|\xi| + |\sigma|)^{2(m-1)}$$

whenever $p(x, \xi + i|\sigma|\, \nabla\psi) = 0$, so we pass to the limit as $|\sigma| \to 0$. Since $A \to p(x, \xi)$ as $|\sigma| \to 0$, we get $\{p, H_p\varphi\} \geq C|\xi_0|^{2(m-1)}$, when $p(x_0, \xi_0) = 0$, provided there are nearby points (x, ξ) and small σ such that $p(x, \xi + i|\sigma|\, \nabla\psi) = 0$. This yields, in view of (2.31) and Lemma 2.4,

$$H_p^2 \varphi \geq C|\xi|^{2(m-1)} \text{ if } p(x, \xi) = 0 \text{ and } H_p\varphi = 0. \tag{2.32}$$

Thus (2.20) implies $\{\psi \leq 0\}$ is convex with respect to the null bicharacteristics of $p(x, \xi)$. Thus, near x_0, so is $\{\varphi \leq 0\}$ convex. By propagation of singularities results, we see that $p(x, D)u = 0$, $u = 0$ on $\varphi > 0$ implies $u \in C^\infty$ near x_0. Thus Theorem 2.5 is automatically strengthened to the following.

COROLLARY 2.6. *In Theorem 2.5, we can replace the hypothesis $u \in C^\infty$ by $u \in \mathcal{D}'$.*

We now show that, for second order differential operators with real principal part, hypothesis (2.20) is *equivalent* to the convexity of $\{\psi \leq 0\}$ with respect to the null bicharacteristics. Indeed, in this case, $p_2(x, \xi) = T(x)\xi \cdot \xi$ for a matrix valued function $T(x)$, and we have

$$\begin{aligned} A = Re\, p(x, \xi + i|\sigma|\, \nabla\psi) &= T(x)\xi \cdot \xi - |\sigma|^2 T(x)\, \nabla\psi \cdot \nabla\psi \\ &= p(x, \xi) - |\sigma|^2 p(x, \nabla\psi) \end{aligned} \tag{2.33}$$

and

$$B = Im\, p(x, \xi + i|\sigma|\,\nabla\psi) = 2|\sigma|T(x)\xi \cdot \nabla\psi \\ = |\sigma| H_p\psi. \tag{2.34}$$

Consequently,

$$\{A, B\} = |\sigma| H_p^2\psi - |\sigma|^3\{p(x, \nabla\psi), H_p\psi\}.$$

If $\{\psi \le 0\}$ is convex with respect to the null bicharacteristics of $p(x, \xi)$, it follows that $p(x, \xi) = 0$, $H_p\psi = 0$ implies $H_p^2\psi \ge 0$. Now for (2.33) and (2.34), we see that (2.20) is equivalent to the hypothesis

$$p(x, \xi) = |\sigma|^2 p(x, \nabla\psi), \qquad H_p\psi = 0 \Rightarrow \{A, B\} \ge C|\sigma|(|\xi|^2 + |\sigma|^2). \tag{2.35}$$

In particular, it is convenient to have $\{p(x, \nabla\psi), H_p\psi\} \le -C_1 < 0$. Note that this quantity in a function of x alone for second order p.

LEMMA 2.7. *Without changing the zero set $\{\psi = 0\}$ or the set $\{\psi > 0\}$ or $\{\psi < 0\}$, we can alter ψ so that $\{p(x, \nabla\psi), H_p\psi\} \le -C_1 < 0$.*

Proof. Since $p(x, \xi) = T(x)\xi \cdot \xi$, setting $q(x) = p(x, \nabla\psi)$, we see that $\{p(x, \nabla\psi), H_p\psi\} = \nabla_x q \cdot T(x)\, \nabla_x\psi$. Letting ν be the normal to the surface $\Sigma = \{\psi = 0\}$ pointing into $\{\psi > 0\} = \Sigma^+$, specify a function $q(x)$ such that $\nabla_x q \cdot T(x)\nu < 0$. Now let ψ solve the eikonal equation

$$p(x, \nabla\psi) = q \qquad \text{and} \\ \psi|_\Sigma = 0$$

such that $\nabla_x\psi$ points into Σ^+. This does the trick.

The conclusion of Lemma 2.7 having been arranged, it follows that (2.35) is equivalent to

$$p(x, \xi) = |\sigma|^2 p(x, \nabla\psi), \qquad H_p\psi = 0 \\ \Rightarrow H_p^2\psi + C_1|\sigma|^2 \ge C(|\xi|^2 + |\sigma|^2). \tag{2.36}$$

Here any $C > 0$ will do; presumably $C \ll C_1$, so (2.36) can be rewritten as

$$p(x, \xi) = |\sigma|^2 p(x, \nabla\psi), \qquad H_p\psi = 0 \\ \Rightarrow H_p^2\psi + C_2|\sigma|^2 \ge C|\xi|^2. \tag{2.37}$$

Clearly the implication need only be checked for $|\sigma| \ll |\xi|$, so (2.37) follows from the convexity of $\{\psi \le 0\}$ with respect to the bicharacteristics of $p(x, \xi)$, tangent to $\{\psi = 0\}$, which are close to the null bicharacteristics. Since convexity is a stable property under small perturbations, we see that (2.37) follows from the convexity of $\{\psi \le 0\}$ with respect to the null bicharacteristics of $p(x, \xi)$. Noting that, by virtue of (2.33), hypothesis (2.19) follows from (2.18), we can summarize our result on second order differential operators.

THEOREM 2.8. *Let $P = p(x, D)$ be a second order differential operator whose principal symbol $p(x, \xi)$ has real coefficients. Suppose the complex characteristics $p(x, \zeta) = 0$ are simple, or more generally assume* (2.18) *holds. Suppose $u \in \mathscr{D}'$, $u = 0$ on $\varphi > 0$, and $Pu = 0$. Then $u = 0$ in a neighborhood of $\{\varphi = 0\}$, provided $\{\varphi \leq 0\}$ is convex with respect to the null bicharacteristics of $p(x, \xi)$.*

Proof. The convexity of $\{\varphi \leq 0\}$ implies that of $\{\psi \leq 0\}$ near the point of contact x_0.

In the special case where P is elliptic, the geometric condition on $\{\varphi \leq 0\}$ trivializes, and we recover the classical result of Aronszajn [1] and Cordes [1]:

COROLLARY 2.9. *If $P = p(x, D)$ is a second order elliptic differential operator with real principal symbol, then unique continuation holds across any hypersurface. Thus if $Pu = 0$ on a connected domain Ω, $u = 0$ on any open subset $\mathcal{O}$ of Ω, it follows that $u = 0$ on Ω.*

In fact, Aronszajn [1] and Cordes [1] prove stronger results, and we refer the reader to their papers.

Even though we have the strongest positive results for second order differential operators, this is a situation that seems to lack counterexamples. For example, if P is a second order strictly hyperbolic operator, we have no reason to think that unique continuation should fail across any noncharacteristic hypersurface. Clarification of the situation lies in the future.

Exercises

2.1. Prove (2.14) and (2.23).

2.2. Analyze the operator $W(y)$ defined by (2.26) and prove the appropriate analogue of Egorov's theorem.

2.3. Prove (2.29).

2.4. Prove a quantitative version of Lemma 2.2, relating ϵ to s, δ, and the size of supp f.

2.5. If P is a second order elliptic differential operator with real principal symbol, show that the complex characteristics must be simple.

2.6. Let A be a second order elliptic differential operator on a compact connected manifold M, assumed to be bounded from above. Let $u(t, x)$ solve the hyperbolic equation

$$\left(\frac{\partial^2}{\partial t^2} - A\right)u = 0, \qquad \text{on } \mathbf{R} \times M.$$

Suppose $u = 0$ on $\mathbf{R} \times \Omega$, Ω an open subset of M. Prove that $u \equiv 0$. Hint:

Let

$$u_\epsilon(t) = \frac{1}{\sqrt{\pi\epsilon}} \int_{-\infty}^{\infty} e^{-(t-\tau)^2/\epsilon} u(\tau)\, d\tau,$$

entire analytic in $t = t_1 + it_2$. Hence $v_\epsilon(y) = u_\epsilon(iy)$ solves the elliptic *PDE*

$$\left(\frac{\partial^2}{\partial y^2} + A\right) v_\epsilon = 0.$$

§3. UCP, Global Solvability, and All That

There are some important connections between uniqueness in the Cauchy problem and global solvability of a partial differential equation $Pu = f$ on a region Ω, discussed in Malgrange [1], and Hörmander [6], which we briefly describe, referring to these sources for the proofs. Fundamental to these results is the notion of P-convexity of a domain Ω, which could be any paracompact C^∞ manifold.

DEFINITION 3.1. *The region Ω is P-convex if, and only if, for any compact $K \subset \Omega$, there exists a compact $K' \subset \Omega$ such that $\varphi \in \mathscr{E}'(\Omega)$,* supp $P^*\varphi \subset K$ *implies* supp $\varphi \subset K'$.

As shown in Hörmander [6], if $\Omega \subset \mathbf{R}^n$ and P has constant coefficients, $P: C^\infty(\Omega) \to C^\infty(\Omega)$ is surjective if, and only if, Ω is P-convex.

For an operator with variable coefficients, one needs an additional condition. One way to phrase it is in terms of semiglobal solvability.

DEFINITION 3.2. *The equation $Pu = f$ is semi-globally solvable in $C^\infty(\Omega)$ if, and only if, for all $\Omega' \subset\subset \Omega$, and all $f \in C^\infty(\Omega)$, there is a $u \in C^\infty$ such that $Pu = f$* on Ω'.

We then have the following theorem (see Hörmander [6]):

THEOREM 3.3. *The operator $P: C^\infty(\Omega) \to C^\infty(\Omega)$ is surjective if, and only if, Ω is P-convex and $Pu = f$ is semiglobally solvable.*

For some purposes, a neater statement of the result is the following, due to C. Harvey [1].

THEOREM 3.4. *The operator $P: C^\infty(\Omega) \to C^\infty(\Omega)$ is surjective if, and only if, for each compact $K \subset \Omega$ and real s, there exist compact $K' \subset \Omega$ and $C, \sigma \in \mathbf{R}$ such that*

$$u \in \mathscr{E}'(\Omega), \qquad P^*u \in H^s, \qquad \text{supp } P^*u \subset K \Rightarrow \text{supp } u \in K', \qquad u \in H^\sigma,$$

$$\text{and } \|u\|_{H^\sigma} \leq C\|P^*u\|_{H^s}.$$

In particular, if P is elliptic, we can deduce from Theorem 3.4 that $P: C^\infty(\Omega) \to C^\infty(\Omega)$ is surjective if, and only if, Ω is P-convex, provided P^* annihilates no element of $\mathscr{E}'(\Omega)$. More generally, if the principal symbol of P is real, we can deduce the following from Theorem 3.4.

THEOREM 3.5. *Let $P = p(x, D)$ be a differential operator with real principal symbol. Suppose no null bicharacteristic of $p_m(x, \xi)$ lies over a compact subset of Ω, and that P^* annihilates no element of $\mathscr{E}'(\Omega)$. Then $P: C^\infty(\Omega) \to C^\infty(\Omega)$ is surjective if, and only if, Ω is P-convex.*

Proof. We claim Theorem 3.4 applies, with $\sigma = s+m-1$. Indeed, the propagation of singularities results for P^* imply that if $u \in \mathscr{E}'(K)$ and $P^*u \in H^s$, then $u \in H^{s+m-1}$. We need only show that

$$\|u\|_{H^{s+m-1}} \le C\|P^*u\|_{H^s}, \qquad u \in \mathscr{E}'(\Omega), \qquad P^*u \in \mathscr{E}'(K), \tag{3.1}$$

assuming that Ω is P-convex and P^* annihilates no element of $\mathscr{E}'(\Omega)$. Indeed, fix K and pick K' such that $P^*u \in H^s_K(\Omega)$, $u \in \mathscr{E}'(\Omega)$ implies $u \in H^{s+m-1}_{K'}(\Omega)$, where $H^s_K(\Omega) = \{u \in \mathscr{E}'(K): u \in H^s(\Omega)\}$. Consider the unbounded operator

$$P^* = H^{s+m-1}_{K'} \to H^s_K$$

with domain $\mathscr{D}(P^*) = \{u \in H^{s+m-1}_{K'}: P^*u \in H^s_K\}$. Clearly P^* is a closed operator. Our hypotheses imply, via the closed graph theorem, that

$$\|u\|_{H^{s+m-1}} \le C\|P^*u\|_{H^s} + C\|u\|_{H^{s+m-2}}, \qquad u \in \mathscr{D}(P^*)$$

so P^* has closed range. We are also assuming P^* is injective, so by the open mapping theorem P^* is an isomorphism of $\mathscr{D}(P^*)$ onto its range in H^s_K, from which (3.1) is an immediate consequence.

In view of our unique continuation theorem for solutions to second order elliptic equations, we have the following simple corollary.

COROLLARY 3.6. *Let $P = p(x, D)$ be a second order elliptic operator with real principal symbol. Then $P: C^\infty(\Omega) \to C^\infty(\Omega)$ is surjective, provided that Ω has no compact connected component.*

Proof. Theorem 3.5 applies, where K' is the complement of the unbounded component of $\Omega - K$.

As a simple consequence of the corollary, one can show that the top dimensional de Rham cohomology $H^n(\Omega, \mathbf{C})$ vanishes, provided Ω has no compact connected components ($n = \dim \Omega$). Here $H^j(\Omega, \mathbf{C}) = \ker d_j / \operatorname{im} d_{j-1}$ where $d_j: C^\infty(\Omega, \Lambda^j) \to C^\infty(\Omega, \Lambda^{j+1})$ is the exterior derivative, acting on smooth j-forms. Indeed, we are to show that $d_{n-1}: C^\infty(\Omega, \Lambda^{n-1}) \to C^\infty(\Omega, \Lambda^n)$ is surjective, which follows from the surjectivity of the Laplace-Beltrami operator $\Delta_j = d_j^* d_j + d_{j-1} d_{j-1}^*$; note that $\Delta_n = d_{n-1} d_{n-1}^*$. Since Δ_j is a second order elliptic operator whose principal symbol is a real scalar, Corollary 3.6 shows it is surjective. In a similar fashion, if Ω is an open subset of $\mathbf{C}^n$, and we define the Dolbeault cohomology group $H^{p,q}(\Omega, \mathcal{O}) = \ker \bar\partial_{p,q} / \operatorname{im} \bar\partial_{p,q-1}$ where $\bar\partial_{p,q}: C^\infty(\Omega, \Lambda^{p,q}) \to C^\infty(\Omega, \Lambda^{p,q+1})$, we see that $H^{p,n}(\Omega, \mathcal{O}) = 0$. This result is fundamental in Sato's theory of hyperfunctions (see Sato, Kawai, and Kashiwara [1]).

CHAPTER XV

Operators with Double Characteristics

In previous chapters we have mainly treated operators with simple characteristics. Here we give some results on the theory of operators with multiple characteristics, a theory that is still in its infancy. The main examples of such operators are the following:

(1) *Hypoelliptic operators.* Operators arising in the treatment of the $\bar{\partial}$-Neumann problem, which arises in several complex variable theory, form the core of this theory. Such problems were first solved by Kohn, using energy inequality methods, and we follow his method in Section 1. Other methods include construction of parametrices, by Boutet de Monvel, Grigis, and Helffer [1], and by Stein, Folland, and Rothschild (see Folland and Stein [1] and Rothschild and Stein [1]), using two somewhat different points of view. Also Hörmander [19] and Menikoff [3] have strong results using energy estimates.

(2) *Hyperbolic operators.* The particular examples of greatest interest here are the equations of crystal optics, discussed in Section 5, and the equations of an anisotopic vibrating elastic solid, which have an analogous behavior. Results here are more primitive than in the case of hypoelliptic operators, and we discuss only special cases.

For multiple characteristics, there is a far richer class of model equations than for simple characteristics. In particular, the so-called subprincipal symbol plays an important role. We give a brief discussion in Section 2 of the problem of reducing an operator to some sort of standard form.

At this point it is safe to say that some of the most important problems in the immediate future of linear partial differential equations center around operators with multiple characteristics.

§1. Hypoelliptic Operators

Consider a second order differential operator of the form

$$P = \sum_{j=1}^{k} X_j^2 + X_0 \tag{1.1}$$

where X_j are real vector fields on $\Omega \subset \mathbf{R}^n$. In Hörmander [11] Hörmander proved the following.

THEOREM 1.1. *Suppose the Lie algebra over $C^\infty(\Omega)$ generated by $(X_0, \ldots, X_k)$ is the space of all vector fields on Ω. Then P is hypoelliptic, and there is an $\epsilon > 0$ such that*

$$Pu \in H^s_{\text{loc}}(\Omega) \Rightarrow u \in H^{s+\epsilon}_{\text{loc}}(\Omega). \tag{1.2}$$

Thus P is hypoelliptic, with loss of $2-\epsilon$ derivatives. Kohn [2] and Radkevic [1] gave a simpler proof of Theorem 1.1, by a different method. Their method does not lead to the best value of ϵ in (1.2), but it does apply to a more general class of operators, of the form

$$P = \sum_{j=1}^{k} A_j^2 + iA_0 + B \tag{1.3}$$

where we assume all operators are properly supported and

(1.4) $A_j \in OPS^1$ have real principal symbol

(1.5) $B = B_1 + \sum_{j=1}^{k} C_j A_j$, where $C_j \in OPS^0$, and $B_1 \in OPS^1$ has principal symbol:

$$b_1(x, \xi) \geq 0. \tag{1.5A}$$

Note that (1.1) is a special case of (1.2), with $A_j = iX_j$, $B = 0$, and a sign change. Theorem 1.1 is contained in the following.

THEOREM 1.2. *Suppose the Lie algebra over OPS^0 generated by $(A_0, A_1, \ldots, A_k)$ and OPS^0 is the space OPS^1. Let P be given by* (1.3), *assume* (1.4), (1.5), (1.5A) *and if A_0 or commutators involving A_0 are needed to generate OPS^1, strengthen* (1.5A) *to $b_1(x, \xi) = 0$. Then P is hypoelliptic and there is an $\epsilon > 0$ such that*

$$Pu \in H^s_{\text{loc}}(\Omega) \Rightarrow u \in H^{s+\epsilon}_{\text{loc}}(\Omega). \tag{1.6}$$

As the proof will show, the theorem is microlocalizable in the obvious fashion.

We remark that the hypothesis of Theorem 1.2 is equivalent to the following. Let $a_j(x, \xi)$ be the principal symbol of A_j, homogeneous of degree 1 in ξ. We are assuming that the Lie algebra (over the ring of functions homogeneous of degree zero) generated by $(a_0, a_1, \ldots, a_k)$, using the Poisson bracket, includes all smooth functions homogeneous of degree one in ξ.

Before we prove Theorem 1.2, let us pursue some of its implications. Let $P = p(x, D) \in OPS^2$, and say its expansion into homogeneous terms is

$$p(x, \xi) \sim p_2(x, \xi) + p_1(x, \xi) + \cdots. \tag{1.7}$$

Suppose $p_2(x, \xi) \geq 0$, and p_2 vanishes to exactly second order on a smooth conic manifold $\Sigma \subset T^*(\Omega)$, of codimension k. Then, by the Morse lemma, you can write, locally,

$$p_2(x, \xi) = \sum_{j=1}^{k} a_j(x, \xi)^2 \tag{1.8}$$

where each $a_j(x, \xi)$ is homogeneous of degree 1 in ξ, and the a_j have linearly independent gradients on Σ. Let $A_j = a_j(x, D)$. Then

$$P = \sum_{j=1} A_j^2 + B \tag{1.9}$$

with $B = B(x, D)$; on the set Σ we have

$$\begin{aligned} B(x, \xi) &= p_1(x, \xi) - \frac{1}{i} \sum_{|\alpha|=1} a_j^{(\alpha)}(x, \xi) a_{j(\alpha)}(x, \xi) \bmod S^0 \\ &= p_1(x, \xi) + \frac{i}{2} \sum_{\nu} \frac{\partial^2}{\partial x_\nu \, \partial \xi_\nu} \left(\sum_j a_j(x, \xi)^2 \right) \\ &= p_1(x, \xi) + \frac{i}{2} \sum_{\nu} \frac{\partial^2}{\partial x_\nu \, \partial \xi_\nu} p_2(x, \xi). \end{aligned} \tag{1.10}$$

The expression (1.10) is an especially significant quantity in the study of operators with multiple characteristics, and is called the subprincipal symbol:

$$\operatorname{sub} \sigma(P) = p_1(x, \xi) + \frac{i}{2} \sum_{\nu} \frac{\partial^2}{\partial x_\nu \, \partial \xi_\nu} p_2(x, \xi). \tag{1.11}$$

What (1.10) shows is that $B(x, \xi) = \operatorname{sub} \sigma(P) \bmod S^0$ on Σ. Since theorem 1.2 be microlocalized, a corollary of Theorem 1.2 in the following.

THEOREM 1.3. *Suppose $p(x, \xi)$ is given by* (1.7), (1.8), *and suppose the Lie algebra over S^0 generated by $(a_1, \ldots, a_k)$, using the Poisson bracket, is all of S^1. Then P is hypoelliptic, with loss of $2-\epsilon$ derivatives, provided*

$$\operatorname{sub} \sigma(P) = \alpha_1(x, \xi) + \sum_{j=1}^{k} b_j(x, \xi) a_j(x, \xi) \tag{1.12}$$

with $b_j \in S^0$ and $\alpha_1 \in S^1$ with

$$Re\, \alpha_1(x, \xi) \geq 0 \tag{1.13}$$

or, what is equivalent to (1.12), (1.13), *provided*

$$\text{(1.14)} \qquad Re \text{ sub } \sigma(P) \geq 0 \textit{ on } \Sigma.$$

Now the property that $(a_1, \ldots, a_k)$ and *single* Poisson brackets generate all of S^1 has a simple geometric interpretation. In this case, we require that, at each point $(x_0, \xi_0) \in \Sigma$, $\{a_j, a_\ell\}(x_0, \xi_0) \neq 0$ for some j, ℓ. Since $\{p, q\} = \sigma(H_p, H_q) = H_p q$, this says $H_{a_j} a_\ell(x_0, \xi_0) \neq 0$, so some H_{a_j} is transversal to Σ at each point. Since a vector X is tangent to Σ at (x_0, ξ_0) if, and only if, $Xa_j = 0$ at (x_0, ξ_0), $j = 1, \ldots, k$, and $Xa_j = \sigma(X, Ha_j)$, we see that

$$T_{(x_0,\xi_0)}\Sigma = (H_{a_1}, \ldots, H_{a_k})^\sigma$$

the space of vectors orthogonal to the linear span of $(H_{a_1}, \ldots, H_{a_k})$ with respect to the bilinear form σ. The assertion that some H_{a_j} is transversal to Σ at (x_0, ξ_0) means the linear span $(H_{a_1}, \ldots, H_{a_k})$ is not contained in $(H_{a_1}, \ldots, H_{a_k})^\sigma$, at (x_0, ξ_0), i.e., that $T_{(x_0,\xi_0)}\Sigma$ does not contain $(T_{(x_0,\xi_0)}\Sigma)^\sigma$. A vector subspace of a symplectic space which contains its orthogonal complement with respect to the symplectic form is called involutive; otherwise it is called not involutive. We use the terminology "not involutive" rather than "non involutive," since the latter phrase has become synonymous with "symplectic" in the literature. Consequently, we have the following.

THEOREM 1.4. *Suppose $p(x, \xi)$ is given by* (1.7) *and suppose $p_2(x, \xi) \geq 0$ vanishes to exactly second order in the smooth conic submanifold Σ of $T^*(\Omega)$. Suppose $T_{(x_0,\xi_0)}\Sigma$ is not involutive at each $(x_0, \xi_0) \in \Sigma$. Then P is hypoelliptic, with loss of* 1 *derivative, provided*

$$\text{(1.15)} \qquad Re \text{ sub } \sigma(P) \geq 0 \textit{ on } \Sigma.$$

The fact that only one derivative is lost in this case will follow from the method of proof, which yields the best value of ϵ if only single commutators are needed.

We turn now to the proof of Theorem 1.2. The proof we give is a straightforward generalization of Kohn's proof [2] of Theorem 1.1.

First we remark that each A_j may be assumed to be self-adjoint, since replacing A_j by its self-adjoint part amounts to changing B_1 to an operator which still satisfies (1.5), (1.6). Now we have a first simple energy estimate, under the hypotheses of Theorem 1.2.

LEMMA 1.5. *There exists a constant C such that, for any $u \in C_0^\infty(U)$, $U \subset\subset \Omega$,*

$$\text{(1.16)} \qquad \sum_{j=1}^{k} \|A_j u\|_{L^2}^2 \leq C\, Re(Pu, u) + C\|u\|_{L^2}^2.$$

Proof. We have

$$(Pu, u) = \sum_1^k \|A_j u\|_{L^2}^2 + i(A_0 u, u) + (B_1 u, u) + \sum_1^k (A_j u, C_j^* u).$$

Using the sharp Gårding inequality, which says $Re(B_1 u, u) \geq -C\|u\|_{L^2}^2$, we have

$$Re(Pu, u) \geq \sum_1^k \|A_j u\|_{L^2}^2 - C\|u\|_{L^2}^2 - C\|u\| \sum_1^k \|A_j u\|_{L^2}$$

$$\geq \frac{1}{2}\sum_1^k \|A_j u\|_{L^2}^2 - C^1\|u\|_{L^2}^2.$$

This proves (1.16).

Our objective will be to establish the following a priori inequality:

(1.17) $$\|u\|_{H^\epsilon} \leq C\|Pu\|_{L^2} + C\|u\|_{L^2}, \qquad u \in C_0^\infty(U).$$

For general second order operators, (1.17) does not imply hypoellipticity. For example, (1.17) holds for lots of hyperbolic operators, which are not hypoelliptic. However, the special form (1.3) of P will allow us to deduce hypoellipticity from (1.17). We shall do this now, and then prove (1.17). First we show that (1.17) is localizable. We shall suppose, without loss of generality, that the distribution kernel of P (and of any other pseudodifferential operator which crops up) is contained in a sufficiently small neighborhood of the diagonal in $\Omega \times \Omega$.

LEMMA 1.6. *If P is of the form* (1.3) *and if* (1.17) *holds, then for any $\zeta, \zeta_1 \in C_0^\infty(U)$ with $\zeta_1 = 1$ on a neighborhood of* supp ζ, *we have*

(1.18) $$\|\zeta u\|_{H^\epsilon} \leq C\|\zeta_1 Pu\|_{L^2} + C\|\zeta_1 u\|_{L^2}, \qquad u \in C^\infty(\Omega).$$

Proof. The equation (1.17) yields $\|\zeta u\|_{H^\epsilon} \leq C\|P\zeta u\|_{L^2} + C\|\zeta_1 u\|_{L^2}$, so it suffices to obtain

(1.19) $$\|[P, \zeta]u\|_{L^2} \leq C\|\zeta_1 Pu\|_{L^2} + C\|\zeta_1 u\|_{L^2}.$$

Now

(1.20) $$[P, \zeta]u = 2\sum_1^k [A_j, \zeta]A_j u + \sum_1^k [A_j, [A_j, \zeta]]u + [B, \zeta]u.$$

The last two terms are easy to estimate since they involve operators of order zero. Thus to prove (1.19) it suffices to estimate $[A_j, \zeta]A_j u$. We have

$$\|[A_j, \zeta]A_j u\|_{L^2}^2 \leq C\|A_j\tilde{\zeta}_1 u\|_{L^2}^2 + C\|\tilde{\zeta}_1 u\|_{L^2}^2$$
$$\leq C\, Re(P\tilde{\zeta}_1 u, \tilde{\zeta}_1 u) + C'\|\tilde{\zeta}_1 u\|_{L^2}^2,$$

by Lemma 1.5. Here we have picked $\tilde{\zeta}_1 = 1$ near supp ζ, but $\zeta_1 = 1$ near supp $\tilde{\zeta}_1$. Thus it remains only to show that

$$Re(P\tilde{\zeta}_1 u, \tilde{\zeta}_1 u) \le C\|\zeta_1 Pu\|_{L^2}^2 + C\|\zeta_1 u\|_{L^2}^2,$$

and to do this we need only show that

$$|Re([P, \tilde{\zeta}_1]u, \tilde{\zeta}_1 u)| \le C\|\zeta_1 Pu\|_{L^2}^2 + C\|\zeta_1 u\|_{L^2}^2. \tag{1.21}$$

The analogue of (1.20) shows that

$$[P, \tilde{\zeta}_1]u = 2\sum_1^k [A_j, \tilde{\zeta}_1]A_j u$$

plus terms which are easy to estimate. But

$$|Re([A_j, \tilde{\zeta}_1]A_j u, \tilde{\zeta}_1 u)| \le C\|\zeta_1 u\|_{L^2}^2$$

since $[A_j, \tilde{\zeta}_1]A_j \in OPS^1$ has pure imaginary symbol. This proves (1.21) and hence the lemma.

Lemma 1.6 generalizes in a straightforward fashion to pairs of pseudo differential operators $\zeta(x, D)$, $\zeta_1(x, D)$ in $OPS^0_{1,0}$, with the property that $\zeta_1(x, \xi) = 1$ on a conic neighborhood of supp $\zeta(x, \xi)$. This microlocalizes the estimate. More generally still, we can use two families of operators, $\zeta(\delta, x, D)$, $\zeta_1(\delta, x, D)$, $0 < \delta < 1$, with the properties

(1.22) $\zeta(\delta)$, $\zeta_1(\delta)$ bounded in OPS^0_1, $\quad 0 < \delta < 1$, self-adjoint;

(1.23) $\zeta(\delta)A\zeta_1(\delta) = \zeta(\delta)A + R_1(\delta)$ for any $A \in OPS^m$, with $R_1(\delta) \in OPS^{-\infty}$, bounded;

(1.24) $\zeta_1(\delta)A\zeta(\delta) = A\zeta(\delta) + R_2(\delta)$ for such A, with $R_2(\delta) \in OPS^{-\infty}$ bounded;

(1.25) $[Q_\ell[\cdots[Q_1, \zeta(\delta)]]\cdots] = A_\ell(\delta)\zeta_1(\delta) + R_3(\delta)$, for any $Q_\nu \in OPS^1$, with $A_\ell(\delta) \in OPS^0_{1,0}$ bounded.

If $\zeta(\delta)$, $\zeta_1(\delta) \in OPS^{-\infty}$ for each $\delta \in (0, 1)$, we furthermore have the analogue of (1.18), valid for any distribution u,

$$\|\zeta(\delta)u\|_{H^\epsilon} \le C\|\zeta_1(\delta)Pu\|_{L^2} + C\|\zeta_1(\delta)u\|_{L^2} \qquad u \in \mathscr{D}'(\Omega) \tag{1.26}$$

where the constant is independent of δ. As an example, let

$$\zeta(\delta) = \psi(2\delta D)\zeta(x), \qquad \zeta_1(\delta) = \psi(\delta D)\zeta_1(x), \tag{1.27}$$

where $\psi(\xi) \in C_0^\infty(|\xi| \le 3/2)$, $\psi(\xi) = 1$ for $|\xi| \le 1$. One easily verifies (1.22) through (1.25) in this case.

Finally, we replace the L^2 norms by Sobolev norms. If Λ^s and Λ^{-s} are the usual elliptic operators of order s and $-s$, altered to be properly supported on Ω, note that the symbol of $\Lambda^s P \Lambda^{-s}$ is

$$p(x, \xi) + \frac{1}{i} \sum_{\ell} \varphi_{\ell,s}(\xi) D_{x_\ell} p_2(x, \xi) \bmod S^0$$

$$= p(x, \xi) + \frac{2}{i} \sum \varphi_{\ell,s}(\xi)(D_{x_\ell} a_j) a_j(x, \xi),$$

where $\varphi_{\ell,s}(\xi) \in S^0$. Thus $\Lambda^s P \Lambda^{-s}$ satisfies the hypotheses of Theorem 1.2 if P does. Now if we apply (1.26), with u replaced by $\Lambda^s u$ and P by $\Lambda^s P \Lambda^s$, and $\zeta(\delta)$ replaced by $\zeta_s(\delta) = \Lambda^s \zeta(\delta) \Lambda^{-s}$, $\zeta_1(\delta)$ by $\zeta_{1,s}(\delta) = \Lambda^s \zeta_1(\delta) \Lambda^{-s}$, we see that

$$\begin{aligned} \|\zeta(\delta)u\|_{H^{s+\epsilon}} &= \|\Lambda^s \zeta(\delta) u\|_{H^\epsilon} \\ &= \|\zeta_s(\delta) \Lambda^s u\|_{H^\epsilon} \\ &\le C\|\zeta_{1,s}(\delta)(\Lambda^s P \Lambda^{-s})\Lambda^s u\|_{L^2} + C\|\zeta_{1,s}(\delta)\Lambda^s u\|_{L^2} \\ &= C\|\Lambda^s \zeta_1(\delta) P u\|_{L^2} + C\|\Lambda^s \zeta_1(\delta) u\|_{L^2}. \end{aligned}$$

Hence

$$(1.28) \quad \|\zeta(\delta)u\|_{H^{s+\epsilon}} \le C\|\zeta_1(\delta)Pu\|_{H^s} + C\|\zeta_1(\delta)u\|_{H^s}, \qquad \forall u \in \mathcal{D}'(\Omega)$$

with C independent of δ. Now suppose $Pu \in H^s_{\text{loc}}(\Omega)$, $u \in \mathcal{D}'(\Omega)$. Shrinking Ω slightly, we may suppose $u \in H^{-N}_{\text{loc}}(\Omega)$. Then (1.28) implies $u \in H^{-N+\epsilon}_{\text{loc}}(\Omega)$, and this argument can be iterated to yield $u \in H^{s+\epsilon}_{\text{loc}}(\Omega)$. Thus we shall have proved Theorem 1.2 once we establish the a priori inequality (1.17). We turn to this.

By hypothesis, we can express Λ as a linear combination of commutators of order $\le L$:

$$\sum_{1 \le \ell \le L+1} a_j \, ad\, A_{j_\ell} \cdots ad\, A_{j_2} \cdot A_{j_1} \bmod OPS^0$$

where $a_j = a_j(x, D) \in OPS^0$, $ad\, X \cdot Y = [X, Y]$, $j = (j_1, \ldots, j_\ell)$ is a multi-index. Denote by F^ℓ a typical operator of the form ad $A_{j_\ell} \cdots ad\, A_{j_2} \cdot A_{j_1}$, and similarly write $F^\ell = [A, F^{\ell-1}]$, to streamline notation. It suffices to obtain the bound, for $\ell \le L+1$,

$$(1.29) \qquad \|F^\ell u\|_{H^{\epsilon-1}} \le C\|Pu\|_{L^2} + C\|u\|_{L^2}, \qquad u \in C_0^\infty(U)$$

in order to prove (1.17). Now, with $T^{2\epsilon-1} = \Lambda^{2\epsilon-2} F^\ell \in OPS^{2\epsilon-1}$, we have

$$\begin{aligned} (1.30) \qquad \|F^\ell u\|^2_{H^{\epsilon-1}} &= (F^\ell u, T^{2\epsilon-1}u) \\ &= (AF^{\ell-1}u, T^{2\epsilon-1}u) - (F^{\ell-1}Au, T^{2\epsilon-1}u). \end{aligned}$$

We shall estimate each term on the right separately, and in the following $T^{2\epsilon-1}$ will often denote any element of $OPS^{2\epsilon-1}$. There are two cases to consider.

First, suppose $A = A_j$, $1 \le j \le k$. Then

$$\begin{aligned}(A_jF^{\ell-1}u, T^{2\epsilon-1}u) &= (F^{\ell-1}u, A_jT^{2\epsilon-1}u)\\ &= (F^{\ell-1}u, T^{2\epsilon-1}A_ju) + 0(\|F^{\ell-1}u\|_{H^{2\epsilon-1}}\|u\|_{L^2})\\ &= 0(\|F^{\ell-1}u\|_{H^{2\epsilon-1}}(\|A_ju\|_{L^2} + \|u\|_{L^2})).\end{aligned}$$

Thus, using (1.16),

$$(1.31)\quad |(A_jF^{\ell-1}u, T^{2\epsilon-1}u)| \le C\|F^{\ell-1}u\|^2_{H^{2\epsilon-1}} + C\|Pu\|^2_{L^2} + C\|u\|^2_{L^2}.$$

Similarly we have

$$(F^{\ell-1}A_ju, T^{2\epsilon-1}u) = (-1)^{\ell-1}(A_ju, T^{2\epsilon-1}F^{\ell-1}u) + 0(\|A_ju\|_{L^2}\|u\|_{H^{2\epsilon-1}}).$$

Thus, in this first case, we get

$$(1.32)\qquad \|F^{\ell}u\|_{H^{\epsilon-1}} \le C\|F^{\ell-1}u\|_{H^{2\epsilon-1}} + C\|Pu\|_{L^2} + C\|u\|_{L^2}.$$

Before we turn to the case where $A = A_0$, we remark that, if only commutators of $A_1, \dots, A_k$ are needed to generate Λ, we get the a priori inequality (1.17), with $\epsilon = 2^{-L}$, where commutators of order $\le L$ are required. At this point, the proof of Theorem 1.3 is complete. If only single commutators of $A_1, \dots, A_k$ are required, it is a simple matter to improve the estimate (1.32), and get hypoellipticity with loss of 1 derivative, instead of with loss of 3/2 derivatives. We briefly discuss this. Suppose

$$\Lambda = \sum_{j,\ell=1}^{k} b_{j\ell}(x, D)[A_j, A_\ell] + \sum_{m=1}^{j} b_m(x, D)A_m$$

with $b_{j\ell}(x, D), b_m(x, D) \in OPS^0$. Then, for $u \in C_0^\infty(\Omega)$,

$$\begin{aligned}(\Lambda u, u) &= \sum (b_{j\ell}A_jA_\ell u, u) - \sum (b_{j\ell}A_\ell A_ju, u) + \sum (b_m(x, D)A_mu, u)\\ &\le C\sum_{j=1}^{k} \|A_ju\|^2_{L^2} + C\|u\|^2_{L^2}.\end{aligned}$$

By Lemma 1.5 we deduce that

$$(\Lambda u, u) \le C\, Re(Pu, u) + C\|u\|^2_{L^2},$$

so

$$\|u\|^2_{H^{1/2}} \le C\|Pu\|_{H^{-1/2}}\|u\|_{H^{1/2}} + C\|u\|^2_{L^2}$$

and hence

$$\|u\|^2_{H^{1/2}} \le C\|Pu\|^2_{H^{-1/2}} + C\|u\|^2_{L^2}.$$

Replacing u *by* $\Lambda^{s+1/2}u$ and P by $\Lambda^{s+1/2}P\Lambda^{-(s+1/2)}$ in this estimate, we get

$$(1.33)\qquad \|u\|_{H^{s+1}} \le C\|Pu\|_{H^s} + C\|u\|_{H^2}, \qquad u \in C_0^\infty(\Omega).$$

We can localize this estimate and insert Friedrichs mollifiers, as before, to conclude that, if $Pu \in H^s_{\text{loc}}$, then $u \in H^{s+1}_{\text{loc}}$. In particular, the proof of Theorem 1.4 is complete. We can improve the conclusion a bit by observing that Lemma 1.5 implies

$$\sum_1^k \|A_j u\|^2_{L^2} \leq C\|Pu\|_{H^{-1/2}}\|u\|_{H^{1/2}} + C\|u\|^2_{L^2},$$

or more generally

$$\sum_1^k \|A_j u\|^2_{H^{s+1/2}} \leq C\|Pu\|_{H^s}\|u\|_{H^{s+1}} + C\|u\|^2_{H^s},$$

and estimating $\|u\|_{H^{s+1}}$ by (1.33) and adding the two estimates, we get the following improvement on (1.31), valid when $A_1, \ldots, A_k$ and single commutators of these operators generate OPS^1:

$$\|u\|_{H^{s+1}} + \sum_{j=1}^k \|A_j u\|_{H^{s+1/2}} \leq C\|Pu\|_{H^s} + C\|u\|_{H^s}. \tag{1.34}$$

We now complete the proof of inequality (1.17), via the estimate of (1.30). Our final case is $A = A_0$. First write

$$-iA_0 = P - \sum_1^k A_j^2 - \sum C_j A_j \bmod OPS^0$$

since in case this analysis is used we assume $b_1(x, \xi) = 0$. Thus

$$iA_0 = P^* - \sum_1^k A_j^2 - \sum_1^k A_j C_j^* \bmod OPS^0.$$

We can estimate the first term on the right in (1.30) by

$$\begin{aligned} (1.35)\quad & i(A_0 F^{\ell-1}u, T^{2\epsilon-1}u) \\ &= (P^*F^{\ell-1}u, T^{2\epsilon-1}u) - \sum_{j=1}^k (A_j^2 F^{\ell-1}u, T^{2\epsilon-1}u) \\ &\quad - \sum_{j=1}^k (A_j C_j^* F^{\ell-1}u, T^{2\epsilon-1}u) + 0(\|F^{\ell-1}u\|_{H^{2\epsilon-1}}\|u\|_{L^1}). \end{aligned}$$

Now $(P^*F^{\ell-1}u, T^{2\epsilon-1}u) = (F^{\ell-1}u, T^{2\epsilon-1}Pu) + (F^{\ell-1}u, [P, T^{2\epsilon-1}]u)$ and since

$$[P, T^{2\epsilon-1}] = \sum_{j=1}^k T_j^{2\epsilon-1}A_j + T_{k+1}^{2\epsilon-1}, \qquad T_\nu^{2\epsilon-1} \in OPS^{2\epsilon-1},$$

we see that

$$(1.36)\quad |(P^*F^{\ell-1}u, T^{2\epsilon-1}u)| \le C\|F^{\ell-1}u\|^2_{H^{2\epsilon-1}} + C\|Pu\|^2_{L^2} + C\|u\|^2_{L^2} + C\sum_{j=1}^{k}\|A_ju\|_{L^2}\|F^{\ell-1}u\|_{H^{2\epsilon-1}} \le C'\|F^{\ell-1}u\|_{H^{2\epsilon-1}} + C'\|Pu\|^2_{L^2} + C'\|u\|^2_{L^2}$$

by (1.16). Next,

$$(1.37)\quad (A_j^2F^{\ell-1}u, T^{2\epsilon-1}u) = (A_jF^{\ell-1}u, A_jT^{2\epsilon-1}u) = (A_j(T^{2\epsilon-1})^*F^{\ell-1}u, A_ju) + ([(T^{2\epsilon-1})^*, A_j]F^{\ell-1}u, A_ju) + (A_jF^{\ell-1}u, [A_j, T^{2\epsilon-1}]u).$$

Furthermore,

$$|(A_jT^{2\epsilon-1}F^{\ell-1}u, A_ju)| \le \|A_jT^{2\epsilon-1}F^{\ell-1}u\|^2_{L^2} + \|A_ju\|^2_{L^2} \le C\,Re(PT^{2\epsilon-1}F^{\ell-1}u, T^{2\epsilon-1}F^{\ell-1}u) + C\|F^{\ell-1}u\|^2_{H^{2\epsilon-1}} + C\,Re(Pu, u) + C\|u\|^2_{L^2}$$

and, setting $T^{2\epsilon-1}F^{\ell-1} = T^{2\epsilon}$, we have

$$(PT^{2\epsilon}u, T^{2\epsilon-1}F^{\ell-1}u) = ([P, T^{2\epsilon}]u, T^{2\epsilon-1}F^{\ell-1}u) + (Pu, T^{4\epsilon-1}F^{\ell-1}u).$$

Since

$$[P, T^{2\epsilon}] = \sum_{j=1}^{k} T_j^{2\epsilon}A_j + T_{k+1}^{2\epsilon},$$

using (1.16) as before, we see that

$$(1.38)\quad |(A_jT^{2\epsilon-1}F^{\ell-1}u, A_ju)| \le C\|F^{\ell-1}u\|_{H^{4\epsilon-1}} + C\|Pu\|^2_{L^2} + C\|u\|^2_{L^2}.$$

Furthermore,

$$(1.39)\quad |([T^{2\epsilon-1}, A_j]F^{\ell-1}u, A_ju)| \le C\|F^{\ell-1}u\|_{H^{2\epsilon-1}} + C\|Pu\|^2_{L^2} + C\|u\|^2_{L^2},$$

and (1.31) yields

$$(1.40)\quad |(A_jF^{\ell-1}u, [A_j, T^{2\epsilon-1}]u)| \le C\|F^{\ell-1}u\|^2_{H^{2\epsilon-1}} + C\|Pu\|^2_{L^2} + C\|u\|^2_{L^2}.$$

Using (1.38) through (1.40), we can estimate the right side of (1.37), to obtain

$$(1.41)\quad |(A_j^2F^{\ell-1}u, T^{2\epsilon-1}u)| \le C\|F^{\ell-1}u\|^2_{H^{4\epsilon-1}} + C\|Pu\|^2_{L^2} + C\|u\|^2_{L^2}$$

and hence the second term on the right of (1.35) is estimated. It remains to estimate $|(A_jC_j^*F^{\ell-1}u, T^{2\epsilon-1}u)|$, which using the same reasoning as in (1.31) we can bound by

$$(1.42)\quad |(A_jC_j^*F^{\ell-1}u, T^{2\epsilon-1}u)| \le C\|F^{\ell-1}u\|^2_{H^{2\epsilon-1}} + C\|Pu\|^2_{L^2} + C\|u\|^2_{L^2}.$$

Substituting (1.36), (1.41) and (1.42) into (1.35), we obtain

$$(1.43)\quad |(A_0F^{\ell-1}u, T^{2\epsilon-1}u)| \le C\|F^{\ell-1}u\|^2_{H^{4\epsilon-1}} + C\|Pu\|^2_{L^2} + C\|u\|^2_{L^2},$$

so the first term on the right in (1.30) is estimated. To deal with the second term, write

$$-iA_0 = P - \sum_1^k A_j^2 - \sum_1^k C_jA_j \bmod OPS^0,$$

so

$$(1.44)\quad -i(F^{\ell-1}A_0u, T^{2\epsilon-1}u)$$
$$= (F^{\ell-1}Pu, T^{2\epsilon-1}u) - \sum_{j=1}^k (F^{\ell-1}A_j^2u, T^{2\epsilon-1}u)$$
$$- \sum_{j=1}^k (F^{\ell-1}C_jA_ju, T^{2\epsilon-1}u) + 0(\|F^{\ell-1}u\|_{H^{2\epsilon-1}}\|u\|).$$

Noting that $(F^{\ell-1})^* = (ad\, A_{j_{\ell-1}} \cdots ad\, A_{j_2} \cdot A_{j_1})^* = ad\, A_{j_1} \cdots ad\, A_{j_{\ell-2}} \cdot A_{j_{\ell-1}} = F^{\ell-1}$, we see that

$$|(F^{\ell-1}Pu, T^{2\epsilon-1}u)| = |(Pu, F^{\ell-1}T^{2\epsilon-1}u)|$$
$$\le |(Pu, T^{2\epsilon-1}F^{\ell-1}u)| + |(Pu, [T^{2\epsilon-1}, F^{\ell-1}]u)|,$$

so if $2\epsilon \le 1$,

$$(1.45)\quad |(F^{\ell-1}Pu, T^{2\epsilon-1}u)| \le C\|Pu\|^2_{L^2} + C\|F^{\ell-1}u\|^2_{H^{2\epsilon-1}} + C\|u\|^2_{L^2}.$$

Meanwhile

$$(1.46)\quad (F^{\ell-1}A_j^2u, T^{2\epsilon-1}u)$$
$$= (A_j^2F^{\ell-1}u, T^{2\epsilon-1}u) + ([F^\ell - 1, A_j^2]u, T^{2\epsilon-1}u),$$

and since $[F^{\ell-1}, A_j^2] = [F^{\ell-1}, A_j]A_j + A_j[F^{\ell-1}, A_j] = -\tilde{F}^\ell A_j + A_j\tilde{F}^\ell$ ($\tilde{F}^\ell$ denoting another commutator of order ℓ) we can estimate $([F^{\ell-1}, A_j^2]u, T^{2\epsilon-1}u)$ by the method of case 1 (since $1 \le j \le k$). Thus the last term in (1.46) is bounded by

$$C\|\tilde{F}^\ell u\|^2_{H^{2\epsilon-1}} + C\sum_{j=1}^k \|A_ju\|^2_{L^2} + C\|u\|^2_{L^2}$$

and applying (1.32), with F^ℓ replaced by $\tilde{F}^\ell$, ϵ replaced by 2ϵ, we get

$$\|\tilde{F}^\ell u\|^2_{H^{2\epsilon-1}} \le C\|F^{\ell-1}u\|^2_{H^{4\epsilon-1}} + C\|Pu\|^2_{L^2} + C\|u\|^2_{L^2}.$$

Since the first term on the right in (1.46) was estimated by (1.41), we see that

$$(1.47)\quad |(F^{\ell-1}A_j^2u, T^{2\epsilon-1}u)| \le C\|F^{\ell-1}u\|^2_{H^{4\epsilon-1}} + C\|Pu\|^2_{L^2} + C\|u\|^2_{L^2},$$

and we have estimated the second term on the right in (1.44). As for the third term,

$$\begin{aligned} &|(F^{\ell-1}C_jA_ju, T^{2\epsilon-1}u)| \\ &\qquad = |(C_jA_ju, F^{\ell-1}T^{2\epsilon-1}u)| \\ &\qquad \le C\|A_ju\|_{L^2}(\|T^{2\epsilon-1}F^{\ell-1}u\|_{L^2} + \|[T^{2\epsilon-1}, F^{\ell-1}]u\|_{L^2}) \\ &\qquad \le C\|Pu\|^2_{L^2} + C\|F^{\ell-1}u\|^2_{H^{2\epsilon-1}} + C\|u\|^2_{L^2}, \end{aligned}$$

if $2\epsilon \le 1$, so putting together (1.45), (1.47) and (1.44) we have the estimate for (1.44):

$$(1.49)\quad |(F^{\ell-1}A_0u, T^{2\epsilon-1}u)| \le C\|F^{\ell-1}u\|^2_{H^{4\epsilon-1}} + C\|Pu\|^2_{L^2} + C\|u\|^2_{L^2}.$$

From (1.43) and (1.49) we get our estimate on (1.30) in this case:

$$(1.50)\quad \|F^{\ell}u\|_{H^{\epsilon-1}} \le C\|F^{\ell-1}u\|_{H^{4\epsilon-1}} + C\|Pu\|_{L^2} + C\|u\|_{L^2}, \qquad \text{if } \epsilon \le \tfrac{1}{2}.$$

Note that (1.50) is a bit weaker than (1.32).

Inductively we obtain

$$(1.51)\qquad \|F^{\ell}u\|_{H^{\epsilon-1}} \le C\sum_{j=0}^{k} \|A_ju\|_{H^{4^{\ell-1}\epsilon-1}} + C\|Pu\|_{L^2} + C\|u\|_{L^2}.$$

Now we claim that

$$(1.52)\qquad \|A_ju\|_{H^{-1/2}} \le C\|Pu\|_{L^2} + C\|u\|_{L^2}, \qquad 0 \le j \le k.$$

Granted this, we obtain Theorem 1.2, with $\epsilon = 2 \cdot 4^{-L}$. If $1 \le j \le k$, (1.52) is weaker than (1.16). For $j = 0$, write

$$\begin{aligned} \|A_0u\|^2_{H^{-1/2}} &= (-iA_0u, T^0u) \\ &= (Pu, T^0u) - \sum_{j=1}^{k} (A_j^2u, T^0u) - \sum_{j=1}^{k} (C_jA_ju, T^0u). \end{aligned}$$

Clearly $|(Pu, T^0u)| \le C\|Pu\|_{L^2}\|u\|_{L^2}$. Meanwhile

$$\sum_1^k |(A_j^2u, T^0u)| \le C\sum_1^k \|A_ju\|^2_{L^2} + C\|u\|^2_{L^2} \le C\|Pu\|^2_{L^2} + C\|u\|^2_{L^2}.$$

Finally,

$$\sum_1^k |(C_jA_ju, T^0u)| \le \sum_1^k \|A_ju\|^2_{L^2} + C\|u\|^2_{L^2} \le C\|Pu\|^2_{L^2} + C\|u\|^2_{L^2}.$$

Thus (1.52) is proved, and the proof of Theorem 1.2 is complete.

We now want to sharpen up Theorem 1.4 a bit, relaxing the hypothesis (1.15). It is clear that the conclusion of the theorem holds if (1.15) is replaced by any other assumption on sub $\sigma(P)$ that yields (1.16). In particular, we may expect to get a better result if appeal to the sharp Gårding inequality is replaced by use of the Hörmander-Melin inequality, discussed in Chapter XIII, Section 4. So, to re-examine (1.16), write, with $0 < \gamma < 1$,

$$(Pu, u) = (1-\gamma)\sum_1^k \|A_j u\|^2 + i(A_0 u, u) + \left(\left(\gamma\sum_1^k A_j^2 + B_1\right)u, u\right) + \sum_1^k (A_j u, C_j^* u).$$

We see that (1.16) will hold provided

$$Re\left(\left(\gamma\sum_1^k A_j^2 + B_1\right)u, u\right) \geq -C\|u\|_{L^2}^2. \tag{1.53}$$

According to Theorem 4.1 of Chapter XIII, (1.53) holds provided

$$Re\, b_1 + \gamma\, tr^+ F \geq 0 \text{ on } \Sigma \tag{1.54}$$

and

$$\text{the symplectic form has constant rank on } \Sigma \tag{1.55}$$

where we recall that F is the fundamental matrix of

$$p_2 = \sum_1^k a_j^2 \text{ on } \Sigma,$$

defined by

$$\sigma(u, Fv) = Q(u, v),$$

where Q is the hessian of $p_2(x, \xi)$. The nonzero eigenvalues of F are of the form $\pm i\mu_\nu (\mu_\nu > 0)$, and $tr^+ F = \Sigma \mu_\nu$. We remark that, if $tr^+ F > 0$ on Σ, then $T_{(x_0, \xi_0)}\Sigma$ is not involutive. Indeed, since

$$T_{(x_0,\xi_0)}\Sigma = \ker Q = \ker F \quad \text{and} \tag{1.56}$$

$$T^\sigma_{(x_0,\xi_0)}\Sigma = \sigma^{-1} \text{ range } Q = \text{range } F, \tag{1.57}$$

we see that

$$\begin{aligned} T_{(x_0,\xi_0)}\Sigma \text{ not involutive} &\Leftrightarrow T^\sigma_{(x_0,\xi_0)}\Sigma \not\subset T_{(x_0,\xi_0)}\Sigma \\ &\Leftrightarrow \text{range } F \not\subset \ker F \\ &\Leftrightarrow F^2 \neq 0. \end{aligned} \tag{1.58}$$

In fact the proof of Lemma 4.3 of chapter XIII shows that the converse is true; $F^2 \neq 0$ implies $tr^+ F > 0$. Also from (1.56) and (1.57) we see that

$$(1.59) \quad T_{(x_0,\xi_0)}\Sigma \text{ is symplectic} \Leftrightarrow \ker F \cap \text{range } F = 0$$
$$\Leftrightarrow F \text{ has no nonzero nilpotent part.}$$

Note that

$$\text{ranks } \sigma\big|_{T_{\zeta_0}\Sigma} = \dim T_{\zeta_0} - \dim T_{\zeta_0} \cap T^{\sigma}_{\zeta_0},$$

which is $>(2n-k)-k$ precisely when $T^{\sigma}_{\zeta_0} \not\subset T_{\zeta_0}$.

We have proved the following.

THEOREM 1.7. *Suppose $p(x, \xi)$ is given by* (1.7) *and suppose $p_2(x, \xi) \geq 0$ vanishes to exactly second order on the smooth conic submanifold Σ, of codimension k. Suppose*

(1.60) *the symplectic form restricted to $T\Sigma$ has constant rank, $>2n-2k$.*

(1.61) $\quad Re \text{ sub } \sigma(P) + \gamma\, tr^+ F \geq 0$ *on* Σ, *for some* $\gamma \in (0, 1)$.

Then P is hypoelliptic, with loss of one derivative.

Note that (1.61) is a consequence of the assumption

$$(1.62) \qquad Re \text{ sub } \sigma(P) + tr^+ F > 0 \text{ on } \Sigma.$$

If Σ is symplectic, hypotheses (1.61) and (1.62) are equivalent. We can relax (1.60) when (1.61) is replaced by (1.62), dropping the requirement that σ have constant rank on $T\Sigma$.

THEOREM 1.8. *Suppose $p_2(x, \xi) \geq 0$ vanishes to exactly second order on Σ, as above, with $T\Sigma$ not involutive at each point of Σ, and suppose* (1.62) *holds. Then P is hypoelliptic, with loss of one derivative.*

Proof. The proof is the same as before, where (1.53) is justified (for $\gamma < 1$ close to 1) by the variant of Hörmander's inequality, Theorem 4.2 of Chapter XIII, which says (1.53) holds provided p_2 vanishes to precisely second order on Σ and

$$(1.63) \qquad Re\, b_1 + \gamma\, tr^+ F > 0 \text{ on } \Sigma$$

with no assumption that σ have constant rank on $T\Sigma$.

If $p(x, D) \in OPS^2$ is a *system*, with scalar principal part $p_2(x, \xi) \geq 0$ as above, but $p_1(x, \xi)$ a matrix, the above argument goes straight through, and Theorem 1.8 immediately generalizes, with (1.62) replaced by

$$(1.62') \qquad \tfrac{1}{2}(\text{sub } \sigma(P) + \text{sub } \sigma(P)^*) + tr^+ F > 0 \text{ on } \Sigma.$$

It is more interesting to derive hypoellipticity from conditions on the eigenvalues of the matrix sub $\sigma(P)$, rather than from conditions on the numerical range of this matrix. Indeed, we easily obtain the following.

THEOREM 1.9. *Suppose* $p(x, \xi)$ *is given by* (1.7) *with positive scalar principal symbol* $p_2(x, \xi)$, *vanishing to exactly second order on* Σ, *of codimension* k. *Suppose that* $T\Sigma$ *is not involutive, at each point, and that, for all eigenvalues* S_ν *of the matrix* sub $\sigma(P)$,

$$Re\, S_\nu + tr^+ F > 0 \text{ on } \Sigma. \tag{1.64}$$

Then P *is hypoelliptic, with loss of one derivative.*

Proof. It follows from (1.64) that there is an elliptic $e_0(x, \xi) \in S^0$ such that, on Σ, $e_0(\text{sub } \sigma(P))e_0^{-1} + tr^+ F$ has numerical range in the open right-half plane. Let $E = e_0(x, D)$ and let E^{-1} be a paramatrix for E. Consider $\tilde{P} = EPE^{-1}$. $\tilde{P}$ has principal symbol $p_2(x, \xi)$ and

$$\text{sub } \sigma(\tilde{P}) = e_0(\text{sub } \sigma(P))e_0^{-1} \text{ on } \Sigma.$$

By Theorem 1.8, trivially generalized to systems, $\tilde{P}$ is hypoelliptic. Thus so is P.

Hörmander [19], Menikoff [3] and Boutet de Monvel, Grigis, and Helffer [1] have obtained necessary and sufficient conditions that a scalar operator P of the form (1.7) with $p_2(x, \xi) \geq 0$ vanishing precisely to second order on a manifold Σ, be hypoelliptic with loss of one derivative. The condition is that, for $(x_0, \xi_0) \in \Sigma$, $\alpha_\nu = 0, 1, 2, \ldots$

$$\text{sub } \sigma(P) + Q_{(x_0,\xi_0)}(v, \bar{v}) + \Sigma(2\alpha_\nu + 1)\mu_\nu \neq 0, \qquad v \in V_0 \tag{1.65}$$

where V_0 is the space of generalized eigenvectors of F belonging to the eigenvalue 0, μ_ν the positive eigenvalues of $(1/i)F$. As before, Q is the hessian of $p_2(x, \xi)$. In fact, these authors consider more general situations, where $p_2(x, \xi)$ takes values in a proper cone and Σ need not be a manifold (in which case (1.65) is replaced by something more complicated). We refer the reader to these papers for a proof. Note that, when Σ is a symplectic manifold, $V_0 = 0$ and $Q_{(x_0,\xi_0)}(v, \bar{v})$ drops out of (1.65). Certainly (1.65) is more general than the conditions of theorem 1.7. See also Sjöstrand [1].

As an example of how the above results apply, consider the following second order operator on $\mathbf{H}_n = \mathbf{C}^n \times \mathbf{R}$, with coordinates $(z_1, \ldots, z_n, t)$:

$$\mathcal{L}_\alpha = -\frac{1}{2}\sum_{j=1}^{n} (Z_j\bar{Z}_j + \bar{Z}_jZ_j) + i\alpha T \tag{1.66}$$

where

$$Z_j = \frac{\partial}{\partial z_j} + i\bar{z}_j\frac{\partial}{\partial t}, \qquad T = \frac{\partial}{\partial t}.$$

Here

$$p_2(x, y, t; \xi, \eta, \tau) = \sum_{j=1}^{n} |\xi_j + i\eta_j - \bar{z}_j\tau|^2,$$

vanishing on $\Sigma = \{\xi_j = x_j\tau, \eta_j = y_j\tau, \tau \in \mathbf{R}\}$, which is a symplectic submanifold of $T^*(\mathbf{H}_n)\backslash 0$ of dimension $2n + 1$. Indeed, $[Z_j, \bar{Z}_k] = \delta_{jk}T$, and $\tau \neq 0$ on Σ. One can compute that $(1/i)F$ has n positive eigenvalues, each one being $|\tau|$. Also

$$Re \text{ sub } \sigma(\mathscr{L}_\alpha) = -\alpha\tau,$$

so

$$Re \text{ sub } \sigma(\mathscr{L}_\alpha) + tr^+ F = n|\tau| - \alpha\tau. \tag{1.67}$$

Consequently Theorem 1.7 implies that $\mathscr{L}_\alpha$ is hypoelliptic (with loss of one derivative) provided $|\alpha| < n$, whereas the condition (1.65) leads to hypoellipticity of $\mathscr{L}_\alpha$ whenever $n \pm \alpha$ is not an even integer ≤ 0. The necessity and sufficiency of this condition was established by Folland and Stein [1], who wrote down an explicit fundamental solution for the permissible α and a singular element of ker $\mathscr{L}_\alpha$ for the exceptional α.

There is a natural way of giving $\mathbf{H}_n$ the structure of a CR manifold (i.e., identifying $\mathbf{H}_n$ with the surface $Im\, z_{n+1} = |z_1|^2 + \cdots + |z_n|^2$ in $\mathbf{C}_{n+1}$) such that the Levi form is positive definite. If $\bar{\partial}_b$ denotes the induced Cauchy-Riemann operator on $\mathbf{H}_n$, acting on $(0, p)$ forms, and if $\bar{\partial}_b^*$ is the adjoint with respect to a "Levi metric," we form the Kohn Laplacian

$$\Box_b = \bar{\partial}_b^*\bar{\partial}_b + \bar{\partial}_b\bar{\partial}_b^*. \tag{1.68}$$

As shown in Folland-Stein

$$\Box_b = \mathscr{L}_\alpha \text{ with } \alpha = n - 2p, \qquad \text{on } (0, p) \text{ forms.} \tag{1.69}$$

Theorem 1.7 implies that $\Box_b$ is hypoelliptic on $(0, p)$ forms, for $p \neq 0$ on n.

For a general hypersurface S in $\mathbf{C}^{n+1}$, $\bar{\partial}_b$ and $\Box_b$ are defined. $\Box_b$ has a scalar principal symbol, and its characteristic variety Σ, which is the line bundle iN, N being the normal bundle to S, is symplectic if, and only if, the Levi form is nondegenerate on S. If S is strongly pseudoconvex, it has been shown (see Folland and Stein [1]) in the case when S has a "Levi metric," that on Σ, Re sub $\sigma(\Box_b) + tr^+ F = (n \pm (n - 2p))\Xi$, where the factor Ξ is positive and homogeneous of degree 1. (In general, the complete symbol of $\Box_b$ is not a scalar, but sub $\sigma(\Box_b)|_\Sigma$ is a scalar.) It follows that, on general strictly pseudoconvex surfaces, $\Box_b$ is hypoelliptic on $(0, p)$ forms for $p \neq 0$ or n.

The study of $\Box_b$ is closely related to the study of the $\bar{\partial}$-Neumann problem for $\Box = \bar{\partial}^*\bar{\partial} + \bar{\partial}\bar{\partial}^*$ on a region Ω in $\mathbf{C}^{n+1}$ with boundary $S = \partial\Omega$.

This is a boundary value problem of the form

$$\Box u = 0,$$
$$[vu]_S = f, \qquad [v\bar{\partial}u]_S = g.$$

This boundary value problem is not coercive, but Kohn [1] showed that it is hypoelliptic, with loss of one derivative, on $(0, p)$ forms for $p \neq 0, n$, provided Ω is strongly pseudoconvex. This result, for (0, 1) forms, leads to the demonstration that such Ω is a domain of holomorphy (see Folland and Kohn [1]). Kohn also showed that $\Box_b$ is hypoelliptic, by similar methods, and it was suspected that these two results were equivalent. This equivalence was demonstrated for (0, 1) forms by Greiner and Stein in [1]. In fact, define a first order pseudodifferential operator $\Box^+$ on $\partial\Omega$ by $g = \Box^+ h$ where u solves the Dirichlet problem $\Box u = 0$, $[vu]_S = 0$, $[u_t]_S = h$, and $g = [v\bar{\partial}u]_S$. $\Box^+$ is doubly characteristic on half the line bundle Σ, and there is another operator $\Box^-$, characteristic on the other half of Σ but elliptic on the characteristics of $\Box^+$, such that $\Box^-\Box^+ - \Box_b = R$ is a first order operator whose principal symbol vanishes on Σ. Thus hypoellipticity of $\Box_b$ is equivalent to hypoellipticity of $\Box^-\Box^+$ (and hence leads to hypoellipticity of $\Box^+$).

For detailed studies of $\Box_b$ and the $\bar{\partial}$ Neumann problem, the reader is referred to Folland and Kohn [1], Boutet de Monvel [2], and especially Greiner and Stein [1].

Finally, we remark that Theorem 1.3 can be generalized along the lines of Theorem 1.9, to yield the following.

THEOREM 1.10. *Suppose $p(x, \xi)$ is given by (1.7) with positive scalar principal part $p_2(x, \xi)$, vanishing to exactly second order on Σ, and suppose that, if Σ is defined by $a_1(x, \xi) = \cdots = a_k(x, \xi) = 0$, then some Poisson bracket $\{a_{j_s}, \ldots, \{a_{j_2}, a_{j_1}\} \cdots\}$ is nonvanishing, at each $(x_0, \xi_0) \in \Sigma$, so Σ is symplectically flat to only finite order, at each point. Suppose also that, for all eigenvalues S_ν of the matrix* sub $\sigma(P)$,

$$\text{(1.70)} \qquad Re\, S_\nu + tr^+ F > 0 \text{ on } \Sigma.$$

Then P is hypoelliptic, with loss of $2-\epsilon$ derivatives, for some $\epsilon > 0$.

The proof proceeds from Theorem 1.3 precisely as the proof of Theorem 1.9 proceeds from Theorem 1.4.

We remark that Theorem 1.10 is not really a satisfactory result. In fact, the results of Boutet de Monvel et al. [1] and Hörmander [12] apply in this case, yielding hypoellipticity with loss of one derivative. However, it seems likely that an appropriate improvement of the Hörmander-Melin inequality could be obtained and used to weaken hypothesis (1.70), producing a new result.

Exercises

1.1. Let $q(x, \xi) = a(|\xi|^\delta x) r(|\xi|^{-\rho}\xi')$ where $a \in C_0^\infty(\mathbf{R}^n)$, $a(x) = 1$ for $|x| \le 1$, and $r \in C_0^\infty(\mathbf{R}^{n-1})$, $r(\xi') = 1$ for $|\xi'| \le 1$. Show that $q(x, \xi) \in S^0_{\rho,\delta}$. Take $\delta < 1/2 < \rho$, δ, ρ close to 1/2. If $s(x, \xi) = \tilde{a}(|\xi|^\delta x)\tilde{r}(|\xi|^{-\rho}\xi')$, $\tilde{a} \in C_0^\infty(|x| \le 1/2)$, $\tilde{a}(0) = 1$, and $\tilde{r} \in C_0^\infty(|\xi'| \le 1/2)$, $\tilde{r}(0) = 1$, show that

$$q(x, D)s(x, D) = s(x, D) \bmod OPS^{-\infty}.$$

Since $s(x, D)$ is not smoothing, deduce that, for any $s \in \mathbf{R}$, $\epsilon > 0$, there exists $u \in H^s \backslash H^{s+\epsilon}$ such that

$$q(x, D)u = u \bmod C^\infty. \tag{1.70}$$

1.2. Suppose $p(x, \xi) \in S^m$ with $p_m(x_0, \xi_0) = 0$. Say $x_0 = 0$, $\xi_0 = (1, 0, \ldots, 0)$. Show that

$$p(x, D)q(x, D) \in OPS^{m-\sigma}_{\rho,\delta}, \qquad \sigma = \tfrac{1}{2} - \epsilon \text{ if } \rho, \delta \text{ close to } \tfrac{1}{2}.$$

Suppose furthermore that $\nabla_{x,\xi} p_m(x_0, \xi_0) = 0$. Then show that

$$p(x, D)q(x, D) \in OPS^{m-\tau}_{\rho,\delta}, \qquad \tau = 1 - \epsilon \text{ if } \rho, \delta \text{ close to } \tfrac{1}{2}.$$

1.3. Suppose $p(x, D) \in OPS^m$ is characteristic at (x_0, ξ_0) as above. Show there is $u \in H^s - H^{s+\epsilon}$ such that $p(x, D)u \in H^{s+1/2-m-\epsilon}$. If $\nabla_{x,\xi} p(x_0, \xi_0) = 0$ also, show there is a $u \in H^s \backslash H^{s+\epsilon}$ such that $p(x, D)u \in H^{s+1-m-\epsilon}$. Thus, if $p(x, D)$ has a double characteristic, one cannot have hypoellipticity with loss of less than one derivative.

1.4. Let $P \in OPS^m$ on a compact manifold M, and suppose $P: \mathcal{D}'(M) \to \mathcal{D}'(M)$ is invertible, with

$$P^{-1}: H^s(M) \to H^{s+m-1}(M).$$

Show that $P^{-1} \in OPS^{-m+1}_{1/2,1/2}$. Hint: Use the characterization of $OPS^{-m+1}_{1/2,1/2}$ given in Chapter VIII, section 8. See Beals [3] for further results along these lines.

§2. The Subprincipal Symbol and Microlocal Equivalence of Operators

In the previous section we introduced the subprincipal symbol of the operator $P \in OPS^m$, with symbol $p(x, \xi) \sim p_m(x, \xi) + p_{m-1}(x, \xi) + \cdots$,

$$\text{sub}\,\sigma(P) = p_{m-1}(x, \xi) - \frac{1}{2i}\sum_{j=1}^n \frac{\partial^2 p_m}{\partial x_j\, \partial \xi_j}. \tag{2.1}$$

We shall now make a brief examination of the invariance of sub $\sigma(P)$ on the set of double characteristics of P, given by

$$\Sigma_2 = \{(x, \xi): p_m(x, \xi) = 0, \nabla_{x,\xi} p_m(x, \xi) = 0\}. \tag{2.2}$$

First, consider the case where $P = p(x, D) \in OPS^2$ is of the form

$$P = \sum_{j=1}^{k} A_j B_j + C \tag{2.3}$$

generalizing (1.3). We suppose A_j, B_j, $C \in OPS^1$ with symbols

$$a_j(x, \xi) \sim \sum_{\nu \leq 1} a_{j\nu}(x, \xi),$$

etc., and that $a_{j1}(x, \xi) = b_{j1}(x, \xi) = 0$ on Σ_2. We see that

$$p(x, \xi) = \sum_{j=1}^{k} a_j(x, \xi) b_j(x, \xi) + \frac{1}{i} \sum_{j=1}^{k} \sum_{|\alpha|=1} (D_\xi^\alpha a_j)(D_x^\alpha b_j) + c(x, \xi) \bmod S^0$$

and hence

$$\text{sub } \sigma(P) = c_1(x, \xi) + \frac{1}{2i} \sum_j \{a_{j1}, b_{j1}\}(x, \xi) \text{ on } \Sigma_2, \tag{2.4}$$

generalizing (1.10). From this it easily follows that, if P is of the form (2.3) and if J is an elliptic Fourier integral operator of the sort studied in Chapter VIII, with associated canonical transformation $\mathcal{J}$, then

$$\text{sub } \sigma(JPJ^{-1})(\mathcal{J}(x, \xi)) = \text{sub } \sigma(P) \text{ on } \Sigma_2. \tag{2.5}$$

Indeed, we need only apply Egorov's theorem to (2.4). This argument is used in Hörmander [20].

To obtain (2.5) in general, we reason as follows. First note that if $A \in OPS^\mu$, then

$$\begin{aligned} \text{sub } \sigma(PA) &= \text{sub } \sigma(P) a_\mu + p_m \text{ sub } \sigma(A) + \frac{1}{2i} \{p_m, a_\mu\}, \\ \text{sub } \sigma(AP) &= \text{sub } \sigma(A) p_m + a_\mu \text{ sub } \sigma(P) + \frac{1}{2i} \{a_\mu, p_m\} \end{aligned} \tag{2.6}$$

and in particular sub $\sigma(PA) = \text{sub } \sigma(P) \cdot a_\mu$ and sub $\sigma(AP) = a_\mu \text{ sub } \sigma(P)$ on Σ_2. Hence, if A is elliptic,

$$\text{sub } \sigma(APA^{-1}) = \text{sub } \sigma(P) \text{ on } \Sigma_2. \tag{2.7}$$

Thus, to establish (2.5), we need only consider $P(t) = S(t, 0)PS(0, t)$, which is doubly characteristic at $C(t)\Sigma_2$, where $C(t)$ is the canonical teanformation corresponding to the solution operator $S(t, 0)$ to the hyperbolic equation $(\partial/\partial t - i\lambda)u = 0$. Now we can use a variant of the method used in Chapter VIII, Section 1, to prove Egorov's theorem. This method was used by Guillemin.

Indeed, one sees from (2.6) that

$$\text{sub}\,\sigma([\lambda, P]) = \frac{1}{i}\{\lambda_1, \text{sub}\,\sigma(P)\} + \frac{1}{i}\{\text{sub}\,\sigma(\lambda_1), p_m\}. \tag{2.8}$$

Hence the operator equation

$$\frac{d}{dt}P(t) = i[\lambda, P(t)]$$

yields

$$\frac{\partial}{\partial t}\text{sub}\,\sigma(P(t)) = \{\lambda_1, \text{sub}\,\sigma(P)\} + \{\text{sub}\,\sigma(\lambda_1), p_m\}, \tag{2.9}$$

and since $\{\text{sub}\,\sigma(\lambda_1), p_m\} = 0$ on $C(t)\Sigma_2$, this implies

$$\left(\frac{\partial}{\partial t} - H_{\lambda_1}\right)\text{sub}\,\sigma(P(t)) = 0 \text{ on } C(t)\Sigma_2$$

which proves (2.5) in the general case.

We now examine the question of microlocal equivalence of two operators $P = p(x, D)$, $\tilde{P} = \tilde{p}(x, D) \in OPS^m$. We say these two operators are microlocally equivalent near (x_0, ξ_0) if there exist two operators $A, B \in OPS^0$, *elliptic* in a conic neighborhood of (x_0, ξ_0), such that

$$APB = \tilde{P}, \qquad \text{mod } OPS^{-\infty}. \tag{2.10}$$

More generally, if in (2.10) equality holds modulo OPS^{m-k}, we say P and $\tilde{P}$ are microlocally equivalent mod OPS^{m-k} near (x_0, ξ_0). We assume

$$p_m(x, \xi) = \tilde{p}_m(x, \xi), \qquad \text{real valued.} \tag{2.11}$$

The point of this is that, conjugating by a Fourier integral operator, we may put $p_m(x, \xi)$ in a standard form, and then we would like to put the whole operator P in a standard form, if possible, via (2.10). We suppose $A = a(x, D)$, $B = b(x, D)$, with

$$a(x, \xi) \sim \sum_{j \le 0} a_j(x, \xi), \qquad \text{etc.}$$

We may as well suppose

$$b_0(x, \xi) = a_0(x, \xi)^{-1}, \qquad \text{near } (x_0, \xi_0). \tag{2.12}$$

Now to make P and $\tilde{P}$ microlocally equivalent mod OPS^{m-2}, following Melrose (unpublished) we try $A = a_0(x, D)$, $B = A^{-1}(1 + C)$, $C = c_{-1}(x, D) \in OPS^{-1}$. We shall restrict our attention here to the scalar case. We see that the principal symbol (in S^{m-1}) of $APB - \tilde{P}$ is

$$a_0^{-1}H_{p_m}a_0 + p_m c_{-1} + (p_{m-1} - \tilde{p}_{m-1}), \tag{2.13}$$

so we want to make this vanish, near (x_0, ξ_0). To guarantee that a_0 is nonvanishing, write $a_0 = e^{\alpha_0}$, so the vanishing of (2.13) is equivalent to

$$\frac{1}{i} H_{p_m}\alpha_0 + p_m c_{-1} = \tilde{p}_{m-1} - p_{m-1}. \tag{2.13}$$

Thus the condition is that $\tilde{p}_{m-1} - p_{m-1}$ belong to the sum of the range of the vector field H_{p_m} and the ideal generated by p_m, near (x_0, ξ_0). Clearly this requires that $p_{m-1} = \tilde{p}_{m-1}$, on Σ_2, which in view of (2.11) is equivalent to sub $\sigma(P) =$ sub $\sigma(\tilde{P})$ on Σ_2. Thus equality of subprincipal symbols on Σ_2 is a necessary condition for microlocal equivalence modulo OPS^{m-2} As we shall see, for certain $p_m(x, \xi)$ this will be a sufficient condition, though for other $p_m(x, \xi)$ this condition will not suffice.

Suppose we can solve (2.13). Then P is microlocally equivalent to APB, which we now denote by P, and we have $P - \tilde{P} \in OPS^{m-2}$, near (x_0, ξ_0). We next would like to find conditions under which the difference can be lowered another degree, after altering P again. In particular, we look for operators $A_{-1} \in OPS^{-1}$, $C_{-2} \in OPS^{-2}$ such that, near (x_0, ξ_0),

$$(I + A_{-1})P(I - A_{-1} + C_{-2}) = \tilde{P} \bmod OPS^{m-3}.$$

More generally, if P has been altered so that $P - \tilde{P} = R_{j+1} \in OPS^{m-j-1}$ near (x_0, ξ_0), $j \geq 1$, we look for $A_{-j} \in OPS^{-j}$, $C_{-j-1} \in OPS^{-j-1}$ such that, near (x_0, ξ_0),

$$(I + A_{-j})P(I - A_{-j} + C_{-j-1}) = \tilde{P} \bmod OPS^{m-j-2}. \tag{2.14}$$

Evaluating the principal symbol of the difference of these two expressions, we see that this identity requires

$$\frac{1}{i} H_{p_m} a_{-j} + p_m(c_{-j-1} - [a_{-j}^2]) = r_{j+1} \tag{2.15}$$

where the term in brackets occurs only for $j = 1$. Again, the condition that (2.15) be solvable is that r_{j+1} belong to the sum of the span of the vector field H_{p_m} and the ideal generated by p_m, near (x_0, ξ_0).

Obviously P and $\tilde{P}$ are microlocally equivalent at points of ellipticity. Furthermore, suppose p_m vanishes to precisely first order at (x_0, ξ_0), so $\nabla_{x,\xi} p_m(x_0, \xi_0) \neq 0$ there. If $\nabla_{x,\xi} p_m(x_0, \xi_0)$ is not parallel to $\Sigma \xi_j \, dx_j$, so that H_{p_m} is not parallel to the cone axis, clearly H_{p_m} is onto, as a map of S^μ to $S^{\mu+m-1}$. We refer to Guillemin and Schaeffer [1] for some results where H_{p_m} is parallel to the cone axis. The next situation is the case of double characteristics. We consider now several typical special cases.

The simplest special case is when p_m is characteristic of precisely second order on a hypersurface Σ. We can suppose $m = 2$ and $p_2 = a_1^2$ where

$a_1 = 0$ on Σ, $\nabla_{x,\xi}a_1 \neq 0$ on Σ. Also assume $\nabla_{x,\xi}a_1$ not proportional to $\Sigma\xi_j\,dx_j$, near a point $(x_0, \xi_0) \in \Sigma$. The task of making P equivalent to $\tilde{P}$ modulo OPS^0 near (x_0, ξ_0) is equivalent to solving

$$\frac{2}{i}\,a_1 H_{a_1}\alpha_0 + a_1^2 c_{-1} = p_1 - \tilde{p}_1. \tag{2.16}$$

As we have seen, it is necessary that $p_1 - \tilde{p}_1 = 0$ on Σ. Conversely, if $p_1 - \tilde{p}_1 = 0$ on Σ, then $q_0 = a_1^{-1}(p_1 - \tilde{p}_1)$ is smooth, and solving (2.16) is equivalent to solving

$$\frac{2}{i}\,H_{a_1}\alpha_0 + a_1 c_{-1} = q_0$$

which is possible; you can even take $c_{-1} = 0$. Consequently, in this case, microlocal equivalence mod OPS^{m-2} of P and $\tilde{P}$ is equivalent to the identity of their subprincipal symbols, on Σ. Thus, if one were first to conjugate P by a Fourier integral operator to change its principal symbol to ξ_1^2, we see that P would be made equivalent to $-(\partial^2/\partial x_1^2)$, mod OPS^0 (near some point where $\xi_1 = 0$) if, and only if, the subprincipal symbol of P vanished on Σ. There will be further obstruction to getting equivalence modulo still lower order terms, which we can leave as an exercise.

The next case we consider is when p_m vanishes on the union of two hypersurfaces, Σ_a and Σ_b, which intersect transversally, at $\Sigma_2 = \Sigma_a \cap \Sigma_b$. Say $p_m = ab$ where a vanishes to precisely first order on Σ_a, etc. We suppose that da, db, and $\Sigma\xi_j dx_j$ are linearly independent at $(x_0, \xi_0) \in \Sigma_2$. We consider two cases, either Σ_2 symplectic, so $\{a, b\}(x_0, \xi_0) \neq 0$, or Σ_2 involutive, so $\{a, b\} = 0$ on Σ_2, near (x_0, ξ_0). We shall not consider in this section any cases where the symplectic form changes rank on Σ_2.

First, consider the case where Σ_2 is symplectic. We can suppose $a \in S^0$, $b \in S^1$. Then $\{a, b\} = r \in S^0$ is nonvanishing near (x_0, ξ_0). We can suppose $\{a, b\} = 1$ near (x_0, ξ_0). Indeed, we would have $\{a, eb\} = 1$ near (x_0, ξ_0) provided

$$bH_a e + er = 1, \qquad e = r^{-1} \text{ on } \{b = 0\}. \tag{2.17}$$

Since H_a is transversal to $\{b = 0\}$, it is easy to solve (2.17) to infinite order at $\{b = 0\}$. With $\tilde{b} = eb$, we have $\{a, \tilde{b}\} - 1$ vanishing to infinite order at $\{\tilde{b} = 0\}$. Now $\tilde{\tilde{b}}$, defined by $H_a\tilde{\tilde{b}} = 1$, $\tilde{\tilde{b}} = 0$ on $\{\tilde{b} = 0\}$, must agree with $\tilde{b}$ to infinite order, so we have $\tilde{\tilde{b}} = (\tilde{\tilde{b}}\tilde{b}^{-1})eb$, a smooth nonvanishing multiple of b. Now, by Darboux' theorem, making a canonical transformation we can suppose $a = x_1$, $b = \xi_1$. Thus

$$H_{p_1} = \xi_1\frac{\partial}{\partial \xi_1} - x_1\frac{\partial}{\partial x_1},$$

and solving (2.13) becomes

$$\frac{1}{i}\left(\xi_1\frac{\partial}{\partial\xi_1} - x_1\frac{\partial}{\partial x_1}\right)\alpha_0 + x_1\xi_1 c_{-1} = p_0 - \tilde{p}_0. \tag{2.18}$$

Granted that $p_0 - \tilde{p}_0 = 0$ on $\{x_1 = \xi_1 = 0\}$, it is easy to construct α_0 so that

$$\frac{1}{i}\left(\xi_1\frac{\partial}{\partial\xi_1} - x_1\frac{\partial}{\partial x_1}\right)\alpha_0 = p_0 - \tilde{p}_0$$

on $\{x_1 = 0\} \cup \{\xi_1 = 0\}$, i.e., so that

$$\frac{1}{i}\xi_1\frac{\partial}{\partial\xi_1}\alpha_0 = p_0 - \tilde{p}_0$$

on $\{x_1 = 0\}$ and

$$-\frac{1}{i}x_1\frac{\partial}{\partial x_1}\alpha_0 = p_0 - \tilde{p}_0$$

on $\{\xi_1 = 0\}$. Then the difference

$$\frac{1}{i}\left(\xi_1\frac{\partial}{\partial\xi_1} - x_1\frac{\partial}{\partial x_1}\right)\alpha_0 - (p_0 - \tilde{p}_0)$$

is divisible by $x_1\xi_1$. Thus (2.18) can be solved if, and only if, $p_0 - \tilde{p}_0 = 0$ on $\{x_1 = \xi_1 = 0\}$.

The case when $p_m = ab$ and Σ_2 is involutive will provide our first example where identity of the subprincipal symbols on Σ_2 does not guarantee that P and $\tilde{P}$ are equivalent mod OPS^{m-2}. In this case, we can suppose $a, b \in OPS^1$, $\{a, b\} = 0$ on $\{a = b = 0\}$. We claim that, upon multiplying by an elliptic factor, we can arrange $\{a, b\} = 0$ on a whole conic neighborhood of a point $(x_0, \xi_0) \in \Sigma_2$. Indeed, with $\alpha, \beta \in S^0$, consider

$$\{e^\alpha a, e^\beta b\} = e^{\alpha+\beta}[ab\{\alpha, \beta\} + a\{\alpha, b\} + b\{a, \beta\} + \{a, b\}].$$

The hypothesis $\{a, b\} = 0$ on Σ_2 implies $\{a, b\} = \lambda a + \mu b$, with $\lambda, \mu \in S^0$, so

$$e^{-\alpha-\beta}\{e^\alpha a, e^\beta b\} = [\lambda - H_b\alpha]a + [\mu + H_a\beta + aH_\alpha\beta]b. \tag{2.19}$$

If we want this to vanish on a conic neighborhood Γ of $(x_0, \xi_0) \in \Sigma_2$, first arrange that α satisfies

$$H_b\alpha = \lambda \text{ on } \Gamma \tag{2.20}$$

and then that β satisfies

$$(H_a + aH_\alpha)\beta = -\mu \text{ on } \Gamma.$$

Now relabel $e^\alpha a$, $e^\beta b$, calling them a and b, respectively. Again, by Darboux' theorem, upon performing a canonical transformation we can suppose $a = \xi_1, b = \xi_2$. Thus

$$H_{p_2} = \xi_1 \frac{\partial}{\partial x_2} + \xi_2 \frac{\partial}{\partial x_1},$$

and (2.13) becomes

$$\frac{1}{i}\left(\xi_1 \frac{\partial}{\partial x_2} + \xi_2 \frac{\partial}{\partial x_1}\right)\alpha_0 + \xi_1\xi_2 c_{-1} = p_1 - \tilde{p}_1. \tag{2.22}$$

Of course, to solve this, it is necessary to have $p_1 - \tilde{p}_0 = 0$ on $\{\xi_1 = \xi_2 = 0\}$. It suffices to find $\alpha_0 \in S^0$ such that

$$\frac{1}{i}\left(\xi_1 \frac{\partial}{\partial x_2} + \xi_2 \frac{\partial}{\partial x_1}\right)\alpha_0 = p_1 - \tilde{p}_1 \text{ on } \{\xi_1 = 0\} \cup \{\xi_2 = 0\},$$

i.e.

$$\xi_2 \frac{\partial}{\partial x_1} \alpha_0 = i(p_1 - \tilde{p}_1) \text{ on } \xi_1 = 0 \qquad \text{and}$$

$$\xi_1 \frac{\partial}{\partial x_2} \alpha_0 = i(p_1 - \tilde{p}_1) \text{ on } \xi_2 = 0$$

or

$$\begin{aligned} &\frac{\partial}{\partial x_1} \alpha_0 = i\xi_2^{-1}(p_1 - \tilde{p}_1) \text{ on } \xi_1 = 0 \qquad \text{and} \\ &\frac{\partial}{\partial x_2} \alpha_0 = i\xi_1^{-1}(p_1 - \tilde{p}_1) \text{ on } \xi_2 = 0. \end{aligned} \tag{2.23}$$

Now, this system is overdetermined. It can be solved if, and only if,

$$\frac{\partial}{\partial x_2}\frac{\partial}{\partial \xi_2}(p_1 - \tilde{p}_1) = \frac{\partial}{\partial x_1}\frac{\partial}{\partial \xi_1}(p_1 - \tilde{p}_1) \text{ on } \xi_1 = \xi_2 = 0. \tag{2.24}$$

(Of course, we have just gone through a special case of Frobenius' theorem.) This is a fairly stringent condition for microlocal equivalence mod OPS^{m-2}, but as we shall see in Section 3, it allows some flexibility.

We now consider the case where $p_m = a_1^2 + a_2^2$, with $a_1 = a_2 = 0$ on Σ, where ∇a_1, ∇a_2, and $\Sigma \xi_j \, dx_j$ are independent. We suppose Σ is symplectic, so $\{a_1, a_2\} \neq 0$ on Σ. We may as well suppose $m = 1$, so $a_1, a_2 \in S^{1/2}$. Then $\{a_1, a_2\}$ is of degree 0. We claim there exists an elliptic $b \in S^0$ such that

$$bp_2 = \tilde{a}_1^2 + \tilde{a}_2^2 \text{ with } \{\tilde{a}_1, \tilde{a}_2\} = 1 \tag{2.25}$$

near a point $(x_0, \xi_0) \in \Sigma$. Indeed, if $b = \{a_1, a_2\}^{-1/2}$ on Σ, we see that $\{ba_1, ba_2\} = 1$ on Σ. Relabel ba_1, and ba_2, calling them a_1 and a_2, respectively. We have arranged that $\{a_1, a_2\} = 1$ on Σ. More generally, suppose $\{a_1, a_2\} - 1$ vanishes to order $k \geq 1$ on Σ. Then, following an argument of

Duistermaat and Sjöstrand [1], we write

$$\{a_1, a_2\} = 1 + \sum_{j=0}^{k} c_j p^j \bar{p}^{k-j}$$

where $p = (1/\sqrt{2})(a_2 + ia_1)$, $c_j \in S^{-k/2}$, $c_j = \bar{c}_{k-j}$. Putting

$$b = 1 - \frac{1}{k+2} \sum_{j=0}^{k} c_j p^j \bar{p}^{k-j},$$

we find that $\{ba_1, ba_2\} - 1$ vanishes to order $k+1$ on Σ. Continuing this argument and applying Borel's theorem, we obtain $b_0 \in S^0$ such that $\{b_0 a_1, b_0 a_2\} - 1$ vanishes to infinite order on Σ. Again, relabel these quantities, calling them a_1, a_2. Now define $\tilde{a}_2$ by

$$H_{a_1}\tilde{a}_2 = 1, \qquad \tilde{a}_2 = 0 \text{ on } \{a_2 = 0\}.$$

It follows that $a_2 - \tilde{a}_2$ vanishes to infinite order on Σ. Thus

$$a_1^2 + a_2^2 = \left(\frac{a_1^2 + a_2^2}{a_1^2 + \tilde{a}_2^2}\right)(a_1^2 + \tilde{a}_2^2) \quad \text{and the factor} \quad \frac{a_1^2 + a_2^2}{a_1^2 + \tilde{a}_2^2} \quad \text{is smooth}$$

near Σ. This establishes our claim (2.25).

By Darboux's theorem, given any pair $b_1, b_2 \in S^{1/2}$ with $\{b_1, b_2\} = 1$ near $\{b_1 = b_2 = 0\}$, assuming (2.25), there exists a homogeneous canonical transformation taking a_j to b_j. An associated Fourier integral operator J conjugates an operator with principal symbol $a_1^2 + a_2^2$ to one with principal symbol $b_1^2 + b_2^2$. Consequently, given any two operators $P \in OPS^m$, $Q \in OPS^\mu$, with principal symbols $p_m \geq 0$, $q_\mu \geq 0$, vanishing to precisely second order on *codimension two* varieties Σ and S, which are symplectic, there exists an elliptic *FIOP* J and an elliptic $B \in OPS^{m-\mu}$ such that, near some $(x_0, \xi_0) \in S$,

$$JPJ^{-1} = BQ \bmod OPS^{m-1}.$$

A model example is $m = 2$, $p_2 = \xi_1^2 + x_1^2\xi_2^2$, near $x_1 = \xi_1 = 0$, $\xi_2 \neq 0$. Given P and $\tilde{P}$ with this principal symbol, we claim that P and $\tilde{P}$ are microlocally equivalent mod OPS^0, provided their subprincipal symbols agree on $\{x_1 = \xi_1 = 0\}$. Indeed, in this case

$$\tfrac{1}{2}H_{p_2} = x_1\xi_2^2 \frac{\partial}{\partial \xi_1} - x_1^2\xi_2 \frac{\partial}{\partial x_2} - \xi_1 \frac{\partial}{\partial x_1},$$

and (2.13) becomes

$$\text{(2.26)} \quad \frac{1}{i}\left(x_1\xi_2^2 \frac{\partial}{\partial \xi_1} - x_1^2\xi_2 \frac{\partial}{\partial x_2} - \xi_1 \frac{\partial}{\partial x_1}\right)\alpha_0 + (x_1^2\xi_2^2 + \xi_1^2)c_{-1} = p_1 - \tilde{p}_1 = r_1.$$

We work on the hyperplane $\xi_2 = 1$, and later extend α_0 and c_{-1} to have the correct degree of homogeneity, in a conic neighborhood of $\xi_2 = 1$, $\xi_1 = \xi_3 = \cdots = \xi_n = 0$.

To solve (2.26), we first work on the ring of formal power series in x_1 and ξ_1 with coefficients in the ring of smooth functions of $x_2, x_3, \xi_3, \ldots, x_n, \xi_n$. Let E_μ denote the submodule spanned by $x_1^k \xi_1^\ell$, $k+\ell = \mu$. Write

$$X_0 = x_1 \frac{\partial}{\partial \xi_1} - \xi_1 \frac{\partial}{\partial x_1} \qquad \text{and} \qquad X_2 = -x_1^2 \frac{\partial}{\partial x_2}.$$

We see that (on $\xi_2 = 1$), $H_{p_2} = X_0 + X_2$ and $X_0 : E_\mu \to E_\mu$, $X_2 : E_\mu \to E_{\mu+2}$. Also, if $Mu = (x_1^2 + \xi_1^2)u$, then $M : E_{\mu-2} \to E_\mu$. On $\xi_2 = 1$, (2.26) becomes $(X_0 + X_2)\alpha_0 + Mc_{-1} = r_1$. Thus, letting R_μ be the components of r_1 in E_μ, and noting that $r_1 = 0$ on $\{x_1 = \xi_1 = 0\}$, we want to solve, for $A_\mu \in E_\mu$, $C_\mu \in E_\mu$,

$$\begin{aligned} X_0 A_1 &= R_1 \\ X_0 A_2 + MC_0 &= R_2 \\ X_0 A_3 + MC_1 &= R_3 - X_2 A_1 \\ &\vdots \\ X_0 A_\mu + MC_{\mu-2} &= R_\mu - X_2 A_{\mu-2}. \end{aligned} \tag{2.27}$$

The first equation of (2.27) is clearly solvable, for given $R_1 = \rho_1 x_1 + \sigma_1 \xi_1$, set $A_1 = \rho_1 \xi_1 - \sigma_1 x_1$. To solve the rest of the equations, it suffices to consider the vector space situation (where the variables $x_2, \ldots, \xi_n$ are omitted); $T_\mu = X_0 \oplus M : E_\mu \oplus E_{\mu-2} \to E_\mu$, with $E_\mu \approx \mathbf{C}^{\mu+1}$. We look for a right inverse of T_μ, which merely involves checking that T_μ is surjective. Note that, if we use polar coordinates on $x_1 - \xi_1$ space, so $x_1 = r\cos\theta$, $\xi_1 = r\sin\theta$, then $X_0 = r(\partial/\partial\theta)$, so $\ker X_0$ is spanned by $r^{2j}, j \geq 0$; of course, $r^2 = x_1^2 + \xi_1^2$. In particular, if μ is odd, $X_0 : E_\mu \to E_\mu$ is injective, hence surjective. If μ is even, $\ker X_0$ in E_μ is one dimensional, being spanned by $(x_1^2 + \xi_1^2)^{\mu/2}$, so the range of X_0 is E_μ has codimension 1. If we can find an element of E_μ which is in the range of M but not in the range of X_0, it will follow that $X_0 \oplus M$ maps onto E_μ. Since elements of the range of X_0 have mean value 0 with respect to rotations, it is clear that $(x_1^2 + \xi_1^2)x_1^{\mu-2}$ will do (for μ even). This establishes the surjectivity of T_μ.

It follows from the above that (2.26) can be solved in such a ring of formal power series. By Borel's theorem, we can pick smooth $\alpha_0 \in S^0$, $\tilde{c}_{-1} \in S^{-1}$, homogeneous, such that, on $\xi_2 = 1$,

$$\sum_{k+\ell=\mu} \frac{x_1^j \xi_1^\ell}{k!\ell!} D_{x_1}^k D_{\xi_1}^\ell \alpha_0 = A_\mu, \qquad \sum_{k+\ell=\mu} \frac{x_1^k \xi_1^\ell}{k!\ell!} D_{x_1}^k D_{\xi_1}^\ell \tilde{c}_{-1} = C_\mu.$$

Thus, with $\tilde{c}_{-1}$ in place of c_{-1}, (2.26) is satisfied to infinite order at $x_1 = \xi_1 = 0$. Call the remainder term $\tilde{r}_{-1}$. Then let $c_{-1} = \tilde{c}_{-1} - (x_1^2\xi_2^2 + \xi_1^2)^{-1}\tilde{r}_1$, which is smooth, and (2.26) is satisfied exactly. We now summarize what we have just proved.

PROPOSITION 2.1. *Let $P, \tilde{P} \in OPS^m$ both have the same principal symbol $p_m \geq 0$, vanishing to second order on a variety Σ_2, assumed to have codimension 2 and be symplectic. Then P and $\tilde{P}$ are microlocally equivalent* mod OPS^{m-2}, *if, and only if,* sub $\sigma(P) =$ sub $\sigma(\tilde{P})$ *on Σ_2.*

As a simple corollary of our discussion of such operators as in Proposition 2.1, we mention the following, whose proof we leave as an exercise.

COROLLARY 2.2. *Let P satisfy the hypotheses of Proposition 2.1, and assume also that P operates on $\mathscr{D}'(\Omega)$ where Ω is a 3-dimensional region. Then for any $(x_0, \xi_0) \in \Sigma_2$ there is a conic neighborhood Γ and an elliptic operator Fourier integral operator J defined on Γ, and an elliptic operator $B \in OPS^{2-m}$, such that $BJPJ^{-1}$ is microlocally equal,* mod OPS^0, *to an operator of the form*

$$\mathscr{L}_0 + i\alpha(x, y, t)T \tag{2.28}$$

where $\mathscr{L}_0 = -(1/2)(Z_1\bar{Z}_1 + \bar{Z}_1 Z_1)$ as in (1.66).

We remark that, with a little more work, one can eliminate the B, and get P microlocally *conjugate* to (2.28), mod OPS^0, provided $m = 2$. In fact, in Proposition 2.1 one can strengthen "microlocally equivalent" to "microlocally conjugate."

Exercises

In 2.1 through 2.4, assume $p_2 = a_1^2 + a_2^2$ where $a_j \in S^1$, $\{a_1, a_2\} = 0$ on $\Sigma_2 = \{a_1 = a_2 = 0\}$. Suppose ∇a_1, ∇a_2, and $\Sigma \xi_j \, dx_j$ are linearly independent.

2.1. Show that there is a positive $c \in S^0$ such that $cp_2 = b_1^2 + b_2^2$ with $b_j \in S^1$, $\{b_1, b_2\} = 0$ in a conic neighborhood of Σ_2. Hint: Look at Duistermaat and Hörmander [1], Lemma 7.2.3.

2.2. Deduce that there exists an elliptic $FIOP$ J and an elliptic $B \in OPS^0$ such that, microlocally,

$$BJPJ^{-1} = Q = q(x, D)$$

with $q_2 = \xi_1^2 + \xi_2^2$.

2.3. Suppose Q, $\tilde{Q}$ both have principal symbol $\xi_1^2 + \xi_2^2$. If we try to make Q and $\tilde{Q}$ microlocally equivalent mod OPS^0 by the methods of this section, producing α_0 and c_{-1} such that

$$H_{q_2}\alpha_0 + q_2 c_{-1} = q_1 - \tilde{q}_1, \tag{2.29}$$

show that a necessary condition for doing this is

$$(2.30)\qquad \left(\frac{\partial}{\partial x_2}\frac{\partial}{\partial \xi_1} - \frac{\partial}{\partial x_1}\frac{\partial}{\partial \xi_2}\right)(q_1 - \tilde{q}_1) = 0 \text{ on } \xi_1 = \xi_2 = 0,$$

and, of course, $q_1 - \tilde{q}_1 = 0$ on $\xi_1 = \xi_2 = 0$.

2.4. Conversely, if (2.30) is satisfied, show that Q and $\tilde{Q}$ are microlocally equivalent, mod OPS^0. Hint: First find β_0 with $H_{q_2}\beta_0 = q_1 - \tilde{q}_1$ on $\{\xi_1 = 0\} \cup \{\xi_2 = 0\}$, i.e., $H_{q_2}\beta_0 = q_1 - \tilde{q}_1 + \xi_1\xi_2\rho_{-1}$. With $\alpha_0 = \beta_0 + \gamma_0$, you need $H_{q_2}\gamma_0 + q_2 c_{-1} = -\xi_1\xi_2\rho_{-1}$, or, setting $\gamma_0 = \xi_1\sigma + \xi_2\delta$, you need

$$(\xi_1^2 + \xi_2^2)\left(\frac{\partial\sigma}{\partial x_1} + c_{-1}\right) + \xi_2^2\left(\frac{\partial\delta}{\partial x_2} - \frac{\partial\sigma}{\partial x_1}\right) = \xi_1\xi_2\left(-\rho_{-1} - \frac{\partial\delta}{\partial x_1} - \frac{\partial\sigma}{\partial x_2}\right).$$

Pick σ and δ to solve the Cauchy-Riemann system

$$\frac{\partial\delta}{\partial x_2} - \frac{\partial\sigma}{\partial x_1} = 0,$$

$$\frac{\partial\delta}{\partial x_1} + \frac{\partial\sigma}{\partial x_2} = -\rho_{-1}$$

or

$$\left(\frac{\partial}{\partial x_1} - i\frac{\partial}{\partial x_2}\right)(\delta + i\sigma) = -\rho_{-1}.$$

Let $c_{-1} = -(\partial\sigma/\partial x_1)$.

§3. Characteristics with Involutive Self-intersection

We devote this section to a study of operators $P \in OPS^m$ with $p = ab$, where a and b vanish simply on two varieties, Σ_1 and Σ_2, which intersect transversally, and we assume $\Sigma_1 \cap \Sigma_2$ is involutive, i.e., $\{a, b\} = 0$ on $\Sigma_1 \cap \Sigma_2$. We also assume ∇a, ∇b and $\Sigma \xi_j\, dx_j$ are linearly independent. As shown in the previous section, P can be microlocally conjugated by an elliptic Fourier integral operator and multiplied by an elliptic pseudodifferential operator, of order $2-m$, to yield $D_{x_1}D_{x_2} + A(x, D_x)$. We are working near $\{\xi_1 = \xi_2 = 0\}$, say in a small conic neighborhood of a point where $\xi_1 = \xi_2 = \cdots = \xi_{n-1} = 0$, $\xi_n = 1$. On such a neighborhood, the results of Section 2 show that $D_{x_1}D_{x_2} + A(x, D_x)$ is microlocally equivalent to an operator with a simpler form.

Indeed, we can replace $A_1(x, \xi)$ by any $\tilde{A}_1$ such that $A_1 = \tilde{A}_1$ on $\{\xi_1 = \xi_2 = 0\}$ and such that (2.24) holds, i.e.,

$$(3.1)\qquad \frac{\partial}{\partial x_2}\frac{\partial}{\partial \xi_2}(A_1 - \tilde{A}_1) = \frac{\partial}{\partial x_1}\frac{\partial}{\partial \xi_1}(A_1 - \tilde{A}_1) \text{ on } \xi_1 = \xi_2 = 0.$$

We claim we can take $\tilde{A}_1 = \tilde{A}_1(x, \xi')$ independent of ξ_1. Indeed, in such a case, we require (with $\xi = (\xi_1, \xi'), \xi' = (\xi_2, \xi''))$

$$\tilde{A}_1(x, 0, \xi'') = A_1(x, 0, 0, \xi'') \tag{3.2}$$

and

$$-\frac{\partial}{\partial x_2}\frac{\partial}{\partial \xi_2}\tilde{A}_1(x, 0, \xi'') = -\frac{\partial}{\partial x_2}\frac{\partial}{\partial \xi_2}A_1(x, 0, 0, \xi'') + \frac{\partial}{\partial x_1}\frac{\partial}{\partial \xi_1}A_1(x, 0, 0, \xi''). \tag{3.3}$$

Since we know the right side of (3.3), we can find a solution for $(\partial/\partial\xi_1)\tilde{A}_1(x, 0, \xi'')$ and then take any extension $\tilde{A}_1(x, \xi')$ of (3.2) such that $(\partial/\partial\xi_1)\tilde{A}_1$ is so given at $\xi_1 = 0$. We can repeat this reasoning for all lower order terms in $A(x, \xi)$.

Consequently, we are reduced to studying operators of the form

$$Pu = D_{x_1}D_{x_2}u + \tilde{A}(x, D_{x'})u \tag{3.4}$$

with $\tilde{A} \in OPS^1$. Our next step will be to change this to a first order system, and then we shall construct a parametrix for that system. In doing this, we follow Uhlmann [1]. Now, let

$$v_1 = u, \qquad v_2 = D_1 u. \tag{3.5}$$

Then the equation $Pu = f$ becomes

$$D_1 v_1 = v_2,$$
$$D_2 v_2 = -\tilde{A}v_1 + f$$

or

$$\begin{pmatrix} D_1 & 0 \\ -\tilde{A} & D_2 \end{pmatrix}\begin{pmatrix} v_1 \\ v_2 \end{pmatrix} = \begin{pmatrix} v_2 \\ f \end{pmatrix}. \tag{3.6}$$

In order to construct a parametrix for (3.6), we make the hypothesis that sub $\sigma(P) = 0$ on $\Sigma_1 \cap \Sigma_2$, which implies that $\tilde{A}_1 = 0$ on $\xi_2 = 0$. This is known as the "Levi condition" in the literature. Thus we are supposing that

$$-\tilde{A}_1(x, \xi') = \xi_2 B_0(x, \xi'). \tag{3.7}$$

We can rewrite (3.6) as

$$\begin{pmatrix} D_1 & 0 \\ B_0 D_2 & D_2 \end{pmatrix}\begin{pmatrix} v_1 \\ v_2 \end{pmatrix} = \begin{pmatrix} 0 & 1 \\ C & 0 \end{pmatrix}\begin{pmatrix} v_1 \\ v_2 \end{pmatrix} + \begin{pmatrix} 0 \\ f \end{pmatrix} \tag{3.8}$$

where $C = C(x, D_{x'}) \in OPS^0$.

LEMMA 3.1. *There exist homogeneous* $M_{ij} \in S^0$ *such that, with* $M = (M_{ij})$

$$\begin{pmatrix} \xi_1 & 0 \\ B_0\xi_2 & \xi_2 \end{pmatrix} M = \begin{pmatrix} \xi_1 & 0 \\ 0 & \xi_2 \end{pmatrix}. \tag{3.9}$$

Proof. Take

$$M = \begin{pmatrix} 1 & 0 \\ B_0 & 1 \end{pmatrix}^{-1}.$$

Thus $M = M(x, \xi')$.

If we make a change of dependent variables, setting

$$v = M(x, D_{x'})w$$

(2.8) becomes

$$\begin{pmatrix} D_1 & 0 \\ 0 & D_2 \end{pmatrix} w = B(x, D_{x'})w + g \tag{3.10}$$

where $B(x, D_{x'}) \in OPS^0$.

We now make a change of variables, setting $t = x_1$, $y_1 = x_2 - t$, $y_j = x_{j+1}$, $2 \le j \le n-1$. Then (3.10) becomes

$$\begin{pmatrix} D_t & 0 \\ 0 & D_t + D_{y_1} \end{pmatrix} w = \tilde{\tilde{B}}w + g$$

where $\tilde{\tilde{B}} = \tilde{\tilde{B}}(t, y, D_{t,y}) \in OPS^0$ in a conic neighborhood of the set Σ_2 of double characteristics $\{\tau = \tau + \eta_1 = 0\}$. Our next goal is to replace $\tilde{\tilde{B}}$ by an operator only involving D_y.

LEMMA 3.2. *There exists an elliptic* $C \in OPS^0$ *such that*

$$\begin{pmatrix} D_t & 0 \\ 0 & D_t + D_{y_1} \end{pmatrix} - \tilde{B} = \left[\begin{pmatrix} D_t & 0 \\ 0 & D_t + D_{y_1} \end{pmatrix} - \tilde{\tilde{B}}\right] C \tag{3.11}$$

microlocally in a conic neighborhood of a point

$$\zeta_0 = ((t_0, y_0), (\tau_0, \eta_0)) \in \Sigma_2, \qquad \text{with } \tilde{B} = \tilde{B}(t, y, D_y) \in OPS^0.$$

Proof. Set

$$C \sim \sum_{j \ge 0} C_j, \qquad \tilde{B} \sim \sum_{j \ge 0} \tilde{B}_j,$$

C_j and $\tilde{B}_j$ of order $-j$. We determine these terms recursively. Take $C_0 = 1$, so C is elliptic. Suppose C_j and $\tilde{B}_{j-1}$ have been chosen for $j \le k$, so that the difference between the left side of (3.11) and the right, with $\tilde{B}$ replaced by $\tilde{B}_0 + \cdots + \tilde{B}_{k-1}$ and with C replaced by $C_0 + \cdots + C_k$, belongs to OPS^{-k}, microlocally near ζ_0. Denote by g_k the principal symbol of this difference. We want to find $\tilde{B}_k$ and C_{k+1} such that the new difference belongs to OPS^{-k-1}, microlocally near ζ_0. Break up g_k as

$$g_k(t, y, \tau, \eta) = \begin{pmatrix} g_k^{11}(t, y, 0, \eta) & g_k^{12}(t, y, 0, \eta) \\ g_k^{21}(t, y, \eta_1, \eta) & g_k^{22}(t, y, \eta_1, \eta) \end{pmatrix}$$

$$+ \begin{pmatrix} \tau \int_0^1 \frac{\partial}{\partial \tau} g_k^{11}(t, y, s\tau, \eta)\, ds & \tau \int_0^1 \frac{\partial}{\partial \tau} g_k^{12}(t, y, s\tau, \eta)\, ds \\ (\tau - \eta_1) \int_0^1 \frac{\partial}{\partial \tau} & (\tau - \eta_1) \int_0^1 \frac{\partial}{\partial \tau} \\ \times\, g_k^{21}(t, y, s\tau + (1-s)\eta_1, \eta)\, ds & \times\, g^{22}(t, y, s\tau + (1-s)\eta_1, \eta)\, ds \end{pmatrix}$$

Now choose for the symbol c_{k+1}^{ij} of C_{k+1}:

$$c_{k+1}^{1j}(t, y, \tau, \eta) = \int_0^1 \frac{\partial}{\partial \tau} g_k^{1j}(t, y, s\tau, \eta)\, ds \quad (j = 1, 2) \qquad \text{and}$$

$$c_{k+1}^{2j}(t, y, \tau, \eta) = \int_0^1 \frac{\partial}{\partial \tau} g_k^{2j}(t, s\tau + (1-s)\eta_1, \eta)\, ds \quad (j = 1, 2)$$

on a conic neighborhood of ζ_0; conveniently extend c_{k+1}^{ij}, off this neighborhood, to belong to S^{-k-1}. Next, choose

$$\tilde{B}_k^{ij}(t, y, \eta) = g_k^{1j}(t, y, 0, \eta) \qquad \text{and}$$
$$\tilde{B}_k^{2j}(t, y, \eta) = g_k^{2j}(t, y, \eta_1, \eta).$$

It follows that the difference between the two sides of (3.12), with $\tilde{B}$ replaced by $B_0 + \cdots + \tilde{B}_k$ and with C replaced by $C_0 + \cdots + C_{k+1}$, belongs to OPS^{-k-1} microlocally in a conic neighborhood of ζ_0. An inductive procedure completes the proof.

Relabeling variables, we are left with studying

$$\frac{\partial}{\partial t} w = \begin{pmatrix} 0 & 0 \\ 0 & -\frac{\partial}{\partial y_1} \end{pmatrix} w + \tilde{B}(t, y, D_y) w + g, \tag{3.12}$$

$\tilde{B}(t, y, D_y) \in OPS^0$. We construct a parametrix for (3.13), with initial data given:

$$w(0, x) = h(x). \tag{3.13}$$

We may as well take $g = 0$, since the general case can be obtained using Duhamel's principle.

The approximate solutions to (3.12), (3.13), with $g = 0$, will be of the form $E_1 h + E_2 h + E_3 h$, where E_1 and E_2 are Fourier integral operators (in fact, E_1 is pseudodifferential) and E_3 is a slightly more complicated operator. Our approach to this follows Uhlmann [1]; previous work was done by Granoff and Ludwig [1]. The operators E_j will have the following form.

$$E_1 h(t, y) = \int e^{iy\cdot\xi} e_1(t, y, \xi)\hat{h}(\xi)\, d\xi, \tag{3.14}$$

$$E_2 h(t, y) = \int e^{i(y_1 - t)\xi_1 + iy'\cdot\xi'} e_2(t, y, \xi)\hat{h}(\xi)\, d\xi, \quad \text{and} \tag{3.15}$$

$$E_3 h(t, y) = \int_{-t}^{t}\int e^{i\left(y_1 - \frac{s+t}{2}\right)\xi_1 + iy'\cdot\xi'} e_3(s, t, y, \xi)\hat{h}(\xi)\, d\xi\, ds. \tag{3.16}$$

Here we have set $y = (y_1, y')$, $\xi = (\xi_1, \xi')$. We shall pick

$$e_1, e_2 \in S^0, \quad e_3 \in S^1 \text{ such that } \frac{\partial}{\partial t} + \begin{pmatrix} 0 & 0 \\ 0 & \dfrac{\partial}{\partial y_1} \end{pmatrix} - \tilde{B} \text{ annihilates}$$

$(E_1 + E_2 + E_3)h$ and such that $E_1 + E_2 + E_3 = I$ at $t = 0$. The transport equations for the terms in the asymptotic expansion of e_3 will turn out to be symmetric hyperbolic equations, for the 2×2 matrix valued unknowns. Denote by φ_i the phase functions in (3.14)–(3.16), respectively, e.g., $\varphi_3(s, t, y, \xi) = (y_1 - (s + t)/2)\xi_1 + y'\cdot\xi'$.

A short calculation yields

$$\begin{aligned}
&\left(\frac{\partial}{\partial t} + \begin{pmatrix} 0 & 0 \\ 0 & \dfrac{\partial}{\partial y_1} \end{pmatrix} - \tilde{B}\right)(E_1 + E_2 + E_3)h \\
&= \int e^{iy\cdot\xi}\left[\frac{\partial e_1}{\partial t} - e^{-iy\cdot\xi}\tilde{B}(e_1 e^{iy\cdot\xi})\right]\hat{h}\, d\xi \\
&\quad + \int e^{i\varphi_2}\left[\left(\frac{\partial}{\partial t} + \frac{\partial}{\partial y_1}\right)e_2 - e^{-i\varphi_2}\tilde{B}(e_2 e^{i\varphi_2})\right]\hat{h}\, d\xi \\
&\quad + \int_{-t}^{t}\int e^{i\varphi_3}\left[\begin{pmatrix} \dfrac{\partial}{\partial t} - \dfrac{\partial}{\partial s} & 0 \\ 0 & \dfrac{\partial}{\partial t} + \dfrac{\partial}{\partial s} + \dfrac{\partial}{\partial y_1} \end{pmatrix} e_3 - e^{i\varphi_3}\tilde{B}(e_3 e^{i\varphi_3})\right]\hat{h}\, d\xi\, ds \\
&\quad + \int e^{iy\cdot\xi}\left[e^{-iy\cdot\xi}\begin{pmatrix} 0 & 0 \\ 0 & \dfrac{\partial}{\partial y_1} \end{pmatrix}(e^{iy\cdot\xi}e_1) - 2\begin{pmatrix} 0 & 0 \\ 0 & 1 \end{pmatrix} e_3(-t, t, y, \xi)\right]\hat{h}\, d\xi \\
&\quad + \int e^{i\varphi_2}\left[e^{-i\varphi_2}\begin{pmatrix} -\dfrac{\partial}{\partial y_1} & 0 \\ 0 & 0 \end{pmatrix}(e_2 e^{i\varphi_2}) - 2\begin{pmatrix} 1 & 0 \\ 0 & 0 \end{pmatrix} e_3(t, t, y, \xi)\right]\hat{h}\, d\xi.
\end{aligned}$$

In requiring each of these five terms to vanish formally, we shall obtain the transport equations and other conditions on the amplitudes e_j. Write

$$e_j \sim \sum_{v \le 0} e_{j,v}, \qquad j = 1, 2, \qquad e_3 \sim \sum_{v \le 1} e_{3,v},$$

where $e_{j,v}$ is homogeneous of degree v in ξ. Requiring the first two terms to vanish leads to the "usual" transport equations for e_1 and e_2:

$$\begin{aligned} &\frac{\partial}{\partial t} e_{1v} - \tilde{b}_0(t, y, \xi) e_{1v} = f_{1v}(t, y, \xi) \quad (v \le 0) \qquad \text{and} \\ &\left(\frac{\partial}{\partial t} + \frac{\partial}{\partial y_1}\right) e_{2v} - \tilde{b}_0 e_{2v} = f_{2v} \end{aligned} \tag{3.18}$$

where f_{jv} are determined by $e_{j0}, \ldots, e_{j,v+1}$; $f_{j0} = 0$; $\tilde{b}_0$ is the principal symbol of $\tilde{B}_0$. The initial conditions are

$$e_{10}(0, y, \xi) = \begin{pmatrix} 1 & 0 \\ 0 & 0 \end{pmatrix}, \qquad e_{20}(0, y, \xi) = \begin{pmatrix} 0 & 0 \\ 0 & 1 \end{pmatrix} \tag{3.19}$$

and $e_{jv}(0, y, \xi) = 0$, $v \le -1$. Note that $E_1 + E_2 = I$ at $t = 0$.

Making the third term have order $-\infty$ requires

$$\begin{pmatrix} \dfrac{\partial}{\partial t} - \dfrac{\partial}{\partial s} & 0 \\ 0 & \dfrac{\partial}{\partial t} + \dfrac{\partial}{\partial s} + \dfrac{\partial}{\partial y_1} \end{pmatrix} e_{3v} - \tilde{b}_0 e_{3v} = \ell_v(s, t, y, \xi) \qquad (v \le 1) \tag{3.20}$$

where $\ell_1 = 0$ and ℓ_v is determined by $\ell_1, \ldots, \ell_{v+1}$, for $v \le 0$. This is clearly a symmetric hyperbolic system for (each column of) e_{3v}. The boundary conditions for e_{3v} are of Goursat type and are obtained by requiring the fourth and fifth terms to formally vanish. Indeed, we obtain

$$\begin{pmatrix} 0 & 0 \\ e_{3v}^{21}(-t, t, y, \xi) & e_{3v}^{22}(-t, t, y, \xi) \end{pmatrix} = \tfrac{1}{2} w_v(t, y, \xi) \tag{3.21}$$

where

$$e^{iy\cdot\xi} \begin{pmatrix} 0 & 0 \\ 0 & \dfrac{\partial}{\partial y_1} \end{pmatrix} (e_1 e^{iy\cdot\xi}) \sim \sum_{v \le 1} w_v,$$

and

$$\begin{pmatrix} e_{3v}^{11}(t, t, y, \xi) & e_{3v}^{12}(t, t, y, \xi) \\ 0 & 0 \end{pmatrix} = \tfrac{1}{2} v_v(t, y, \xi) \tag{3.22}$$

where

$$e^{-i\varphi_2}\begin{pmatrix} -\dfrac{\partial}{\partial y_1} & 0 \\ 0 & 0 \end{pmatrix}(e_2 e^{i\varphi_2}) \sim \sum_{\nu \le 1} v_\nu.$$

The existence of a unique smooth solution to such a characteristic boundary value problem as (3.20) through (3.22) is classical; see Garabedian [1].

If we write

$$E = E(t) = \sum_{j=1}^{3} E_j(t),$$

we can read off $WF(E(t)h)$ in terms of $WF(h)$. The only problem is to describe $WF(E_3(t)h)$, and this is accomplished as a special case of the following result (with $\lambda(x, D) = (1/i)(\partial/\partial y_1)$).

LEMMA 3.2. *Let $\lambda(x, D) \in OPS^1$ have real principal symbol, and let $P(t) \in OPS^m$ be a smooth family of operators. Let*

$$A = \int_a^b P(t)e^{it\lambda(x,D)}\, dt. \tag{3.23}$$

Then

$$\begin{gathered} WF(Au) \subset C(a)WF(u) \cup C(b)WF(u), \\ \bigcup_{a \le s \le b} C(s)(\text{char. } \lambda \cap WF(u)). \end{gathered} \tag{3.24}$$

Proof. Write $u = u_1 + u_2$ where $WF(u_1)$ belongs to a small conic neighborhood of char $\lambda = \{\lambda_1(x, \xi) = 0\}$, and $WF(u_2) \cap \text{char } \lambda = \varnothing$. Then, with $Q(t) = e^{-it\lambda}P(t)e^{it\lambda}$, and λ^{-1} denoting a parametrix for λ on a conic neighborhood of $WF(u_2)$,

$$\begin{aligned} Au = \int_a^b P(t)e^{it\lambda}u_1\, dt + \sum_{j=1}^{N} e^{ib\lambda}\lambda^{-j}Q^{(j)}(b)u_2 - \sum_{j=1}^{N} e^{ia\lambda}\lambda^{-j}Q^{(j)}(a)u_2 \\ + \int_a^b e^{it\lambda}\lambda^{-N}Q^{(N)}(t)u_2\, dt \bmod C^\infty. \end{aligned} \tag{3.25}$$

Application of Egorov's theorem, as in the proof of Theorem 1.2 of Chapter VIII, shows that the wave front set of the right side of (3.25) is contained in an arbitrarily small neighborhood of the set described in (3.24), which completes the proof.

§4. Characteristics with Noninvolutive Self-intersection

We consider here operators with principal symbol $p_m = ab$ where $\{a = 0\}$ and $\{b = 0\}$ intersect transversally with $\Sigma = \{a=b=0\}$ non-involutive, i.e. $\{a, b\} \ne 0$ on Σ. Suppose ∇a, ∇b, and $\Sigma \xi_j\, dx_j$ independent on Σ. As shown in Section 2, there is no loss of generality in taking $m = 1$, $a \in S^0$, $b \in S^1$, and we can assume $\{a, b\} = 1$ on a conic neighborhood of a point $(x_0, \xi_0) \in \Sigma$. Thus, conjugating by an appropriate Fourier integral operator, we can suppose $P = x_1 D_{x_1} + A(x, D_x)$. Furthermore, we can simplify further. For example, P is microlocally equivalent mod OPS^{-1} to any $\tilde{P} = x_1 D_{x_1} + \tilde{A}$ where $\tilde{A}_0 = A_0(x, \xi)$ on Σ. In particular, we can take $\tilde{A}_0 = \tilde{A}_0(x', \xi') = A_0(0, x', 0, \xi')$. Continuing this argument, we can make P microlocally equivalent to

$$x_1 D_{x_1} + \tilde{A}(x', D_{x'}). \tag{4.1}$$

Let us relabel coordinates, setting $y = x_1$ and denoting x' by x, so our work reduces to the study of

$$P = y\frac{\partial}{\partial y} + A(x, D_x) \tag{4.2}$$

with $A \in OPS^0$. Results on parametrices and propagation of singularities for (4.2) were obtained by Ivrii [1], Melrose (unpublished), and Hanges [1]. We follow Hanges, in spirit, if not in detail. We differ from these authors, in that we construct parametrices as pseudodifferential-operator-valued distributions, rather than as Fourier integral distributions with singular amplitudes.

To treat (4.2), we put x_0 in a compact manifold M and extend $A(x, \xi)$, so in particular A is a bounded operator on $H^s(M)$. We consider (4.2) as an operator equation, and define an operator valued function, holomorphic in $w, z \in \mathbf{C}\backslash i\mathbf{R}^-$, by

$$\begin{aligned} w\frac{\partial}{\partial w} a(z, w) &= Aa, \\ a(w, w) &= \frac{1}{w} I. \end{aligned} \tag{4.3}$$

The solution to (4.3) is

$$a(z, w) = z^{-1-A} w^A. \tag{4.4}$$

Here, for $w \in \mathbf{C}\backslash i\mathbf{R}^-$, $w^A = e^{A \log w}$, and $e^{\zeta A}$ is defined by

$$e^{\zeta A} = \sum_{j=0}^{\infty} \frac{\zeta^j}{j!} A^j = (2\pi i)^{-1} \int_\gamma e^{\zeta\lambda}(\lambda - A)^{-1}\, d\lambda$$

where γ is any simple curve encircling the spectrum of A. It is easy to see that a is in fact a holomorphic function of z, w taking values in the Frechet space $OPS^0_{1,0}$. Now let

$$F(y, y') = H(y' - y)(y' + i0)^{-1-A}(y + i0)^A. \tag{4.5}$$

where $H(y)$ is the Heaviside function, $H(y) = 1$ for $y \geq 0$, 0 for $y < 0$. F is well defined as an operator valued distribution, and we have

$$\left(y\frac{\partial}{\partial y} - A\right)F = \delta(y-y')I. \tag{4.6}$$

We note that, considering $F(y, y')$ as an operator valued distribution, with values in $OPS^0_{1,0}$, we have

$$\begin{aligned} WF(F) \subset &\{(y, y', \eta, \eta'): y = 0, \eta > 0, \eta' = 0, y' \leq 0\} \\ &\cup \{y' = 0, \eta' > 0, \eta = 0, y \geq 0\} \\ &\cup \{y = y' = 0, \eta' \geq -\eta\} \\ &\cup \{y = y', \eta = -\eta'\} \\ &= W_1 \cup W_2 \cup W_3 \cup W_4. \end{aligned} \tag{4.7}$$

Now we claim that if F is considered as a distribution in y, y', x, x', its wave front set is contained in the set which describes the wave front set of a distribution of the form $w(y, y') \otimes \delta(x-x')$, where $WF(w) \subset W_1 \cup W_2 \cup W_3 \cup W_4$. To be precise, we have the following.

LEMMA 4.1. *Let $w(z)$ be a distribution taking values in $OPS^m_{1,0}$ and define $\tilde{w}(z, x, x')$ by $\int \varphi(z)w(z)f(x)\,dz = \int \varphi(z)\tilde{w}(z, x, x')f(x')\,dx'\,dz$. Then*

$$\begin{aligned} WF(\tilde{w}) \subset &\{(z, x, x', \zeta, \xi, \xi'):(z, \zeta) \in WF(w) \text{ and } x = x', \xi = -\xi'\} \\ &\cup \{(z, \zeta) \in WF(w) \text{ and } \xi = \xi' = 0\} \\ &\cup \{x = x', \xi = -\xi' \text{ and } \zeta = 0 \text{ and } z \in \operatorname{supp} w\}. \end{aligned}$$

This lemma is a special case of the following more general assertion, whose proof we leave as an exercise. Pick a real number σ and a closed convex cone $\Gamma \subset T^*(X)\backslash 0$; where X is a compact manifold, for example $X = M \times M$. Let $H^\sigma_\Gamma = \{u \in H^\sigma(X): WF(u) \subset \Gamma\}$; H^σ_Γ has a natural Fréchet space topology. (In Lemma 4.1, Γ is the normal bundle to the diagonal in $M \times M$.)

LEMMA 4.2. *Let $w(z)$ be a distribution taking values in H^σ_Γ, and define $\tilde{w} \in \mathcal{D}'(Z \times X)$ by $\int \varphi(z)\langle w(z), \psi\rangle\,dz = \int \varphi(z)w(z, x)\psi(x)\,dz\,dx$. Then*

$$\begin{aligned} WF(\tilde{w}) \subset &\{(z, x, \zeta, \xi) \in T^*(Z \times X)\backslash 0:(z, \zeta) \in WF(w) \text{ and } (x, \xi) \in \Gamma\} \\ &\cup \{(z, \zeta) \in WF(w) \text{ and } \xi = 0\} \\ &\cup \{(x, \xi) \in \Gamma \text{ and } \zeta = 0, z \in \operatorname{supp} w\}. \end{aligned}$$

Note that, although F is a right fundamental solution of $y(\partial/\partial y) - A$, it is not defined on all compactly supported distributions in (y, x). It is naturally defined on all distributions u which vanish for y sufficiently large and positive, and such that

$$WF(u) \cap \{(y, x, \eta, \xi): y = 0, \xi = 0, \eta > 0\} = \emptyset. \tag{4.8}$$

Note that the negative adjoint $-F^*$ is a left fundamental solution for $y(\partial/\partial y) - B$, where $B = -A^* - I$:

$$-F^*\left(y\frac{\partial}{\partial y} - B\right) = \delta(y-y')I. \tag{4.9}$$

This identity is valid as long as both sides are applied to a distribution v vanishing for y large and negative, and satisfying

$$WF(v) \cap \{y = 0, \xi = 0, \eta < 0\} = \varnothing. \tag{4.10}$$

In particular, (4.9) does not contradict the example

$$\left(y\frac{\partial}{\partial y} - 1\right)\delta(y) = 0.$$

Our interest, recall, is focused on a small conic neighborhood of a point $(0, x_0, 0, \xi_0)$, $\xi_0 \neq 0$.

From the parametrix F^* of (4.9) we can obtain the following information on propagation of singularities.

PROPOSITION 4.3. *Suppose* $u \in \mathscr{D}'(\mathbf{R}^n)$ *with*

$$(0, x_0, 0, \xi_0) \notin WF\left(\left(y\frac{\partial}{\partial y} - B\right)u\right),$$

and with

$$\mathscr{L}_{ij}(x_0, \xi_0) = \{(y, x_0, \eta, \xi_0) \in T^*(\mathbf{R}^n)\backslash 0 : y\eta = 0 \text{ and either } (-1)^i y > 0 \text{ or } (-1)^j \eta > 0\} \tag{4.11}$$

Suppose

$$\mathscr{L}_{11}(x_0, \xi_0) \cap WF(u) = \varnothing. \tag{4.12}$$

Then

$$(0, x_0, 0, \xi_0) \notin WF(u). \tag{4.13}$$

Proof. Pick $\chi \in OPS^0$ compactly supported, with symbol 1 in a small conic neighborhood of $(0, x_0, 0, \xi_0)$ and vanishing outside a slightly larger conic neighborhood, such that χu has compact support. Then

$$(0, x_0, 0, \xi_0) \notin WF\left(\left[\chi, y\frac{\partial}{\partial y} - B\right]u\right) = WF\left(\left(y\frac{\partial}{\partial y} - B\right)\chi u\right). \tag{4.14}$$

Now since χu satisfies (4.10), we have

$$-F^*\left(y\frac{\partial}{\partial y} - B\right)\chi u = \chi u. \tag{4.15}$$

Applying (4.7) and Lemma 4.1 to F^*, we deduce (4.13) from (4.14) and (4.12).

We remark that, if A were replaced by an operator A' differing from A by an element of $OPS^{-\infty}$, the corresponding distribution $F'(y, y') = H(y'-y)(y'+i0)^{-1-A'}(y+i0)^{A'}$, would differ from $F(y, y')$ by a distribution with wave front set described by (4.7), taking values in $OPS^{-\infty}$ (acting on distribution in x). Therefore,

$$\left(y\frac{\partial}{\partial y} - A\right)F' = \delta(y-y')I + R$$

where R is not necessarily a smoothing operator, though $\xi = 0$ on $WF(Ru)$, and consequently such a "parametrix" would suffice to prove Proposition 4.3.

We can construct three other parametrices similar to F, replacing $H(y'-y)$ by $H(y-y')$ and/or replacing

$$(y'+i0)^{-1-A}(y+i0)^{A} \quad \text{by} \quad (y'-i0)^{-1-A}(y-i0)^{A}.$$

These give rise to three other regularity theorems, analogous to Proposition 4.3. We record them all.

THEOREM 4.4. *Suppose $u \in \mathscr{D}'(\mathbf{R}^n)$ with*

$$(0, x_0, 0, \xi_0) \notin WF\left(\left(y\frac{\partial}{\partial y} - B\right)u\right).$$

If there exists (i, j) such that $\mathscr{L}_{ij}(x_0, \xi_0) \cap WF(u) = \varnothing$, then

$$(0, x_0, 0, \xi_0) \notin WF(u).$$

We now look into some propagation of singularity theorems in which properties of the lower order term $A(x, D_x)$ play a role. We start off with a construction of a right parametrix for $y(\partial/\partial y) - A$ which requires $Re\, A_0(x, \xi) \le a_0 < 0$. It follows from Gårding's inequality that, after perhaps altering A by an element of $OPS^{-\infty}$,

$$Re(Au, u) \le -C\|u\|^2_{L^2}, \qquad \forall u \in L^2(M). \tag{4.16}$$

Consequently, if $f \in C(\mathbf{R}, L^2(M))$, we can define a solution to

$$y\frac{\partial}{\partial y}u - Au = f \tag{4.17}$$

by

$$\begin{aligned} u(y) &= \int_0^1 \rho^{-1-A} f(y\rho)\, d\rho \\ &= |y|^A \int_0^{|y|} \tau^{-1-A} f((\operatorname{sgn} y)\tau)\, d\tau \\ &= \int G(y, y') f(y')\, dy' \end{aligned} \tag{4.18}$$

where

(4.19) $$G(y, y') = \begin{cases} |y|^A |y'|^{-1-A} & \text{if } 0 \le y' \le y \text{ or } y \le y' \le 0 \\ 0 & \text{otherwise.} \end{cases}$$

Indeed, (4.18) gives the unique solution $u \in C(\mathbf{R}, L^2(M))$, since any other solution must differ from u, on $y > 0$, by something of the form $y^A v_0$, which blows up as $y \downarrow 0$ unless $v_0 = 0$ (see Exercise 4.1); similarly for $y < 0$. More generally, for any real s, with $(u, v)_{H^s} = (\Lambda^s u, \Lambda^s v)_{L^2}$ an inner product on $H^s(M)$, we can alter A by an element of $OPS^{-\infty}$ to get

$$Re(Au, u)_{H^s} \le C\|u\|_{H^s}^2, \qquad u \in H^s(M),$$

and then, if $f \in C(\mathbf{R}, H^s(M))$, the unique $u \in C(\mathbf{R}, H^s(M))$ solving (4.17) is given by (4.18).

We now describe the wave front set of the $OPS^0_{1,0}$ valued distribution $G(y, y')$.

(4.20) $$WF(G) \subset \{(y, y', \eta, \eta') : y' = \eta = 0\} \cup \{y = y', \eta = \eta'\} \cup \{y = y' = 0 \text{ and } \exists \rho \in [0, 1] \text{ such that } -\eta = \rho\eta'\}.$$

Indeed, the description of $WF(G)$ is clear except for the points lying over $y = y' = 0$. To get the complete description we need to know that

(4.21) if Γ is a closed cone in $\mathbf{R}^2$ on which $|\eta + \rho\eta'| \ge C(|\eta| + |\eta'|)$, $\rho \in [0, 1]$, then $(0, 0, \eta, \eta') \notin WF(G)$, for $(\eta, \eta') \in \Gamma$.

Indeed, let $\chi \in C_0^\infty(\mathbf{R}^2)$ be equal to 1 near $(y, y') = (0, 0)$, and use

$$\widehat{\chi G}(\eta, \eta') = \langle G, \chi(y, y') e^{i(y\eta + y'\eta')} \rangle$$
$$= \iint_0^1 \rho^{-1-A} \chi(y, y\rho) e^{iy(\eta + \rho\eta')} \, d\rho \, dy.$$

Thus, if Π_j is a seminorm on $OPS^0_{1,0}$, for $(\eta, \eta') \in \Gamma$,

$$\Pi_j(\widehat{\chi G}(\eta, \eta')) \le C_N(|\eta| + |\eta'|)^{-N} \int_0^1 \Pi_j(\rho^{-1-A}) \, d\rho \int |D_y^N \chi(y, \rho y)| \, dy$$
$$\le C_N'(|\eta| + |\eta'|)^{-N},$$

which establishes (4.21) and hence (4.20). From (4.20) and Lemma 4.1 one can read off $WF(\tilde{G}(y, x, y', x'))$ where $\tilde{G}$ is the kernel determined by G.

We can now get a propagation of singularities result for $(y(\partial/\partial y) - A)u$ when $Re\, A_0(x, \xi) \le -a_0 < 0$.

PROPOSITION 4.5. *Let $(y(\partial/\partial y) - A)u = f$ and suppose $Re\, A_0(x, \xi) \le -a_0 < 0$. Suppose $(0, x_0, 0, \xi_0) \notin WF(f)$, and also suppose*

(4.22) $$\{(y, x_0, 0, \xi_0) : y \ne 0\} \cap WF(u) = \emptyset.$$

Then

$$(0, x_0, 0, \xi_0) \notin WF(u). \tag{4.23}$$

Proof. Take a compactly supported $\chi \in OPS^0$ whose symbol equals 1 near $(0, x_0, 0, \xi_0)$ and vanishes outside a small conic neighborhood of this point, such that χu has compact support. Note that (4.22) holds with u replaced by χu. Also $\chi u \in C(\mathbf{R}, H^s(M))$ for some s, since $\xi \neq 0$ on $WF(\chi u)$. Also, since $(0, x_0, 0, \xi_0) \notin WF([y(\partial/\partial y) - A, \chi]u)$, we have

$$(0, x_0, 0, \xi_0) \notin WF\left(\left(y\frac{\partial}{\partial y} - A\right)(\chi u)\right) = WF(g), \tag{4.24}$$

where we have set $g = (y(\partial/\partial y) - A)(\chi u)$. From (4.22) and (4.24) we see that

$$\{(y, x_0, 0, \xi_0): y \in \mathbf{R}\} \cap WF(g) = \varnothing. \tag{4.25}$$

Now $\chi u = Gg$, so we can apply (4.20) and Lemma 4.1 to (4.25), which yields (4.23), proving the proposition.

From the equation

$$\left(y\frac{\partial}{\partial y} - A\right)G = \delta(y-y')I, \tag{4.26}$$

valid when both sides are applied to a distribution $u(y, x)$ with $\xi \neq 0$ on $WF(u)$, we deduce that

$$-G^*\left(y\frac{\partial}{\partial y} - B\right) = \delta(y-y')I, \qquad B = -A^* - I \tag{4.27}$$

valid when both sides are applied to u with $\xi \neq 0$ on $WF(u)$. Note that the hypothesis $Re\, A_0(x, \xi) < 0$ is equivalent to $Re\, B_0(x, \xi) > -1$. The preceding analysis yields the following result.

PROPOSITION 4.6. *Let* $(y(\partial/\partial y) - B)u = f$ *and suppose* $Re\, B_0(x, \xi) > -1$. *Suppose* $(0, x_0, 0, \xi_0) \notin WF(f)$ *and also suppose*

$$\{(0, x_0, \eta, \xi_0): \eta \neq 0\} \cap WF(u) = \varnothing. \tag{4.28}$$

Then

$$(0, x_0, 0, \xi_0) \notin WF(u). \tag{4.29}$$

Our next goal is to construct a right parametrix for $y(\partial/\partial y) - A$ whenever $A_0(x, \xi)$ avoids the discrete set $\{0, 1, 2, \ldots\} = \mathbf{Z}^+ \cup \{0\}$.

PROPOSITION 4.7. *Let $A_0(x, \xi) \notin \{0, 1, 2, \ldots\}$. Then there exists an operator H such that*

$$\left(y\frac{\partial}{\partial y} - A\right)H = \delta(y-y')I + R \tag{4.30}$$

where $Ru \in C^\infty$, provided both sides are applied to u with $\xi \neq 0$ on $WF(u)$.

Proof. We shall construct H. Thus its wave front set will be apparent. Choose a positive integer k such that $A_0(x, \xi) - k < 0$. We have seen that for a certain G_k,

$$\left(y\frac{\partial}{\partial y} - (A-k)\right)G_k = \delta(y-y')I. \tag{4.31}$$

Since $A_0(x, \xi)$ avoids $\{0, 1, 2, \ldots\}$, $A - j$ is elliptic for each $j = 0, 1, 2, \ldots$. Take $S_j \in OPS^0$ such that $S_j(A - j) = I + R_j$, $R_j \in OPS^{-\infty}$ on M. Define operators $\gamma_j: C^\infty(\mathbf{R}, \mathcal{D}'(M)) \to \mathcal{D}'(M)$ and $M_k: C^\infty(\mathbf{R}, \mathcal{D}'(M)) \to C^\infty(\mathbf{R}, \mathcal{D}'(M))$ by $\gamma_j u = (\partial^j/\partial y^j)u(0, x)$,

$$u = \sum_{j=0}^{k-1} \gamma_j u + y^k M_k u.$$

Now set

$$H = \sum_{j=0}^{k-1} \frac{y^j}{j!} S_j\gamma_j + y^k G_k M_k. \tag{4.32}$$

A short calculation yields

$$\left(y\frac{\partial}{\partial t} - A\right)H = I + \sum_{j=0}^{k-1} \frac{y^j}{j!} R_j\gamma_j$$

which implies (4.30).

We remark that, for the operator valued distributions $y^j\gamma_j$ and M_k, we have

$$\begin{aligned} &WF(y^j\gamma_j) \subset \{(y, y', \eta, \eta'): y' = 0, \eta = 0\}, \\ &WF(M_k) \subset \{y' = 0, \eta = 0\} \cup \{y = y', \eta = -\eta'\}. \end{aligned} \tag{4.33}$$

Consequently, as with $WF(G_k)$, we have

$$\begin{aligned} WF(H) \subset \{y = y', \eta = -\eta'\} &\cup \{y' = 0, \eta = 0\} \\ &\cup \{y = y' = 0 \text{ and } \exists \rho \in [0, 1] \text{ with } -\eta = \rho y'\}. \end{aligned} \tag{4.34}$$

From (4.30) we deduce

$$-H^*\left(\frac{\partial}{\partial y} - B\right) = I + R^*, \qquad B = -A^* - I. \tag{4.35}$$

Note that $WF(R^*) \subset \{y = 0, \eta' = 0\}$. Consequently, since R^* takes values is $OPS^{-\infty}$, on $WF(R^*)$ one has $y = 0$ and $\xi = 0$. In view of (4.34), the same reasoning as used to prove Proposition 4.6 yields the following.

THEOREM 4.8. *Let* $(y(\partial/\partial y) - B)u = f$ *and suppose* $B_0(x, \xi)$ *avoids the set* $\{-1, -2, -3, \ldots\} = \mathbf{Z}^-$. *Suppose* $(0, x_0, 0, \xi_0) \notin WF(f)$ *and also suppose*

$$\{(0, x_0, \eta, \xi_0) : \eta \neq 0\} \cap WF(u) = \emptyset.$$

Then

$$(0, x_0, 0, \xi_0) \notin WF(u).$$

In passing from Theorems 4.4 and 4.8 to our final results, we note that, if $p \sim p_m + \cdots$ with $p_m = ab$, and if $p(x, D)$ is converted to $y(\partial/\partial y) - B(x, D_x)$ as indicated, then $\mathcal{J}$ denoting the canonical transformation associated with J, which takes a to an elliptic multiple of y and b to an elliptic multiple of η,

$$\frac{i \text{ sub } \sigma(P)}{\{a, b\}} \cdot \mathcal{J}^{-1} = B_0(x, \xi) + \tfrac{1}{2}. \tag{4.36}$$

If $a(x_0, \xi_0) = b(x_0, \xi_0) = 0$, we define four half-rays starting at (x_0, ξ_0). γ_{11} (respectively γ_{12}) is the forward (respectively backward) integral half ray of H_a starting at (x_0, ξ_0), except $(x_0, \xi_0) \notin \gamma_{1j}$, and γ_{21}, γ_{22} are the analogous half rays corresponding to H_b. Theorem 4.4 immediately translates into the following.

THEOREM 4.9. *Let* $u \in \mathcal{D}'$ *with* $(x_0, \xi_0) \notin WF(Pu)$. *Suppose that, for each* $j \in \{1, 2\}$, *there is a* $k \in \{1, 2\}$ *such that* $\gamma_{jk} \cap WF(u) = \emptyset$. *Then* $(x_0, \xi_0) \notin WF(u)$.

From Theorem 4.8 we obtain the following.

THEOREM 4.10. *Suppose* $u \in \mathcal{D}'$, $(x_0, \xi_0) \notin WF(Pu)$, *and suppose*

$$\frac{(-1)^{j+1} i \text{ sub } \sigma(P)}{\{a, b\}} + \frac{1}{2} \notin \{0, -1, -2, \ldots\}.$$

If $(\gamma_{j1} \cup \gamma_{j2}) \cap WF(u) = \emptyset$, *then* $(x_0, \xi_0) \notin WF(u)$.

Indeed, the case $j = 1$ follows directly from (4.36) and Theorem 4.8, and the case $j = 2$ follows from interchanging the roles of a and b.

Exercises

4.1. Suppose A is a continuous linear operator on a Hilbert space H and $Re(Au, u) \leq -C_0\|u\|^2$, $u \in H$. Suppose $e^{-\sigma A}u_0$ is bounded, $1 \leq \sigma < \infty$. Show that $u_0 = 0$. Conclude that, if $y^A u_0$ is bounded, $0 < y \leq 1$, then $u_0 = 0$.

4.2. Suppose $A \in OPS^0$ on M, compact manifold, and $Re(Au, u) \le -C_0\|u\|^2$. Show that, for any seminorm Π_j on $OPS^0_{1,0}$,

$$\Pi_j(e^{\sigma A}) \le C_j e^{-(C_0/2)\sigma}, \ 1 \le \sigma < \infty.$$

(Hint: $\Pi_j(e^{\sigma A}) \le (1/2\pi) \int_\gamma |e^{\sigma\lambda}| \Pi_j((\lambda - A)^{-1}) |d\lambda|$ where γ is an appropriate curve.) Deduce that $\Pi_j(\rho^{-1-A}) \le C_j' \rho^{-1+(C_0/2)}$, $0 < \rho \le 1$.

4.3. Work out a basic theory of wave front sets of vector valued distributions. Discuss the following three cases of increasing generality: distributions with values in a Hilbert space, in a projective limit of a sequence of Hilbert spaces, and in a Frechet space. Prove Lemma 4.2.

§5. Characteristics with Conical Singularities, and Conical Refraction

Let $p_m(x, \xi)$ vanish on Σ, as before. We say $(x_0, \xi_0) \in \Sigma$ is a conical point if, locally, there exist functions $a_j(x, \xi)$, homogeneous of order 1, with linearly independent gradients (also independent of $\Sigma \xi_j dx_j$) such that Σ is given by

$$a_1(x, \xi)^2 + a_2(x, \xi)^2 = a_3(x, \xi)^2.$$

and $a_j(x_0, \xi_0) = 0$. In such a case, the variety Σ is singular precisely along $\Sigma_c = \{a_j(x, \xi) = 0, j = 1, 2, 3\}$, which has codimension three. We suppose p_m vanishes simply on $\Sigma - \Sigma_c$. It follows that, with q_{m-2} a homogeneous elliptic factor,

$$p_m = q_{m-2}(a_1^2 + a_2^2 - a_3^2).$$

In this section we shall restrict our attention to a very simple special case, where the Poisson brackets $\{a_j, a_k\}$ all vanish identically and there are no lower order terms. This is sufficient to handle some problems in crystal optics involving the phenomenon of conical refraction. The idea for treating this special case arose during a conversation between G. Uhlmann and the author in the Courant Institute's coffee lounge.

A. The Equations of Crystal Optics

We consider Maxwell's equations for propagation of light in a crystal:

$$\frac{\partial}{\partial t} D = c \operatorname{curl} H,$$

$$\frac{\partial}{\partial t} B = c \operatorname{curl} E, \tag{5.1}$$

$$\operatorname{div} B = 0, \qquad \operatorname{div} D = 0.$$

We are assuming there are no currents or charged particles. Here E and B are the electric and magnetic fields, respectively, and

$$D = \epsilon E, \qquad B = \mu H \tag{5.2}$$

where ϵ and μ are 3×3 symmetric, positive definite matrices. In an isotropic medium like glass they are both scalar, but in a crystal this is not so. We shall suppose however, that μ is a scalar, which is the case for so-called nonmagnetic media. However, we will assume ϵ has three distinct positive eigenvalues.

We want to treat reflection of singularities of solutions to (5.1) in the special case when ϵ, μ, c are all constant and the boundary is a flat plane. Choosing units appropriately, we may suppose $\mu = c = 1$. The first part of this section will discuss generalities about the system (5.1), well posedness, boundary conditions and the geometry of the characteristics. Further material on this may be found in Courant and Hilbert [1].

We begin this half with a brief discussion of well posedness of (5.1) given appropriate boundary conditions. Using (5.2), we can construct second order equations for E and B, namely

$$\epsilon \frac{\partial^2}{\partial t^2} E = -\operatorname{curl} \operatorname{curl} E, \tag{5.3}$$

$$\frac{\partial^2}{\partial t^2} B = -\operatorname{curl} \epsilon^{-1} \operatorname{curl} B. \tag{5.4}$$

From now on we will restrict our discussion to a treatment of E. A similar discussion can be given for B. Using $\operatorname{div} \epsilon E = 0$, we can obtain from (5.3) that

$$\epsilon \frac{\partial^2}{\partial t^2} E = (-\operatorname{curl} \operatorname{curl} + A\epsilon \operatorname{grad} \operatorname{div} \epsilon)E, \tag{5.5}$$

where $A > 0$ is any conveniently picked positive real number. Now the symbol of $\tilde{\tilde{\Delta}} = -\operatorname{curl} \operatorname{curl} + A\epsilon \operatorname{grad} \operatorname{div} \epsilon$ is

$$\sigma_{\tilde{\tilde{\Delta}}}(x, \xi) = -|\xi|^2(\tilde{P}_\xi + A\epsilon P_\xi \epsilon)$$

where P_ξ is the orthogonal projection onto the linear space spanned by ξ, i.e., $|\xi|^2 P_\xi v = (\xi \cdot v)\xi$, and $\tilde{P}_\xi = I - P_\xi$ is the complementary projection. Note that $((\tilde{P}_\xi + A\epsilon P_\xi \epsilon)v, v) = \|\tilde{P}_\xi v\|^2 + A\|P_\xi \epsilon v\|^2$; if this vanishes, $v \in \ker \tilde{P}_\xi$ and $\epsilon v \perp \ker \tilde{P}_\xi$, so $v \cdot \epsilon v = 0$, which implies $v = 0$, since ϵ is assumed to be positive definite. Thus $\tilde{\tilde{\Delta}}$ is a strongly elliptic operator.

It is convenient to set $F = \sigma E$, with the positive definite matrix $\sigma = \epsilon^{1/2}$. Thus (5.5) becomes

$$\frac{\partial^2}{\partial t^2} F = (-\sigma^{-1} \operatorname{curl} \operatorname{curl} \sigma^{-1} + A\sigma \operatorname{grad} \operatorname{div} \sigma)F \tag{5.6}$$

and $\tilde{\Delta} = -\sigma^{-1} \operatorname{curl} \operatorname{curl} \sigma^{-1} + A\sigma \operatorname{grad} \operatorname{div} \sigma$ is also strongly elliptic, with symbol

$$\sigma_{\tilde{\Delta}}(x, \xi) = -|\xi|^2(\sigma^{-1}\tilde{P}_\xi\sigma^{-1} + A\sigma P_\xi\sigma). \tag{5.7}$$

Note that $\operatorname{div} \sigma F = 0$.

Now suppose $\mathbf{R}^3$ is divided into two regions by a smooth hypersurface Σ, and suppose ϵ assumes two different values on these two regions. The physical requirement on light waves propagating across Σ are that E and H be continuous across Σ. Denoting the two regions separated by Σ, by $\Omega_\pm$, letting $E_\pm = E|\Omega_\pm$, we have

$$E_+ = E_- \text{ on } \Sigma. \tag{5.8}$$

This yields three boundary conditions, but (5.6) requires six. We can get two more from $\operatorname{div} \epsilon E = 0$.

$$\operatorname{div} \epsilon E_+ = 0 \quad \text{and} \quad \operatorname{div} \epsilon E = 0 \text{ on } \Sigma. \tag{5.9}$$

We obtain one more from the continuity of H across Σ, which implies the continuity of $\operatorname{curl} E$ across Σ.

$$\operatorname{curl} E_+ = \operatorname{curl} E_- \text{ on } \Sigma. \tag{5.10}$$

Now (5.8) through (5.10) yield eight boundary conditions, which are too many. We pare them down to

$$\left.\begin{aligned} \nu \times E_+ &= \nu \times E_-, \\ \operatorname{div} \epsilon E_+ &= 0 = \operatorname{div} \epsilon E_-, \\ \nu \times \operatorname{curl} E_+ &= \nu \times \operatorname{curl} E_- \end{aligned}\right\} \text{on } \Sigma. \tag{5.11}$$

This translates into the following boundary conditions for F:

$$\left.\begin{aligned} \nu \times \sigma^{-1}F_+ &= \nu \times \sigma^{-1}F_-, \\ \operatorname{div} \sigma F_+ &= 0 = \operatorname{div} \sigma F_-, \\ \nu \times \operatorname{curl} \sigma^{-1}F_+ &= \nu \times \operatorname{curl} \sigma^{-1}F_- \end{aligned}\right\} \text{on } \Sigma. \tag{5.12}$$

In order to make natural our choice of these six boundary conditions from among the available eight, we show that they specify a domain for $\tilde{\Delta}$ making the operator symmetric. Since this computation involves the use of Stoke's formula, it is convenient to identify the vector field F with a 1-form φ, via the Euclidean metric on $\mathbf{R}^3$. Then curl becomes $*d$, grad becomes d, and div becomes $\delta = *d*$. Thus

$$\tilde{\Delta} = \sigma^{-1}\delta\, d\sigma^{-1} + A\sigma\, d\delta\sigma. \tag{5.13}$$

The natural inner product on 1-forms defined on a region Ω is $(\varphi, \psi) = \int_\Omega \varphi \wedge *\psi$. Stoke's formula implies

$$(\tilde{\Delta}\varphi, \psi)_\Omega = (d\sigma^{-1}\varphi, d\sigma^{-1}\psi)_\Omega + \int_{\partial\Omega} (*d\sigma^{-1}\varphi) \wedge \sigma^{-1}\psi \tag{5.14}$$
$$+ A(\delta\sigma\varphi, \delta\sigma\psi)_\Omega + A\int_{\partial\Omega} (\delta\sigma\varphi) \wedge *\sigma\psi.$$

In our case, $\Omega = \Omega_+ \cup \Omega_-$, $\partial\Omega = \Sigma_+ \cup \Sigma_-$ (the two sides of Σ). If φ and ψ are smooth with bounded support, and possible jumps across Σ, we apply (5.14), and ask that the boundary integrals vanish. Thus we want

$$\int_\Sigma \delta\sigma\varphi_+ \wedge \sigma\psi_+ = \int_\Sigma \delta\sigma\varphi_- \wedge *\sigma\psi_-, \tag{5.15}$$

$$\int_\Sigma *d\sigma^{-1}\varphi_+ \wedge \sigma^{-1}\psi_+ = \int_\Sigma *d\sigma^{-1}\varphi_- \wedge \sigma^{-1}\psi_-. \tag{5.16}$$

Indeed, div $\sigma F_+ = 0 =$ div σF_- on Σ yields $\delta\sigma\varphi_\pm = 0$ on Σ, so (5.15) holds. Meanwhile $\nu \times \sigma^{-1}F_+ = \nu \times \sigma^{-1}F_-$ on Σ is equivalent to the identity of $\sigma^{-1}\psi_+$ and $\sigma^{-1}\psi_-$, pulled back to Σ, so then (5.16) is equivalent to

$$\int_\Sigma *d\sigma^{-1}(\varphi_+ - \varphi_-) \wedge \sigma^{-1}\psi = 0$$

for any smooth compactly supported 1-form ψ on Σ, which is equivalent to $*d\sigma^{-1}\varphi_+$ and $*d\sigma^{-1}\varphi_-$ having identical pull backs to Σ. This is in turn equivalent to $\nu \times \operatorname{curl} \sigma^{-1}F_+ = \nu \times \operatorname{curl} \sigma^{-1}F_-$ on Σ.

Consequently, for smooth F, G with bounded support, satisfying the conditions (5.12), we have

$$(\tilde{\Delta}F, G)_\Omega = -(\operatorname{curl} \sigma^{-1}F, \operatorname{curl} \sigma^{-1}G)_\Omega - A(\operatorname{div} \sigma F, \operatorname{div} \sigma G)_\Omega. \tag{5.17}$$

Furthermore, the boundary conditions (5.12) satisfy the coerciveness condition, leading to the appropriate elliptic regularity estimates for $\tilde{\Delta}$ (see Exercise 5.1). In particular, the Friedrichs extension method provides a natural self-adjoint extension for $\tilde{\Delta}$, which we also denote $\tilde{\Delta}$, with domain $\mathcal{D}(\tilde{\Delta})$ contained in $H^2(\Omega_+) \oplus H^2(\Omega_-)$. Thus the wave equation (5.6) is well posed, with Cauchy data $F(0, x) \in \mathcal{D}(\tilde{\Delta}^{1/2})$, $(\partial/\partial t)F(0, x) \in L^2(\mathbf{R}^3)$, and the results of the introduction to Chapter IX hold. The fact that $\mathbf{R}^3$ is unbounded can be circumvented by finite propagation speed, but we will omit the details.

We now examine the characteristics of the operator $(\partial^2/\partial t^2) - \tilde{\Delta}$, whose symbol is $-\tau^2 + |\xi|^2\sigma^{-1}\tilde{P}_\xi\sigma^{-1} + A|\xi|^2\sigma P_\xi\sigma = -\tau^2 + A(\xi) + B(\xi)$. Note that $A(\xi)$ and $B(\xi)$ are both self-adjoint semidefinite matrices, whose sum is positive definite, and $A(\xi)B(\xi) = 0 = B(\xi)A(\xi)$; in particular, $A(\xi)$ and $B(\xi)$ commute. We compute the eigenvalues of $A(\xi)$ and $B(\xi)$ separately. A straightforward calculation shows that

$$\det(\tau^2 - A(\xi)) = \tau^2(\tau^4 - \tau^2\Psi(\xi) + |\xi|^2\Phi(\xi)) \tag{5.18}$$

where, if we choose coordinates so that

$$\sigma = \begin{pmatrix} \sigma_1 & & \\ & \sigma_2 & \\ & & \sigma_3 \end{pmatrix},$$

then,

$$\Psi(\xi) = \sigma_1^{-2}(\xi_2^2+\xi_3^2) + \sigma_2^{-2}(\xi_1^2+\xi_3^2) + \sigma_3^{-2}(\xi_1^2+\xi_2^2),$$
$$\Phi(\xi) = (\sigma_2\sigma_3)^{-2}\xi_1^2 + (\sigma_1\sigma_3)^{-2}\xi_2^2 + (\sigma_1\sigma_2)^{-1}\xi_3^2.$$

Similarly, we have

$$\det(\tau^2 - B(\xi)) = \tau^4(\tau^2 - A\sigma_1^2\xi_1^2 - A\sigma_2^2\xi_2^2 - A\sigma_3^2\xi_3^2). \tag{5.19}$$

So $A(\xi)$ has 0 as a simple eigenvalue, and two positive eigenvalues, which may coincide, while $B(\xi)$ has 0 as a double eigenvalue, and one positive eigenvalue. Consequently, $\sigma_{\tilde{\Delta}} = A(\xi) + B(\xi)$ has for its three eigenvalues, the two roots τ^2 of the equation

$$\tau^4 - \tau^2\Psi(\xi) + |\xi|^2\Phi(\xi) = 0 \tag{5.20}$$

and the eigenvalue $\tau^2 = A(\sigma_1^2\xi_1^2 + \sigma_2^2\xi_2^2 + \sigma_3^2\xi_3^2)$.

We now examine the geometry of the set of solutions to (5.20). The zeros τ^2, as functions of ξ, are given by

$$\tau^2 = \tfrac{1}{2}\Psi(\xi) \pm \tfrac{1}{2}\sqrt{\Psi(\xi)^2 - 4|\xi|^2\Phi(\xi)}. \tag{5.21}$$

If the diagonal entries in σ are assumed to satisfy $\sigma_1 < \sigma_2 < \sigma_3$, a simple calculation shows that, with $\alpha_j = \sigma_j^{-2}$, (so $\alpha_1 > \alpha_2 > \alpha_3$),

$$\begin{aligned} D(\xi) &= \Psi(\xi)^2 - 4|\xi|^2\Phi(\xi) \\ &= P(\xi)^2 + 4(\alpha_1 - \alpha_3)(\alpha_1 - \alpha_2)\xi_2^2\xi_3^2 \end{aligned} \tag{5.22}$$

where $P(\xi) = (\alpha_2 - \alpha_3)\xi_1^2 + (\alpha_1 - \alpha_3)\xi_2^2 - (\alpha_1 - \alpha_2)\xi_3^2$. Thus we have written the discriminant $D(\xi)$ as a sum of squares. Note that, for $\xi \neq 0$, $D(\xi)$ vanishes precisely when

$$(\alpha_2 - \alpha_3)\xi_1^2 = (\alpha_1 - \alpha_2)\xi_3^2 \quad \text{and} \quad \xi_2 = 0. \tag{5.23}$$

It follows that the variety defined by (5.20) is a four sheeted cover of the ξ-space, except along this singular set, where it exhibits a conical singularity, which we define in the following fashion.

DEFINITION. *Let V be a conic variety in $\mathbf{R}^n$*, i.e. $\zeta \in V \Rightarrow r\zeta \in V$, $r > 0$. *Let $\zeta = (\zeta_1, \zeta')$. Suppose $\omega = (\omega_1, \omega') \in V$. We say V has a conical singularity at ω, with respect to the ζ' variables, if there exist smooth functions $f(\zeta')$, $g(\zeta')$, $h(\zeta')$, homogeneous of degree 1 in ζ', such that ∇g and ∇h are linearly independent at $\zeta' = \omega'$, and such that, on a conic neighborhood of ω, V is given by*

$$\zeta_1 = f(\zeta') \pm \sqrt{g(\zeta')^2 + h(\zeta')^2}. \tag{5.24}$$

By this definition, we can show that the variety defined by (5.21) has a conical singularity with respect to the ξ variables, at a point (τ, ξ) in the variety with ξ satisfying (5.23). Since (5.21) is an equation for τ^2 rather than for τ, the verification requires a little computation. In fact,

$$\begin{aligned}(\tfrac{1}{2}\Psi \pm \tfrac{1}{2}\sqrt{D})^{1/2} &= (\tfrac{1}{2}\Psi)^{1/2}(1 \pm \Psi^{-1}\sqrt{D})^{1/2}\\ &= (\tfrac{1}{2}\Psi)^{1/2}(1 \pm a_1\Psi^{-1}\sqrt{D} + a_2\Psi^{-2}D \pm a_3\Psi^{-3}\sqrt{D^3} + \cdots)\\ &= (\tfrac{1}{2}\Psi)^{1/2}[(1 + a_2\Psi^{-2}D + a_4\Psi^{-4}D^2 + \cdots)\\ &\quad \pm (a_1\Psi_1^{-1} + a_3\Psi^{-3}D + a_5\Psi^{-5}D^2 + \cdots)\sqrt{D}],\end{aligned}$$

so (4.21) assumes the form (4.24), with $\zeta_1 = \tau$, $\zeta' = \xi$, $f(\zeta') = (\tfrac{1}{2}\Psi)^{1/2}(1 + a_2\Psi^{-2}D + \cdots)$, $g(\zeta') = (\tfrac{1}{2}\Psi)^{1/2}(a_1\Psi^{-1} + a_3\Psi^{-3}D + \cdots)P(\xi)$, and $h(\zeta') = 2(\tfrac{1}{2}\Psi)^{1/2}(\alpha_1 - \alpha_3)^{1/2}(\alpha_1 - \alpha_2)^{1/2}(a_1\Psi^{-1} + a_3\Psi^{-3}D + \cdots)\xi_2\xi_3$. (Note: $a_1 = 1/2$.)

We need to know under what circumstances does the variety V defined by (5.21) have a conical singularity with respect to the (τ, ζ') variables, where $\zeta = (\zeta_1, \zeta')$ is some orthogonal coordinate system on ξ space, not necessarily (ξ_1, ξ_2, ξ_3). Rewrite (5.21) as an equation for $\tilde{\xi} = \tau^{-1}\xi$, which is appropriate since $\tau \neq 0$ on V, except at the origin. We have, with $\rho = |\tilde{\xi}|$, $\tilde{\xi} = \rho\alpha$, $\alpha \in S^2 \subset \mathbf{R}^3$,

$$\rho^4\Phi(\alpha) - \rho^2\Psi(\alpha) + 1 = 0.$$

Therefore,

$$\begin{aligned}\rho^2 &= \tfrac{1}{2}\Phi(\alpha)^{-1}(1 \pm \sqrt{\Psi(\alpha)^2 - 4\Phi(\alpha)})\\ &= \tfrac{1}{2}\Phi(\alpha)^{-1}(1 \pm \sqrt{D(\alpha)})\end{aligned} \tag{5.25}$$

with $D(\alpha)$ given by (5.22). From this representation one can deduce that, if $\zeta_1, \zeta_2, \zeta_3$ are orthogonal linear coordinates on ξ-space, then the variety V defined by (5.21) has conical singularities at a point $(\tau_0, \xi_0) \in V$ with respect to the (τ, ζ') variables, provided the vector $\nabla\zeta_1 \in \mathbf{R}^3$ is time-like with respect to the tangent cone to the variety defined by (5.25), at the singular point $\tau_0^{-1}\xi_0$.

B. Conical Refraction at a Flat Boundary

In this section we shall show how to construct parametrices for (5.6), (5.12). We suppose $\Omega = \mathbf{R}^3$ is divided into two half spaces by a plane, $y = 0$. On one side, Ω_+, we suppose

$$\epsilon = \begin{pmatrix} \epsilon_1 & & \\ & \epsilon_2 & \\ & & \epsilon_3 \end{pmatrix}$$

with $0 < \epsilon_1 < \epsilon_2 < \epsilon_3$. On the other side, Ω_-, we suppose

$$\epsilon = \begin{pmatrix} 1 & & \\ & 1 & \\ & & 1 \end{pmatrix}.$$

Thus we suppose Ω_- is an isotropic medium, e.g., a vacuum, and Ω_+ is a crystal. It is equivalent to construct parametrices for solutions to

$$\begin{aligned} &\frac{\partial^2}{\partial t^2} F_+ = (-\sigma^{-1} \text{ curl curl } \sigma^{-1} + A\sigma \text{ grad div } \sigma)F_+ \\ &\frac{\partial^2}{\partial t^2} F_- = \Delta F_- \end{aligned} \tag{5.26}$$

which are outgoing, i.e., $F_+ = F_- = 0$ for $t \ll 0$, and which satisfy on the boundary $\Sigma = \{y = 0\}$, the conditions

$$\left.\begin{aligned} \nu \times \sigma^{-1} F_+ - \nu \times F_- &= f \\ \text{div } \sigma F_+ &= g \\ \text{div } F_- &= h \\ \nu \times \text{curl } \sigma^{-1} F_+ - \nu \times \text{curl } F_- &= k \end{aligned}\right\} \text{ on } \mathbf{R} \times \Sigma \tag{5.27}$$

Here f, g, h, k are distributions with compact support on Σ. f and k take values in $\mathbf{C}^2$ (complex tangent vectors to Σ); g and h take values in $\mathbf{C}$.

We make the following assumptions on f, g, h, k, namely that their wave front sets are all contained in a conic set $\Gamma \in T^*(\mathbf{R} \times \Sigma)$ over which no grazing rays for (5.21) pass, and that, if $(t_0, x_0', \tau_0, \zeta_0') \in \Gamma$, and if $(\tau_0, \zeta_1, \zeta_0')$ is a singular point on the variety V for some $\zeta_1 \in \mathbf{R}$, then this point is a conical point with respect to the (τ, ζ') variables. If this latter assumption fails to hold, then rays along which conical refraction occur will include some grazing rays, and the analysis is more complicated. Without loss of generality, we assume the cone Γ is a small conic neighborhood of a single point $(t_0, x_0', \tau_0, \zeta_0') \in T^*(\mathbf{R} \times \Sigma) - 0$. We also suppose some $(\tau_0, \zeta_1, \zeta_0') \in V$ is a conical singular point of V, with respect to the (τ, ζ') variables, since the case when all such $(\tau_0, \zeta_1, \zeta_0') \in V$ are regular points is handled by the discussion in Chapter IX. At such a point, det $(\tau_0^2 - |\zeta_0|^2\sigma^{-1}\tilde{P}_{\zeta_0}\sigma^{-1}) = 0$, where $\zeta_0 = (\zeta_1, \zeta_0')$. For simplicity, we assume det $(\tau_0^2 - A|\zeta_0|^2\sigma P_{\zeta_0}\sigma) \neq 0$, which can certainly be arranged if A is chosen small enough (or large enough).

If we rewrite the system (5.26) as a first order system of the type

$$\frac{\partial}{\partial y} u = K(y, y', D_{y'})u \tag{5.28}$$

where $y' = (t, x') \in \Sigma$, the above hypotheses imply that, for $(y'\ \eta') = (t, x', \tau, \xi') \in \Gamma$, $K_1(y, y', \eta')$ can be put in a block diagonal form

$$K_1(y, y', \eta') \sim \begin{pmatrix} i\lambda_1 & & & & & \\ & \ddots & & & & \\ & & i\lambda_j & & & \\ & & & E_0 & & \\ & & & & E_+ & \\ & & & & & E_- \end{pmatrix} \tag{5.29}$$

where $\lambda_\nu(y, y', \eta')$ are real $(0 \leq j \leq 4)$, $E_\pm(y, y', \eta')$ have spectrum in the half spaces with either strictly positive or strictly negative real part, respectively, and $E_0(y, y', \eta')$ is a 2×2 matrix, homogeneous of degree 1 in η', whose eigenvalues $i\lambda_\pm(y, y', \eta')$ are pure imaginary and of the form

$$\lambda_\pm(y, y', \eta') = p(\eta') \pm \sqrt{q(\eta')^2 + r(\eta')^2} \tag{5.30}$$

for certain real valued smooth functions p, q, r, homogeneous of degree 1 in η', with $\nabla q, \nabla r$ linearly independent at $\eta' = (\tau_0, \xi_0')$. Since the complete symbol of $K(y, y', D_{y'})$ is homogeneous of degree 1 and depends only on η', $K(y, y', \eta') = K_1(\eta')$, we can assume K_1 is exactly in the form (4.29), by a change of dependent variable u. No appeal to the total decoupling procedure of Chapter IX, Section 1 is required. Of course, the entries of $E_0(y, y', \eta') = E_0(\eta')$ are smooth. Now write

$$E_0(\eta') = p(\eta')I + F_0(\eta') \tag{5.31}$$

where $F_0(\eta')$ is a 2×2 matrix, with smooth entries, with eigenvalues $\pm\sqrt{q(\eta')^2 + r(\eta')^2}$. Solving

$$\begin{aligned} &\frac{\partial}{\partial t} v_0 = ip(D_{y'})v_0 + iF_0(D_{y'})v_0 \qquad \text{and} \\ &v_0(0) = \varphi_0 \end{aligned} \tag{5.32}$$

is equivalent to solving

$$\frac{\partial}{\partial y} w_0 = iF_0(D_{y'})w_0 \quad \text{and} \tag{5.33}$$
$$w_0(0) = \varphi_0$$

with $v_0(y) = e^{iyp(D_{y'})}w_0(y)$, i.e., v_0 obtained from w_0 via a Fourier integral operator. Thus it suffices to analyze solutions to (5.33). Multiplying $(\partial/\partial y) - iF_0(D_{y'})$ by its cofactor matrix, we obtain a scalar second order equation for w_0:

$$\left(\frac{\partial^2}{\partial y^2} + [q(D_{y'})^2 + r(D_{y'})^2]\right)w_0 = 0, \tag{5.34}$$
$$w_0(0) = \varphi_0, \qquad \frac{\partial}{\partial y} w_0(0) = iF_0(D_{y'})\varphi_0.$$

Now, (5.34) is a constant coefficient problem, but to analyze it, it is most convenient to apply a Fourier integral operator which is not translation invariant, to get a simpler constant coefficient problem. Indeed, one can choose a canonical transformation taking $q(\eta')$ into ξ_1 and $r(\eta')$ into ξ_2, and implement this by a Fourier integral operator J such that

$$Jq(D_{y'})J^{-1} = \frac{1}{i}\frac{\partial}{\partial x_1}, \qquad Jr(D_{y'})J^{-1} = \frac{1}{i}\frac{\partial}{\partial x_2}, \qquad \text{mod } OPS^{-\infty}. \tag{5.35}$$

Thus under the transformation $w_0(y, y') \to \tilde{w}_0 = Jw_0(y, x')$, we have

$$\left(\frac{\partial^2}{\partial y^2} - \left(\frac{\partial^2}{\partial x_1^2} + \frac{\partial^2}{\partial x_2^2}\right)\right)\tilde{w}_0 = 0, \tag{5.36}$$
$$\tilde{w}_0(0) = \psi_0 = J\varphi_0, \qquad \frac{\partial}{\partial y}\tilde{w}_0 = iJF_0(D_{y'})\varphi_0 = \psi_1.$$

Now (5.36) is simply the classical wave equation in three variables (2 space and 1 time), the difference being that $\tilde{w}_0$ is a distribution on $\mathbf{R}^4$. If $E(y, x_1, x_2, y', x_1', x_2')$ is the distribution kernel for the fundamental solution of (5.36) on distributions on $\mathbf{R}^3$, the fundamental solution to (5.36) is

$$F = \delta(x_3 - x_3') \otimes E(y, x_1, x_2, y', x_1', x_2'). \tag{5.37}$$

Note that E is singular on the cone $(x_1 - x_1')^2 + (x_2 - x_2')^2 = (y - y')^2$ and in C^∞ but nonvanishing inside this cone. This gives rise to extra singularities for F, which accounts for the conical refraction phenomenon. Namely, the wave front sets of ψ_0, ψ_1 away from $\xi_1 = \xi_2 = 0$ get propagated at two speeds, while the wave front sets in $\xi_1 = \xi_2 = 0$ get smeared out. We leave it as an exercise to read off the wave front set of v_0 from (5.32) through (5.37).

Now analyses of reflection and refraction of singularities for various boundary value problems follow along the same lines as in Chapter IX.

For a general study of operators whose characteristics have conical singularities, in the involutive case, see Melrose and Uhlmann [2].

Bibliography

Abraham, R. and Marsden, J.
[1] *Foundations of Mechanics*. 2nd ed. Benjamin/Cummings, Reading, 1978.
Adams, R.
[1] *Sobolev Spaces*. Academic Press, New York, 1975.
Agmon, S.
[1] "On the eigenfunctions and on the eigenvalues of general elliptic boundary value problems." *Comm. Pure Appl. Math. 15* (1962), 119–147.
———.
[2] *Lectures on Elliptic Boundary Value Problems*. Van Nostrand, New York, 1964.
Agmon, S., Douglis, A., and Nirenberg, L.
[1] "Estimates near the boundary for solutions of elliptic differential equations satisfying general boundary conditions." *Comm. Pure Appl. Math. 12* (1959), 623-727.
Agranovich, M.
[1] "Boundary value problems for first order pseudodifferential operators." *Russian Math. Surveys 24* (1969), 59–126.
S. Alinhac.
[1] "Paramétrixe pour un système hyperbolique à multiplicité variable." *Comm. PDE 2* (1977), 251–296.
Amrein, W., Jauch, J., and Sinha, K.
[1] *Scattering Theory in Quantum Mechanics*. Benjamin, Reading, 1977.
Andersson, K. and Melrose, R.
[1] "The propagation of singularities along gliding rays." *Invent. Math. 41* (1977), 197–232.
Arnold, V.
[1] "On a characteristic class entering in the quantization conditions." *Funct. Anal. Appl. 1* (1967), 1–13.
———.
[2] *Mathematical Methods of Classical Mechanics*. Springer, New York, 1978.
Aronszajn, N.
[1] "A unique continuation theorem for solutions of elliptic partial differential equations or inequalities of second order." *J. Math. Pure Appl. 36* (1957), 235–249.

Aronszajn, N. and Smith, K.

[1] "Theory of Bessel Potentials I." *Ann. Inst. Fourier Grenoble.* *11* (1961), 385–475.

Beals, R.

[1] "Spatially inhomogeneous pseudo-differential operators, II." *Comm. Pure Appl. Math.* *27* (1974), 161–205.

———.

[2] "A general calculus of pseudo-differential operators." *Duke Math. J.* *42* (1975), 1–42.

———.

[3] "Characterization of pseudodifferential operators and applications." *Duke Math. J.* *44* (1977), 45–57; correction 46 (1979), 215.

———.

[4] "On the boundedness of pseudo-differential operators." *Comm. PDE* *2* (1977), 1063–1070.

———.

[5] "L^p and Hölder estimates for pseudo-differential operators." To appear.

———.

[6] "Weighted distribution spaces associated to pseudo-differential operators." To appear.

Beals, R., and Fefferman, C.

[1] "On local solvability of linear partial differential equations." *Ann. of Math.* *97* (1973), 482–498.

[2] "Spatially inhomogeneous pseudodifferential operators." *Comm. Pure Appl. Math.* *27* (1974), 1–24.

Berger, M., Gauduchon, P., and Mazet, E.

[1] *Le Spectre d'une Variété Riemannienne.* Lecture Notes in Math. 194, Springer, New York, 1971.

Bergh, J. and Löfström, J.

[1] *Interpolation Spaces, an Introduction.* Springer, New York, 1976.

Bers, L., John, F., and Schechter, M.

[1] *Partial Differential Equations.* Wiley, New York, 1964.

Birkhoff, G.

[1] "Quantum mechanics and asymptotic series." *Bull. A.M.S.* *39* (1933), 681–700.

Bona, J. and Scott, R.

[1] "Solutions of the Korteweg-de Vries equation in fractional order Sobolev spaces." *Duke Math. J.* *43* (1976), 87–99.

Bony, J.

[1] "Propagation des singularités differéntiables pour une classe d'operateurs differentials à coefficients analytiques." *Astérisque* *34–35* (1976), 43–91.

———.

[2] "Equivalence des diverses notions de spectre singulier analytique." Séminaire Goulaouic-Schwartz, no. 3. Ecole Polytechnique, 1976–1977.

Bony, J. and Schapira, P.

[1] "Propagation des singularités analytiques pour les solutions des équations aux dérivées partielles." *Ann. Inst. Fourier 26* (1976), 81–140.

Boutet de Monvel, L.

[1] "Boundary value problems for pseudodifferential operators." *Acta. Math. 126* (1971), 11–51.

———.

[2] "Hypoelliptic operators with double characteristics and related pseudodifferential operators." *Comm. Pure Appl. Math. 27* (1974), 585–639.

Boutet de Monvel, L., Grigis, A., and Helffer, B.

[1] "Paramétrixes d'operateurs pseudodifferentiels à caracteristiques multiple." *Asterisque 34–35* (1976), 93–121.

Boutet de Monvel, L. and Sjöstrand, J.

[1] "Sur la singularité des noyaux de Bergmann et de Szegö." *Asterisque 34–36* (1976), 123–164.

Boutet de Monvel, L. and Trèves, F.

[1] "On a class of pseudodifferential operators with double characteristics." *Inventiones Math. 24* (1974), 1–34.

———.

[2] "On a class of systems of pseudodifferential equations with double characteristics." *Comm. Pure Appl. Math. 27* (1974), 59–89.

Browder, F.

[1] "Estimates and existence theorems for elliptic boundary value problems." *Proc. Nat. Acad. Sci. USA. 45* (1959), 365–372.

Butzer, P. and Berens, H.

[1] *Semi-groups of Operators and Approximation.* Springer, New York, 1967.

Butzer, P. and Nessel, R.

[1] *Fourier Analysis and Approximation.* Birkhäuser, Basel, 1971.

Cahn, R.

[1] "The *K*-types of principal series representations." To appear.

Cahn, R. and Taylor, M.

[1] "Asymptotic behavior of multiplicities of representations of compact groups." *Pacific J. Math. 83* (1979).

Calderon, A.

[1] "Uniqueness in the Cauchy problem of partial differential equations." *Amer. J. Math. 80* (1958), 16–36.

———.

[2] "Intermediate spaces and interpolation, the complex method." *Studia Math. 24* (1964), 113–190.

Calderon, A. and Vaillancourt, R.

[1] "On the boundedness of pseudodifferential operators." *J. Math. Soc. Japan 23* (1971), 374–378.

———.

[2] "A class of bounded pseudodifferential operators." *Proc. Nat. Acad. Sci. USA, 69* (1972), 1185–1187.

Calderon, A. and Zygmund, A.
[1] "Singular integral operators and differential equations." *Amer. J. Math. 79* (1957), 901–921.
Caratheodory, C.
[1] *Calculus of Variations and Partial Differential Equations of the First Order.* Holden-Day, San Francisco, 1965.
Carleman, T.
[1] "Sur un probleme d'unicite pour les systèmes d'équations aux dérivées partielles à deux variables indépendantes." *Ark. Mat. Astr. Fys. 26B, no. 17* (1939), 1–9.
Carleson, L.
[1] "An interpolation problem for bounded analytic functions." *Amer. J. Math. 80* (1958), 921–930.
Chazarain, J.
[1] "Construction de la paramétrix du problème mixte hyperbolique pour l'equation des ondes." *C.R. Acad. Sci. Paris, 267* (1973), 1213–1215.
———.
[2] "Formule de Poisson pour les variétés riemanniennes." *Invent. Math. 24* (1974), 65–82.
———.
[3] "Propagation des singularities pour une classe d'opérateurs a caracteristiques multiplies et résobulité locale." *Ann. Inst. Fourier Grenoble 24* (1974), 203–223.
Cheeger, J.
[1] "Spectral geometry of spaces with cone-like singularities." *Proc. Nat. Acad. Sci. 76* (1979), 2103–2106.
Cheeger, J. and Taylor, M.
[1] "Diffraction of waves by conical singularities." *Comm. Pure Appl. Math.* To appear.
Clerc, J.
[1] "Sommes de Riesz sur un groupe de Lie compact." *Comptes Rendus 275* (1972), 591–593.
Chevelley, C.
[1] *Theory of Lie Groups.* Princeton Univ. Press, Princeton, 1946.
Cohen, P.
[1] "The non-uniqueness of the Cauchy problem." Office of Naval Research Technical Report *93*, Stanford, 1960.
Coifman, R. and Meyer, Y.
[1] "Au dela des opérateurs pseudodifferentiels." *Asterisque, 57* (1979), 1–184.
Coifman, R. and Weiss, G.
[1] *Analyze Harmonique Non-commutative sur Certains Espaces Homogenes.* Lecture Notes in Math. 242, Springer, New York, 1971.
Cordes, H.
[1] "Uber die eindeutige Bestimmtheit der Lösungen elliptischen Differentialgleichungen durch Anfangsvorgaben." *Nachr. Akad. Wiss. Göttingen Math.-Phys. Kl. IIa, no. 11* (1956), 239–258.

———.

[2] "On compactness of commutators of multiplications and convolutions, and boundedness of pseudo-differential operators." *J. Funct. Anal. 18* (1975), 115–131.

Cordes, H. and Herman, E.

[1] "Gelfand theory of pseudo-differential operators." *Amer. J. Math. 90* (1968), 681–717.

Cotlar M.

[1] "A unified theory of Hilbert transforms and ergodic theory." *Rev. Mat. Cuyana 1* (1955), 105–167.

Courant, R. and Friedrichs, K.

[1] *Supersonic Flow and Shock Waves.* Springer, New York, 1976.

Courant, R. and Hilbert, D.

[1] *Methods of Mathematical Physics II.* J. Wiley, New York, 1966.

Courant, R. and Lax, P.

[1] "The propagation of discontinuities in wave motion." *Proc. Nat. Acad. Sc. USA 42* (1956), 872–876.

Dieudonné, J.

[1] *Foundation of Modern Analysis.* Academic Press, New York, 1964.

Dionne, P.

[1] "Sur les problèmes de Cauchy hyperboliques bien posés." *J. Anal. Math. 10* (1962–1963), 1–90.

DiPerna, R.

[1] "Singularities of solutions of nonlinear hyperbolic systems of conservation laws." *Arch. Rat. Mech. Anal. 60* (1976), 75–100.

Dixmier, J.

[1] *Les C* algebres et leurs représentations.* Gauthier Villars, Paris, 1964.

Donoghue, W.

[1] *Distributions and Fourier Transforms.* Academic Press, New York, 1969.

Dugundji, J.

[1] *Topology.* Allyn and Bacon, New York, 1966.

Duistermaat, J.

[1] *Fourier Integral Operators.* Courant Institute Lecture Notes, New York, 1974.

———.

[2] "Oscillatory integrals, Lagrange immersions and unfolding of singularities." *Comm. Pure Appl. Math.* 27 (1974), 207–281.

Duistermaat J. and Guillemin, V.

[1] "The spectrum of positive elliptic operators and periodic bicharacteristics." *Invent. Math. 29* (1975), 39–79.

Duistermaat, J. and Hörmander, L.

[1] "Fourier integral operators II." *Acta Math. 128* (1972), 183–269.

Duistermaat, J. and Sjöstrand, J.

[1] "A global construction for pseudo-differential operators with non-involutive characteristics." *Invent. Math. 20* (1973), 209–225.

Dunford, N. and Schwartz, J.

[1] *Linear Operators.* Interscience, New York, vol. 1, 1958; vol. 2, 1963.

Dynin, A.

[1] "Pseudodifferential operators on Heisenberg groups." Sem. C.I.M.E., Bressanone, Italy, June 1977.

Ehrenpreis, L.

[1] "Solutions of some problems of division." *Amer. J. Math. 76* (1954), 883–903; *78* (1956), 685–715; *82* (1960), 522–588.

Egorov, Yu. V.

[1] "On canonical transformations of pseudo-differential operators." *Uspehi Mat. Nauk. 24* (1969), 235–236.

Erdelyi, A.

[1] *Asymptotic Expansions.* Dover, New York, 1965.

Farris, M.

[1] "Egorov's theorem on manifolds with diffractive boundary." *Comm. PDE,* to appear.

Fedoryuk, M.

[1] "The stationary phase method and pseudodifferential operators." *Russian Math. Surveys 26* (1971), 65–115.

Fefferman, C.

[1] "L^p bounds for pseudo-differential operators." *Israel J. Math. 14* (1973), 413–417.

Fefferman, C. and Phong, D.

[1] "On positivity of pseudo-differential operators." *Proc. Nat. Acad. Sci. USA 75* (1978), 4673–4674.

Flashka, H. and Strang, G.

[1] "The correctness of the Cauchy problem." *Adv. in Math. 6* (1971), 347–379.

Folland, G. and Kohn, J.

[1] *The Neumann Problem for the Cauchy-Riemann Complex.* Ann. of Math. Studies 75, Princeton Univ. Press, Princeton, 1972.

Folland, G. and Stein, E.

[1] "Estimates for the $\bar{\partial}_b$ complex and analysis on the Heisenberg group." *Comm. Pure Appl. Math. 27* (1974), 429–522.

Friedlander, F.

[1] "The wavefront set of a simple initial-boundary value problem with glancing rays." *Math. Proc. Camb. Phil. Soc. 79* (1976), 145–149.

Friedlander, F. and Melrose, R.

[1] "The wave front set of a simple initial-boundary value problem with glancing rays, II." *Math. Proc. Camb. Phil. Soc. 81* (1977), 97–120.

Friedman, A.

[1] *Generalized Functions and Partial Differential Equations.* Prentice-Hall, Englewood Cliffs, 1963.

———.

[2] *Stochastic Differential Equations and Applications, I and II.* Academic Press, New York, 1975.

Friedrichs, K.

[1] "Differentiability of solutions of elliptic partial differential equations." *Comm. Pure Appl. Math. 5* (1953), 299–326.

———.

[2] "Symmetric positive systems of differential equations." *Comm. Pure Appl. Math. 11* (1958), 333–418.

Fujiwara, D.

[1] "Concrete characterization of the domains of fractional powers of some elliptic differential operators of the second order." *Proc. Japan Acad. 43* (1967), 82–86.

Garabedian, P.

[1] *Partial Differential Equations.* Wiley, New York, 1964.

Gårding, L.

[1] "Dirichlet's problem for linear elliptic partial differential equations." *Math. Scand. 1* (1953), 55–72.

———.

[2] "Solution directe du problème de Cauchy pour les équations hyperboliques." Centre national de la recherche scientifique, Colloques internationaux, *71* (1956), 71–90.

Garsia, A.

[1] *Topics in Almost Everywhere Convergence.* Markham, Chicago, 1970.

Gelfand, I. and Shilov, I.

[1] *Generalized Functions. Vol.* 1–3. Moscow, 1958.

Gelfand, I. and Vilenkin, N.

[1] *Generalized Functions. Vol.* 4. Moscow, 1961.

Gilbarg, D. and Trudinger, N.

[1] *Elliptic Partial Differential Equations of Second Order.* Springer, New York, 1977.

Gilkey, P.

[1] *The Index Theorem and the Heat Equation.* Publish or Perish, Boston, 1974.

Glimm, J.

[1] "Solutions in the large for nonlinear hyperbolic systems of equations." *Comm. Pure Appl. Math. 18* (1965), 697–715.

Glimm, J. and Lax, P.

[1] "Decay of solutions of systems of nonlinear hyperbolic conservation laws." Memoirs A.M.S. *101*, Providence, 1970.

Goldstein, H.

[1] *Classical Mechanics.* Addison-Wesley, New York, 1950.

Golubitsky, M. and Guillemin, V.

[1] *Stable Mappings and their Singularities.* Springer, New York, 1973.

Granoff, B. and Ludwig, D.

[1] "Propagation of singularities along characteristics with non-uniform multiplicity." *J. Math. Anal. Appl. 21* (1968), 556–574.

Greiner, P. and Stein, E.

[1] *Estimates for the $\bar{\partial}$-Neumann Problem.* Math. Notes No. 19, Princeton Univ. Press, Princeton 1977.

Grusin, V.

[1] "On a class of hypoelliptic pseudodifferential operators degenerate on a submanifold." *Mat. Sbornik 84* (1971), 111–134.

Guillemin, V.

[1] "Clean intersection theory and Fourier integrals." In *Fourier Integral Operators and Partial Differential Equations*, pp. 23–35. Lecture Notes in Math., 459, Springer, New York, 1975.

———.

[2] "Lectures on spectral theory of elliptic operators." *Duke Math. J. 44* (1977), 485–518.

Guillemin, V. and Melrose, R.

[1] "Poisson summation formula on a manifold with boundary." *Adv. in Math. 32* (1979), 204–232.

Guillemin, V. and Schaeffer, D.

[1] "Remarks on a paper of D. Ludwig." *Bull. A.M.S., 79* (1973), 382–385.

Guillemin, V. and Sternberg, S.

[1] *Geometric Asymptotics.* A.M.S., Providence, 1977.

Gunning, R. and Rossi, H.

[1] *Analytic Functions of Several Complex Variables.* Prentice Hall, Englewood Cliffs, 1965.

Hadamard, J.

[1] *Le Problème de Cauchy et les Equations aux Dérivées Partielles Linéaires Hyperboliques.* Hermann, Paris, 1932.

Hamilton, R.

[1] *Harmonic Maps of Manifolds with Boundary.* Lecture Notes, 471, Springer, New York, 1975.

Hanges, N.

[1] "Parametrices and local solvability for a class of singular hyperbolic operators." *Comm. PDE 3* (1978), 105–152.

———.

[2] "Propagation of singularities for a class of operators with double characteristics." In *Seminar on Singularities*, pp. 113–126. Ann. of Math. Studies 91, Princeton Univ. Press, Princeton, 1979.

Harvey, C.

[1] "On domination estimates and global existence." *J. Math. Mech. 16* (1967), 675.

Helffer, B.

[1] "Sur une class d'opérateurs hypoelliptiques à caractéristiques multiple." *J. Math. Pure Appl. 55* (1976), 207–215.

Herman, E.

[1] "The symbol of the algebra of singular integral operators." *J. Math. Mech. 15* (1966), 147–156.

Hersch, R.

[1] "Mixed problems in several variables." *J. Math. Mech. 12* (1963), 317–334.

Hewitt, E. and Ross, K.

[1] *Abstract Harmonic Analysis I.* Springer, New York, 1963.

Hille, E. and Phillips, R.

[1] *Functional Analysis and Semi-groups.* Colloq. Publ. vol. 31. A.M.S., Providence, 1957.

Hörmander, L.

[1] "On the theory of general partial differential operators." *Acta. Math. 94* (1955), 161–248.

———.

[2] "Local and global properties of fundamental solutions." *Math. Scand. 5* (1957), 27–39.

———.

[3] "On the interior regularity of the solutions of partial differential equations." *Comm. Pure Appl. Math. 5* (1958), 197–215.

———.

[4] "Estimates for translation invariant operators in L^p spaces." *Acta. Math. 104* (1960), 93–140.

———.

[5] "Linear partial differential equations without solutions." *Math. Ann. 140* (1960), 169–173.

———.

[6] *Linear Partial Differential Operators*, Springer, New York, 1964.

———.

[7] "Pseudo-differential operators." *Comm. Pure Appl. Math. 18* (1965), 501–517.

———.

[8] *Introduction to Complex Analysis in Several Variables.* Van Nostrand, Princeton, 1966.

———.

[9] "On the Riesz means of spectral functions and eigenfunction expansions for elliptic differential operators." In *Recent Advances in the Basic Sciences*, pp. 155–202. Yeshiva Univ. Conference, Nov. 1966.

———.

[10] "Pseudodifferential operators and non-elliptic boundary problems." *Ann. of Math. 83* (1966), 129–209.

———.

[11] "Hypoelliptic second order differential equations." *Acta Math. 119* (1967), 147–171.

———.

[12] "Pseudo-differential operators and hypoelliptic equations." In *Proc. Sym. Pure Math. X (Singular Integrals)*, pp. 138–183. A.M.S., Providence, 1967.

———.

[13] "The spectral function of an elliptic operator." *Acta Math. 121* (1968), 193–218.

———.

[14] "The calculus of Fourier integral operators." In *Prospects in Mathematics*, pp. 33–57. Ann. of Math. Studies 70, Princeton Univ. Press, Princeton 1971, 33–57.

———.

[15] "Fourier integral operators I." *Acta Math. 127* (1971), 79–183.

———.

[16] "On the existence and the regularity of solutions of linear pseudo-differential equations." *L'Enseign. Math. 17* (1971), 99–163.

———.

[17] "On the L^2 continuity of pseudo-differential operators." *Comm. Pure Appl. Math. 24* (1971), 529–535.

———.

[18] "Uniqueness theorems and wave front sets for solutions of linear differential equations with analytic coefficients." *Comm. Pure Appl. Math. 24* (1971), 671–703.

———.

[19] "A class of pseudodifferential operators with double characteristics." *Math. Ann. 217* (1975), 165–188.

———.

[20] "The Cauchy problem for differential equations with double characteristics." *J. Anal. Math. 32* (1977), 118–196.

———.

[21] "Spectral analysis of singularities," In *Seminar on Singularities*, pp. 3–49. Ann. of Math. Studies 91, Princeton Univ. Press, Princeton, 1979.

———.

[22] "Subelliptic operators." In *Seminar on Singularities*, pp. 127–208. Ann. of Math. Studies 91, Princeton Univ. Press, Princeton, 1979.

———.

[23] "The Weyl calculus of pseudodifferential operators." *Comm. Pure Appl. Math. 32* (1979), 355–443.

Hughes, T., Kato, T., and Marsden, J.

[1] "Well posed quasi-linear second order hyperbolic systems with applications to nonlinear elastodynamics and general relativity." *Arch. Rat. Mech. Anal. 63* (1976), 273–294.

Illner, R.

[1] "A class of L^p bounded pseudo-differential operators." *Proc. A.M.S. 51* (1975), 347–355.

Ivrii, V.

[1] "Wave fronts of solutions of symmetric pseudodifferential systems." *Soviet Math. Dokl. 18* (1977), 540–544.

Ivrii, V. and Petkov, V.

[1] "Necessary conditions for the Cauchy problem for non-strictly hyperbolic equations to be well-posed." *Uspehi Mat. Nauk. 29* (1974), 5.

Jackson, D.

[1] *The Theory of Approximation*. Colloq. Publ. vol. 11, A.M.S., Providence, 1930.

Jackson, J.

[1] *Classical Electrodynamics*. J. Wiley, New York, 1962.

John, F.

[1] *Partial Differential Equations*, Springer, New York, 1975.

Johnson, J. and Smoller, J.

[1] "Global solutions for an extended class of hyperbolic systems of conservation laws." *Arch. Rat. Mech. Anal. 32* (1969), 169–189.

Jones, B. F.

[1] "A class of singular integrals." *Amer. J. Math. 86* (1964), 441–462.

Kagan, V.

[1] "Boundedness of pseudo-differential operators in L^p." *Izv. Vyss. Ucebn. Zaurl Mat. 73* (1968) no. 6, 35–44. (In Russian).

Kannai, Y.

[1] "Off diagonal short time asymptotics for fundamental solutions of diffusion equations." Preprint.

Kashiwara, M. and Kawai, T.

[1] "Microhyperbolic pseudodifferential operators, I." *J. Math. Soc. Japan 27* (1975), 359–404.

Kashiwara, M. and Schapira, P.

[1] "Micro-hyperbolic systems." *Acta Math. 142* (1979), 1–55.

Kato, T.

[1] "Boundedness of some pseudo-differential operators." *Osaka J. Math. 13* (1976), 1–9.

Keller, J.

[1] "Corrected Bohr-Sommerfeld quantum conditions for non-separable systems." *Ann. Phys. 4* (1958), 180–188.

Kirchhoff, G.

[1] *Vorlesungen über math. Physik* 2 (Optik), Leipzig, 1981.

Knapp, A. and Stein, E.

[1] "Intertwining operators on semi-simple Lie groups." *Ann. Math. 93* (1971), 489–578.

Kohn, J.

[1] "Harmonic integrals on strongly pseudo convex manifolds," I and II. *Ann of Math. 78* (1963), 112–147; *79* (1964), 450–472.

———.

[2] "Pseudo-differential operators and hypoellipticity." In *Proc. Symp. Pure Math.* vol. 23, pp. 61–69. A.M.S., Providence, 1973.

———.

[3] "Subellipticity of the $\bar{\partial}$-Neumann problem on pseudoconvex domains: sufficient conditions." *Acta Math. 142* (1979), 79–122.

Kohn, J., and Nirenberg, L.

[1] "An algebra of pseudo-differential operators." *Comm. Pure Appl. Math. 18* (1965), 269–305.

———.

[2] "Noncoercive boundary value problems." *Comm. Pure Appl. Math. 18* (1965), 443–492.

Korevaar, J.

[1] "Distribution proof of Wiener's Tauberian theorem." *Proc. Amer. Math. Soc. 16* (1965), 353–355.

Kotlow, D.
[1] "Quasilinear parabolic equations and first order quasilinear conservation laws with bad Cauchy data." *J. Math. Anal. App. 35* (1971), 563–576.
Kreiss, H.
[1] "Initial boundary value problems for hyperbolic systems." *Comm. Pure Appl. Math. 23* (1970), 277–298.
Krushkov, S.
[1] "First order quasi-linear equations in several independent variables." *Math. USSR Sbornik 10* (1970), no. 2, 217–243.
Kumano-go, H.
[1] "Pseudo differential operators and the uniqueness in the Cauchy problem." *Comm. Pure Appl. Math. 22* (1969), 73–129.
———.
[2] "Algebras of pseudo-differential operators." *J. Fac. Sci. Univ. Tokyo. 17* (1970), 31–50.
———.
[3] *Pseudo Differential Operators.* Iwanami Shoten, Japan, 1974. (In-Japanese)
———.
[4] "A calculus of Fourier integral operators on R^n and the fundamental solution for an operator of hyperbolic type." *Comm. PDE 1* (1976), 1–44.
———.
[5] "Factorizations and fundamental solutions for differential operators of elliptic-hyperbolic type." *Proc. Japan Acad. 52* (1976), 480–483.
Kumano-go, H. and Nagase, M.
[1] "L^p theory of pseudo-differential operators." *Proc. Japan Acad. 46* (1970), 138–142.
———.
[2] "Approximation theorems in the theory of pseudo-differential operators and sharp Gårding's inequality." To appear.
Kumano-go, H. and Taniguchi, K.
[1] "Multi-products of phase functions of Fourier integral operators with an application." *Comm. PDE 3* (1978), 349–380.
Lascar, R.
[1] "Parametrices microlocales de problèmes aux limites pour une clase d'equations pseudodifferentielles à caracteristiques de multiplicité variable." Seminaire Goulaouic-Schwartz, no. 4. Ecole Polytechnique, 1978–1979.
Lax, A.
[1] "On Cauchy's problem for partial differential equations with multiple characteristics." *Comm. Pure Appl. Math 9* (1956), 135–169.
Lax, P.
[1] "Asymptotic solutions of oscillatory initial value problems." *Duke Math. J. 24* (1957), 627–646.
———.
[2] "Hyperbolic systems of conservation laws, II." *Comm. Pure Appl. Math. 10* (1957), 537–566.

Lax, P. and Nirenberg, L.
[1] "On stability for difference schemes; a sharp form of Gårding's inequality." *Comm. Pure Appl. Math. 19* (1966), 473–492.
Lax, P. and Phillips, R.
[1] *Scattering Theory.* Academic Press, New York, 1967.
Lebedev, N.
[1] *Special Functions and their Applications.* Dover, New York, 1972.
Leray, J.
[1] *Hyperbolic Differential Equations.* Princeton Univ. Press, Princeton, 1952.
Lewy, H.
[1] "An example of a smooth linear partial differential equation without solutions." *Ann. Math. 66* (1957), 155–158.
Lions, J.
[1] *Equations Différentielles Operationnelles,* Springer, New York, 1961.
———.
[2] "Espaces d'interpolation et domaines de puissance fractionaires d'opérateurs." *J. Math. Soc. Japan 14* (1962), 233–241.
Lions, J. and Magenes, E.
[1] *Non-Homogeneous Boundary Value Problems and Applications, I and II.* Springer, New York, 1972.
Lions, J. and Peetre, J.
[1] "Sur une classe d'espaces d'interpolation." Publ. Math. Inst. Hautes Etudes Sci. no. 19, pp. 5–68. Paris 1964.
Lopatinski, Y.
[1] "On a method of reducing boundary problems for a system of differential equations of elliptic type to regular equations." *Ukrain. Math. Ž. 5* (1953), 123–151.
Ludwig, D.
[1] "Uniform asymptotic expansions at a caustic." *Comm. Pure Appl. Math. 19* (1966), 215–250.
———.
[2] "Uniform asymptotic expansions of the field scattered by a convex object at high frequencies." *Comm. Pure Appl. Math 19* (1967), 103–138.
McKean, H.
[1] *Stochastic Integrals.* Academic Press, New York, 1969.
McKean, H. and Singer, I.
[1] "Curvature and the eigenvalues of the Laplacian." *J. Diff. Geom. 1* (1967), 43–69.
Majda, A.
[1] "High frequency asymptotics for the scattering matrix and the inverse problem of acoustical scattering." *Comm. Pure Appl. Math. 29* (1976), 261–291.
Majda, A. and Osher, S.
[1] "Reflection of singularities at the boundary." *Comm. Pure Appl. Math. 28* (1975), 479–499.

Majda, A. and Ralston, J.
[1] "An analogue of Weyl's theorem for unbounded domains." *Duke Math. J. 45* (1978), 513–536.
Majda, A., and Taylor, M.
[1] "The asymptotic behavior of the diffraction peak in classical scattering." *Comm. Pure Appl. Math. 30* (1977), 639–669.
———.
[2] "Inverse scattering problems for transparent obstacles, electromagnetic waves, and hyperbolic systems." *Comm. PDE 2* (1977), 395–438.
Malgrange, B.
[1] "Existence et approximation des solutions des équations aux deriveés partielles et des équations de convolutions." *Ann Inst. Fourier Grenoble 6* (1955–6), 271–355.
———.
[2] Ideals of Differentiable Functions. Oxford Univ. Press, Oxford, 1966.
Marcinkiewicz, J.
[1] "Multiplicateurs des séries de Fourier." *Studia Math. 8* (1939), 78–91.
Maslov, V.
[1] *Theorie des Perturbations et Methodes Asymptotiques* Dunod, Paris, 1972.
Mather, J.
[1] "Differentiable invariants." *Topology 16* (1977), 145–155.
Melin, A.
[1] "Lower bounds for pseudo-differential operators." *Ark. Mat. 9* (1971), 117–140.
Melin, A. and Sjöstrand, J.
[1] "Fourier integral operators with complex valued phase functions." In *Fourier Integral Operators and Partial Differential Equations*, p.p. 120–223. Lecture Notes in Math., 459, Springer, New York, 1975.
———.
[2] "A calculus for Fourier integral operators in domains with boundary and applications to the oblique derivative problem." *Comm. PDE 2* (1977), 857–935.
Melrose, R.
[1] "Local Fourier-Airy integral operators." *Duke Math. J. 42* (1975), 583–604.
———.
[2] "Microlocal parametrices for diffractive boundary value problems." *Duke Math. J. 42* (1975), 605–635.
———.
[3] "Equivalence of glancing hypersurfaces." *Invent. Math. 37* (1976), 165–191.
———.
[4] "Airy operators." *Comm. PDE 3* (1978), 1–76.
———.
[5] "Equivalence of glancing hypersurfaces, II." To appear.
———.
[6] "Transformation of boundary value problems." To appear.

———.
[7] "Weyl's conjecture for manifolds with concave boundary." To appear.
Melrose, R. and Sjöstrand, J.
[1] "Singularities of boundary value problems, I." *Comm. Pure Appl. Math. 31* (1978), 593–617.
Melrose, R. and Taylor, M.
[1] "The corrected Kirchoff approximation and uniform analysis of the amplitude in classical scattering." To appear.
Melrose, R., and Uhlmann, G.
[1] "Lagrangian intersections and the Cauchy problem." *Comm. Pure Appl. 32* (1979), 483–519.

———.
[2] "Microlocal structure of involutive conical refraction." *Duke Math. J. 46* (1979), 571–582.
Menikoff, A.
[1] "Carleman estimates for partial differential operators with real coefficients." *Arch. Rat. Mech. Anal. 54* (1974), 118–133.

———.
[2] "Parametrices for subelliptic operators." *Comm. PDE 2* (1977), 69–108.

———.
[3] "Subelliptic estimates for pseudo-differential operators with double characteristics." Manuscript.
Menikoff, A. and Sjöstrand, J.
[1] "On the eigenvalues of a class of hypoelliptic operators." *Math. Anallen 235* (1978), 55–85.
Mikhlin, S.
[1] *Multidimensional Singular Integral Equations.* Pergammon Press, New York, 1965.
Milnor, J.
[1] *Topology from the Differential Viewpoint.* Univ. Press of Virginia, Charlottesville, 1965.
Minakshishundaram, S. and Pleijel, A.
[1] "Some properties of the eigenfunctions of the Laplace operator on Riemannian manifolds." *Canad. J. Math. 1* (1949), 242–256.
Mizohata, S.
[1] *The Theory of Partial Differential Equations.* Cambridge Univ. Press, Cambridge, 1973.
Morawetz, C.
[1] "Exponential decay of solutions of the wave equation." *Comm. Pure Appl. Math. 19* (1966), 439–444.
Morawetz, C. and Ludwig, D.
[1] "The generalized Huyghens' principle for reflecting bodies." *Comm. Pure Appl. Math. 22* (1969), 189–205.
Morrey, C.
[1] *Multiple Integrals and the Calculus of Variations.* Springer, New York, 1966.

Muskheleshvili, N.
[1] *Singular Integral Equations*. P. Nordhoff, Groningen, 1953.
Nagase, M.
[1] "The L^p boundedness of pseudo-differential operators with non-regular symbols." *Comm. PDE 2* (1977), 1045–1061.
Nagel, A. and Stein, E.
[1] "A new class of pseudo-differential operators." *Proc. Nat. Acad. Sci. U.S.A.* 75 (1978), 582–585.
Nelson, E.
[1] "Operants, a functional calculus for non-commuting operators." In *Functional Analysis and Related Topics*, pp. 172–187. Springer, New York, 1970.
Neri, U.
[1] "Singular integral operators on manifolds." In *Proc. Symp. Pure Math.* vol. 10, pp. 232–242. A.M.S., Providence, 1967.
Nirenberg, L.
[1] *Lectures on Linear Partial Differential Equations*. Reg. Conf. Ser. in Math., no. 17. A.M.S., Providence, 1972.
Nirenberg, L. and Treves, F.
[1] "On local solvability of linear partial differential equations." *Comm. Pure Appl. Math. 23* (1970), 459–509; *24* (1971), 279–288.
Nosmas, J.
[1] "Parametrix du problème de Cauchy pour une classe de systèmes hyperboliques à caracteristiques réelles involutive de multiplicité variable." Seminaire Goulaouic-Schwartz, no. 8. Ecole Polytechnique, 1978–1979.
Nussensweig, H.
[1] "High frequency scattering by an impenetrable sphere." *Ann. Phys. 34* (1965), 23–95.
Oleinik, O. and Radkevic, E.
[1] "On the local smoothness of weak solutions and the hypoellipticity of differential second order equations." *Uspehi Math. Nauk. 26* (1971), 265–281.
Olver, F.
[1] "The asymptotic expansion of Bessel functions of large order." *Phil Trans. Roy. Soc. London Ser. A.* 257 (1954), 328–368.
———.
[2] *Asymptotics and Special Functions*. Academic Press, New York, 1974.
Palais, R. ed.
[1] *Seminar on the Atiyah-Singer Index Theorem*. Ann. of Math. Studies 57, Princeton Univ. Press, Princeton, 1963.
Peetre, J.
[1] "Interpolation of Lipschitz operators and metric spaces." *Mathematica* (Cluj) *12* (1970), 325–334.
Phillips, R.
[1] "Dissipative operators and hyperbolic systems of partial differential equations." *Trans. A.M.S.* 90 (1959), 193–254.

Pliś, A.

[1] "Non-uniqueness in the Cauchy problem for differential equations of elliptic type." *J. Math. Mech.* 9 (1960), 557–562.

Polking, J.

[1] "Boundary value problems for parabolic systems of partial differential equations." In *Proc. Symp. Pure Math.* vol. 10, pp. 243–274. A.M.S., Providence, 1967.

Povzner, A. and Sukarevskii, K.

[1] "Discontinuities of the Green's function of a mixed problem for the wave equation." *Mat. Sbornik 51* (1960), 3–26.

Protter, M. and Weinberger, H.

[1] *Maximum Principles in Differential Equations.* Prentice Hall, Englewood Cliffs, 1967.

Radkevic, E.

[1] "Hypoelliptic operators with multiple characteristics." *Math. USSR Sbornik, 8* (1969), 181–206.

Ralston, J.

[1] "Solutions of the wave equation with localized energy." *Comm. Pure Appl. Math. 22* (1969), 807–824.

———.

[2] "Difficiency indices of symmetric operators with elliptic boundary conditions." *Comm. Pure Appl. Math. 23* (1970), 221–232.

———.

[3] "Diffraction by convex bodies." Seminaire Goulaouic-Schwartz, no. 23. Ecole Polytechnique, 1978–1979.

Rauch, J.

[1] "L^2 is a continuable initial condition for Kreiss' mixed problems." *Comm. Pure Appl. Math. 25* (1972), 265–286.

Rauch, J. and Massey, F.

[1] "Differentiability of solutions to hyperbolic initial-boundary value problems." *Trans. A.M.S. 189* (1974), 303–318.

Rauch, J. and Taylor, M.

[1] "Penetration into shadow regions and unique continuation for solutions to hyperbolic mixed problems." *Indiana J. Math. 22* (1972), 277–285.

———.

[2] "Exponential decay of solutions to symmetric hyperbolic equations in bounded domains." *Indiana J. Math. 24* (1974), 79–86.

———.

[3] "Decay of solutions to nondissipative hyperbolic equations on compact domains." *Comm. Pure Appl. Math. 28* (1975), 501–523.

Lord Rayleigh.

[1] "On waves propagated along the plane surface of an elastic solid." *Proc. London Math. Soc. 17* (1885), 4–11.

Reed, M. and Simon, B.

[1] *Methods of Mathematical Physics.* Academic Press, New York. Vol. 1,

Functional Analysis, 1975; Vol. 2, Fourier Analysis and Self Adjointness, 1975; Vol. 3, Scattering Theory, 1978; Vol. 4, Spectral Theory, 1978.

Richtmyer, R. and Morton, K.

[1] *Difference Methods for Initial-value Problems*. J. Wiley, New York, 1967.

Rothschild, L.

[1] "A criterion for hypoellipticity of operators constructed from vector fields." *Comm. PDE 4* (1979), 645–699.

———.

[2] "Local solvability of left invariant differential operators on the Heisenberg group." *Proc. A.M.S. 74* (1979), 383–388.

Rothschild, L. and Stein, E.

[1] "Hypoelliptic differential operators and nilpotent groups." *Acta Math.* 137 (1976), 247–320.

Sakamoto, R.

[1] "Mixed problems for hyperbolic equations." *J. Math. Kyoto Univ. 10* (1970), 349–373; 403–417.

Sarason, L.

[1] "On weak and strong solutions of boundary value problems." *Comm. Pure Appl. Math. 15* (1962), 237–288.

Sato, M., Kawai, T., and Kashiwara, M.

[1] "Microfunctions and pseudo differential equations." In *Hyperfunctions and Pseudo-Differential Equations*, pp. 265–529. Lecture Notes In Math, 287. Springer, New York, 1973.

Schapira, P.

[1] "Propagation at the boundary and reflection of analytic singularities of solutions of linear partial differential equations I." *Publ. Res. Inst. Math. Sci. Kyoto 12* (1977), 441–453.

Schechter, M.

[1] *Modern Methods in Partial Differential Equations, an Introduction*. McGraw-Hill, New York, 1977.

Schwartz, L.

[1] *Théorie des Distributions*. Herman, Paris, 1950.

Seeley, R.

[1] "Refinement of the functional calculus of Calderon and Zygmund." *Proc. Acad. Wet. Ned., Ser. A, 68* (1965), 521–531.

———.

[2] "Singular integrals and boundary value problems." *Amer. J. Math. 88* (1966), 781–809.

———.

[3] "Complex powers of an elliptic operator." In *Proc. Symp. Pure Math.* vol 10, pp. 288–307. A.M.S., Providence, 1967.

———.

[4] "The resolvent of an elliptic boundary value problem." *Amer. J. Math. 91* (1969), 889–920.

———.

[5] "A sharp asymptotic remainder estimate for the eigenvalues of the Laplacian in a domain of $\mathbf{R}^3$." *Adv. in Math. 29* (1978), 144–169.

———.

[6] "An estimate near the boundary for the spectral function of the Laplacian." To appear.

Shubin, M.

[1] *Pseudodifferential Operators and Spectral Theory*. Moscow, 1978. (In Russian)

Sjöstrand, J.

[1] "Parametrices for pseudodifferential operators with multiple characteristics." *Ark. för Math. 12* (1974), 85–130.

———.

[2] "Propagation of singularities for operators with multiple involutive characteristics." *Ann. Inst. Fourier Grenoble 26* (1976), 141–155.

———.

[3] "Propagation of analytic singularities for second order Dirichlet problems." *Comm. PDE 5* (1980), 41–94.

Smoller, J. and Taylor, M.

[1] "Wave front sets and the viscosity method." *Bull. A.M.S. 79* (1973), 431–436.

Sobolev, S.

[1] "Sur un théorème d'analyse fonctionelle." *Math. Sbornik 45* (1938), 471–496.

Sommerfeld, A. and Runge, J.

[1] "Anwendung der Vektorrechnung auf die Grundlagen der geometrischen Optik." *Ann. Phys. 35* (1911), 277–298.

Stakgold, I.

[1] *Boundary Value Problems of Mathematical Physics*, vol. 2. Macmillan, New York, 1968.

Stanton, R.

[1] "Mean convergence of Fourier series on compact Lie groups." *Trans. Amer. Math. Soc. 217* (1976), 61–87.

Stanton, R. and Tomas, P.

[1] "Polyhedral summability of Fourier series on compact Lie groups." *Amer. J. Math. 100* (1978), 477–493.

Stein, E.

[1] *Singular Integrals and Differentiability Properties of Functions*. Princeton Univ. Press, Princeton, 1970.

———.

[2] *Topics in Harmonic Analysis Related to the Littlewood-Paley Theory*. Ann. of Math. Studies, Princeton Univ. Press, Princeton, 1970.

Stein, E. and Weiss, G.

[1] *Introduction to Fourier Analysis on Euclidean Spaces*. Princeton Univ. Press, Princeton, 1971.

Sternberg, S.

[1] *Lectures on Differential Geometry*. Prentice Hall, Englewood Cliffs, 1964.

Strichartz, R.

[1] "Multipliers on fractional Sobolev spaces." *J. Math. Mech. 16* (1967), 1031–1060.

———.

[2] "A functional calculus for elliptic pseudodifferential operators." *Amer. J. Math. 94* (1972), 711–722.

Taibleson, M.

[1] "On the theory of Lipschitz spaces of distributions on Euclidean n-space." I, *J. Math. Mech. 13* (1964), 407–479; II, *J. Math. Mech. 14* (1965), 821–839.

Tartakoff, D.

[1] "Regularity of solutions to boundary value problems for first order systems." Thesis, Univ. of Calif., Berkeley, 1969.

———.

[2] "Local analytic hypoellipticity for $\Box_b$ on nondegenerate Cauchy Riemann manifolds." *Proc. Nat. Acad. Sci. 75* (1978), 3027–3028.

Tartar, L.

[1] "Interpolation non linéaire et régularité." *J. Funct. Anal. 9* (1972), 469–489.

Taylor, M.

[1] "Gelfand theory of pseudodifferential operators and hypoelliptic operators." *Trans. Amer. Math. Soc. 153* (1971), 495–510.

———.

[2] "Vector valued analytic functions and applications." *Mich. Math. J. 19* (1972), 353–356.

———.

[3] *Pseudodifferential operators.* Lecture Notes in Math. 416. Springer-Verlag, New York, 1974.

———.

[4] "Reflection of singularities of solutions to systems of differential equations." *Comm. Pure Appl. Math. 28* (1975), 457–478.

———.

[5] "Grazing rays and reflection of singularities of solutions to wave equations." *Comm. Pure Appl. Math. 29* (1976), 1–38.

———.

[6] "Grazing rays and reflection of singularities of solutions to wave equations. Part II (systems). "*Comm. Pure Appl. Math. 29* (1976), 463–481.

———.

[7] "First order hyperbolic systems with a small viscosity term." *Comm. Pure Appl. Math. 31* (1978), 707–786.

———.

[8] "Propagation, reflection, and diffraction of singularities of solutions to wave equations." *Bull. A.M.S. 84* (1978), 589–611.

———.

[9] "Fourier integral operators and harmonic analysis on compact manifolds." In *Proc. Symp. Pure Math.*, vol. 35, pt. 2, pp. 115–136, A.M.S., Providence, 1979.

———.
[10] "Rayleigh waves in linear elasticity as a propagation of singularities phenomenon." In *Proc. Conf. on PDE and Geometry*, pp. 273–291. Marcel Dekker, 1979.

Taylor, M., Farris, M. and Yingst, D.
[1] "Grazing rays and reflection of singularities of solutions to wave equations, part III," To appear.

Titchmarsh, E.
[1] *Introduction to the Theory of Fourier Integrals*. Oxford Univ. Press, Oxford, 1937.

———.
[2] *Eigenfunction Expansions associated with Second-order Differential Equations*. Oxford Univ. Press, Oxford, 1946.

Trèves, F.
[1] *Linear Partial Differential Equations with Constant Coefficients*. Gordon and Breach, New York, 1966.

———.
[2] *Locally Convex Spaces and Linear Partial Differential Equations*. Springer, New York, 1967.

———.
[3] "Hypoelliptic partial differential equations of principal type. Sufficient conditions and necessary conditions." *Comm. Pure Appl. Math. 24* (1971), 631–670.

———.
[4] "A new method of proof of the subelliptic estimates." *Comm. Pure Appl. Math. 24* (1971), 71–115.

———.
[5] "A link between solvability of pseudodifferential equations and uniqueness in the Cauchy problem." *Amer. J. Math. 94* (1972), 267–288.

———.
[6] "On the existence and regularity of solutions of linear partial differential equations." In *Proc. Symp. Pure Math.*, vol. 23, pp. 33–60. A.M.S., Providence, 1973.

———.
[7] "Analytic hypoellipticity for a class of pseudodifferential operators." *Comm. PDE 3* (1978).

Uhlmann, G.
[1] "Pseudo-differential operators with involutive double characteristics." *Comm. PDE 2* (1977), 713–779.

Unterberger, A.
[1] "Résolution d'équations aux derivées partielles dans des espaces de distributions d'ordre de regularité variable." *Ann. Inst. Fourier 21* (1971), 85–128.

Vaillancourt, R.
[1] "Pseudo-translation operators." Thesis, N.Y.U., 1969.

———.

[2] "A simple proof of the Lax-Nirenberg theorem." *Comm. Pure Appl. Math.* *23* (1970), 151–163.

Visik, M. and Eskin, G.

[1] "Normally solvable problems for elliptic systems of equations in convolutions." *Mat. Sbornik 78* (1967), 326–356.

Vol'pert, A.

[1] "The spaces BV and quasilinear equations." *Math. USSR Sbornik 2* (1967), 225–267.

Wainger, S.

[1] "Special trigonometric series in k dimensions." *Mem. A.M.S.* no. 59 (1965).

Wallach, N.

[1] *Harmonic Analysis on Homogenous Spaces.* Marcel Dekker, New York, 1973.

———.

[2] "An asymptotic formula of Gelfand and Gangolli for the spectrum of $\Gamma\backslash G$." *J. Diff. Geom. 11* (1976), 91–101.

Weinstein, A.

[1] "Fourier integral operators, quantization, and the spectra of riemannian manifolds." Centre national de la recherche scientifique, Colloques Internationaux *237* (1976), 289–298.

———.

[2] "Asymptotics of eigenvalue clusters for the Laplacian plus a potential." *Duke Math. J. 44* (1977), 883–892.

Weiss, N.

[1] "L^p estimates for bi-invariant operators on compact Lie groups." *Amer. J. Math. 94* (1972), 103–118.

Whitham, G.

[1] *Linear and Nonlinear Waves.* Wiley Interscience, New York, 1974.

Wiener, N.

[1] "Tauberian theorems." *Ann. of Math. 33* (1932), 1–100.

Yamaguti, M. and Nogi, T.

[1] "An algebra of pseudo-difference schemes and its application." *Publ. RIMS 3* (1967), 151–166.

Yingst, D.

[1] "The Kirchoff approximation for the Neumann problem." *Indiana Math. J.* To appear.

Yosida, K.

[1] *Functional Analysis.* Springer, New York, 1965.

Zelobenko, D.

[1] *Compact Lie Groups and their Representations.* Transl. Math. Mono. 40. Amer. Math. Soc., Providence, 1973.

Zygmund, A.

[1] *Trigonometrical Series.* 2nd ed., Cambridge Univ. Press, Cambridge, 1959.

General Index

Index of Symbols

Library of Congress Cataloging in Publication Data

Taylor, Michael Eugene, 1946–
Pseudodifferential operators.

(Princeton mathematical series; 34)
Bibliography
Includes index.
1. Differential equations, Partial.
2. Pseudodifferential operators. I. Title.
II. Series.
QA374.T38 1981 515.3′52 80-8580
ISBN 0-691-08282-0 AACR2

Michael E. Taylor is Professor of Mathematics at Rice University

GPSR Authorized Representative: Easy Access System Europe - Mustamäe tee 50, 10621 Tallinn, Estonia, gpsr.requests@easproject.com

www.ingramcontent.com/pod-product-compliance
Ingram Content Group UK Ltd.
Pitfield, Milton Keynes, MK11 3LW, UK
UKHW021824060425
457147UK00006B/132

* 9 7 8 0 6 9 1 6 1 5 0 3 5 *